SOUND VISUALIZATION AND MANIPULATION

This edition first published 2013

Registered office

John Wiley & Sons Singapore Pte. Ltd., 1 Fusionopolis Walk, #07-01 Solaris South Tower, Singapore 138628

For details of our global editorial offices, for customer services and for information about how to apply for permission to reuse the copyright material in this book please see our website at www.wiley.com.

Library of Congress Cataloging-in-Publication Data

Kim, Yang-Hann.
Sound visualization and manipulation / Yang-Hann Kim, Jung-Woo Choi.
pages cm
Includes bibliographical references and index.
ISBN 978-1-118-36847-3 (cloth)
1. Sound-waves–Mathematical models. 2. Helmholtz equation. I. Choi, Jung-Woo. II. Title.
QC243.K46 2013
534.01′5153533–dc23

2013025393

Set in 9/11pt Times by Laserwords Private Limited, Chennai, India.
Printed and bound by CPI Group (UK) Ltd, Croydon, CR0 4YY

C9781118368473_020123

Contents

sound/noise control over elected regions based on the concept of acoustic contrast, which has been widely adopted for the implementation of personal sound zones. Recently, his research has extended to interactive 3D sound/multi-channel audio systems that can be manipulated in real-time by exploiting the beauty of direct integral formulas.

Dr Choi received his B.Sc., M.Sc., and Ph.D. degrees in Mechanical Engineering from the Korea Institute of Science and Technology (KAIST), Korea, in 1999, 2001, and 2005, respectively. He was a Postdoctoral Research Associate with the Center for Noise and Vibration Control (NOVIC), KAIST, Korea, 2005–06. From 2006 to 2007, he was a Visiting Postdoctoral Researcher at the Institute of Sound and Vibration Research (ISVR), University of Southampton, UK. From 2007 to 2011, he was with Samsung Electronics at the Samsung Advanced Institute of Technology (SAIT) in Korea, working on array-based audio systems as a Research & Development staff member and a Senior Engineer. In 2011, he joined the Department of Mechanical Engineering, KAIST, Korea, and has since been a Research Professor there. He is the author of more than 50 papers/conference articles and 15 patent applications, including five registered patents on loudspeaker array systems.

Preface

If only we could see sound propagation in space with our eyes, and if only the sound could be created in any desired shape! Such a fantastic concept is being realized. New approaches to acoustics and noise engineering have allowed innovative changes in these fields.

So far, extensive efforts have been made using various methods to explain how a medium changes as sound propagates in space or how the shape of the sound propagation changes depending on its frequency and wavelength. There are two main approaches being employed to resolve these questions: theoretical and experimental. The theoretical approach is adopted to develop an understanding of the phenomena of sound propagation in acoustic waves and, through this understanding, to attempt to find a solution. The characteristics of acoustic wave equations in certain cases are interpreted by numerically solving the so-called linear acoustic wave equations. Popular numerical techniques are the finite element method and boundary element method; both have achieved incredible developments owing to continuous evolutions in their background theories and improvements in the arithmetic capacity of computers. The experimental approach has also seen rapid improvements. Developments in semiconductor technologies have reduced the microphone size to eliminate unnecessary scattering induced by them, and the reduction in cost allows for tens and hundreds of microphones to be used at the same time. We can now sample, record, and analyze signals from hundreds of microphones in almost real time. These developments allow us to "visualize" sound using our eyes in the real world, which is something that human beings have long dreamt of. The first half of this book explains various methods to visualize sound.

From a mathematical point of view, sound visualization can be regarded as an exploration of methods to transform measured data into information that is visible to the human eyes. Most information transformation is determined during selection of the desired basis function because information transformation can produce different results depending on its mapping functions, as shown in the figure below. Thus, we need to deal with problems such as selecting a basis function and expressing a sound field as a visible image using the selected basis function. This book explains the planar, cylindrical, and spherical basis functions used in acoustic holography and the functions employed for beamforming methods. Their advantages and disadvantages as well as practical applicability are addressed. The advantage of the acoustic holography method includes visualization of information with great physical significance, such as acoustic pressure, velocity, intensity, and energy. On the other hand, the beamforming method can provide a variety of visualization information depending on the type of basis function used for the beamformer.

The concept that visualization results vary significantly depending on the basis functions can be reasonably expanded to an idea of sound manipulation where arbitrary or desired forms of sound can be created in space. Desired sounds can be produced by selecting basis functions so that the sounds generated from the sound sources arranged in the space are of these types of basis functions, as shown in the figure. Well-known methods include wave field synthesis (WFS) and Ambisonics. WFS is a representative method based on the so-called Kirchhoff–Helmholtz integral

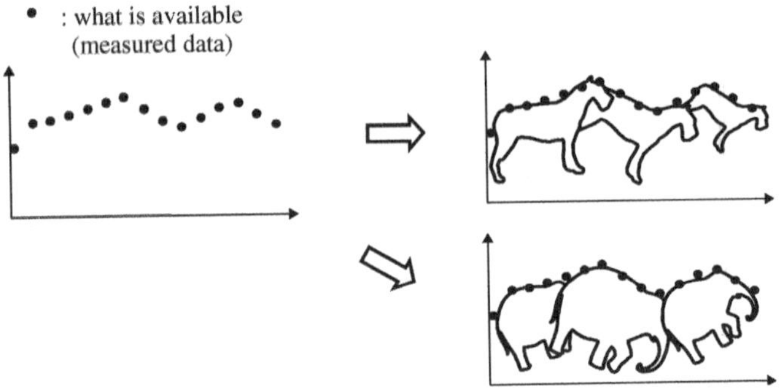

equation, whereas Ambisonics is a technique that expresses sound fields using spherical harmonics and embodies the desired shapes of sound in space using such expressions. From a unifying point of view, the manipulation of a sound field is an issue in obtaining the desired output using the available sound sources; accordingly, we can select the best basis function depending on the definition of the desired function. Based on this idea, the sound focusing problem of concentrating the sound in desired areas or dividing an area into acoustically bright and dark zones and maximizing the ratio of sound energies between the two areas can also be explained.

Both the sound visualization method and sound manipulation method demand considerable theoretical knowledge of mathematics and acoustics as well as knowledge of signal processing to understand their principles and realize their practical applications. To aid potential readers who want to understand the basic concepts or those who will practically apply the methods, the simplest one-dimensional theories are introduced in this book, and their mathematical and theoretical explanations are presented in every chapter. Chapter 1 is intended to aid understanding of the basic physical quantities in acoustics. Part I consists of two chapters. Chapter 1 explains three physical quantities in acoustics using one-dimensional examples: interrelationships among acoustic pressure, particle velocity, and acoustic density. This approach is justified in that the principle of superposition holds for a linear system; hence, most of the concepts explained in one dimension can be extended to multidimensional cases.

Part II introduces the sound visualization methods and explains how their basic principles can be varied depending on certain basis functions. Accordingly, basis functions and approaches for the acoustic holography and the beamforming method are introduced. An appropriate basis function should be used depending on what we want to visualize; depending on this basis function, the information of visualized sound fields can be varied.

In Part III, we deal with sound manipulation techniques. Sound manipulation is carried out using two main methods; both are discussed with respect to how they are embodied in one-dimensional situations. Sound manipulation involves a sound focusing technique that concentrates the sound in specific areas in space and a sound field reproduction method that generates a wave front in the desired forms. For realizing these two methods, unique inputs to generate sound fields in the desired forms need to be determined. Therefore, the sound focusing and reproduction problems are defined as inverse problems corresponding to the beamforming and acoustic holography methods, respectively. Thus, the sections on acoustic holography and sound field reproduction are organized to complement each other. The sections on beamforming and sound focusing address similar issues but explain them from different points of view. The chapter on beamforming focuses on a signal processing technique to extract the parameter determining the locations of sound sources, whereas the chapter on sound focusing explains resolution variations depending on the geometric configuration of arrays and beam pattern variations depending on the basic aperture functions. Thus, Parts II and III address different and similar issues from complementary points of view; readers interested in visualization are strongly recommended to read the manipulation part. It would be efficient for readers of this book to use Part I as a reference when they need to know more about acoustics while reading Parts II and III.

In conclusion, this book introduces and explains methods for sound visualization and manipulation. The book is organized such that readers can gain a profound understanding of basic concepts and theoretical approaches from the one-dimensional case. The methods of visualization and manipulation are explained as a unifying approach for creating certain assumed or desired shapes in space based on the measured or available information using basis functions.

Yang-Hann Kim
Jung-Woo Choi

Acknowledgments

It was around 1990 that the first author had an idea about the basis function described in this book. He visited his old friend, Prof. J. K. Hammond of the Institute of Sound and Vibration Research (ISVR), University of Southampton, who was giving a lecture on nonlinear signal processing for a group of people in the industry at the time. The first page of the handout made for that class included a primitive version of the picture that is published in this book's preface. In fact, this picture originated and evolved from the image in *Science with a Smile* (Robert L. Weber, Institute of Physical Publishing, Bristol and Philadelphia, 1992, pp. 111–12). The moment he looked at this picture, it occurred to him that a part of this picture could be used to explain the processing of every signal. Signal processing essentially involves finding desired information using available data. Thus, this picture symbolically shows that the quality of information obtained eventually depends on how well the processing method represents substantial, physical, or mathematical situations. Ultimately, the result is fully dependent on the processing method one has chosen, that is, a basis function. If so, how do we select a basis function? Although it is a very basic question, it is self-evident that if we can suggest the best method to be selected, it would serve as a very innovative and useful method in this discipline.

In fact, since the first author was at the time working on issues such as mechanical noise diagnosis and fault detection using signal processing, he looked at the picture that Prof. Hammond had used in a symbolic manner and gained an idea to view various problems in a unifying manner. He came to realize that both sound visualization techniques – acoustic holography and beamforming – eventually produced different results with regard to visualization owing to differences in the basic functions that were used. If so, in-depth knowledge of whether a basis function used has a mathematical function to perform something well would allow one to have a good understanding of the result of the sound visualization, that is, the picture. Thus, to clearly interpret the visualized information gained through the acoustic holography method and to accurately analyze the desired information, it is necessary to analyze how well the basis function expresses the desired visual information. Similarly, the following questions can be approached from an understanding of the basis function that was used: what is the specific information that can be gained from the spatial distribution of the beamforming power obtained from the beamforming method? Does the maximum value of the beaming power correctly describe the locations of sound or noise sources? What properties of the sources does the spatial distribution of the beaming power represent? In this regard, the visualization described in this book was greatly inspired by the discussions with Prof. Hammond.

The sound manipulation study described in the second half of this book started, it is recalled, around 1999 when the first author thought he had found some improvements in the study of visualization or was getting bored with studying sound visualization. The second author was studying how to focus sounds in an arbitrary space as part of the work for a master's degree, and based on this, he started a full-scale study on sound manipulation. The first result aimed to practically implement a system that allows one to hear a desired sound without disturbing others by focusing the sound in a specific space. Fortunately, the experiment was successful, and in 2000, the authors succeeded in focusing sound in a specific space by using six loudspeakers. Later on, this study was expanded to a study of a home/mobile speaker array system by the second author in the industry and to another study of monitor speaker array development by the first author at KAIST. The monitor speaker array system led to the implementation of a personal audio system that focuses sound using nine speakers. Among those who participated in the theoretical development and the experiment are Chan-hee Lee, currently working at Hyundai Heavy Industries Co., Ltd.; Dr Ji-ho Chang, currently at DTU after completing his doctoral degree; and Jin-young Park, currently pursuing his doctoral degree. At that time, the result attracted so much attention that it was broadcast on national TV. Thanks to this, the first author was granted an unexpected research fund, and his team could build a set of experimental equipment consisting of 32 speakers supported by KAIST's HRHR project. Using this experimental equipment, the research team implemented a method for focusing sounds on specific spots in various ways, and the effects were found to be better than expected. One day, a question was raised about what results would be produced if the focusing point was moved to an arbitrary location. Min-ho Song developed an interface using an iPhone, in which as a finger moved to a location, the location at which sound was focused was also changed, making the sound audible in real time. In fact, from a theoretical viewpoint, they knew that the sound focusing solution had nothing to do with 3D sound. However, the listener could feel the effect of the location of the sound source moving through the sound focusing solution only. As a matter of fact, studies have reported on a focused source using the time-reversed driving function in wave field synthesis, but these did not have sufficient theoretical basis, and no perfect integral equation form was available for the array in the form of surrounding the listener. The theoretical explanation of the experimental results was completed by the second author in 2011, and it was proved that a general solution can be drawn by combining Porter-Bojarski integral with a multipole virtual source. The first work to create sounds using this solution aimed to relocate a mosquito's sound to a desired space, and it was a great success. This substantial success became a motivation for the book's third part. The doctoral students, Jeong-min Lee and Dong-su Kang, made substantial contributions to developing a speaker system that implemented a sound ball. In addition, the authors would like to acknowledge Dr Min-Ho Song, who performed great research while completing his doctoral degree at the Graduate School of Cultural Technology and who contributed to developing one particular interface.

In fact, the graduates' wonderful studies were greatly helpful to the authors in writing the sound visualization part. In particular, studies by Dr Jae-Woong Choi, who has made great achievements in the areas of spherical beamforming and music, and Dr Young-Key Kim, who founded a company and has been disseminating sound visualization technology, were very helpful in writing the beamforming chapter. Dr Hyu-sang Kwon developed moving frame acoustic holography (MFAH), and he is expected to realize great achievements as an expert in this area. Furthermore, Dr Soon-hong Park of the Korea Aerospace Research Institute has made a great contribution to the method by applying MFAH to moving sound sources. The authors also want to acknowledge Dr Sea-Moon Kim of the Korea Institute of Ocean Science and Technology who successfully lead the acoustic holography experiment on the King Seong-deok Bell; Dr Kyung-Uk Nam of Hyundai Motor Company who greatly contributed to the partial field acoustic holography; and Dr Chun-Su Park who developed the time domain acoustic holography technique using the spatio-temporal complex envelope.

Credit for Chapter 4 of this book also belongs to Ku-Hwan Kim, who programmed most of the codes for beamforming simulations. The authors would like to express their appreciation to all the laboratory members – Jung-Min Lee, Dong-Soo Kang, Dae-Hoon Seo, Ki-Won Kim, Myung-Ryun Lee, Seong-Woo Jung – for their enthusiasm in correcting errors and giving advice to improve the content of this book.

The second author also wants to thank his former advisors at KAIST – Yoon-Sik Park, Chong-Won Lee, Jeong-Guon Lee, and Young-Jin Park – for teaching him the fundamentals of sound and vibration. Special thanks must be given to Prof. P. A. Nelson and S. J. Elliott and Dr F. M. Fazi of ISVR for many hours of fruitful discussions with him regarding the sound field reproduction and sound focusing projects. The experiences with his former colleagues at Samsung Electronics – Youngtae Kim, Jungho Kim, Sang-Chul Ko, and Seoung-Hun Kim – were greatly helpful in summarizing the techniques discussed in Chapters 4–6.

Finally, the authors would like to express their special thanks to James Murphy and Clarissa Lim of John Wiley & Sons for their consistent help and cooperation with regard to editing this book. Without their encouragement, this book would not have been possible.

Yang-Hann Kim
Jung-Woo Choi

Part I

Essence of Acoustics

Sound is an important part of our lives. Even in the womb, human beings are capable of detecting sounds. We create and enjoy sounds, and we can identify information conveyed by sound. We live with sound and are familiar with the fundamental concepts associated with it. Fundamental concepts of sound visualization and manipulation can also be explained on the basis of the mechanisms by which a sound wave is generated, propagated, and decayed by various internal and external disturbances.

Acoustics is a vast field of study that explains the propagation of waves in different media, and it cannot be completely covered in merely the first two chapters of this book. In this book, however, we limit the scope by focusing on the general idea of acoustics in terms of its essential physical measures. This part of the book discusses essential measures that refer to the primary measures or physical variables used in acoustics, such as acoustic pressure, velocity, intensity, and energy, which can be used to describe sound propagation. Various impedances, radiations, scatterings, surfaces, and so on, are also considered as important measures that affect wave propagation in space. In order to uniquely and conveniently explain the physics of acoustics, this part of the book relies heavily on the concept of impedance as a window to study sound propagation in time and space.

Chapter 1 introduces the essential physical parameters used in acoustics measurements. The significance of physical parameters other than impedance, such as sound pressure, speed, energy, power, and intensity, are explained (Figure I.1). It is emphasized that these parameters form the fundamental concepts required for understanding the propagation of sound waves. The aforementioned parameters are explained by using a one-dimensional approach. The Euler equation is used to describe the relation between the sound pressure and the particle velocity in a given medium. The state equation, on the other hand, is used to evaluate the relation between the acoustic density and the fluctuating pressure, which is the acoustic pressure that causes the sound propagation. The third equation used is the law of conservation of mass for the compressible fluid, which defines how the fluctuating density and the fluid particle velocity are associated with each other. Therefore, the three essential variables: sound pressure, particle velocity, and fluctuating density, follow these three equations. This enables the derivation of the acoustic wave equation that governs all the parameters associated with acoustic wave propagation. Chapter 1 discusses the two different approaches that can be used to solve this acoustic wave equation. The first one is based on the eigenfunction analysis, in which the solution is determined as the superposition of eigenmodes. Another approach uses Green's function, which describes how a sound field is constructed when the field has a monopole source at an arbitrary position in space. This approach leads to the Kirchhoff–Helmholtz equation.

Sound Visualization and Manipulation, First Edition. Yang-Hann Kim and Jung-Woo Choi.

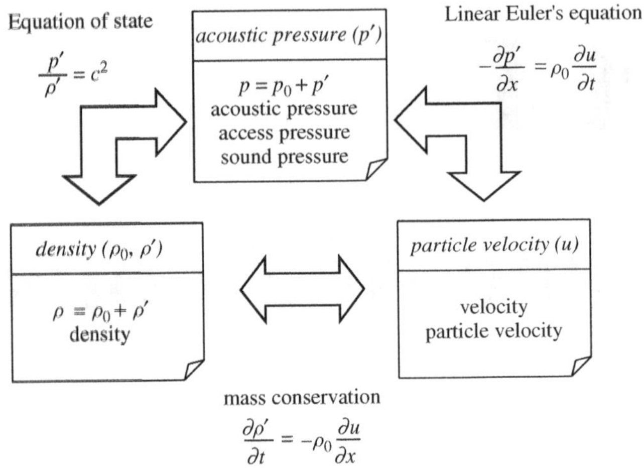

Figure I.1 Pictorial relation between three variables that govern acoustic wave propagation (p_0 and ρ_0 express the mean pressure and static density, respectively; p' and ρ' denote acoustic pressure and fluctuating density, respectively; c denotes the speed of propagation, and u is the velocity of the fluctuating medium)

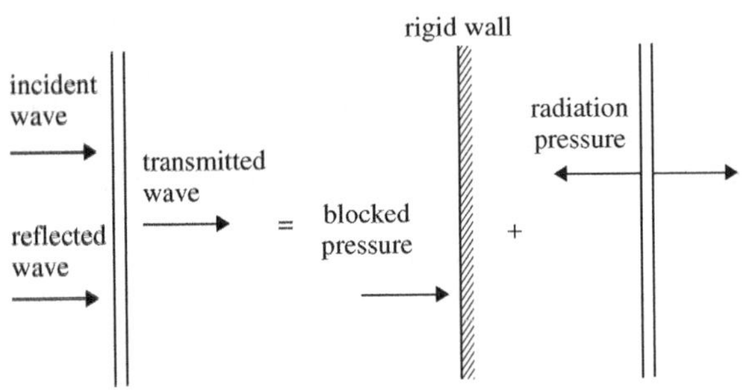

Figure I.2 Reflection and transmission phenomena using the principle of superposition

Chapter 2 takes a rather ambitious route to describe how sound wave reacts under impedance mismatch in space and time using the concept of radiation, scattering, and diffraction. It is believed that the scattering and the diffraction of sound can be explained by acoustic radiation. For instance, a scattered sound field is a result of radiation scattering (Figure I.2), whereas diffraction is a result of the radiation from an object that has spatial impedance mismatch.

An understanding of the first two chapters is expected to help in analyzing and explaining the results obtained by the sound visualization described in Chapters 3 and 4, and the manipulation described in Chapters 5 and 6.

1

Acoustic Wave Equation and Its Basic Physical Measures[1]

1.1 Introduction

The waves along a string propagate along its length, but the string itself moves perpendicular to the propagation direction. It therefore forms a transverse wave. If the particle of a medium moves in the direction of propagation, we refer to it as a longitudinal wave. The waves in air, water, or any compressible medium are longitudinal waves, which are often referred to as acoustic waves. This chapter explores the underlying physics and sensible physical measures related to acoustic waves, including pressure, velocity, intensity, and energy. Impedance plays a central role with regard to its effect on these measures.

In the area of sound visualization, our objectives are to determine a rational means to convert essential acoustic variables such as pressure, velocity, and density, or other physically sensible acoustic measures such as intensity or energy, into visible representations. One very straightforward way to accomplish these objectives is to express acoustic pressure by using a color code. Notably, there are many ways to visualize a sound field, depending on a mapping or general basis function, which relates acoustic variables to visual expressions. Therefore, this chapter starts with a discussion on the visualization of a one-dimensional acoustic wave.

1.2 One-Dimensional Acoustic Wave Equation

The simplest case is illustrated in Figure 1.1. The end of a pipe or duct which is filled with a homogeneous compressible fluid (air, water, etc.) is excited with a radian frequency ($\omega = 2\pi f$, f: frequency in Hz). If the pipe is semi-infinitely long, then the pressure in the pipe ($p(x,t)$) can be mathematically written as

$$p(x,t) = P_0 \cos(kx - \omega t + \phi) \tag{1.1}$$

where P_0 is the pressure magnitude and ϕ is an initial phase. Here, k represents the spatial frequency ($k = 2\pi/\lambda$, λ: wavelength in m) of the pressure field, which is often called wavenumber.

If the pipe is of finite length L, then the possible acoustic pressure in the pipe can be written as

$$p(x,t) = P_0 \cos k(L - x) \cos \omega t. \tag{1.2}$$

[1] Sections of Chapter 1 have been re-used with permission from Ref. [1].

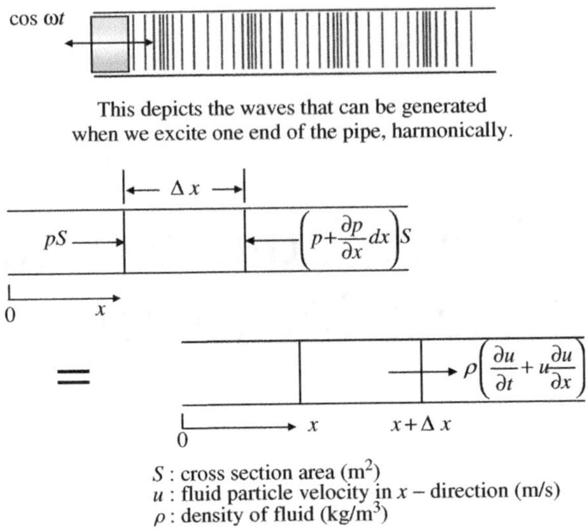

Figure 1.1 Relation between forces and motion of an infinitesimal fluid element in a pipe (expressing momentum balance: the left-hand side shows the forces and the right exhibits the change of momentum)

Equations (1.1) and (1.2) are different simply because of the boundary conditions: the former has no boundary condition prescribed at $x = L$, but the latter has a rigid-wall condition (velocity is zero).

To understand what is happening in the pipe, we have to understand how pressures and velocities of the fluid particles behave and are associated with each other. This motivates us to look at an infinitesimal element of the volume of the fluid in the pipe; specifically, we will investigate the relation between force and motion.

As illustrated in Figure 1.1, the forces acting on the fluid between x and $x + \Delta x$ and its motion will follow the conservation of momentum principle. That is,

$$\text{Sum of the forces acting on the fluid} = \text{momentum change} \tag{1.3}$$

We can mathematically express this equality as

$$(pS)_x - (pS)_{x+\Delta x} = \rho S \frac{du}{dt} \Delta x \tag{1.4}$$

where it has already been assumed that the viscous force, which likely exists in the fluid, is small enough (relative to the force induced by pressure) to be neglected.

The rate of change of velocity (du/dt) can be expressed by

$$\frac{du}{dt} = \frac{\partial u}{\partial t} + \frac{\partial u}{\partial x}\frac{\partial x}{\partial t} \tag{1.5}$$

where u is a function of position (x) and time (t) and velocity is the time rate change of the displacement. Therefore, we can rewrite Equation (1.5) as

$$\frac{du}{dt} = \frac{\partial u}{\partial t} + u\frac{\partial u}{\partial x}. \tag{1.6}$$

If the cross-section between x and $x + \Delta x$ is maintained constant and Δx becomes small ($\Delta x \to 0$), then Equation (1.4) can be expressed as[2]

$$-\frac{\partial p}{\partial x} = \rho\left(\frac{\partial u}{\partial t} + u\frac{\partial u}{\partial x}\right) = \rho\frac{Du}{Dt} \tag{1.7}$$

where

$$p = p_0 + p' \tag{1.8}$$

$$\rho = \rho_0 + \rho' \tag{1.9}$$

$$\frac{D}{Dt} = \frac{\partial}{\partial t} + u\frac{\partial}{\partial x}. \tag{1.10}$$

Note that the pressure (p) is composed of the static pressure (p_0) and the acoustic pressure (p'), which is induced by the small fluctuation of fluid particles. The density also has two components: the static density (ρ_0) and the small fluctuating density (ρ').

Equation (1.10) is the total derivative, and is often called the material derivative. The first term expresses the rate of change with respect to time, and the second term can be obtained by examining the change with respect to space as we move with the velocity u.[3] As can be anticipated, the second term is generally smaller than the first.

If the static pressure (p_0) and density (ρ_0) do not vary significantly in space and time, then Equation (1.7) becomes

$$-\frac{\partial p'}{\partial x} = \rho_0\frac{\partial u}{\partial t} \tag{1.11}$$

where p' is acoustic pressure and is directly related to acoustic wave propagation. As already implied in Equation (1.8), acoustic pressure is considerably smaller than static pressure.[4] Equation (1.11) essentially means that a small pressure change across a small distance (∂x) causes the fluid of mass/unit volume ρ_0 to move with the acceleration of $\partial u/\partial t$. This equation is generally referred to as a linearized Euler equation. Equation (1.7), on the other hand, is an Euler equation.

Equations (1.7) and (1.11) describe three physical parameters, pressure, fluid density, and fluid particle velocity. In other words, they express the relations between these three basic variables. In order to completely characterize the relations, two more equations are needed.

The relation between density and fluid particle velocity can be obtained by using the conservation of mass. Figure 1.2 shows how much fluid enters the cross-section at x and how much exits through the surface at $x + \Delta x$. If we apply the principle of conservation of mass law to the fluid volume between x and $x + \Delta x$, the following equality can be written.

$$\left\langle \begin{array}{l} \text{the rate of mass increase in the infinitesimal element} \\ \quad = \text{the decrease of mass resulting from the fluid that is entering and exiting through} \\ \quad\quad \text{the surface at } x \text{ and } x + \Delta x \end{array} \right\rangle$$

Expressing this equality mathematically leads to

$$\frac{\partial}{\partial t}(\rho S\,\Delta x) = (\rho u S)_x - (\rho u S)_{x+\Delta x} \tag{1.12}$$

[2] Note that we used $\partial x/\partial t = u$ in Equation (1.6). This is the Lagrangian description, which describes the motion of a mass of fluid at Δx. The other method to describe the momentum change through fixed infinitesimal control volume is by using the Euler description (Section 8.1.2). Note also that a more precise momentum balance can be expressed as $-\frac{\partial p}{\partial x} = \frac{D\rho u}{Dt}$.

[3] We assume that the effect of mass transport is negligible.

[4] We also refer to the acoustic pressure as "access pressure" or "sound pressure."

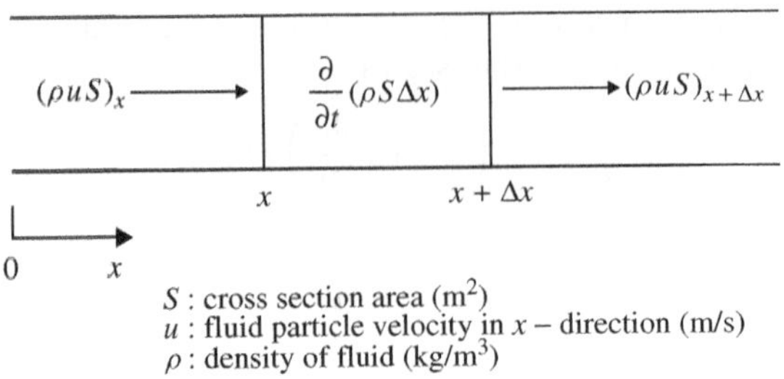

Figure 1.2 Conservation of mass in an infinitesimal element of fluid (increasing mass of the infinitesimal volume results from a net decrease of the mass through the surfaces of the volume)

as illustrated in Figure 1.2. As assumed before, if the area of the cross-section (S) remains constant, then Equation (1.12) can be rewritten as

$$\frac{\partial \rho}{\partial t} = -\frac{\partial}{\partial x}(\rho u). \tag{1.13}$$

We can linearize this equation by substituting Equation (1.9) into Equation (1.13). Equation (1.13) then becomes

$$\frac{\partial \rho'}{\partial t} = -\rho_0 \frac{\partial u}{\partial x}. \tag{1.14}$$

Equations (1.11) and (1.14) express the relation between sound pressure and fluid particle velocity, as well as the relation with fluctuating density and fluid particle velocity, respectively. One more equation is therefore needed to completely describe the relations of the three acoustic variables: acoustic pressure, fluctuating density, and fluid particle velocity. The other equation must describe how acoustic pressure is related to fluctuating density. Recall that a pressure change will induce a change in density as well as other thermodynamic variables, such as entropy. This leads us to postulate that acoustic pressure is a function of density and entropy, that is,

$$p = p(\rho, s) \tag{1.15}$$

where s denotes entropy. We can then write the change of pressure, or fluctuating pressure, dp or p', by modifying Equation (1.15) as follows:

$$dp = \left.\frac{\partial p}{\partial \rho}\right|_s d\rho + \left.\frac{\partial p}{\partial s}\right|_\rho ds. \tag{1.16}$$

This equation simply states that a pressure change causes a density change ($d\rho$) and entropy variation (ds). It is noticeable that the fluid obeys the law of isentropic processes when it oscillates within the range of the audible frequency: 20 Hz to 20 kHz.[5] The second term on the right-hand side of Equation (1.16) is therefore negligible. This implies that the small change of sound pressure with regard to the infinitesimal change of density can be assumed to have certain proportionality. (An alternative way to deduce the same relation can be found in Appendix B, Section B.1.3.) Note that the second relation of Equation (1.16) is mostly found experimentally. This reduces Equation (1.16) to the form

$$\frac{p'}{\rho'} = \frac{B}{\rho_0} = c^2 \tag{1.17}$$

[5] This is possible if the period of oscillation by the fluid particle is much smaller than the time required to dissipate or transfer the heat energy within the wavelength of interest.

where B is the bulk modulus that expresses the pressure required for a unit volume change and c is the speed of sound. We may obtain Equation (1.17) by introducing a gas dynamics model. This equation is an equation of state. Tables 1.1 and 1.2 summarizes the speed of sound in accordance with the state of gas [2]. An alternative method of deducing Equations (1.16) and (1.17) can be found in Appendix B, Section B.1.3.

Table 1.1 The dependency of the speed of sound on temperature

Temperature (°C)	Speed of sound (m/s)	Temperature (°C)	Speed of sound (m/s)	Temperature (°C)	Speed of sound (m/s)
−100	263.5	−35	309.5	30	349.1
−95	267.3	−30	312.7	35	352.0
−90	271.1	−25	315.9	40	354.8
−85	274.8	−20	319.1	45	357.6
−80	278.5	−15	322.3	50	360.4
−75	282.1	−10	325.3	55	363.2
−70	285.7	−5	328.4	60	365.9
−65	289.2	0	331.5	65	368.6
−60	292.7	5	334.5	70	371.3
−55	296.1	10	337.5	75	374.0
−50	299.5	15	340.4	80	376.7
−45	302.9	20	343.4	–	–
−40	306.2	25	346.3	–	–

Table 1.2 The dependency of the speed of sound on relative humidity and on frequency

Relative humidity/ frequency	0%		30%		60%		100%	
	Decay rate (%)	Speed of sound (m/s)	Decay rate (%)	Speed of sound (m/s)	Decay rate (%)	Speed of sound (m/s)	Decay rate (%)	Speed of sound (m/s)
20	0.51	343.477	0.03	343.807	0.02	344.182	0.01	344.685
40	1.07	343.514	0.11	343.808	0.06	344.183	0.04	344.685
50	1.26	343.525	0.17	343.810	0.09	344.183	0.06	344.685
63	1.43	343.536	0.25	343.810	0.15	344.184	0.09	344.685
100	1.67	343.550	0.50	343.814	0.34	344.185	0.22	344.686
200	1.84	343.559	1.01	343.821	0.99	344.190	0.77	344.689
400	1.96	343.561	1.59	343.826	1.94	344.197	2.02	344.695
630	2.11	343.562	2.24	343.827	2.57	344.200	3.05	344.699
800	2.27	343.562	2.85	343.828	2.94	344.201	3.57	344.701
1 250	2.82	343.562	5.09	343.828	4.01	344.202	4.59	344.704
2 000	4.14	343.562	10.93	343.829	6.55	344.203	6.29	344.705
4 000	8.84	343.564	38.89	343.831	18.73	344.204	13.58	344.706
6 300	14.89	343.565	90.61	343.836	42.51	344.204	27.72	344.706
10 000	26.28	343.566	204.98	343.846	101.84	344.206	63.49	344.706
12 500	35.81	343.566	294.08	343.854	155.67	344.208	96.63	344.707
18 000	52.15	343.567	422.51	343.865	247.78	344.211	154.90	344.708
20 000	75.37	343.567	563.66	343.877	373.78	344.215	237.93	344.709

Adapted from CRC Handbook of Chemistry and Physics, 79th ed., 1998, pp. 14–38, CRC Press.: With kind permission of Taylor & Francis Group LLC-Books

Note that Equation (1.17) expresses how the access pressure or acoustic pressure communicates with the fluctuating density. Equations (1.11) and (1.14) completely express the laws that govern the waves in which we are interested. Therefore, we can summarize the relations as

$$-\frac{\partial p'}{\partial x} = \rho_0 \frac{\partial u}{\partial t} \tag{1.11}$$

$$\frac{\partial \rho'}{\partial t} = -\rho_0 \frac{\partial u}{\partial x} \tag{1.14}$$

$$\frac{p'}{\rho'} = c^2. \tag{1.17}$$

Figure 1.3 demonstrates how these equations and physical variables are related. If we eliminate ρ' and u from Equations (1.11), (1.14), and (1.17), then we obtain

$$\frac{\partial^2 p'}{\partial x^2} = \frac{1}{c^2}\frac{\partial^2 p'}{\partial t^2} \tag{1.18}$$

This is a linearized acoustic wave equation.[6]

Equation (1.18) is essentially a general one-dimensional acoustic wave, that is, the waves in compressible fluid. A similar relation can be found from the propagation of a string wave. The only difference between the waves along a string and acoustic waves lies in whether the directions of wave propagation and velocity fluctuation of medium are collinear or perpendicular. Note that the propagation direction of the waves along a string is perpendicular to that of the motion of the string. Conversely, the acoustic wave propagates in the direction of the fluid particle's velocity.[7]

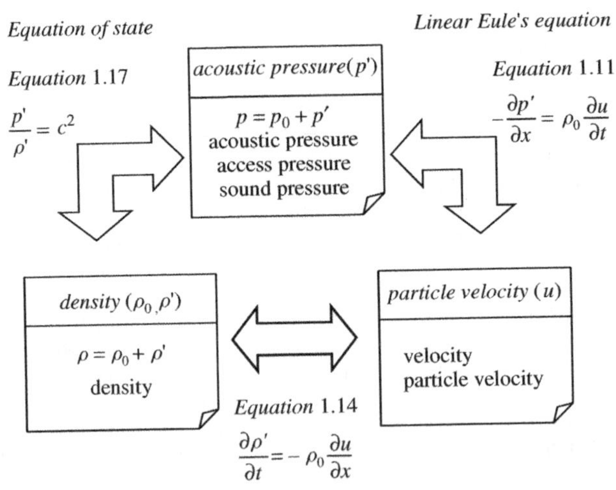

Figure 1.3 Pictorial relation between three variables that govern acoustic wave propagation (p_0 and ρ_0 express the mean pressure and static density, respectively; p' and ρ' denote acoustic pressure and fluctuating density, respectively; c denotes the speed of propagation, and u is the velocity of the fluctuating medium)

[6] If we eliminate p' and u or change to p' and ρ', then we can obtain the equation for ρ' and u, respectively.

[7] The linearized Euler equation (Equation (1.11)) essentially states that the pressure difference induces time rate of velocity u in the x direction.

The former is called a transverse wave, while the latter is considered a longitudinal wave.[8] This means that the acoustic wave, at least a one-dimensional case, is analogous to that of a wave on a string. It therefore appears that a unified concept or theory is very possible.

We can also optimistically consider that any three-dimensional wave can be decomposed into a one-dimensional wave, at least conceptually. The principle of superposition holds for the waves that are governed by the linearized acoustic wave equation. Therefore, any combination of the solutions with respect to each component of the coordinate – for example, (x, y, z) rectangular coordinate, (r, φ, z) cylindrical, and (r, θ, ϕ) spherical coordinates – also satisfies the governing equation. Simply extending this concept leads us to use the Fourier series or transform. This means that we attempt to look at what we are interested in, in terms of the orthogonal series. Many sound visualization and manipulation techniques, such as acoustic holography in Chapter 3, Section 3.3.3 and mode-matching approach in Chapter 6, Section 6.6, utilize orthogonal series to predict or reproduce the sound field.

The second conclusion involves the relations between three acoustic variables, which are illustrated in Figure 1.3. Euler equation (Equation (1.11)), states that the spatial pressure change makes the fluid particle move. Equation (1.14) shows that a higher compression rate in time makes a steeper negative velocity gradient in space. Equation (1.17) indicates that speed of propagation depends on the characteristics of the medium. If we have a smaller density change for a unit pressure change, then we will obtain faster propagation speed (see Tables 1.1 and 1.2). Lastly, the wave equation defines the relation between the spatial and temporal change of sound. From Equations (1.1) and (1.18), we have

$$\omega = ck. \tag{1.19}$$

From Equation (1.19), we can observe that the frequency and wavenumber are bound by the wave equation. This relation, called dispersion relation of one-dimensional acoustic waves, indicates that the possible spatial fluctuation of wave is limited when the frequency of sound is determined.

1.2.1 Impedance

If we use Equation (1.11) to examine the ratio between the pressure acting on the waves and the corresponding particle velocities (their magnitudes), then it leads to the concept of *acoustic impedance*.

$$\frac{p'}{u'} = Z \tag{1.20}$$

We normally regard the acoustic impedance as a complex variable, which allows us to express any possible phase difference between the pressure and particle velocity.

For example, examine the sound propagation along the semi-infinite duct. The pressure field of Equation (1.1) is the real part of the complex exponential

$$\hat{p}(x,t) = \hat{P}e^{i(kx-\omega t)}. \tag{1.21}$$

Note that $\hat{P}$ represents the complex amplitude including both the magnitude and phase of the pressure field ($\hat{P} = P_0 e^{i\phi}$). The real-valued pressure $p(x,t)$ can then be expressed as

$$p(x,t) = \mathrm{Re}[\hat{p}(x,t)].$$

the same token, the particle velocity can be written as

$$\hat{u}(x,t) = \hat{U}e^{i(kx-\omega t)}. \tag{1.22}$$

[8] Depending on whether the direction of wave propagation and that of the medium's oscillation are perpendicular or collinear, the wave can be transverse or longitudinal, respectively. The bending wave essentially makes the change of angle measured from the mid plane.

Since only access pressure and velocity (p', u') are involved with this form of wave propagation, we can derive the acoustic impedance from the ratio of Equation (1.21) to (1.22). That is,

$$Z = \frac{\hat{P}}{\hat{U}} = \rho_0 c. \tag{1.23}$$

The right hand side of Equation (1.23) is composed of the medium density and the speed of sound in the medium. Because it represents the characteristics of the medium, we call this type of impedance as the *characteristic impedance*.

1.3 Three-Dimensional Wave Equation

We now extend Equations (1.11), (1.14), and (1.18) to a three-dimensional case. First, Euler equation can be written as

$$\begin{aligned} -\frac{\partial p'}{\partial x} &= \rho_0 \frac{\partial u}{\partial t} \\ -\frac{\partial p'}{\partial y} &= \rho_0 \frac{\partial v}{\partial t} \\ -\frac{\partial p'}{\partial z} &= \rho_0 \frac{\partial w}{\partial t} \end{aligned} \tag{1.24}$$

where we use (x, y, z) coordinate for convenience. Each equation simply states that the change of pressure with respect to the (x, y, z) direction, that is, the pressure gradient, makes the fluid particle move. u, v, and w denote the velocity with respect to the (x, y, z) coordinate system.

We may use a vector notation to express Equation (1.24), which will yield a more compact form. This gives

$$-\nabla p' = \rho_0 \frac{\partial \mathbf{u}}{\partial t}. \tag{1.25}$$

This is the linearized Euler equation, where

$$\mathbf{u} = (u, v, w). \tag{1.26}$$

Similarly, Equation (1.14) can also be extended to the three-dimensional form, that is,

$$\frac{\partial \rho'}{\partial t} = -\rho_0 \nabla \cdot \mathbf{u} \tag{1.27}$$

The right-hand term of Equation (1.27) represents the net mass flow into the unit volume in space. Appendix B, Section B.1 provides the detailed derivation and theoretical background.

If we eliminate ρ' and $\mathbf{u}$ using Equations (1.25), (1.27), and (1.17), then

$$\nabla^2 p' = \frac{1}{c^2} \frac{\partial^2 p'}{\partial t^2} \tag{1.28}$$

is obtained, which is a three-dimensional form of a wave equation.[9]

[9] We can also express the governing equation by using a potential function, that is, the velocity potential. This is possible because the viscous force is negligible compared to others. This means that $\nabla \times \mathbf{u} = 0$; in other words, the angular deformation can be neglected. Note that $\nabla \Phi = \mathbf{u}$ always satisfies $\nabla \times \mathbf{u} = 0$, where Φ is the velocity potential (Appendix B, Section B.1.4).

To summarize, three physical variables (acoustic pressure, fluid particle velocity, and fluctuating density) govern acoustic waves in a compressible fluid. These variables are not independent. The relations between them are expressed by laws or equations, that is, the linearized Euler equation, conservation of mass, and the state equation of gas. Lastly, we envisage that three-dimensional acoustic waves can also be understood using theory developed to explain one-dimensional acoustic waves.

1.4 Acoustic Intensity and Energy

We acknowledged that acoustic pressure (p'), density (ρ'), and fluid particle velocity (u for one dimension, or $\mathbf{u}$ for the three-dimensional case) are major physical variables that determine everything related to acoustic wave propagation. We also recall that the impedance measures how the waves behave in response to the pressure/velocity perturbation. Power and energy are also major descriptors of wave propagation along a string. Understanding that the waves in a compressible fluid also obey what we have learned in the case of one-dimensional string waves, we explore what is physically meant by the energy and intensity of acoustic waves.

For simplicity, we consider a one-dimensional case (Figure 1.4). We denote acoustic pressure (p') as p, and fluctuating density (ρ') as ρ.

As illustrated in Figure 1.4, there will be a volume change of $\Delta l \cdot S$ because of the pressure difference along the element. The length of the element will be shortened by $-\Delta l$ due to the small pressure change p. The energy stored in the unit volume (potential or elastic energy), can then be written as

$$e_p = \frac{1}{2} p \left(\frac{-\Delta l}{l} \right) \tag{1.29}$$

where Δl has to obey the conservation of mass. We therefore have

$$\rho_0 l S = (\rho_0 + \rho)(l + \Delta l) S. \tag{1.30}$$

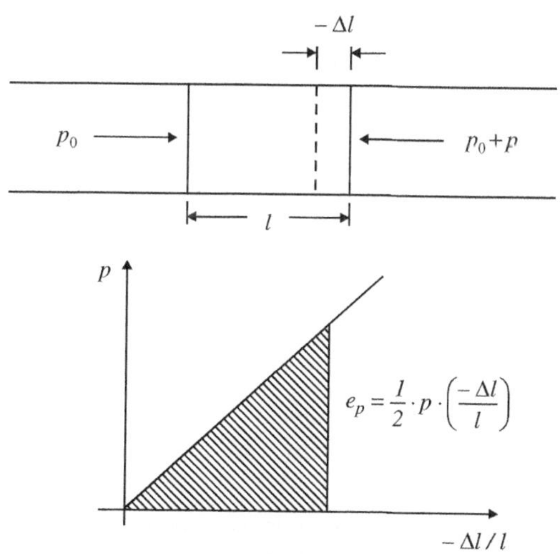

Figure 1.4 Volume change and energy for a one-dimensional element (e_p is potential energy density and p is p' for convenience)

Rearranging this, we obtain

$$\rho_0 l = \rho_0 l + \rho l + \rho_0 \Delta l + \rho \Delta l. \tag{1.31}$$

Note that the last term on the right-hand side is much smaller than the others. Equation (1.31) therefore reduces to

$$-\frac{\Delta l}{l} = \frac{\rho}{\rho_0}. \tag{1.32}$$

This equation simply states that the change of density is proportional to the decrease of volume, $-\Delta l \cdot S$, or the length reduction $(-\Delta l)$.

Substituting Equation (1.32) into (1.29) then gives

$$e_p = \frac{1}{2} p \frac{\rho}{\rho_0}. \tag{1.33}$$

Using the state equation (Equation 1.17), and changing ρ to p, then gives

$$e_p = \frac{1}{2} \frac{p^2}{\rho_0 c^2} \tag{1.34}$$

where e_p denotes the acoustic potential energy.

The kinetic energy per unit volume can be written as

$$e_k = \frac{1}{2} \rho_0 u^2. \tag{1.35}$$

If we assume that the dissipated energy in the fluid is much less than the potential energy or kinetic energy, then the total energy has to be written:

$$e_t = e_p + e_k = \frac{1}{2} \frac{p^2}{\rho_0 c^2} + \frac{1}{2} \rho_0 u^2. \tag{1.36}$$

Note that the potential and kinetic energy are identical if the wave of interest is a plane wave in an infinite domain;[10] in other words, the wave propagates without any reflection. This can be readily understood by recalling the impedance of a plane wave, which is $\rho_0 c$.

The next question then is how acoustic energy changes with time. We can see that the energy per unit volume has to be balanced by the net power flow through the surfaces that enclose the volume of interest, as illustrated in Figure 1.5. This observation can be written conceptually as

$$\begin{aligned} &\text{the rate of increase of energy} \\ &\quad = \text{the power entering through the surface at } (x = x) \\ &\quad - \text{the power exiting through the surface at } (x = x + \Delta x) \end{aligned} \tag{1.37}$$

This can be translated into a mathematical expression as follows:

$$\begin{aligned} \frac{\partial e_t}{\partial t} S\,\Delta x &= (p\,u\,S)_x - (p\,u\,S)_{x+\Delta x} \\ &= -\frac{\partial pu}{\partial x} S\,\Delta x. \end{aligned} \tag{1.38}$$

[10] We have to measure both velocity and pressure to derive sound energy. The sound pressure can be easily measured using a microphone; however, measuring fluid particle velocity is relatively hard. The linearized Euler equation certainly implies that the velocity can be measured by using two pressure sensors. Equation (1.11) states that the velocity can be measured by integrating the signals from the microphones that are separated by a small distance apart.

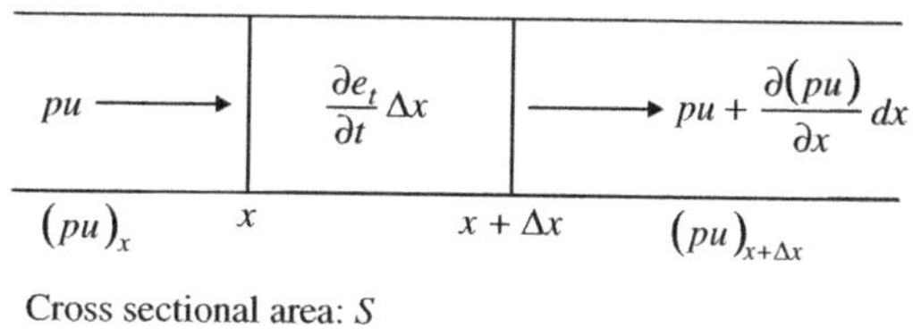

Figure 1.5 Relation between energy (e_t) and one-dimensional intensity (pu) (energy in the volume $S\Delta x$ and the intensity through the surface at x and $x + \Delta x$ must be balanced)

Equation (1.38) can then be reduced to

$$\frac{\partial e_t}{\partial t} + \frac{\partial I}{\partial x} = 0 \tag{1.39}$$

where $I = pu$, which we call "acoustic intensity" or "sound intensity." Acoustic intensity is the acoustic power per unit area.

If we simply extend Equation (1.39) to a three-dimensional case, then

$$\frac{\partial e_t}{\partial t} + \nabla \cdot \mathbf{I} = 0. \tag{1.40}$$

The second term on the left-hand side of Equation (1.40) expresses net outflow power through the fluid surfaces (Figure 1.5). Two major points must be noted in relation to the expression of the intensity. First, intensity is a vector which has direction.[11] Second, intensity is a product of two different physical quantities. Recall that the impedance is also composed of two physical quantities, pressure and velocity, but in a division form; it can therefore be regarded as having an input and output relation. Note, however, that intensity expresses how much power is transported in which direction.

When we have two physical variables, the phase difference between them has significant meaning; for example, the phase relationship between force (pressure) and velocity of impedance: the phase between the force and velocity expresses how well the force generates the velocity (response). In this regard, the intensity can be classified as two different categories: active intensity and reactive intensity. The phase of the former is in-phase and that of the latter has 90° ($\pi/2$) phase difference.

To understand the meaning of the intensities in physical terms, we look again at the simplest case: the intensity of waves propagating in a one-dimensional duct. Figure 1.6 depicts the waves in an infinite-length duct and Figure 1.7 shows the waves for a finite-length L duct.[12] These waves are typical examples of sound visualization. As illustrated in these figures, the visualization can be attempted in various ways, and accordingly all the possible approaches to understanding the propagation of sound waves in space and time can be attained.

When the waves propagate in an infinite duct, where no reflection is possible, the pressure and velocity have the same phase and the frequency of the intensity is double that of the frequency of the pressure and velocity. It can also be observed that the average intensity with respect to time $\langle pu \rangle_{avg}$ is constant, as can be seen in Figure 1.6. The instantaneous intensity oscillates with different phase, however, with regard to the position along the duct.

The general behavior of sound intensity is closely related to the energy transfer at the driving point ($x = 0$ in Figure 1.6). This implies that we can use the impedance at the driving point (*driving*

[11] Strictly speaking, the intensity in one dimension is also indicated in Figure 1.5. The velocity u is in the positive direction of x.

[12] The pressure and velocity of both the infinite as well as the finite duct satisfy the governing equation and the linearized Euler equation. The waves in the finite duct also meet the boundary condition at $x = L$.

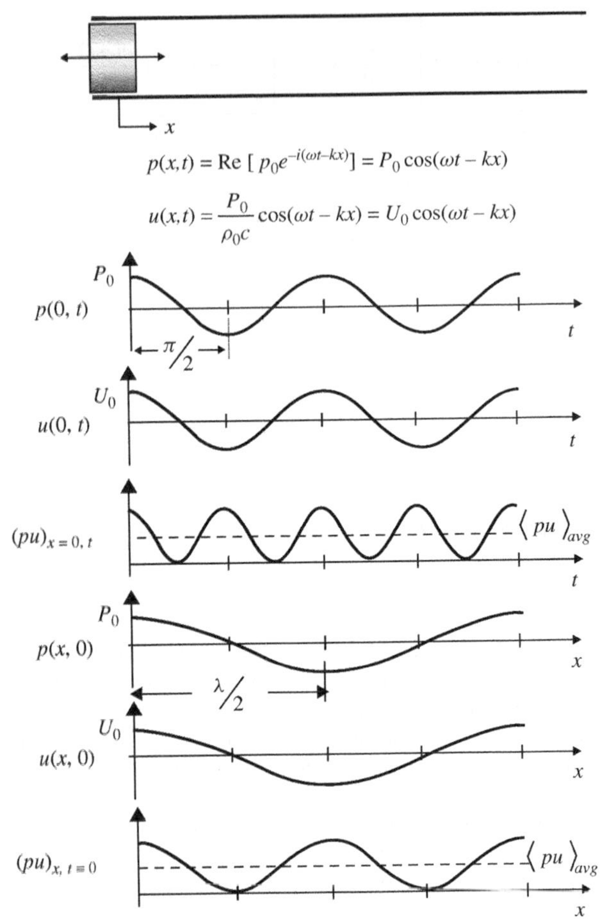

Figure 1.6 The acoustic pressure and intensity in an infinite duct. Note that the pressure and velocity are in phase with each other. Also, the active intensity $\langle pu \rangle_{avg}$ (or average intensity with respect to time) is constant

point impedance) as a representative measure to indicate how well pressure responds to external excitation (velocity). For the case of Figure 1.6, the driving point impedance is given by

$$Z_0 = \frac{p(0,t)}{u(0,t)} = \rho_0 c,$$

which is matched precisely with the characteristic impedance of the medium ($\rho_0 c$). Therefore, the generated waves only propagate in the right-going direction. The fact that the waves all propagate to the right implies that only active intensity is allowed. Accordingly, the excitation effectively supplies energy to the system, which means that the pressure and velocity are in phase. Conversely, if we have the same excitation at one end, the duct has a finite length of L, and a rigid boundary condition exists at the other end $x = L$, then the phase difference between the pressure and velocity will be 90° ($\pi/2$) as depicted in Figure 1.7. Therefore, it is not possible to effectively put energy into the system.

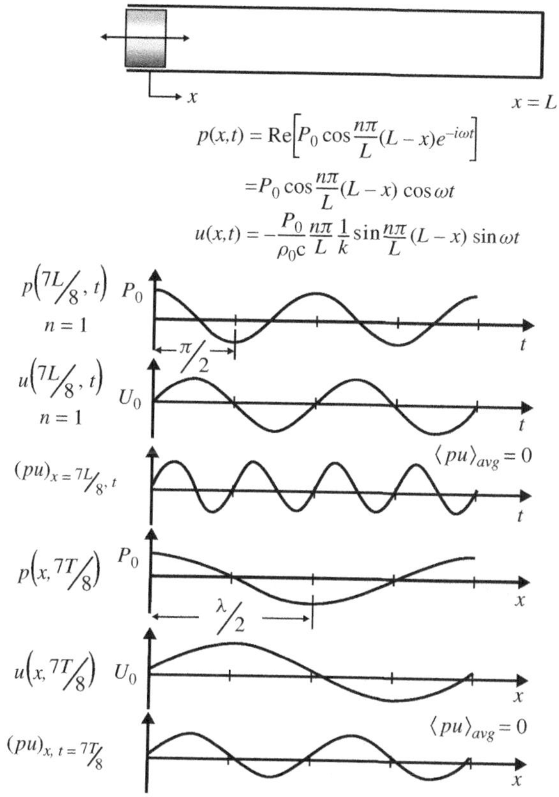

Figure 1.7 The acoustic pressure and intensity in a duct of finite length of L. Note that the phase difference between the pressure and velocity is 90° ($\pi/2$)

Instantaneous acoustic intensity in this case is completely different from what we have for the infinite duct case. The mean intensity is zero, independent of the position of the duct. Also, we see that intensity is always zero at the nodal point of the duct $x = (n + 1)L/4$, but it oscillates between these points where the energy vibrates and does not propagate anywhere. This is a typical characteristic of a finite duct.

We can envisage that a more general case will exhibit behavior which is between those two extreme cases depicted in Figures 1.6 and 1.7. In general, intensity may be both active and reactive.

We have studied fundamental physics and its implications by examining two typical but extreme cases. We now need to explore more specific characteristics of the sound intensity, such as how to calculate and measure the intensity.

The mathematical definition of intensity can be written as

$$\mathbf{I} = p\mathbf{u}. \tag{1.41}$$

The one-dimensional expression is simply

$$I = pu. \tag{1.42}$$

The velocity can be obtained from the Euler equation (Equation (1.11)):

$$u = -\frac{1}{\rho_0}\int \frac{\partial p}{\partial x}\,dt. \tag{1.43}$$

To obtain the derivative with respect to space, we may use two microphones. This means that we approximate the derivative as

$$\frac{\partial p}{\partial x} \approx \frac{p_1 - p_2}{\Delta x}. \tag{1.44}$$

This equation allows us to measure the rate of change of pressure in space. The pressure (p) at the position of the measurement can be approximately obtained as:

$$p \approx \frac{p_1 + p_2}{2} \tag{1.45}$$

where the pressure fluctuates in time and is therefore a dynamic quantity. This means that the microphones that are used to implement Equations (1.44) and (1.45) have to be dynamically identical. In particular, the phase difference has to be treated carefully; this requires accurate calibration for magnitude and phase.

1.4.1 Complex-Valued Pressure and Intensity

We now look at intensity measurement and calculation by considering a plane wave with a radian frequency ω. The expression of pressure as a complex variable can then be written as

$$\begin{aligned}\hat{p}(x,t) &= \hat{p}(x,\omega)e^{-i\omega t}\\ &= \left[P_0(x)\,e^{i\phi_p(x)}\right]e^{-i\omega t}\end{aligned} \tag{1.46}$$

where P_0 denotes the pressure magnitude which has a real value and $\phi_p(x)$ represents the possible phase change in space. This is a more general expression than what we had before, and is devised specially to look at the physics associated with the phase.

To obtain the velocity using the linearized Euler equation a pressure gradient is needed, that is,

$$\frac{\partial \hat{p}}{\partial x} = \left[\frac{dP_0}{dx} + i\left(\frac{d\phi_p}{dx}\right)P_0\right]e^{-i(\omega t - \phi_p)}. \tag{1.47}$$

Equations (1.47) and (1.43) then allow us to obtain the following expression for the complex-valued particle velocity:

$$\hat{u}(x,t) = \frac{1}{i\omega\rho_0}\frac{\partial \hat{p}}{\partial x} = \frac{1}{\omega\rho_0}\left[P_0\frac{d\phi_p}{dx} - i\frac{dP_0}{dx}\right]e^{-i(\omega t - \phi_p)}. \tag{1.48}$$

The first term in Equation (1.48) has the same phase as that of the pressure of Equation (1.46); however, the second has a 90° phase difference. These two different velocities obviously result in very physically distinct intensities.

The intensity generated by the real part of pressure (Equation (1.46)) and the corresponding velocity (Equation (1.48)), which is in phase with the real part of the pressure, can be obtained as

$$\begin{aligned}I_a(x,t) &= P_0\cos(\omega t - \phi_p)\left[\frac{P_0}{\omega\rho_0}\frac{d\phi_p}{dx}\cos\left(\omega t - \phi_p\right)\right]\\ &= \frac{1}{\omega\rho_0}\left[P_0^2\frac{d\phi_p}{dx}\right]\cos^2(\omega t - \phi_p).\end{aligned} \tag{1.49}$$

This is normally referred to as the "active component of sound intensity." The time average of this intensity is often referred to as a mean intensity, or an active intensity, and can be written:

$$I_{avg}(x) = \frac{1}{2\omega\rho_0}\left[P_0^2\frac{d\phi_p}{dx}\right]. \tag{1.50}$$

This intensity can effectively supply power to space, because the velocity and pressure have the same phase.

On the other hand, the multiplication of the real part of the pressure and the imaginary part of the velocity that has 90° phase difference (with respect to the real part of the pressure, Equation (1.46)) will generate the following intensity:

$$\begin{aligned} I_r(x,t) &= P_0\cos(\omega t - \phi_p)\left[-\frac{1}{\omega\rho_0}\frac{dP_0}{dx}\sin\left(\omega t - \phi_p\right)\right] \\ &= -\frac{1}{4\omega\rho_0}\frac{dP_0^2}{dx}\sin 2(\omega t - \phi_p). \end{aligned} \tag{1.51}$$

We refer to this intensity as the "reactive component of sound intensity." The time average of this intensity is 0 and, therefore, there is no net energy transport; it only oscillates.

We now look at the direction of intensity. From Equations (1.49) and (1.51), we see that the active intensity is proportional to the change of phase with space and the reactive intensity is linearly related to the rate of change of pressure squared with space. From these observations, we can conclude that the direction of the active intensity is perpendicular to the wavefront where the phase is constant. In addition, the direction of reactive intensity has to be perpendicular to the surface over which the mean square pressure is constant.[13]

In order to clearly distinguish the intensities at a specific time (Equations (1.49) and (1.51)) from the mean intensity (Equation (1.50)), referred to as the instantaneous active intensity and instantaneous reactive intensity, respectively, are employed. Normally, when we say active intensity, we are referring to a time average of the instantaneous active intensity, that is, Equation (1.50). For the reactive intensity case, we describe its amplitude

$$I_r(x) = -\frac{1}{4\omega\rho_0}\frac{dP_0^2}{dx} \tag{1.52}$$

as reactive.

Note that the instantaneous intensity expressed by Equation (1.42) is composed of two components: the instantaneous active intensity (Equation (1.49)) and instantaneous reactive intensity (Equation (1.51)). We can therefore write them as:

$$I(x,t) = I_{avg}(x)[1+\cos 2(\omega t - \phi_p)] + I_r(x)\sin 2(\omega t - \phi_p). \tag{1.53}$$

Using a complex function, Equation (1.53) can be expressed in simpler form, that is,

$$I(x,t) = \mathrm{Re}\left[C(x)\left(1+e^{-2i(\omega t - \phi_p)}\right)\right] \tag{1.54}$$

where $C(x) = I_{avg}(x) + iI_r(x)$. This is often referred to as complex intensity.

The real part of complex intensity is active intensity (Equation (1.50)) and its imaginary part is the amplitude of reactive intensity (Equation (1.52)). (Appendix B, Section B.1.5 provides details associated with the derivation of intensities.)

---

[13] Depending on the type of exponential function ($\exp(-i\omega t)$ or $\exp(i\omega t)$), the intensity vectors direction may change. Note that the direction of reactive intensity does not matter, but the active intensity direction is in the opposite direction.

To summarize, acoustic energy and intensity represent how acoustic waves propagate in space and time. Acoustic energy is composed of two components: acoustic kinetic energy and acoustic potential energy. The former is due to the motion of the fluid and the latter is induced by compression and expansion of the medium. The rate of propagation of these energies with regard to space and time is known as the intensity. In other words, intensity is energy per unit area in time and space. Intensity is the product of pressure and velocity; it therefore has both magnitude and direction. Because it is a product of two physical quantities, the phase relationship is an important characteristic. Note that impedance is also composed of pressure and velocity but is a division of two physical quantities.

It has been stressed that the phase relation can be simply described by introducing complex notations. In the following sections and chapters, **the complex notation is used as a default descriptor, that is,** p **and** u **denote the complex pressure and velocity** $\hat{p}$ **and** $\hat{u}$.

1.5 The Units of Sound

We have studied how physical quantities, such as pressure, velocity, intensity, energy, and impedance, are related to the propagation characteristics of acoustic waves. We have also seen that a one-dimensional acoustic system such as a duct can express the associated fundamentals.

In considering these physical quantities, it is necessary to know how to communicate them. The units of the physical quantities are the basis of such means. In fact, units are a currency that forms the basis of exchanging these physical quantities. Note also that sound exists before the governing equation is formulated; the measures and units therefore form the essential tenets of acoustics.

The units relevant to sound can be classified into two groups: absolute units and subjective units. The former express the acoustic expression of physical dimensions while the latter take into account the subjective recognition of sound.

Let us first start with the absolute units. The force per unit area, which is pressure, has the units:

$$\text{the unit of pressure} = \text{Pascal(Pa)} = \text{N/m}^2.$$

Velocity, which is defined as the time rate change of displacement, has units:

$$\text{the unit of velocity} = \text{m/s}.$$

Intensity, which is the power per unit area, has units:

$$\text{the unit of intensity} = \text{Pa} \cdot \text{m/s} = \text{W/m}^2.$$

Finally, the unit of energy is defined:

$$\text{the unit of energy} = \text{joule} = \text{W} \cdot \text{s}.$$

Although these absolute units are the necessary parameters that have to be visualized in terms of space and time, it should be noted that subjective units have to be visualized in space and time as well. To understand subjective units, we need to understand how we hear, which means that we need to study our hearing system. Figure 1.8a depicts the human hearing system. Sound arrives at the ear and vibrates the ear drum (Figure 1.8a, external ear, Figure 1.8b). This vibration is amplified by three ossicles before arriving at the cochlea (middle ear, Figure 1.8b). The cochlea has a membrane referred to as the basilar membrane in its middle region, which is filled with fluid (Figure 1.8c). Hair cells are planted on the membrane (Figure 1.8d) and the motion of these cells generates signals that are transported to the brain. The brain processes the signals, feeding them to

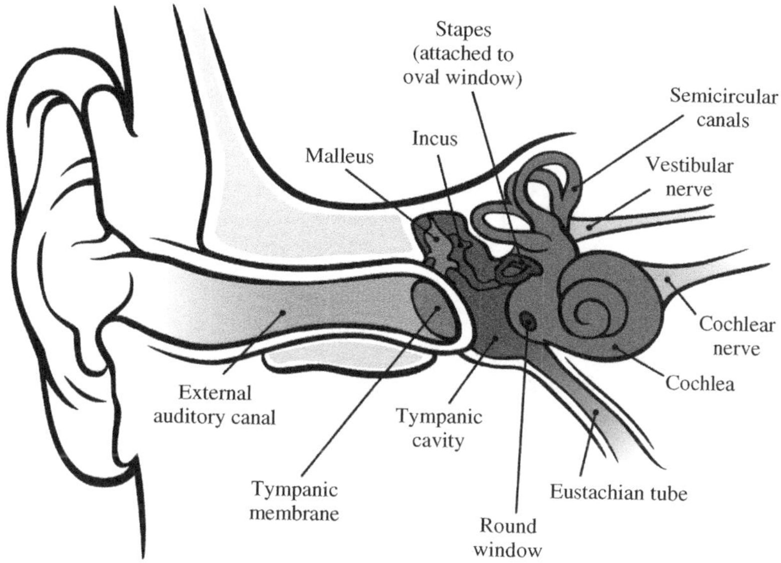

(a)

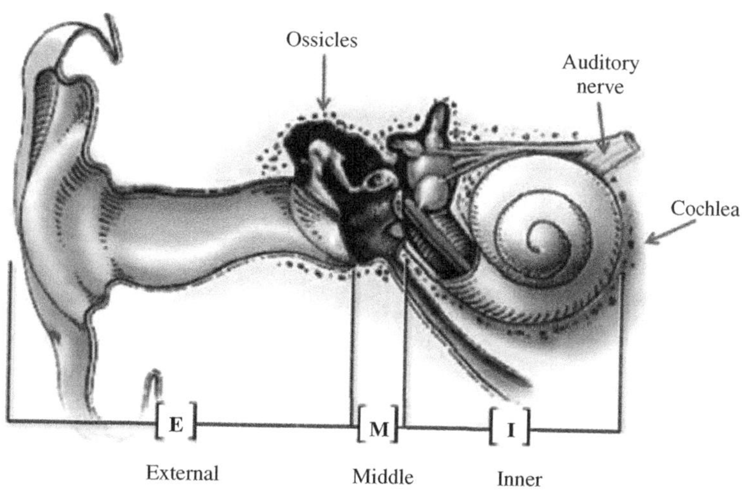

(b)

Figure 1.8 The structure of the ear and its frequency band characteristics. (a) The structure of a human ear (Adapted from [3] with the permission of L. Chittka, this figure was redrawn from the original source: D. Purves et al., Neuroscience, 3rd edition, 2004, pp. 288 (Figure 12.3), Sinauer Associates, Inc., Massachusetts, USA. ©2004 Sinauer Associates, Inc.) (b) external, middle, and inner ear (c) Basilar Membrane and Organ of Corti (d) the cross section of the cochlea shows the sensory cells (located in the organ of Corti) surrounded by the cochlear fluids (e) space-frequency map: moving along the cochlea, difference locations are preferentially excited by different input acoustic frequencies. (f) Tonotopic organization. (Figure 1.8(b–f) Drawings by Stéphan Blatrix, from "Promenade around the cochlea" EDU website http://www.cochlea.org by Rémy Pujol et al. INSERM and University Montpellier.)

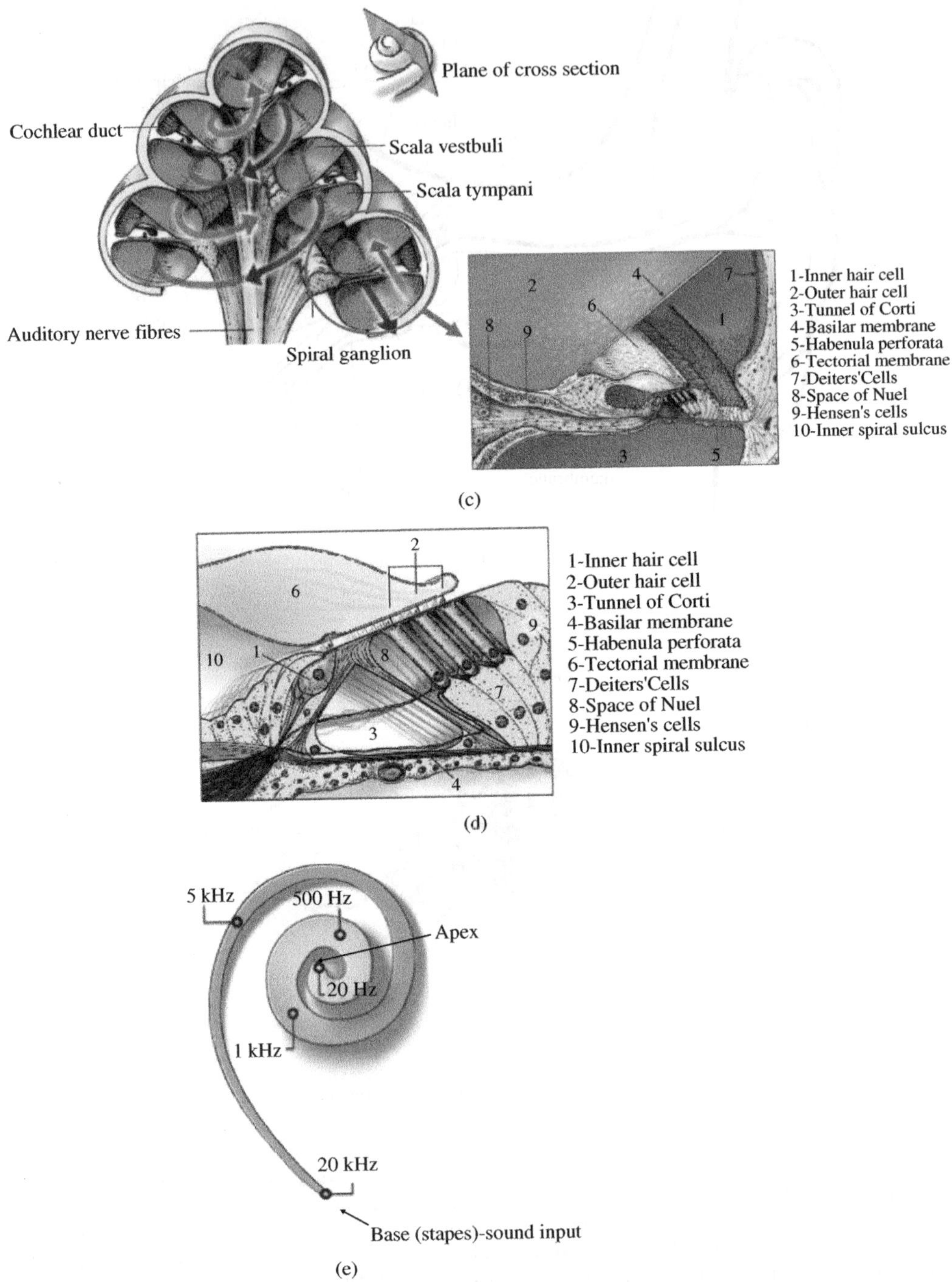

Figure 1.8 (*continued*)

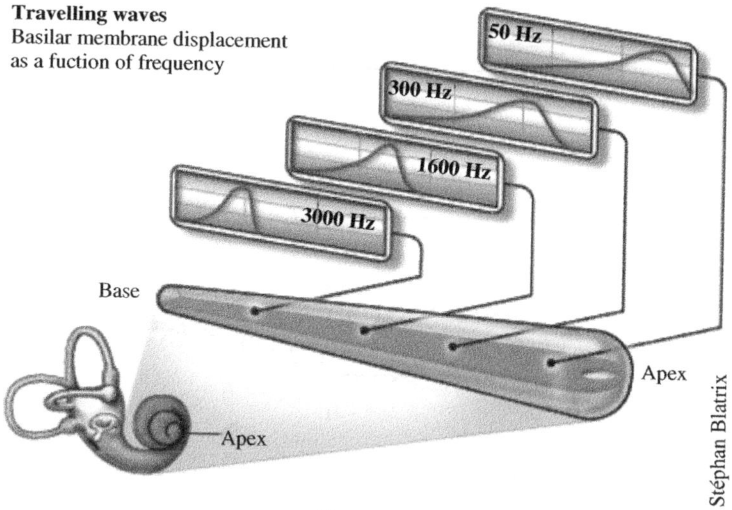

(f)

Figure 1.8 (*continued*)

our cognition system. We recognize the sound and the information that it carries.[14] An interesting characteristic of hair cells is that they behave like a spatially distributed frequency band-pass filter, as illustrated in Figure 1.8e,f.[15]

It is well known that humans do not hear the frequency of sound in absolute scale, but rather relatively.[16] Due to this, we normally use relative units for the frequencies. The octave band is a typical relative scale (Figure 1.9). The band between the reference frequency (f_1) and the frequency that is twice that (f_2) is referred to as an "octave". The 1/3 octave band, which is popular in noise control engineering, has a frequency of $2^{1/3}$ f_1; the reference frequency has the frequency of upper band f_2. The center frequency (f_0) of each band is at the geometrical center of the band.

According to this definition, the center frequency can be any frequency. However, for convenience, we use the standard center frequencies (Table 1.3). The frequency bandwidth of the octave band is about 70% and that of the 1/3 octave band is about 23% of the corresponding band (see Figure 1.9). Therefore, the higher the center frequency, the wider the frequency band.

As we can see in the standard frequency band, humans can hear sound at a broad range of frequencies. The audible frequency range is between 20 Hz and 20 kHz.

[14] The acoustic signal is expected to be effectively transmitted to the fluid of the cochlea. As noted earlier, there is an unavoidable impedance mismatch when sound propagates from air to fluid. To achieve best possible impedance matching our ear attempts, in various ways, to somehow release impedance mismatch. One source is the area ratio between the eardrum size and that of the oval window (about 17 : 1). An additional amplification is due to the lever effect of the middle ear ossicle. Other amplification factors vary with the frequency. For example, the ear canal has maximum amplification of 2.5 kHz. The shoulder, head, and pinna also contribute the amplification. The total of these amplifications is around 30 dB.

[15] Due to the physical arrangement of the hair cell, if a single part is defect then hearing loss corresponding to that frequency band occurs. Because of the characteristics of the hair cell, we often regard it as a spatially distributed spectrum analyzer.

[16] The cognition of frequency or pitch varies with respect to individuals.

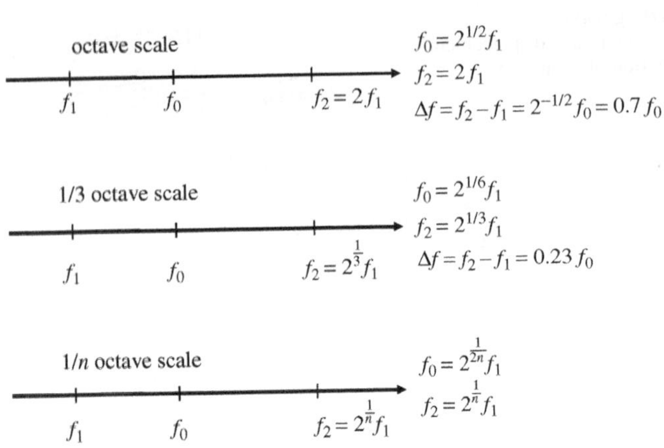

Figure 1.9 Octave, 1/3 octave, and 1/*n* octave

Table 1.3 The center frequency of octave and 1/3 octave

Octave band center frequency	One-third octave band, center frequency	Band limits	
		Lower	Upper
31.5	25	22	28
	31.5	28	35
	40	35	44
63	50	44	57
	63	57	71
	80	71	88
125	100	88	113
	125	113	141
	160	141	176
250	200	176	225
	250	225	283
	315	283	353
500	400	353	440
	500	440	565
	630	565	707
1000	800	707	880
	1000	880	1130
	1250	1130	1414
2000	1600	1414	1760
	2000	1760	2250
	2500	2250	2825
4000	3150	2825	3530
	4000	3530	4400
	5000	4400	5650
8000	6300	5650	7070
	8000	7070	8800
	10 000	8800	11 300
16 000	12 500	11 300	14 140
	16 000	14 140	17 600
	20 000	17 600	22 500

There are two important reasons why we scale or measure the frequency of our interest by an octave or 1/3 of an octave: (i) we hear frequency in a relative scale and (ii) we cannot express our audible frequency range by using a linear frequency scale.

For the amplitude of the sound pressure, we use the sound pressure level (SPL or L_p). It is defined as

$$\mathrm{SPL} = L_p = 10\log_{10}\left(\frac{p_{avg}^2}{p_{ref}^2}\right) \tag{1.55}$$

and is measured in units of decibels (dB); p_{ref} is the reference pressure, p_{avg} is the average pressure, and $\log_{10}$ is a log function that has a base of 10. p_{ref} is 20 μPa (20×10^{-6} N/m^2). The reference pressure is the smallest sound that a human being can hear. The first reason why we use a log scale and do not use a linear scale is that we can hear sound in a range from small amplitude to extremely large amplitude (Figure 1.10). From Figure 1.10, we can see that the human can hear from about 0 dB to somewhere in the range of 130–140 dB [4].

The second reason is that we recognize the level of sound relatively and not in an absolute scale, similar to frequency recognition. Equation (1.55) essentially expresses how large the sound is compared to the reference SPL, and is therefore a good representation of the human hearing

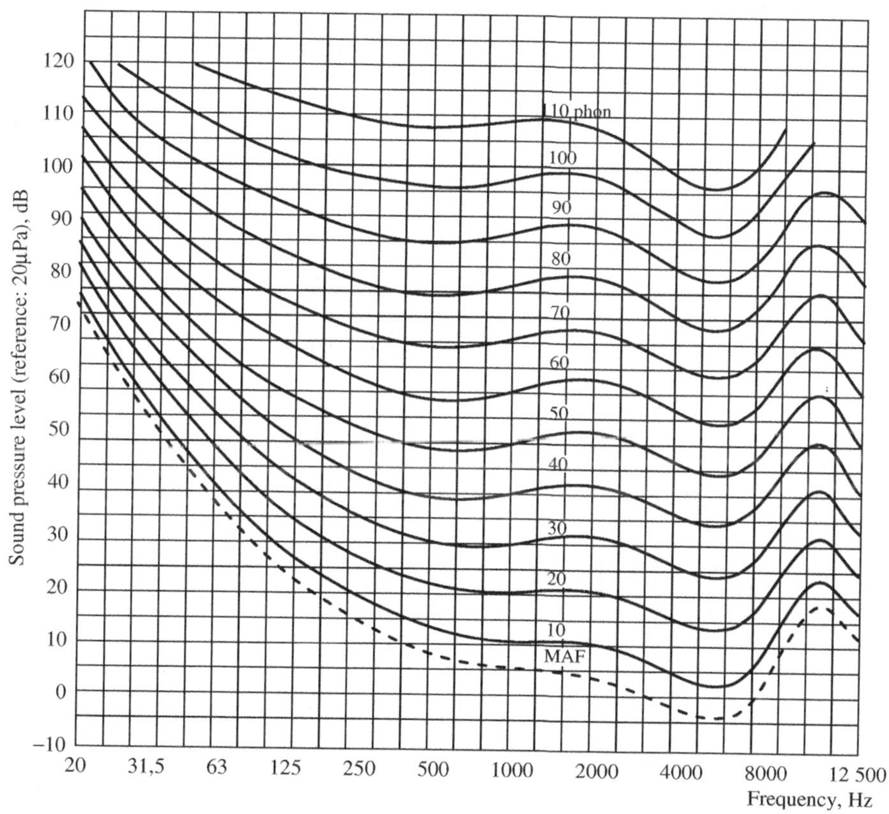

Figure 1.10 Equal-loudness contour: each line shows the SPL with respect to the frequency which corresponds to a loudness (phon) of 1 kHz pure sound. Adapted from ISO 226: 2003 revision "Equal loudness contour"

Table 1.4 Daily life noise level in SPL

SPL (dB ref. 20 μPa)	Example
120	When a jet takes off (60 m)
110	Construction site
100	Loud shouting (1.5 m)
90	Truck noise (15 m)
80	Busy street
70	Moving car interior
60	Normal conversation (1 m)
50	Office
40	Living room
30	Bedroom (night)
20	Broadcasting studio
10	Leaves oscillating by a breeze
0	Minimum audible sound pressure level

Reproduced from Thomas D. Rossing, The science of sound, 2nd Edition, Addison-Wesley, pp 86, 1990

system. Table 1.4 collects some typical samples of sound levels that we can encounter, providing some practical references of the SPL [5].

In order to calculate SPL, we write the mean-squared value of the real-valued pressure $\mathrm{Re}[p(t)]$ as

$$p_{avg}^2 = \frac{1}{T}\int_0^T \mathrm{Re}[p(t)]^2 dt \tag{1.56}$$

where T denotes the measurement time. Equation (1.56) can be expressed as the sum of every frequency component, equivalent to

$$p_{avg}^2 = \frac{1}{T}\int_0^T \sum_n \sum_m \mathrm{Re}\left[p_m e^{-i\omega_m t}\right] \mathrm{Re}\left[p_n e^{-i\omega_n t}\right] dt. \tag{1.57}$$

where p_m is the complex amplitude of the mth frequency component $(p(t) = \sum_m p_m e^{-i\omega_m t})$.

We then use the well-known relation

$$\mathrm{Re}\left[p_m e^{-i\omega_m t}\right] = \frac{1}{2}\left(p_m e^{-i\omega_m t} + p_m^* e^{i\omega_m t}\right) \tag{1.58}$$

where * denotes the complex conjugate. If we rearrange Equation (1.57) using Equation (1.58), then we obtain

$$\begin{aligned} p_{avg}^2 &= \frac{1}{4T}\sum_n \sum_m \int_0^T \left[p_m p_n e^{-i(\omega_m+\omega_n)t} + p_m p_n^* e^{-i(\omega_m-\omega_n)t}\right. \\ &\quad \left. + p_m^* p_n e^{i(\omega_m-\omega_n)t} + p_m^* p_n^* e^{i(\omega_m+\omega_n)t}\right] dt \\ &= \frac{1}{4T}\sum_n \sum_m \int_0^T \{2\mathrm{Re}\left[p_m p_n e^{-i(\omega_m+\omega_n)t}\right] + 2\mathrm{Re}\left[p_m p_n^* e^{-i(\omega_m-\omega_n)t}\right]\} dt \\ &\cong \frac{1}{4T}\sum_m \int_0^T 2\mathrm{Re}\left[|p_m|^2\right] dt \\ &= \sum_m \frac{1}{2}|p_m|^2. \end{aligned} \tag{1.59}$$

Equation (1.59) states that the mean square average of sound pressure is the sum of the mean square average of each frequency's sound pressure. This is only valid if and only if Equation (1.59) has a maximum when $n = m$. Note that when $n \neq m$, the slowly fluctuating terms with frequency $(\omega_m - \omega_n)$ are much greater than those with a frequency of $(\omega_m + \omega_n)$.

Figure 1.11 illustrates the relation between the SPL and the mean square pressure.

Equations (1.55) and (1.59) indicate that it is important to carefully sum the sound levels or the mean square values of different sounds. Let us begin with two sound pressures that have different frequencies, ω_1 and ω_2. According to Equation (1.55), the SPL of each individual tone can then be written as

$$\text{tone 1}: \ \text{SPL} = L_{p_1} = 10\log_{10}\frac{p_{1,avg}^2}{p_{ref}^2} \tag{1.60}$$

$$\text{tone 2}: \ \text{SPL} = L_{p_2} = 10\log_{10}\frac{p_{2,avg}^2}{p_{ref}^2}. \tag{1.61}$$

If these two tones occur at the same time, the SPL can be written

$$\begin{aligned}\text{SPL}_{1+2} = L_{p_{1+2}} &= 10\log_{10}\frac{p_{1,avg}^2 + p_{2,avg}^2}{p_{ref}^2} \\ &= 10\log_{10}(10^{L_{p_1}/10} + 10^{L_{p2}/10}).\end{aligned} \tag{1.62}$$

For example, if each tone has an SPL of 80 dB, that is, $L_{p_1} = 80$ dB and $L_{p_2} = 80$ dB, then the sum of these two must be $L_{p_{1+2}} = 10\log_{10}(10^8 + 10^8)$. This simply means that the SPL increases

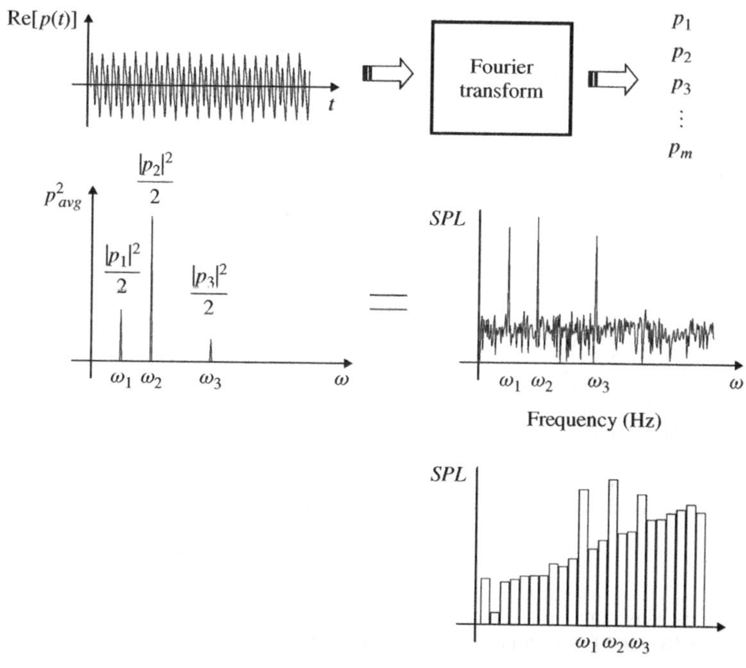

Figure 1.11 Total mean square pressure and the mean square pressure of each frequency band

by 3 dB.[17] This method is valid if we add any sound pressure that has the same level. It also implies that the biggest sound level dominates the total SPL when there are many sounds together. For example, if we have two sounds of SPL 75 and 80 dB, the resulting SPL of the sounds is 81.2 dB.

If we generalize this result to N different pure tone cases, the SPL is defined

$$\mathrm{SPL}_{1+2+\cdots+N} = L_{p_{1+2+\cdots+N}} = 10\log_{10}\left(10^{L_{p1}/10} + \cdots + 10^{L_{pN}/10}\right). \quad (1.63)$$

The SPL can be obtained from Equation (1.63), but it is not yet clear how humans perceive sound. A well-known fact is that we acknowledge sound depending on its frequency content. As illustrated in Figure 1.10, we interpret equal loudness for sounds with different sound pressures. A primary conclusion based on these characteristics of hearing perception is that there is a scaling factor with respect to frequency. This motivates us to introduce a weighting curve to the SPL (Figure 1.12) [9]. The A-weighting is the most widely used weighting curve.

The weighting curve is based on our frequency-dependent sensation. The human ear is normally very sensitive to a frequency of 1–3 kHz in the case of a pure tone, and less sensitive to frequencies lower than this range (Figure 1.10). Recall that the weighting curves are all based on pure tones. For more general sound or noise, the use of these curves is somewhat different from reality. It is therefore reasonable to use the curves as a guideline to considering human perception with regard to frequency.

To summarize, the most commonly used unit to measure the amplitude of sound is sound pressure level (SPL, L_p) which is defined as the ratio of the mean square pressure to the mean square

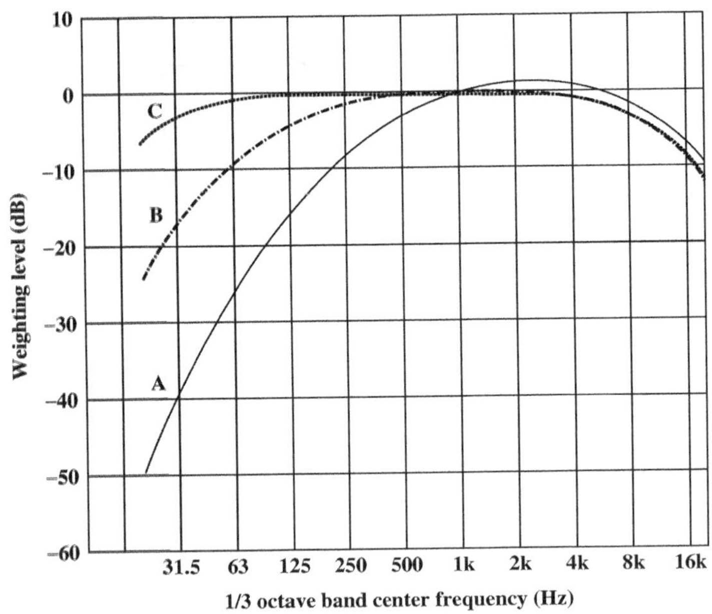

Figure 1.12 Various weighting curves. A-weighting: 40 phon curve (SPL < 55 dB); B-weighting: 70 phon curve (SPL = 55–85 dB); and C-weighting: 100 phon curve (SPL > 85 dB)

[17] The minimum difference in SPL which we can hear is 2 dB. Adapted from Davis A. Bies and Colin H. Hansen, Engineering Noise Control: Theory and Practice, 4th ed., Taylor & Francis, 2009, p. 104.

Table 1.5 The measurement standards

Physical measure	Definition	Measurement reference (SI)
According to ANSI SI.8-1989		
Sound pressure level, dB (gases)	$L_p = 20\ \log_{10}(p_{avg}/p_{ref})$	$p_{ref} = 20\ \mu\text{Pa} = 2 \times 10^{-5}\ \text{N/m}^2$
Sound pressure level, dB (other than gases)	$L_p = 20\ \log_{10}(p_{avg}/p_{ref})$	$p_{ref} = 1\ \mu\text{Pa} = 10^{-6}\ \text{N/m}^2$
Sound power level, dB	$L_W = 10\ \log_{10}(W/W_0)$	$W_0 = 1\ \text{pW} = 10^{-12}\ \text{N m/s}$
Sound power level, bel	$L_W = \log_{10}(W/W_0)$ bel	$W_0 = 1\ \text{pW} = 10^{-12}\ \text{N m/s}$
Sound intensity level, dB	$L_I = 10\ \log_{10}(I/I_0)$	$I_0 = 1\ \text{pW/m}^2 = 10^{-12}\ \text{N/m s}$
Vibratory force level, dB	$L_{Fv} = 20\ \log_{10}(F/F_0)$	$F_0 = 1\ \mu\text{N} = 10^{-6}\ \text{N}$
Frequency level, dB	$N = \log_{10}(f/f_0)$	$f_0 = 1\ \text{Hz}$
Sound exposure level, dB	$L_E = 10\ \log_{10}(E/E_0)$	$E_0 = (20\ \mu\text{Pa})^2\ \text{s} = (2 \times 10^{-5}\ \text{Pa})^2\text{s}$
According to ISO or suggested by the author		
Sound energy level given in ISO 1683;1983	$L_e = 10\ \log_{10}(e/e_0)$	$e_0 = 1\text{pJ} = 10^{-12}\ \text{Nm}$
Sound energy density level given in ISO 1683;1983	$L_D = 10\ \log_{10}(D/D_0)$	$D_0 = 1\ \text{pJ/m}^3 = 10^{-12}\ \text{N/m}^3$
Vibration acceleration level	$L_a = 20\ \log_{10}(a/a_0)$	$a_0 = 10\ \mu\text{m/s}^2 = 10^{-5}\ \text{m/s}^2$
Vibration acceleration level in ISO 1683;1983	$L_a = 20\ \log_{10}(a/a_0)$	$a_0 = 1\ \mu\text{m/s}^2 = 10^{-6}\ \text{m/s}^2$
Vibration velocity level	$L_v = 20\ \log_{10}(v/v_0)$	$v_0 = 10\ \text{nm/s} = 10^{-8}\ \text{m/s}$
Vibration velocity level in ISO 1683; 1983	$L_v = 20\ \log_{10}(v/v_0)$	$v_0 = 1\ \text{nm/s} = 10^{-9}\ \text{m/s}$
Vibration displacement level	$L_d = 20\ \log_{10}(d/d_0)$	$d_0 = 10\ \text{pm} = 10^{-11}\ \text{m}$

Noise and Vibration Control Engineering : Principles and applications, 1992, pp. 20, Reference Quantities for Acoustical Levels from American National Standard ANSI S1.8-1989 (Reaffirmed 2001) and As Preferred by Authors, Leo L. Beranek, István L. Vér, John Wiley & Sons, Inc.: With kind permission of John Wiley & Sons

reference pressure. The acoustic intensity expresses the power transmitted through the unit surface, and its reference intensity (I_0) is defined as $10^{-12}\text{W/m}^2 = 10^{-12}\text{N/(m} \cdot \text{s)}$.

Table 1.5 summarizes the measurement units and its criteria. SPL is the sum of the mean square pressure of individual frequency components of sound [6]. We also see that weighting is necessary to consider the frequency perception of sound. The basic frequency scale has to be an octave or 1/3 octave to take into account human perception. For the same reason, a logarithmic scale is also necessary to measure the SPL.

All of these units can be visualized, depending on the desired use of the outcome. Accordingly, the appropriate measure has to be selected. However, it should be noted that the selected measure has to be understood physically, so that one can interpret the essential meaning of what is being seen.

1.6 Analysis Methods of Linear Acoustic Wave Equation

We have studied the physical variables – pressure, density, and fluid particle velocity – which determine acoustic wave propagation as well as their relations (Section 1.2). The physical implications of intensity and energy on acoustic waves have been also studied in detail (Section 1.4). We have subsequently introduced how these physical quantities, especially sound pressure and intensity, excite our hearing system. What we hear is the output of our hearing system from the input of physical acoustic variables, sound pressure and intensity. The major attributes of our hearing system are how we hear frequency and how we acknowledge SPL. The former is described by the octave band frequency unit while the latter is expressed in units of dB (Section 1.5).

In terms of sound visualization, all measures described thus far, including subjective measures, are all to be visualized in terms of space and time. It is also notable that the expression of these measures is not limited to time and space domains. Depending on the convenience of interpretation, these measures can be expressed in terms of frequency or wavenumber domains as well. As mentioned in this chapter, the primary objectives of visualization are to obtain useful information that cannot be easily obtained by other means. To obtain or to determine what is useful, we have to understand the underlying physics. It is well known that the level of understanding can be increased by understanding the associated analysis method.

This section addresses how we mathematically predict or describe sound in space and time. We do not want to imply that a mathematical approach is the best way to understand acoustic wave propagation, but want to demonstrate that a mathematical approach provides us with a logical way to understand the characteristics of sound propagation. A mathematical approach does not always mean that we will eventually arrive at an exact solution, but is a guide to understanding the fundamentals of all forms of acoustic wave propagation.

1.6.1 Acoustic Wave Equation and Boundary Condition

The acoustic wave equation that we learned in Section 1.2 represents the characteristics of acoustic wave propagation. However, it only expresses that which is related to the medium. It can be applied to every possible wave that satisfies the governing equation. For example, for waves that propagate in a one-dimensional duct, Equations (1.1) and (1.2) satisfy the governing equation. The next point to consider is what makes the solution unique: the answer is the boundary and initial conditions.

We conceive a method to express the sound source. In fact, the governing equation previously studied satisfies waves in time and space without regard to acoustic sources, that is, excitations. The equation is, therefore, a homogeneous governing equation. The inhomogeneous terms that express acoustic sources or excitations include anything that can make acoustic waves; however, for convenience, we consider only pressure and velocity sources. The former creates waves by changing pressure; the latter is related to volume change.

Let us begin with a situation where we have a volume source in one-dimensional infinite space. The volume velocity source makes the mass change by the velocity excitation. We can express this mathematically by modifying Equations (1.12) and (1.13), that is,

$$\frac{\partial \rho}{\partial t} = -\rho_0 \frac{\partial u}{\partial x} + \rho_0 \upsilon, \tag{1.64}$$

where $\upsilon = \upsilon(x, t)$ is the volume velocity at x. Therefore, the last term of the right-hand side of Equation (1.64) expresses the time rate of mass supply to the fluid. In summary, Equation (1.64) states that the time rate of mass change per unit volume is balanced by the net supply of mass and net mass flux through the surface of the fluid.

Substituting Equations (1.17) and (1.11) into this new mass law equation (Equation (1.64)), we obtain the governing equation that includes the acoustic source:

$$\frac{\partial^2 p}{\partial x^2} - \frac{1}{c^2}\frac{\partial^2 p}{\partial t^2} = -\rho_0 \frac{\partial \upsilon}{\partial t}. \tag{1.65}$$

The right-hand side of Equation (1.65) is the multiplication of fluid density and volume acceleration, and therefore expresses the inertia force (mass acceleration) of the source.

We first attempt a harmonic solution, as follows:

$$p(x, t) = p(x, \omega)e^{-i\omega t}. \tag{1.66}$$

Equation (1.65) can then be written as

$$\frac{d^2p(x,\omega)}{dx^2} + k^2p(x,\omega) = -q(x,\omega), \text{ where } \quad q = \rho_0\frac{\partial v}{\partial t} \tag{1.67}$$

where $q(x, \omega)$ represents the right-hand side of Equation (1.65) in an abbreviated form. In fact, we assume that it can be expressed by $q(x, \omega)e^{-i\omega t}$, and for simplicity, the frequency dependency ω will be omitted throughout this section. Equation (1.65) is a typical inhomogeneous governing equation. Note that Equation (1.67) is strictly only valid where the sound source exists; otherwise a homogeneous equation is valid. For example, if there is a point source at x_0, then Equation (1.67) can be rewritten as

$$\frac{d^2p(x)}{dx^2} + k^2p(x) = -q(x_0)\delta(x - x_0), \tag{1.68}$$

where $\delta(x - x_0)$ is a Dirac delta function, that is,

$$\int_{-\infty}^{\infty} \delta(x - x_0)dx = 1;$$

$$\delta(x - x_0) = 0; \quad x \neq x_0. \tag{1.69}$$

If the source exists only in the region L_0, then we can write the governing equation as

$$\frac{d^2p(x)}{dx^2} + k^2p(x) = -\int_{L_0} q(x_0)\delta(x - x_0)dx_0. \tag{1.70}$$

Expanding this equation to a three-dimensional case yields

$$\nabla^2 p(\mathbf{r}) + k^2p(\mathbf{r}) = -\int_{V_0} q(\mathbf{r}_0)\delta(\mathbf{r} - \mathbf{r}_0)dV(\mathbf{r}_0), \tag{1.71}$$

where $\mathbf{r}_0$ and V_0 express the source position and the volume where the source is, respectively. To summarize, if there is a sound source or sources, then the mechanism that we investigated in Section 1.2 is no longer valid at the source position and the governing equation becomes Equation (1.71), that is, the inhomogeneous wave equation.[18]

We now look at how to mathematically express the boundary condition. If a space is closed by a boundary, then the solutions have to satisfy not only the wave equation but also the boundary condition.

We first study the one-dimensional case. The boundary condition can generally be shown as

$$\alpha_0 p(0) + \beta_0 u(0) = \gamma_0, \quad x = 0$$

$$\alpha_L p(L) + \beta_L u(L) = \gamma_L \quad x = L \tag{1.72}$$

where the subscript 0 and L represent boundary value at $x = 0, L$. To understand the boundary conditions expressed by Equation (1.72), let us investigate several typical cases. First, when $\beta_{0,L} = 0$, the condition takes the form

$$p(0) = \frac{\gamma_0}{\alpha_0} \quad p(L) = \frac{\gamma_L}{\alpha_L} \tag{1.73}$$

This type of boundary condition, which describes the pressure on the boundary, is generally known as the Dirichlet boundary condition.

[18] It is possible to express any sound source by pressure source, volume source, or their combination.

On the other hand, if $\alpha_{0,L} = 0$, then the equation becomes

$$u(0) = \frac{\gamma_0}{\beta_0} \quad u(L) = \frac{\gamma_L}{\beta_L} \tag{1.74}$$

Equation (1.74) essentially describes the situation where the boundary moves with the prescribed velocity. If we recollect that the velocity is related to the change of pressure with regard to space (according to Euler equation), then Equation (1.74) describes the boundary condition in terms of pressure derivatives on the boundary. This type of boundary condition is called a Neumann boundary condition.

If $\gamma = 0$, then Equation (1.72) reduces to

$$\begin{aligned}\alpha_0 p(0) + \beta_0 u(0) = 0,\\ \alpha_L p(L) + \beta_L u(L) = 0,\end{aligned} \iff \begin{aligned} p(x) &= -\frac{\beta_0}{\alpha_0}\left(\frac{1}{i\rho_0\omega}\frac{dp(x)}{dx}\right) = \chi_0\frac{dp(x)}{dx}, \quad \text{at } x = 0\\ p(x) &= -\frac{\beta_L}{\alpha_L}\left(\frac{1}{i\rho_0\omega}\frac{dp(x)}{dx}\right) = \chi_L\frac{dp(x)}{dx}, \quad \text{at } x = L.\end{aligned} \tag{1.75}$$

or

$$\frac{p(0)}{u(0)} = -\frac{\beta_0}{\alpha_0}, \quad \frac{p(L)}{u(L)} = -\frac{\beta_L}{\alpha_L}. \tag{1.76}$$

The impedance, which is the ratio between pressure and velocity, is described on the boundary. We refer to the condition where the right-hand side is 0 (as in Equation (1.75)) as a homogeneous boundary condition. If the equation follows the form of Equation (1.73) or (1.74), then we refer to it as an inhomogeneous boundary condition. The latter implies that there is an active element on the boundary while the former means the boundary behaves rather passively.

More generally, the three-dimensional case of Equation (1.75) can be written as

$$\alpha p(\mathbf{r}_0) + \beta u(\mathbf{r}_0) = 0, \quad \text{on} \quad S_0, \tag{1.77}$$

where S_0 is the surface that encloses the space of interest, as depicted in Figure 1.13 where $u(\mathbf{r}_0)$ is the particle velocity that is normal to the surface. Equation (1.72) is the more generally acceptable form in practice, but it must be noted that we can always change the inhomogeneous boundary condition in terms of the homogeneous boundary condition and the inhomogeneous governing equation. Therefore, we will only consider the problem that is governed by the inhomogeneous governing equation (Equation (1.70)) and homogeneous boundary condition (Equation (1.75)). That is,

$$\frac{d^2p(x)}{dx^2} + k^2p(x) = -\int_{L_0} q(x_0)\delta(x - x_0)dx_0$$

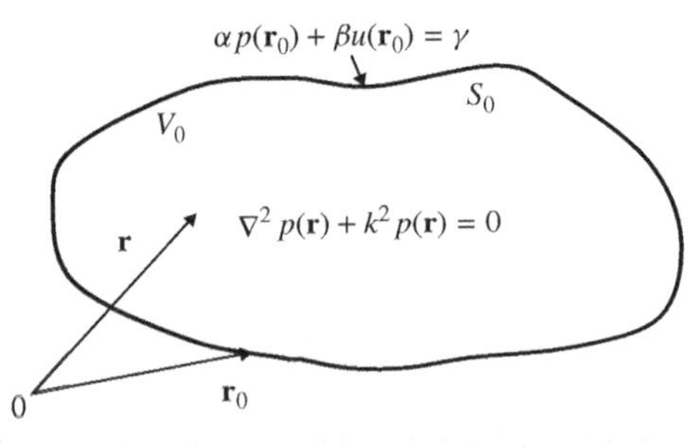

Figure 1.13 General boundary value problem ($p(\mathbf{r})$ is complex amplitude, $u(\mathbf{r})$ is complex velocity, k is the wavenumber, and S_0 expresses the boundary of a volume V_0; $\mathbf{r}$ and $\mathbf{r}_0$ indicate the observation position and boundary, respectively)

and

$$\alpha_0 p(0) + \beta_0 u(0) = 0, \quad x = 0$$
$$\alpha_L p(L) + \beta_L u(L) = 0 \quad x = L$$

1.6.2 Eigenfunctions and Modal Expansion Theory

1.6.2.1 Eigenfunctions

One very well-known method for obtaining solutions that satisfy the wave equation and boundary conditions is to expand the wave field as a sum of linearly independent functions that satisfy the imposed boundary conditions. This kind of approach is commonly referred to as the eigenfunction or modal expansion approach.

To begin with, consider a linear operator, $\mathcal{L}$, that maps a function, $p(x)$, to another function, $-q(x)$:

$$\mathcal{L}\, p(x) = -q(x). \tag{1.78}$$

By comparing Equations (1.78) and (1.70), we can see that the inhomogeneous acoustic wave equation for forced excitation is equivalent to the mapping by the operator $\mathcal{L}$, expressed as follows:

$$\mathcal{L} = [\nabla^2 + k^2] \quad \text{for a 3D wave equation,}$$
$$\mathcal{L} = \left[\frac{d^2}{dx^2} + k^2\right] \quad \text{for a 1D wave equation.} \tag{1.79}$$

In this equation, the linear differential operator $\mathcal{L}$ essentially describes the mapping between the pressure field $p(x)$ and the excitation $q(x)$.

The physical meaning of $\mathcal{L}$ is worth mentioning here. As $\mathcal{L}$ maps the pressure field $p(x)$ to the source excitation $q(x)$, it is possible to estimate $q(x)$ provided that $p(x)$ is known over the region where the excitation exists. This property is beneficial when we attempt to predict the distribution of the excitation from the given $p(x)$, which is exactly what we want to achieve by solving the sound visualization problem. The wave equation itself tells us that such a prediction is theoretically possible.

The solution that fulfills the mapping relation of Equation (1.78) can be efficiently described in terms of eigenfunctions. The eigenfunction or mode $\psi_n(x)$ of $\mathcal{L}$ is defined as the function satisfying the given boundary conditions and the following *characteristic equation*:

$$\mathcal{L}\psi_n(x) = \mu_n \psi_n(x), \tag{1.80}$$

where the constant μ_n is the nth eigenvalue of $\mathcal{L}$ associated with $\psi_n(x)$. For the acoustic wave equation (Equation (1.79)), we can rewrite the characteristic equation (Equation (1.80)) in a more popular form:

$$[\mathcal{L} - \mu_n]\psi_n(x) = 0 \Rightarrow \left[\frac{d^2}{dx^2} + k_n^2\right]\psi_n(x) = 0, \tag{1.81}$$

which now has the form of the homogeneous wave equation. The squared wavenumber $k_n^2 = k^2 - \mu_n$ can be regarded as an eigenvalue, because we can express Equation (1.81) as the characteristic equation for the Laplacian operator $\mathcal{L}_L = d^2/dx^2$:

$$\mathcal{L}_L \psi_n = -k_n^2 \psi_n. \tag{1.82}$$

It is noteworthy that the eigenvector ψ_n of $\mathcal{L}$ also becomes the eigenvector of the Laplacian operator $\mathcal{L}_L$. Accordingly, the eigenvectors that satisfy the homogeneous wave equation (Equation (1.81)) at discrete wavenumbers (k_n^2) can be used to express the solution of the inhomogeneous wave equation (Equation (1.78)). The frequency $\omega_n = ck_n$ associated with the wavenumber k_n is often denoted as the eigenfrequency.

The eigenvalue of Equation (1.82) can have either a real or complex value depending on the boundary condition. To determine the case for which the eigenvalue is complex, we consider the following operation obtained by multiplying Equation (1.82) with the eigenvector and integrating it over $x \in [0, L]$.

$$\int_0^L \psi_n^*(\mathcal{L}_L\psi_n)dx = -k_n^2\int_0^L |\psi_n|^2 dx \tag{1.83}$$

Integration by parts of the left side of Equation (1.83) gives

$$\int_0^L \psi_n^* \frac{d^2\psi_n}{dx^2}dx = \psi_n^*\frac{d\psi_n}{dx}\bigg|_{x=0}^{x=L} - \int_0^L \left|\frac{d\psi_n}{dx}\right|^2 dx. \tag{1.84}$$

The eigenfunction ψ_n satisfies the boundary condition of Equation (1.75); accordingly, we can derive the relation between the eigenvalue of Equation (1.83) and the boundary condition. From Equations (1.75) and (1.83), (1.84) can be written as

$$-k_n^2\int_0^L |\psi_n(x)|^2 dx = \frac{1}{\chi_L}|\psi_n(L)|^2 - \frac{1}{\chi_0}|\psi_n(0)|^2 - \int_0^L \left|\frac{d\psi_n}{dx}\right|^2 dx. \tag{1.85}$$

Since all the integrals are real-valued functions, the eigenvalue k_n^2 becomes complex only when χ_0 or χ_L is complex. Therefore, the eigenvalue is real for the reactive boundary condition (real-valued χ) and is complex if one of the boundaries has a resistive (imaginary χ) part.

1.6.2.2 Self-Adjointness and Orthogonality of Eigenfunctions

Next, we discuss an important property of the operator, of having orthogonality between two different eigenfunctions. If the operator $\mathcal{L}_L$ fulfills

$$\int_0^L f(x)^*(\mathcal{L}_L h(x))dx = \left[\int_0^L \left(\mathcal{L}_L f(x)\right) h(x)^* dx\right]^* \tag{1.86}$$

for arbitrary functions $f(x)$ and $h(x)$ that satisfy the boundary condition, then it is called a *self-adjoint operator*. The self-adjointness property of $\mathcal{L}_L$ or $\mathcal{L}$ can be proved as follows: First, we subtract the right-hand side of Equation (1.86) from its left-hand side, such that

$$\int_0^L \left[f^*\frac{d^2}{dx^2}h - h\frac{d^2}{dx^2}f^*\right]dx = \left[f^*\frac{d}{dx}h - h\frac{d}{dx}f^*\right]_{x=0}^{x=L}. \tag{1.87}$$

If the boundary condition is *homogeneous* (as in Equation (1.75)) and is *purely reactive*, then Equation (1.87) is zero for functions $f(x)$ and $h(x)$. Accordingly, Equation (1.86) holds for the homogeneous and reactive boundary condition. The argument in the integral of Equation (1.87), called the Wronskian of $f(x)$ and $h(x)$, represents the *fitness* of the boundary condition.

From the condition for having a vanishing Wronskian, we can derive the orthogonality of the eigenfunctions. For the self-adjoint operator $\mathcal{L}_L$ fulfilling Equation (1.86) and for the eigenfunctions satisfying the characteristic equation (Equation (1.80)), we can derive the following relation:

$$\int_0^L \psi_n^*(\mathcal{L}_L\psi_m)dx - \left[\int_0^L (\mathcal{L}_L\psi_n)\,\psi_m^* dx\right]^* = (k_n^2 - k_m^2)\int_0^L \psi_n\psi_m^* dx$$
$$= 0, \tag{1.88}$$

which shows the integral of the product of two eigenfunctions as zero if their eigenfrequencies are distinct ($k_n^2 \neq k_m^2$). The integral of the product of two eigenfunctions can then be described as

$$\int_0^L \psi_n(x)\psi_m(x)^* dx = \Lambda_n\delta_{nm}, \tag{1.89}$$

where δ_{nm} is the Kronecker delta given by

$$\delta_{nm} = \begin{cases} 1 & \text{if} \quad m = n \\ 0 & \text{if} \quad m \neq n \end{cases},$$

and Λ_n denotes the total energy of an eigenfunction within the defined interval:

$$\int_0^L \psi_n\psi_n^* dx = \Lambda_n. \tag{1.90}$$

Equation (1.89) essentially describes the mutual orthogonality relation between two different eigenfunctions.

1.6.2.3 Biorthogonality for the Resistive Boundary Condition

Unfortunately, the self-adjointness of $\mathcal{L}_L$ and the orthogonality of the eigenvectors do not hold in many cases. For example, in the case of a resistive boundary condition with imaginary χ, Equation (1.87) is nonzero. In this case, however, a different kind of orthogonality can be defined. Note that the following two different kinds of characteristic equations can be set by applying the complex conjugation operator to Equation (1.82):

$$\mathcal{L}_L\psi_n(x) = -k_n^2\psi_n(x) \text{ and } \mathcal{L}_L^*\varphi_n(x) = -(k_n^2)^*\varphi_n(x) \tag{1.91}$$

with $\psi_n = \varphi_n^*$. Therefore, we have two different kinds of eigenfunction sets corresponding to two operators $\mathcal{L}_L$ and $\mathcal{L}_L^*$. In general, the eigenfunction set consisting of ψ_n is not mutually orthogonal, but ψ_n and φ_n are orthogonal to each other. This kind of orthogonal relationship is called *biorthogonality* [7]. The biorthogonal relation between ψ_n and φ_n can be more simply described by modifying the definition of orthogonality (Equation (1.89)) as

$$\int_0^L \psi_n\psi_m dx = \int_0^L \psi_n(\varphi_m)^* dx = \Lambda_n\delta_{nm}. \tag{1.92}$$

With the definition of Equation (1.92), the eigenfunction set ψ_n is mutually orthogonal with respect to the integration of the Wronskian defined by

$$\int_0^L f(x)(\mathcal{L}_L h(x))dx - \int_0^L (\mathcal{L}_L f(x))h(x)dx = 0. \tag{1.93}$$

Note that the definition of Equation (1.92) is identical to Equation (1.89) for purely reactive boundary conditions.

1.6.2.4 Modal Expansion

With a set of linearly independent eigenfunctions ψ_n, we can expand the pressure field $p(x)$ and the excitation $q(x)$ of Equation (1.78)[19] as

$$p(x) = \sum_{n=0}^{\infty} p_n \psi_n(x), \tag{1.94}$$

$$q(x) = \sum_{n=0}^{\infty} q_n \psi_n(x), \tag{1.95}$$

where p_n and q_n are the modal coefficients of the pressure field and excitation functions, respectively. Then, from Equation (1.80), we can show how the expansion coefficients p_n and q_n are related to each other. Substituting Equations (1.94) and (1.95) into Equation (1.80) yields

$$\begin{aligned} p_n \mathcal{L} \psi_n(x) &= (p_n \mu_n) \psi_n(x) = -q_n \psi_n(x) \\ \Rightarrow p_n \mu_n &= -q_n. \end{aligned} \tag{1.96}$$

Therefore, the modal coefficient of the pressure field can be derived from the knowledge of q_n and the eigenvalue μ_n. This indicates that we can calculate the pressure field generated by the given excitation $q(x)$.

Let us apply the eigenfunction approach for a 1D duct bounded by $x \in [0, L]$. For this specific case, we first try to find the function $\psi_n(x)$ that satisfies

$$\frac{d^2 \psi_n(x)}{dx^2} + k_n^2 \psi_n(x) = 0, \tag{1.97}$$

and also satisfies the boundary condition of Equation (1.75), that is,

$$\begin{aligned} \alpha_0 \psi_n(0) + \frac{\beta_0}{i \rho_0 \omega} \frac{d\psi_n(0)}{dx} &= 0, \quad x = 0 \\ \alpha_L \psi_n(L) + \frac{\beta_L}{i \rho_0 \omega} \frac{d\psi_n(L)}{dx} &= 0 \quad x = L \end{aligned} \tag{1.98}$$

To gain further clarity on this problem, we consider the special case when $\alpha = 0$. In this case, we have a rigid-wall boundary condition, and the eigenmode can be found, intuitively, as:

$$\psi_n(x) = \cos \frac{n\pi}{L} x. \tag{1.99}$$

If $\beta = 0$, which is the case for the pressure release boundary condition, then the solution has to take the form

$$\psi_n(x) = \sin \frac{n\pi}{L} x. \tag{1.100}$$

It is obvious that any combination of these modes (Equation (1.94)) always satisfies the boundary condition of Equation (1.98).

The main obstacle to finding the solution by a linear combination of eigenfunctions is finding the contribution, or weighting, of each mode to the solution. In other words, we must attempt to find p_n that satisfies Equation (1.70). For example, if we have one source at a point where

---

[19] To validate this statement, the completeness of the set of eigenfunctions should be shown. For the discussion on the completeness, refer to the proof of the general Sturm-Liouville theorem (e.g., Ref. [8]).

Equation (1.68) is the governing equation, then we can attempt to construct the solution as given by Equation (1.94). The coefficients can be found using the orthogonality of the eigenfunctions. Using Equations (1.70), (1.92), (1.94), and (1.97), we can obtain the weighting as

$$p_n = -\frac{q_n}{\mu_n} = \frac{1}{(k_n^2 - k^2)\Lambda_n}\int_0^L q(x)\,\psi_n(x)\,dx. \qquad (1.101)$$

This approach entails first finding all possible solutions that satisfy the homogeneous wave equation. We then try to construct the solution for an inhomogeneous wave equation, which has a source excitation on the right-hand side of the equation, from a linear combination of the eigenfunctions. This approach is valid because the equation is linear and, therefore, the principle of superposition holds. In other words, it is akin to spreading a net over a body of water surrounded by a boundary and catching the fish (or coefficients in our case) therein.

1.6.3 Integral Approach Using Green's Function

1.6.3.1 Green's Function

Alternatively, we can try to obtain the solution that satisfies the boundary condition by introducing Green's function. As noted earlier, the sound field induced by the source excitation $q(x)$ can be expressed by Equation (1.67). If we denote the sound pressure due to a unit point source at $x = x_0$ as $g(x|x_0)$, then g has to satisfy the equation:[20]

$$\frac{d^2 g(x|x_0)}{dx^2} + k^2 g(x|x_0) = -\delta(x - x_0) \text{ or } \mathcal{L}g(x|x_0) = -\delta(x - x_0) \qquad (1.102)$$

Since the Green's function is defined as the response to the point excitation in space, it can be regarded as the spatial impulse response function of the linear wave equation. The definition of Equation (1.102), however, does not imply anything about the boundary condition, so there can be various types of Green's functions depending on the imposed boundary condition. For example, the Green's function for the Dirichlet boundary condition is completely different from that for the Neumann boundary condition.

The Green's function can also be regarded as the inverse operator of $\mathcal{L}$, which maps the given excitation function $q(x)$ into the pressure field. The inverse-relation between $\mathcal{L}$ and the Green's function $g(x|x_0)$ can be easily proved from the fact that $\mathcal{L}$ transforms the Green's function into the Dirac delta function. Therefore, we can utilize the Green's function to express the pressure field produced by arbitrary excitation $q(x)$. For example, if the Green's function fulfills the physical boundary condition, the solution of Equation (1.70) can be simply rewritten from Equation (1.102) as

$$p(x) = \int_{L_0} g(x|x_0)q(x_0)dx_0, \text{ for 1D case} \qquad (1.103)$$

$$p(\mathbf{r}) = \int_{V_0} g(\mathbf{r}|\mathbf{r}_0)q(\mathbf{r}_0)dV(\mathbf{r}_0), \text{ for 3D case.} \qquad (1.104)$$

1.6.3.2 Reciprocity Principle

Consider two Green's functions, $g(x|x_0)$ and $g(x|x_1)$, that are the spatial impulse response functions due to the point excitation at x_0 and x_1, respectively. Also assume that two Green's functions satisfy

[20] The solution that satisfies Equation (1.102) is Green's function.

the same boundary condition, that is, the point excitation is applied in the same environments. Then we have two different wave equations:

$$\mathcal{L}g(x|x_0) = -\delta(x - x_0) \text{ and } \mathcal{L}g(x|x_1) = -\delta(x - x_1) \tag{1.105}$$

Multiplying Equation (1.105) by $g(x|x_1)$ and $g(x|x_0)$, respectively, and subtracting two differential equations gives

$$\begin{aligned} &g(x|x_1)(\mathcal{L}g(x|x_0)) - g(x|x_0)(\mathcal{L}g(x|x_1)) \\ &= g(x|x_0)\delta(x - x_1) - g(x|x_1)\delta(x - x_0). \end{aligned} \tag{1.106}$$

The integration of Equation (1.106) with respect to x over the interval $x \in [0, L]$ yields

$$\int_0^L \left\{ g(x|x_1) \frac{d}{dx} g(x|x_0) - g(x|x_0) \frac{d}{dx} g(x|x_1) \right\} dx = g(x_1|x_0) - g(x_0|x_1). \tag{1.107}$$

The left-hand side of Equation (1.107) is zero, as we have derived for Equation (1.93). Therefore, the two Green's functions, $g(x_1|x_0)$ and $g(x_0|x_1)$, are equivalent to each other, and we can write

$$\begin{aligned} &g(x_1|x_0) = g(x_0|x_1) \text{ for the 1D case,} \\ &g(\mathbf{r}_1|\mathbf{r}_0) = g(\mathbf{r}_0|\mathbf{r}_1) \text{ for the 3D case.} \end{aligned} \tag{1.108}$$

This relation is commonly referred to as the *principle of acoustic reciprocity* for a linear acoustic wave. The reciprocity demonstrates that the observed sound signal does not change when the locations of the source and the receiver are interchanged in an unchanging environment.

As can be seen in Equation (1.107), the reciprocity relation holds for the arbitrary Green's function that fulfills the homogeneous boundary condition of Equation (1.75). Note that Equation (1.75) describes the boundaries with the prescribed impedance. This condition differs from the generalized boundary condition that includes the contribution from the active pressure or velocity sources. For the inhomogeneous boundary condition of Equation (1.72), the left-hand side of Equation (1.107) does not vanish and the reciprocity principle may not hold in general.

1.6.3.3 Relation between Green's Function and Eigenfunctions

The representation of sound fields in terms of the Green's function (Equations (1.103) and (1.104)) has the form of the Fredholm's equation of the first kind:

$$p(x) = \int_{L_1}^{L_2} K(x|x_0)q(x_0)dx_0, \tag{1.109}$$

where the integral is taken over the interval $x_0 \in [L_1, L_2]$. The operator $K(x|x_0)$ that maps the function q to another function p is called the *kernel* of the equation. In Equation (1.103), the kernel of the equation is given by the Green's function, which is symmetric with respect to x and x_0 (i.e., a symmetric kernel).

Since the pressure field and excitation function can be expanded in terms of the eigenfunctions (Equations (1.94)–(1.96)), we can rewrite Equation (1.109) as

$$\begin{aligned} &p_n\psi_n(x) = \int_0^L g(x|x_0)q_n\psi_n(x_0)dx_0 \\ &\Rightarrow \psi_n(x) = -\mu_n \int_0^L g(x|x_0)\psi_n(x_0)dx_0, \end{aligned} \tag{1.110}$$

which is another form of the eigenvalue problem defined in Equation (1.80). Equation (1.110) is known as the homogeneous Fredholm's equation of the second kind. From Equation (1.110), we can observe that the eigenfunction of the linear differential equation is equivalent to that of the integral equation.

The kernel of the integral can also be expressed in terms of the eigenfunctions. Since the Green's function is the solution of the inhomogeneous wave equation ((Equation (1.78)) with $q(x) = \delta(x - x_0)$), it can be represented by the modal summation

$$g(x|x_0) = \sum_{n=0}^{\infty} g_n(x)\psi_n(x_0). \tag{1.111}$$

From Equation (1.111), we can derive the expression of the modal coefficient $g_n(x)$. Substituting Equation (1.111) into Equation (1.110) yields

$$\begin{aligned}\psi_n(x) &= -\mu_n \sum_{m=0}^{\infty}\left[g_m(x)\left(\int_0^L \psi_m(x_0)\,\psi_n(x_0)dx_0\right)\right] \quad \Rightarrow \quad g_n(x) = -\frac{\psi_n(x)}{\Lambda_n\mu_n} \\ &= -\mu_n g_n(x)\Lambda_n\end{aligned} \tag{1.112}$$

Therefore, the Green's function can be written as:

$$g(x|x_0) = \sum_{n=0}^{\infty} \frac{\psi_n(x)\psi_n(x_0)}{\Lambda_n(k_n^2 - k^2)}. \tag{1.113}$$

1.6.3.4 Integral Approach

The integral approach attempts to solve the problem using various forms of surface integrals. As we have discussed for the self-adjointness property of the linear acoustic wave equation, the surface integral of two functions (Equation 1.93) is always zero if those functions satisfy the same boundary conditions and are the solution to the wave equation.

However, when the two functions do not satisfy the same boundary condition on the surface, the surface integral has many interesting properties. To formulate such an integral, let us consider the Green's function $g(x|x_0)$ and the pressure field $p(x)$. Multiplying $g(x|x_0)$ by Equation (1.67) and $p(x)$ by Equation (1.102), subtracting the former from the latter and finally integrating with respect to x lead us to

$$\int_0^L \left(p(x)\frac{d^2g(x|x_0)}{dx^2} - g(x|x_0)\frac{d^2p(x)}{dx^2}\right)dx = \int_0^L (g(x|x_0)q(x) - p(x)\delta(x - x_0))\,dx. \tag{1.114}$$

Then, integration by parts yields:

$$p(x)\frac{dg(x|x_0)}{dx}\Bigg]_0^L - g(x|x_0)\frac{dp(x)}{dx}\Bigg]_0^L = \int_0^L g(x|x_0)q(x)dx - p(x_0). \tag{1.115}$$

Exchanging the variable x for x_0 reduces Equation (1.115) to the form[21]

$$p(x) = \int_0^L g(x|x_0)q(x_0)dx_0 + p(x_0)\frac{dg(x|x_0)}{dx_0}\Bigg]_L^0 - g(x|x_0)\frac{dp(x_0)}{dx_0}\Bigg]_L^0. \tag{1.116}$$

[21] The integral sequence is reversed in [0, L] to make the sign of $p(x_0)\frac{\partial g(x|x_0)}{\partial x}$ positive for consistency with 3-D case. In 3-D case, the direction of surface normal vector **n** is inward, and then the sign of $p(x_0)\frac{\partial g(x|x_0)}{\partial x}$ is positive.

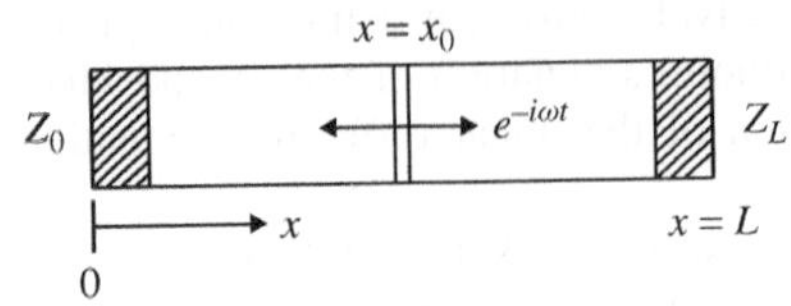

(a) 1 dimensional case

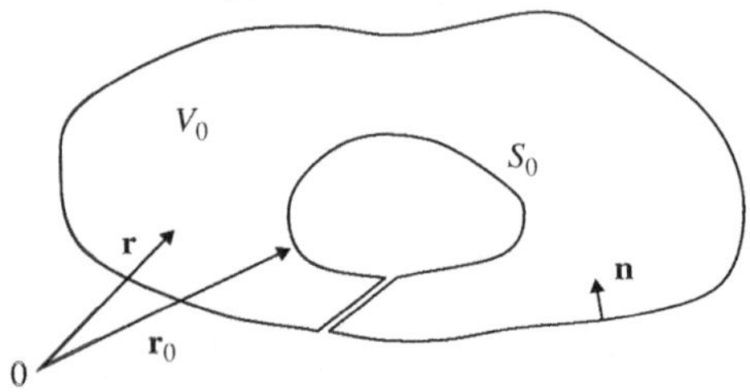

(V: volume, S_0: surface which surrounds the V, **n**: surface normal vector)

(b) 3 dimensional case

Figure 1.14 One-dimensional and three-dimensional boundary value problems

We now investigate how to apply Equation (1.116) when we have a unit amplitude sound source at $x = x_0$, as illustrated in Figure 1.14. This specific case reduces Equation (1.116) to

$$p(x) = g(x|x_0) + p(x_0)\frac{dg(x|x_0)}{dx_0}\Bigg]_L^0 - g(x|x_0)\frac{dp(x_0)}{dx_0}\Bigg]_L^0. \tag{1.117}$$

If the velocity at $(x = 0, L)$ is 0 (rigid-wall boundary condition), or the pressure is 0 (pressure release boundary condition), then Equation (1.117) becomes

$$p(x) = g(x|x_0) + p(x_0)\frac{dg(x|x_0)}{dx_0}\Bigg]_L^0 \tag{1.118}$$

or

$$p(x) = g(x|x_0) - g(x|x_0)\frac{dp(x_0)}{dx_0}\Bigg]_L^0, \tag{1.119}$$

respectively. In other words, the pressure at x is the sum of two pressure components. One is due to the sound source of unit magnitude with radian frequency ω at x_0, and the reflected pressure due to the boundary. Note also that the pressure at the right-hand side is induced by the pressure at the boundary. Equation (1.117) essentially states that the sound pressure in the space of interest is determined by the pressure at the boundary.[22] Interestingly, Green's function $g(x|x_0)$ can be anything that satisfies Equation (1.102). That is, Equation (1.117) is always valid for any function that does satisfy Equation (1.102) (see further details in Chapter 2, Section 2.7.3). We can therefore select $g(x|x_0)$ to make Equation (1.117) simpler; that is, choosing $g(x|x_0)$ that satisfies the Dirichlet boundary condition or Neumann boundary condition.

Equation (1.117) states that the sound pressure at x consists of two components: one is a direct effect from the sound source and the other is due to the reflection from the boundary. In comparison

[22] This type of equation is an integral equation for the one-dimensional case.

to the modal analysis method (discussed earlier in this section), this approach is akin to catching fish from the boundary using a fishing rod, that is, Green's function. Equation (1.117) is, in fact, an integral form and we have to assign or know Green's function in advance.

Expanding Equation (1.117) to a three-dimensional form yields the integral equation

$$p(\mathbf{r}) = \int_{V_0} g(\mathbf{r}|\mathbf{r}_0)q(\mathbf{r}_0)dV + \int_{S_0} (p(\mathbf{r}_0)\nabla_0 g(\mathbf{r}|\mathbf{r}_0) - g(\mathbf{r}|\mathbf{r}_0)\nabla_0 p(\mathbf{r}_0)) \cdot \mathbf{n}\, dS. \tag{1.120}$$

If we do not have the sound source $q(\mathbf{r}_0)$ in the integral volume V_0 (Figure 1.14b), then Equation (1.120) becomes

$$p(\mathbf{r}) = \int_{S_0} (p(\mathbf{r}_0)\nabla_0 g(\mathbf{r}|\mathbf{r}_0) - g(\mathbf{r}|\mathbf{r}_0)\nabla_0 p(\mathbf{r}_0)) \cdot \mathbf{n}\, dS. \tag{1.121}$$

(See Chapter 2, Section 2.7.3 for the detailed derivations.) Equations (1.120) and (1.121) are referred to as Kirchhoff–Helmholtz integral equations. As noted with respect to the one-dimensional case, Green's function g can be anything that satisfies Equation (1.102); we can use any type of fishing rod as long as it is a fishing rod.

Note that Equation (1.121) is essentially the basis of acoustic holography and the boundary element method, which are quite popular in acoustics (Chapter 3). The concept of acoustic holography is to start by constructing the integral surface (S_0) by using any orthogonal coordinate that conveniently expresses what we wish to express. The pressure on the surface is then measured at discrete points. The last step is required to predict the sound pressure at any point ($\mathbf{r} = (x, y, z)$: for the rectangular coordinate) by using the predictor $g(\mathbf{r}|\mathbf{r}_0)$, that is, Green's function. Acoustic holography can therefore provide information that cannot be measured or is not otherwise available on any surface of interest. On the other hand, boundary element method predicts acoustic information (e.g., sound pressure, particle velocity, intensity, etc.) at a position or on a surface that is not measured by constructing the integral surface as an arbitrary shape of surfaces, and dividing the surface to small elements. These two methods use prediction property of Green's function in common.

To summarize, the mathematical methods that can analytically predict acoustic wave propagation in time and space may employ either eigenfunctions or Green's function. The former attempts to describe waves that can be generated by a prescribed acoustic source or sources by seeking the contribution from each eigenfunction; here we are catching fish using an eigenfunction net. The latter attempts to describe the sound field by using Green's function that transforms the information of sound pressure and velocity on the boundary to the point or field of interest.[23] In fact, this approach distributes a function that satisfies the governing equation on the boundary. It can be a singular function or a Green's function. We then try to find the contribution of each function which satisfies the boundary condition; we have many fishing rods on the boundary to catch the fish.

In fact, Equation (1.121) is the tenet to make or visualize sound in space. For a sound focusing problem (Chapter 5), it attempts to maximize sound energy, which is $|p(\mathbf{r})|^2$ in a desired region. One can also try to maximize other acoustic properties: that is, acoustic contrast between selected regions. We can also utilize Equation (1.121) to design virtual sound sources: inside or outside the region of integration (Chapter 6).

1.7 Solutions of the Wave Equation

We have seen that the solutions which satisfy the acoustic wave equation can be obtained using two different approaches: the modal analysis method or Green's function method. The boundary

[23] The ideal technique is to find superposition of the solutions that satisfy the boundary condition and governing equation.

condition and the source determine the solution that expresses the waves in the area of interest and time. However, this statement is too general to grasp the idea of how an acoustic wave would behave in space and time. To be more specific and realistic, we again have to start with the simplest case. We therefore study a representative acoustic wave that satisfies the governing equation.

1.7.1 Plane Wave

As attempted frequently, we will start with a one-dimensional, planar acoustic wave at position x and time t, $p(x, t)$. This can be written as

$$p(x,t) = p(x,\omega)e^{-i\omega t} = A(\omega)e^{i(kx-\omega t)}. \tag{1.122}$$

A wave in a certain direction in space can be expressed as

$$p(\mathbf{r},t) = p(\mathbf{r},\omega)e^{-i\omega t} = A(\omega)e^{i(\mathbf{k}\cdot\mathbf{r}-\omega t)}, \tag{1.123}$$

where A is a complex amplitude. Equation (1.122) obviously satisfies Equation (1.18) as Equation (1.123) satisfies Equation (1.28). The plane wave (Equation (1.123)), as the name implies, essentially has all the same physical properties (i.e., pressure and velocity) at the plane perpendicular to $\mathbf{k}$ at $\mathbf{r}$ (Figure 1.15). Note that its impedance at any position and time is[24]

$$Z_p = \rho_0 c. \tag{1.124}$$

This is precisely the same as the characteristic impedance of the medium. This means that the plane wave in an unbounded fluid propagates in the wavenumber vector direction, independent of the position, frequency, wavenumber, and wavelength. Intensity which expresses the power through the unit area, specifically the average intensity (active intensity), can be expressed as follows (see details in Appendix B, Section B.1.5):

$$I_{avg} = \frac{1}{2}\text{Re}[pu^*] = \frac{1}{2\rho_0 c}|p|^2, \tag{1.125}$$

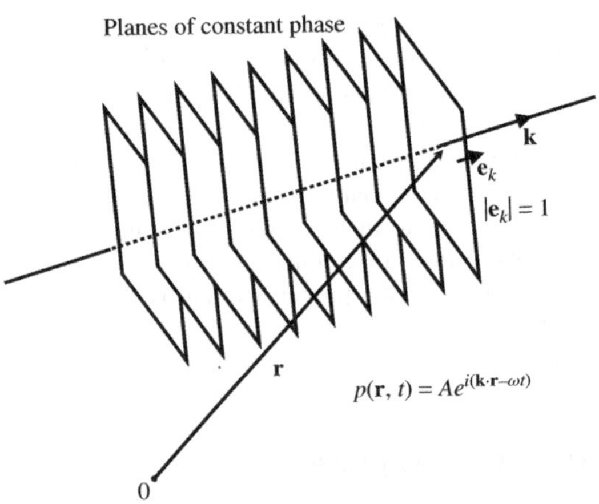

Figure 1.15 A plane wave ($\mathbf{k} = (\omega/c)\mathbf{e}_k$ is normal to the planes of constant phase)

[24] This can be obtained by having the velocity using the one-dimensional Euler equations then finding the impedance.

where u is the velocity in the direction of propagation and $u = p/\rho_0 c$. Therefore, the intensity Equation (1.125) can be written as

$$I_{avg} = \frac{1}{2}\rho_0 c|u|^2. \tag{1.126}$$

1.7.2 *Spherical Wave*

The governing equation that can express this type of wave can also be written in terms of the spherical coordinate. Let us therefore express Equation (1.28) in terms of the spherical coordinate. We assume that the pressure is independent of the polar and azimuth angles and only depends on the distance from the origin (r). Equation (1.28) then becomes

$$\frac{\partial^2}{\partial r^2}(rp) = \frac{1}{c^2}\frac{\partial^2}{\partial t^2}(rp). \tag{1.127}$$

Its solution will be

$$rp = Ae^{i(kr-\omega t)}, \tag{1.128}$$

where A is a complex amplitude. Equation (1.128) can be rewritten as

$$p(r,t) = \frac{A}{r}e^{i(kr-\omega t)} \tag{1.129}$$

which tends to infinity at $r = 0$ and is inversely proportional to the radius r. To assess the velocity, consider Euler equation in the spherical coordinate:

$$-\frac{\partial p}{\partial r} = \rho_0 \frac{\partial u_r}{\partial t}, \tag{1.130}$$

where u_r is the velocity in the radial direction. Equations (1.129) and (1.130) allow us to calculate the velocity in the radial direction, that is,

$$u_r = \frac{A}{r}\cdot\frac{1}{\rho_0 c}\left(1 + \frac{i}{kr}\right)e^{i(kr-\omega t)}. \tag{1.131}$$

Therefore, the impedance at r can be written as

$$Z_r = \frac{p}{u_r} = \rho_0 c\left\{\frac{(kr)^2}{1+(kr)^2} - i\frac{kr}{1+(kr)^2}\right\}. \tag{1.132}$$

Figure 1.16 illustrates the implication of the impedance in terms of the dimensionless distance kr. If the distance from the origin is large compared to the wavelength of interest ($kr \gg 1$), then the impedance approaches the plane wave impedance. This means that the wave propagates like that seen on an infinite string. On the other hand, if the position is very close to the origin with respect to wavelength ($kr \ll 1$), then the pressure and velocity has a $-90°$ phase difference. This means that the fluid particle is accelerated by the acoustic pressure. Interestingly, $kr = 1$ makes the real and imaginary part even. We can therefore argue that kr is a measure of how to share the real and imaginary parts.

Intensity is the product of Equations (1.129) and (1.131). The active intensity, where the pressure and velocity have no phase difference, is inversely proportional to the square of the distance but independent of kr. The reactive intensity on the other hand, where the pressure and velocity have a $90°$ phase difference, is inversely proportional to the square of the distance and also inversely proportional to kr. Therefore, the reactive intensity tends to decrease rapidly as we move away

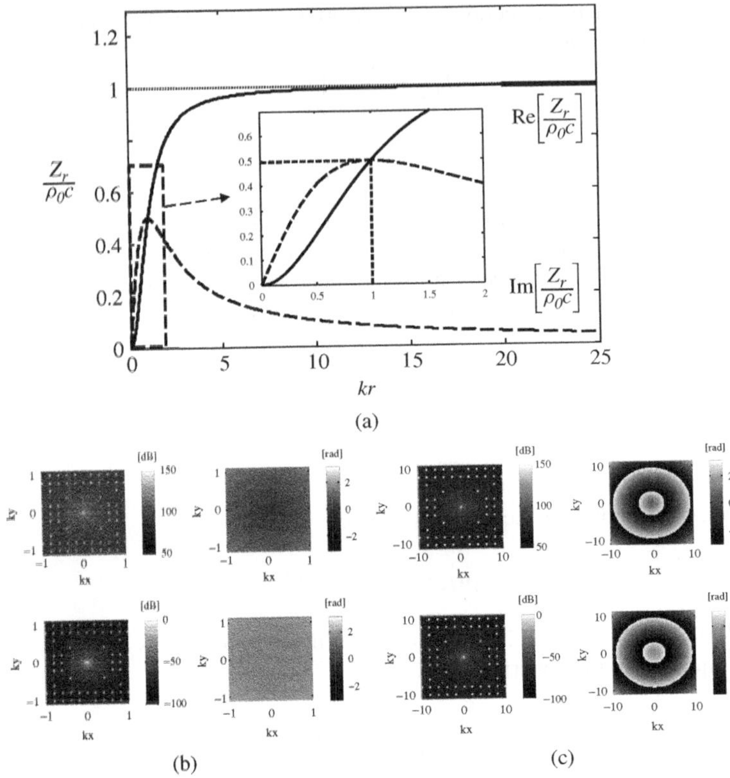

Figure 1.16 Monopole radiation. (a) The monopole's radiation impedance where k is wavenumber, λ indicates the Note is noteworthy that it behaves as a plane wave, as the observation position is far from the origin. (b) Pressure and particle velocity in nearfield (kr is small), magnitude (left), and phase (right) of pressure (top) and particle velocity (bottom); arrows indicate intensity. (c) As for (b) for farfield case

from the origin. This means that, in the vicinity of the origin, the reactive intensity dominates the acoustic behavior in such a way that the waves do not propagate well. We call this region the "nearfield" and this effect the "nearfield effect." The "farfield," therefore, indicates a position that is relatively far from the origin with respect to the wavelength of interest. Note also that the distance is measured in a relative scale with regard to wavelength kr.[25]

The monopole sound source radiating in free-field is defined by Equations (1.129) and (1.131) has a singularity at $r = 0$. This simple solution satisfies the linear wave equation. This implies that superposition of this type of solution also satisfies the governing wave equation. We can therefore attempt to construct any type of wave by using the monopole. This concept is illustrated in Figure 1.17, a graphical expression of Huygens principle. If the two monopoles are close together with the opposite phase, then a dipole is formed. Having 4 monopoles but alternating opposite phases makes a quadrupole. Appendix B, Section B.1.6 explains the various singular sound sources (Figure 1.18).

[25] Note that the information measured at the farfield does not contain the information of the nearfield. It is therefore not possible to predict the sound source based on what is measured in the farfield.

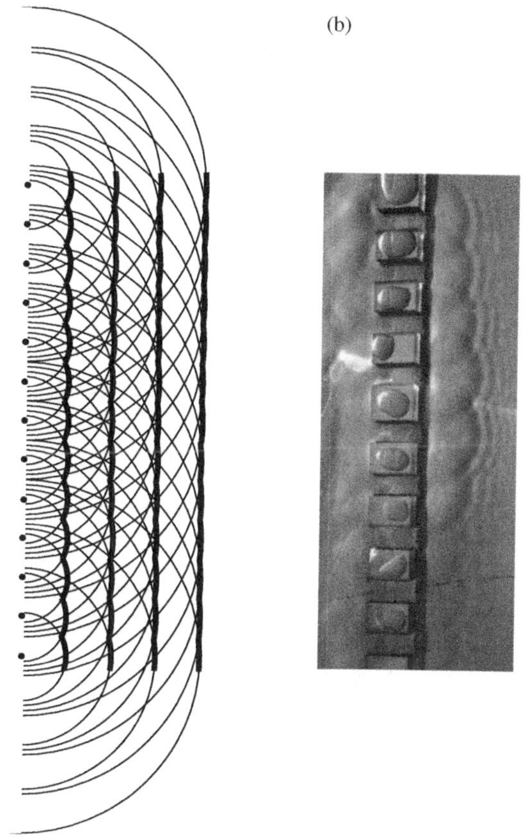

Figure 1.17 Huygens principle. The wavefront constructed by many monopole sound sources: (a) graphical illustration and (b) shallow ripple tank

If we use the monopole and dipole as the Green's functions of Equation (1.121), then the integral equation describes how the resulting sound pressure at the position of interest is caused by monopole propagation of the velocity on the surface and dipole propagation of the surface pressure.

To summarize, the plane wave and monopole sound source satisfy the governing wave equation and represent the simplest but most representative sound sources. The nearfield and farfield are measured by the distance compared to the wavelength of interest. In the nearfield, the monopole source does not effectively drive the sound field, but in the farfield, the monopole behaves as if it were a plane wave. The impedance of plane wave is exactly the same as the characteristic impedance of medium. The impedance of monopole depends on kr, which measures the distance from the origin with regard to the wavelength of interest. In the farfield, the impedance approaches that of a plane wave. In the nearfield, it shares with the real and imaginary part because imaginary part cannot be neglected. Again, we found that impedance describes the underlying physics well. Note also that we can use monopole sources to describe the sound field: this is the foundation of the Kirchhoff–Helmholtz equation and Huygens principle: "Any complicated phenomena can be expressed by the superposition of simple and fundamental elements."

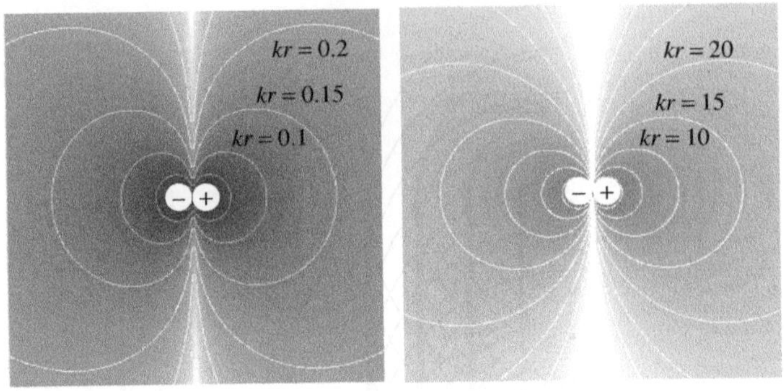

$$|p(r,\theta,\phi=0,\omega)|^2 = k^2|q_{di}|^2\frac{1}{(4\pi r)^2}\cos^2\theta\left\{1+\left(\frac{1}{kr}\right)^2\right\}$$

(a)

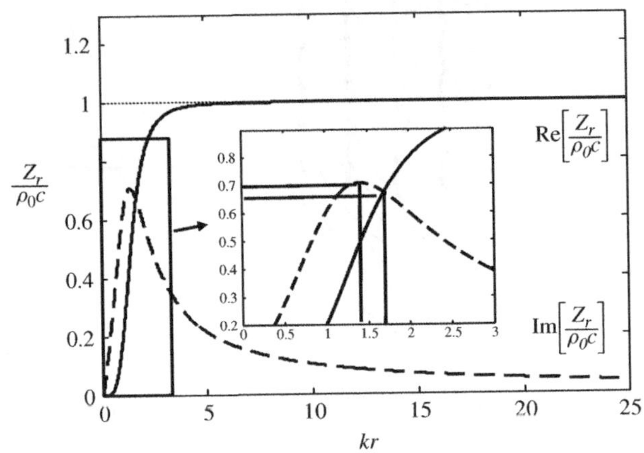

$$Z_r(r,\theta,\phi) = \rho_0 c\,\frac{1+i\left(\frac{1}{kr}\right)}{1+2i\left(\frac{1}{kr}\right)-2\left(\frac{1}{kr}\right)^2}$$

(b)

Figure 1.18 Dipole and quadrupole distributions and their characteristics where (r,θ,ϕ) indicates an arbitrary point in spherical coordinate, k is wavenumber, and q_{di} represents the dipole-moment amplitude (Equation (B59)), and q_{quad} represents the amplitude of quadrupole (Equation (B83)). (a) Pressure of a dipole sound's spatial pattern (b) Impedance of a dipole at r (c) The magnitude (left) and phase (right) of particle velocity of a dipole in nearfield (top) and farfield (bottom), and arrows indicate intensity. (d) Pressure of a quadrupole pattern in space (e) Impedance of a quadrupole at r (f) The magnitude (left) and phase (right) of particle velocity of a quadrupole in nearfield (top) and farfield (bottom), and arrows indicate intensity

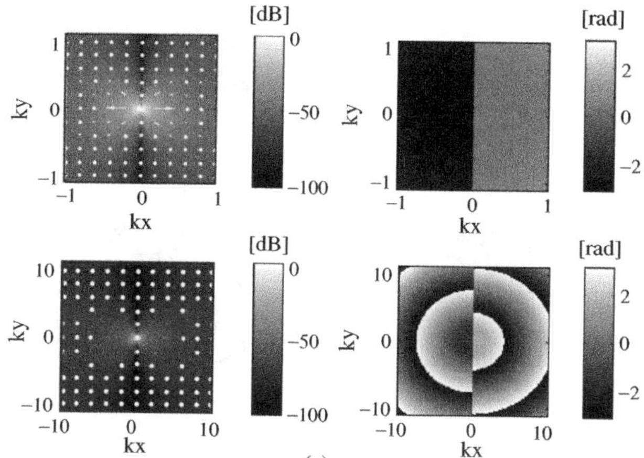

(c)

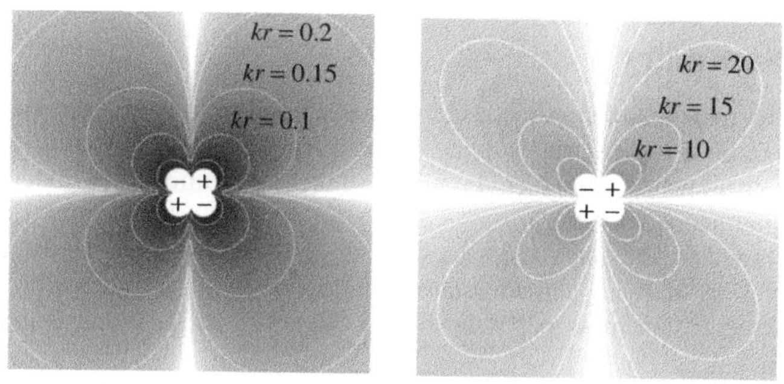

$$|p(r,\theta,\phi = 0 \text{ or } \pi, \omega)|^2 = \frac{k^4|q_{quad}|^2}{4}\frac{1}{(4\pi r)^2}\sin^2 2\theta\cos^2\phi\left[\left\{1-3\left(\frac{1}{kr}\right)^2\right\}^2+9\left(\frac{1}{kr}\right)^2\right]$$

(d)

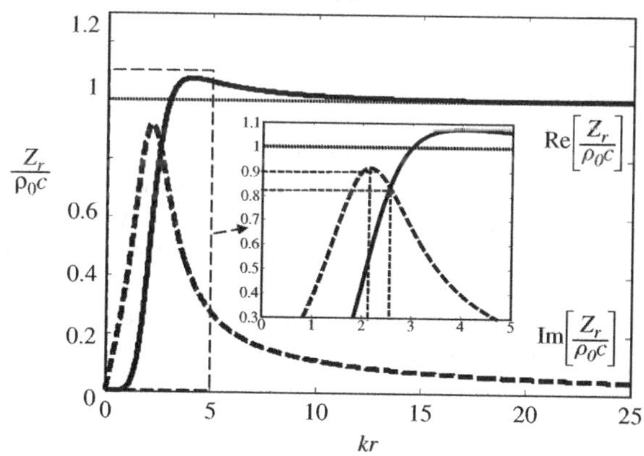

$$Z_r(r,\theta,\phi) = \rho_0 c\,\frac{1+3i\left(\frac{1}{kr}\right)-3\left(\frac{1}{kr}\right)^2}{1+4i\left(\frac{1}{kr}\right)-9\left(\frac{1}{kr}\right)^2-9i\left(\frac{1}{kr}\right)^3}$$

(e)

Figure 1.18 *(continued)*

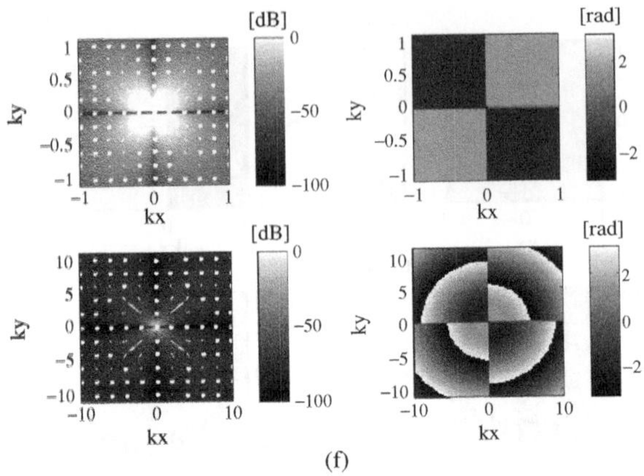

Figure 1.18 *(continued)*

1.8 Chapter Summary

We have attempted to understand how acoustic waves are generated and propagated in a compressible fluid by examining what takes place in a duct or pipe. Generally, the behavior of three-dimensional acoustic waves can be envisaged from the observations and understanding gained from a one-dimensional acoustic wave.

Newton's second law describes how the forces acting on a compressible fluid create the motion of the fluid. Conservation of mass and the state equation of fluid, together with Newton's law, provide three relations between mass per unit volume (density), fluid velocity, and force per unit area (pressure).

We have also learned that acoustic intensity, which is power transmission through a unit area in space, is another representative measure in addition to the three basic physical parameters of density, velocity, and pressure. The intensity is a vector, which expresses not only the power per unit area but also the direction of power propagation.

It was also stressed that acoustic waves are particularly meaningful to our hearing system. We therefore need to establish a way to measure the associated acoustic variables in accordance with human perception. As a result, the primary measure of frequency is the octave or 1/3 octave band and the dB scale is employed to indicate the magnitude of pressure. The A, B, C, and other weightings essentially reflect the dependency of human perception on frequency and magnitude.

Knowing that acoustic waves are governed by the acoustic wave equation and boundary and initial conditions, and recognizing that this can be considered as a typical mathematical problem related to linear partial differential equation, we have investigated possible solution methods that predict how sound waves propagate in space and time. The modal analysis method and Green's function method were extensively studied. The boundary condition determines the solution of an acoustic wave equation. Alternatively, we can also say that the impedance distribution in space certainly determines all possible sound waves.

References

1. Kim, Y.H. (2010) Acoustic wave equation and its basic physical measures, *Sound Propagation: An Impedance Based Approach*, John Wiley & Sons (Singapore) Pte. Ltd, pp. 69–128.
2. Lide D.R. *CRC Handbook of Chemistry and Physics*, 79th edn, CRC Press, 1998, pp. 14–38.
3. Lars Chittka and Axel Brockmann, *Perception Space – The Final Frontier*, *PLoS Biology*, **3** (4):e137 (2005).

4. ISO 226: 2003 revision, *Equal-Loudness Contour*, International Standards Organization.
5. T.D. Rossing, *The Science of Sound*, 2nd edn, Addison-Wesley, 1990, p. 86.
6. L.L. Beranek, *Noise and Vibration Control Engineering: Principles and Applications*, John Wiley & Sons, Inc., p. 20, 1992.
7. Morse, P. M. and Feshbach, H. (1953) *Methods of Theoretical Physics*, McGraw-Hill, New York, pp. 884–886.
8. Al-Gwaiz, M. A. (2008) *Sturm-Liouville Theory and Its Applications*, Springer-Verlag, pp. 79–84.
9. Davis A. Bies and Colin H. Hansen, *Engineering Noise Control: Theory and Practice*, 4th ed., Taylor & Francis, 2009, p. 104.

2

Radiation, Scattering, and Diffraction[1]

2.1 Introduction/Study Objectives

Sound waves are reflected and transmitted when they meet a surface of discontinuity in space. If surface of discontinuity is two dimensional, the discontinuity is expressed by its impedance. The impedance is a function of two spatial variables (e.g., (x, y) for Cartesian coordinate), which means that it is two-dimensional. The reflected and transmitted waves are therefore also two-dimensional. If we have a certain discontinuity that is a function of three spatial variables (e.g., (x, y, z) for Cartesian coordinate), then the waves will have more complicated patterns than those in two- or one-dimensional cases. Figure 2.1 depicts some representative cases. Depending on the type of discontinuities or the type of waves, scattering, diffraction, or radiation may occur.

A mathematical approach to analyze the phenomena can be one of the following two methods: find the solutions that satisfy the linear acoustic wave equation (Equation (1.28)) and boundary conditions, or use the integral equation (Equation (1.121)). These two approaches enable us to understand radiation, scattering, and diffraction.[2]

Scattering is a phenomenon that describes the reflection of waves due to the presence of discontinuities in space. As its name implies, "scattering" typically describes waves scattered by the discontinuities in space. Radiation, on the other hand, describes waves generated by the vibration of structures or by fluid fluctuation (see Figure 2.1a). In this respect, scattering can also be a form of radiation. In particular, scattering occurs when incident waves meet scatterers (e.g., discontinuities) which radiate the scattered sound field. We also refer to "diffraction" as illustrated in Figure 2.1b,d, which exhibits rather different types of waves. The latter are waves which are curved around corners or edges.

Diffraction implies that we cannot see the sound source but can hear the sound. In addition, diffraction introduces the concept of a shadow region where we cannot hear the sound. To understand these rather complicated phenomena, we need to understand sound fields induced by the basic unit sources: the radiation of a breathing sphere and a trembling sphere. These are, in fact, quite similar to what we have learned with regard to a monopole and dipole.

[1] Sections of this chapter have been re-used with permission from [1].

[2] Scattering and diffraction are physical phenomena, which represent waves deflected by characteristics of the discontinuity. Both can therefore be expressed by solutions of the wave equation which satisfy boundary conditions.

Sound Visualization and Manipulation, First Edition. Yang-Hann Kim and Jung-Woo Choi.

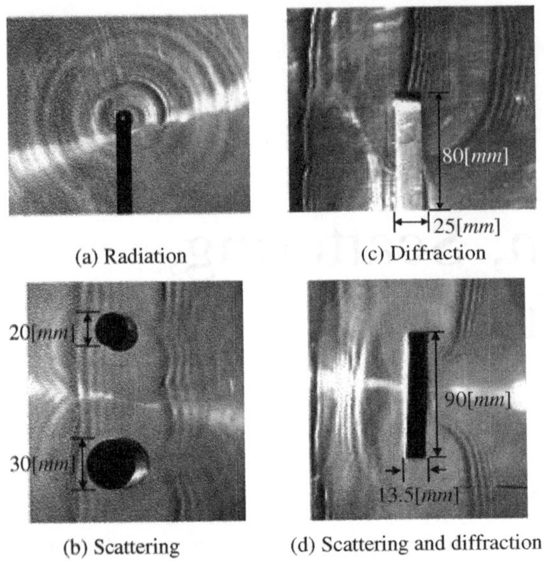

Figure 2.1 (a) Radiation, (b) scattering, and (c) and (d) diffraction: waves are visualized using a ripple tank. Incidence waves for (b–d) are plane waves coming from the left. The depth of the water has to be sufficiently smaller than 1/8th the wavelength to create a nondispersive wave (kh has to be smaller than roughly 0.5, where k is wavenumber and h is water depth). (Photographs taken in Center for Noise and Vibration Control, Korea Advanced Institute of Science and Technology, August 2005.)

What we get by any means of sound visualization, which expresses sound pressure, velocity, intensity, or energy in space and time (sometimes in terms of wavenumber), is due to the radiation, scattering, and diffraction. Therefore the quality of interpretation based on the visualized result is determined by how well we understand radiation, scattering, and diffraction.

2.2 Radiation of a Breathing Sphere and a Trembling Sphere

We first study the radiation due to the vibration of a sphere of radius a (Figure 2.2). The sphere is assumed to vibrate omni-directionally with equal magnitude; we refer to this as a "breathing sphere."

Let us attempt to use the velocity potential to describe the sound radiation from the sphere. The velocity potential can directly provide the velocity and pressure by differentiation with respect to space ($\mathbf{u} = \nabla\Phi$) and time (Equation (B31)), respectively. This is the advantage of using the velocity potential. The velocity potential (Φ) (Appendix B, Section B.1.4), which can effectively describe the acoustic wave field induced by the breathing sphere, must satisfy the following linear differential equation (a linear acoustic wave equation) in polar coordinates,

$$\frac{\partial^2}{\partial r^2}(r\Phi) = \frac{1}{c^2}\frac{\partial^2}{\partial t^2}(r\Phi). \tag{2.1}$$

Equation (2.1) assumes that the velocity potential does not depend on the angle of the polar coordinate because of its symmetry characteristics. The solution of Equation (2.1) can be written, as confirmed in Equation (1.129), in the form

$$\Phi(r, t) = \frac{A}{r}e^{i(kr-\omega t)}, \tag{2.2}$$

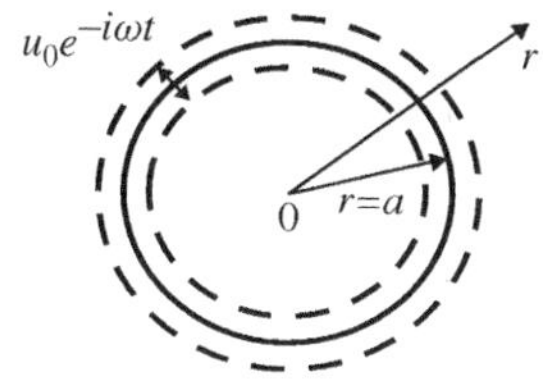

Figure 2.2 A breathing sphere and its radiation pattern: a is the radius, r indicates the radial distance, and u_0 denotes the velocity magnitude

where k is the wavenumber in the r direction. If the surface of the sphere harmonically vibrates, then the velocity on the surface ($u_r(a,t)$) can be written as

$$u_r(a,t) = u_0 e^{-i\omega t}. \tag{2.3}$$

The relation between the two constants A and u_0 in Equations (2.2) and (2.3) can be obtained by recalling that the rate of change of the velocity potential with regard to r has to be the velocity at the sphere (Equation (2.3)). That is,

$$u_r(r,t) = \frac{\partial \Phi(r,t)}{\partial r} = -A\left(\frac{1}{r^2} - \frac{ik}{r}\right)e^{i(kr-\omega t)}. \tag{2.4}$$

If we denote the amplitude of $u_r(r,t)$ at frequency ω as $u_r(r,\omega)$, Equation (2.4) can be written as

$$u_r(r,t) = u_r(r,\omega)e^{-i\omega t}, \quad \text{for } u_r(r,\omega) = -A\left(\frac{1}{r^2} - \frac{ik}{r}\right)e^{ikr}. \tag{2.5}$$

The velocity at $r = a$ will then be

$$u_r(a,t) = -A\left(\frac{1}{a^2} - \frac{ik}{a}\right)e^{i(ka-\omega t)}. \tag{2.6}$$

From Equations (2.3) and (2.6), we can obtain

$$A = -u_0\frac{a^2}{1-ika}e^{-ika}. \tag{2.7}$$

The velocity potential, the velocity, and the acoustic pressure can therefore be written as

$$\Phi(r,t) = \Phi(r,\omega)e^{-i\omega t} = -u_0\frac{a^2}{1-ika}\frac{1}{r}e^{i(k(r-a)-\omega t)}, \tag{2.8}$$

$$u_r(r,t) = u_r(r,\omega)e^{-i\omega t} = u_0\frac{1-ikr}{1-ika}\left(\frac{a}{r}\right)^2 e^{i(k(r-a)-\omega t)}, \tag{2.9}$$

$$p(r,t) = p(r,\omega)e^{-i\omega t} = \rho_0 c u_0\frac{-ika}{1-ika}\left(\frac{a}{r}\right)e^{i(k(r-a)-\omega t)}. \tag{2.10}$$

Note that Equations (2.8)–(2.10) are only valid if r is larger than a. We now have expressions for the velocity potential, radial velocity, and acoustic pressure that satisfy the governing differential equation (Equation (2.1)) and boundary condition (Equation (2.3)). Note that the velocity and pressure depend on the relative scales such as the ratio between the radius of the sphere (a) and the observation position (r): the sphere's radius with respect to the wavelength (ka) and the observation position with regard to the wavelength (kr). In other words, the absolute length scale does not have any significance. For example, the location where we measure the sound pressure should be determined with respect to how far we are in terms of the wavelength and the sphere's diameter.

The significance of the relative scale becomes more important if we look at the acoustic impedance of the sphere (Z_r). This can obtained from Equations (2.9) and (2.10), that is,

$$Z_r = \rho_0 c \left\{ \frac{(kr)^2}{1+(kr)^2} - i\frac{kr}{1+(kr)^2} \right\}$$
$$= \rho_0 c \left\{ \frac{1}{1+\left(\frac{1}{kr}\right)^2} - i\frac{\frac{1}{kr}}{1+\left(\frac{1}{kr}\right)^2} \right\}. \tag{2.11}$$

Figure 2.3a illustrates Z_r with respect to kr, which scales the observation position with respect to the wavelength. From this graph, we can explore many interesting radiation characteristics of the breathing sphere. For example, the real part of the specific acoustic impedance ($Z_r/\rho_0 c$) tends to 1 and its imaginary part is negligible when kr is much larger than 1. This means that we are observing the waves at a distance far from the origin compared to the wavelength of interest. In fact, we often say that we are seeing the waves in a farfield. In this case, the radiation from the breathing sphere resembles a one-dimensional acoustic wave (plane wave), because the acoustic impedance is $\rho_0 c$. We call this distance, or field, the "farfield," as noted in Chapter 1.

In the farfield, we observed that the acoustic impedance approaches the characteristic impedance of the medium, and is purely resistive. This means that the pressure and velocity are in phase with each other. On the other hand, as we approach the nearfield, the real part of the impedance becomes smaller and the contribution of the imaginary part tends to be larger. In particular, when $kr = 1$ (i.e., the wavelength is 2π times the wavelength of interest), then the contribution of the real part and the imaginary part to the radiation is the same. In the region where $kr < 1$, the imaginary part (i.e., the reactive part of the impedance) dominates the radiation characteristics. Therefore, the radiation is not likely to be effective. We generally call this region the nearfield.

It is also interesting to study the radiation impedance of a breathing sphere at $r = a$. This impedance exhibits how well the sphere radiates sound from its surface. This can be obtained from Equation (2.11), that is,

$$Z_{r=a} = \rho_0 c \left\{ \frac{(ka)^2}{1+(ka)^2} - i\frac{ka}{1+(ka)^2} \right\}$$
$$= \rho_0 c \left\{ \frac{1}{1+\left(\frac{1}{ka}\right)^2} - i\frac{\frac{1}{ka}}{1+\left(\frac{1}{ka}\right)^2} \right\}. \tag{2.12}$$

Note that ka determines the size of the sphere with regard to the wavelength. As ka becomes larger, the size of the sphere will therefore be bigger compared to the wavelength. This implies that the size of the radiator has to be larger as the frequency of radiation becomes lower, thus illustrating why a sub-woofer is typically much larger than a mid-frequency speaker. Figure 2.3a illustrates how well the sphere radiates sound with respect to ka, which is the fundamental scale of the radiator. The phenomenon that we observe when ka becomes larger can also be envisaged as many infinite strings attached to a sphere. In this case, we only have an outgoing wave propagating from the surface of the sphere.[3] On the other hand, in the nearfield, the sphere has many strings that have finite length and an impedance boundary condition.

[3] We can also see why the low frequency loudspeaker is louder than the high frequency loudspeaker from this observation.

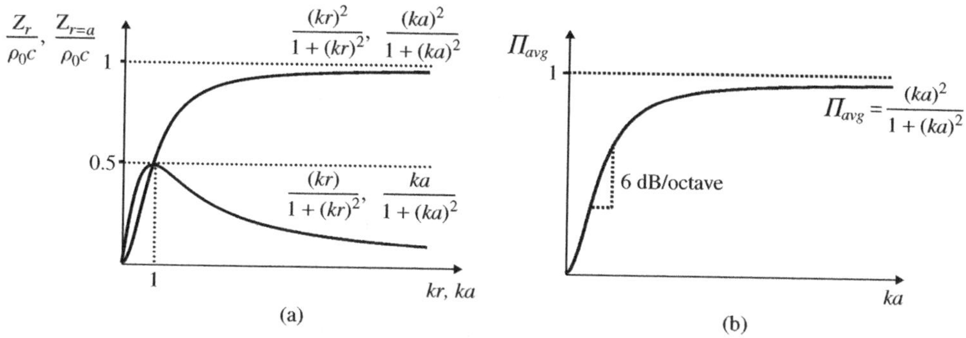

Figure 2.3 Impedances of the breathing sphere: (a) the acoustic impedance and the radiation impedance and (b) the radiation power. (*kr* and *ka* dominate the characteristics of the impedances; as they become larger, the wave behaves as if it is planar.)

We can also define the mechanical impedance (Z_m) by multiplying Equation (2.12) by the surface area ($4\pi a^2$). We then obtain

$$Z_m = 4\pi a^2 \rho_0 c \left\{ \frac{(ka)^2}{1+(ka)^2} - i \frac{ka}{1+(ka)^2} \right\}$$

$$= 4\pi a^2 \rho_0 c \left\{ \frac{1}{1+\left(\frac{1}{ka}\right)^2} - i \frac{\frac{1}{ka}}{1+\left(\frac{1}{ka}\right)^2} \right\}. \tag{2.13}$$

This again allows us to look at the radiation characteristics of the breathing sphere. For example, if *ka* is much larger than 1, then the mechanical impedance will be written as

$$Z_m = 4\pi a^2 \rho_0 c. \tag{2.14}$$

When *ka* is much smaller than 1, the mechanical impedance will be

$$Z_m = -ik4\pi a^3 \rho_0 c. \tag{2.15}$$

Equation (2.15) can be rewritten as

$$Z_m = -3i\omega m_d \tag{2.16}$$

by introducing a new term, $m_d = (4/3)\pi a^3 \rho_0$, which is the mass of the medium occupied by the sphere. Equation (2.16) shows that the pressure simply accelerates the fluid particles when *ka* is much smaller than 1.[4] The imaginary part of the mechanical impedance (Equation (2.13)) using m_d is given by

$$\mathrm{Im}[Z_m] = -3\omega \frac{m_d}{1+(ka)^2}. \tag{2.17}$$

[4] If there is a phase difference of $-i\omega$ in the impedance, then it means that the fluid particle behaves as if it is a mass.

Equation (2.17) implies that the accession to inertia depends on both frequency and ka. Note that the inertia tends to be smaller as the sphere size increases relative to the wavelength of interest. This means that we have a more efficient radiator.

If we calculate the mean intensity (I_{avg}) by using Equations (2.9) and (2.10), then we obtain

$$\begin{aligned} I_{avg} &= \frac{1}{2}\mathrm{Re}[pu_r^*] \\ &= \frac{1}{2}\rho_0 c u_0^2\left(\frac{a}{r}\right)^2 \frac{(ka)^2}{1+(ka)^2} = \frac{1}{2}\rho_0 c u_0^2\left(\frac{a}{r}\right)^2 \frac{1}{1+\left(\frac{1}{ka}\right)^2}, \end{aligned} \tag{2.18}$$

where $p = p(r, \omega)$ and $u_r^* = u_r^*(r, \omega)$ are the sound pressure magnitude (Equation (2.10)) and the complex conjugate of the velocity magnitude (Equation (2.9)). If we want to consider the radiation power, it can be obtained from Equation (2.18) by multiplying the area of interest by

$$I_{avg} \cdot 4\pi r^2 = \frac{1}{2}\rho_0 c 4\pi a^2 u_0^2 \frac{(ka)^2}{1+(ka)^2} = \frac{1}{2}\rho_0 c 4\pi a^2 u_0^2 \frac{1}{1+\left(\frac{1}{ka}\right)^2}. \tag{2.19}$$

Equation (2.18) expresses the intensity at r, and Equation (2.19) denotes the radiation power on the surface of the radius r. Note that the term $(1/2)\rho_0 c 4\pi a^2 u_0^2$ in Equation (2.19) is the radiation power in the case of a plane wave that is radiating from the surface of the area $4\pi a^2$ (Equation (1.126), multiplied by $4\pi a^2$). In other words, the radiation power (Equation (2.19)) of the breathing sphere has the difference with that of the plane wave, which is shown in the last part of Equation (2.19) through the term related with ka.

Consider the normalized radiation power (Π_{avg}) of a circular plate of radius $2a$ with a velocity of u_0 and frequency ω. This is expressed as

$$\Pi_{avg} = \frac{I_{avg} \cdot 4\pi r^2}{\frac{1}{2}\rho_0 c 4\pi a^2 u_0^2} = \frac{(ka)^2}{1+(ka)^2} = \frac{1}{1+\left(\frac{1}{ka}\right)^2}, \tag{2.20}$$

which highlights that the radiation power becomes larger as we increase ka (Figure (2.3b)). It indicates that we can increase the radiation power by 6 dB by doubling the size or dimension of radiator when the wavelength is large relative to the size of the radiator (a). If the wavelength is much smaller than a, then the average power asymptotically approaches that of plane waves, which is the maximum possible radiation.

A second type of basic radiation unit is a trembling sphere (Figure 2.4). Assuming that the sphere vibrates in the z direction with a velocity of u_c, the acoustic waves generated by this trembling sphere would satisfy the acoustic wave equation in the spherical coordinate, that is,

$$\frac{1}{r}\frac{\partial^2}{\partial r^2}(r\Phi) + \frac{1}{r^2\sin\theta}\frac{\partial}{\partial\theta}\left(\sin\theta\frac{\partial\Phi}{\partial\theta}\right) + \frac{1}{r^2\sin^2\theta}\frac{\partial^2\Phi}{\partial\phi^2} = \frac{1}{c^2}\frac{\partial^2\Phi}{\partial t^2}. \tag{2.21}$$

For harmonic wave, the velocity potential is given by $\Phi(\mathbf{r}, t) = \Phi(\mathbf{r}, \omega)\exp[-i\omega t]$, and $\mathbf{r} = (r, \theta, \phi)$ is the directional vector indicating a position in spherical coordinates.

It is obvious that the waves are independent of the ϕ coordinate. In this case, the boundary condition on the surface of the trembling sphere ($r = a$) can written as

$$\begin{aligned} u_r(\mathbf{r}, \omega)|_{r=a} &= u_c \cos\theta \\ &= \left.\frac{\partial\Phi(\mathbf{r}, \omega)}{\partial r}\right]_{r=a}, \end{aligned} \tag{2.22}$$

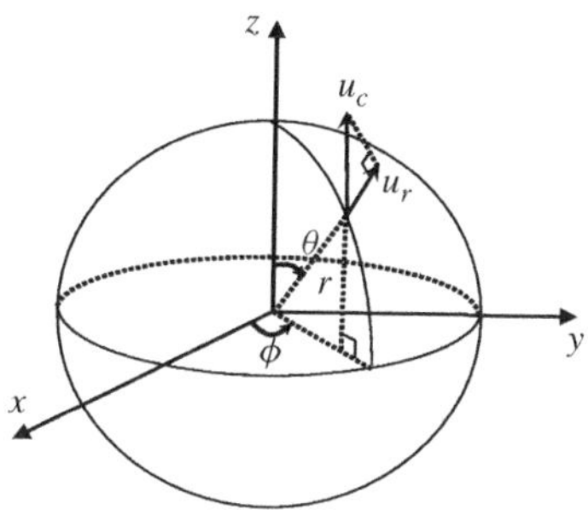

Figure 2.4 The trembling sphere. (The direction of vibration is z, the velocity is u_c, and u_r is the velocity in the r direction. Other symbols represent coordinates.)

The tangential velocity components in the r direction are not considered, because we assume that the fluid is inviscid. In this case, the solution satisfying Equation (2.21) is found to be[5]

$$\Phi(\mathbf{r}, \omega) = A\cos\theta \frac{\partial}{\partial r}\left(\frac{e^{ikr}}{r}\right). \tag{2.23}$$

The pressure in time domain ($p(\mathbf{r}, t)$) can be calculated from Equation (2.23) as

$$\begin{aligned} p(\mathbf{r}, t) &= -\rho_0 \frac{\partial}{\partial t}(\Phi(\mathbf{r}, \omega)e^{-i\omega t}) \\ &= A(i\omega\rho_0)\cos\theta \frac{\partial}{\partial r}\left(\frac{e^{ikr}}{r}\right)e^{-i\omega t}, \end{aligned} \tag{2.24}$$

The velocity (u_r) can be obtained by taking the derivative of the potential function with respect to r as below:

$$u_r(\mathbf{r}, \omega) = \frac{\partial\Phi(\mathbf{r}, \omega)}{\partial r} = A\cos\theta \frac{\partial^2}{\partial r^2}\left(\frac{e^{ikr}}{r}\right). \tag{2.25}$$

Equations (2.22) and (2.25) lead us to write

$$u_c \cos\theta = \left[A\cos\theta \frac{\partial^2}{\partial r^2}\left(\frac{e^{ikr}}{r}\right)\right]_{r=a}, \tag{2.26}$$

and we can obtain A as

$$A = \frac{u_c a^3}{2 - (ka)^2 - 2ika} e^{-ika}. \tag{2.27}$$

The pressure magnitude ($p(\mathbf{r}, \omega)$) can therefore be obtained by using Equations (2.24) and (2.27), that is,

$$p(\mathbf{r}, \omega) = ik\rho_0 c u_c a^3 \cos\theta \frac{1}{2 - (ka)^2 - 2ika}\left(\frac{ik}{r} - \frac{1}{r^2}\right)e^{ik(r-a)}. \tag{2.28}$$

[5] It is known that we can obtain the solution of the trembling sphere by either applying a coordinate transformation or using a separation of variables in the governing equation. However, we can try to use the solution (Equation (2.23)) that satisfies the governing equation and boundary condition. Intuitively, the solution would be of dipole type because the sphere trembles in one direction.

The velocity (u_r) can be calculated using Equations (2.25) and (2.27). This gives us

$$u_r(\mathbf{r}, \omega) = \frac{u_c a^3}{2-(ka)^2-2ika}\cos\theta\left(\frac{2}{r^3}-\frac{2ik}{r^2}-\frac{k^2}{r}\right)e^{ik(r-a)}. \tag{2.29}$$

The acoustic impedance (Z_r) can be obtained from Equations (2.28) and (2.29) as

$$\begin{aligned} Z_r &= \rho_0 c\left\{\frac{(kr)^4}{4+(kr)^4} - i\frac{2kr+(kr)^3}{4+(kr)^4}\right\} \\ &= \rho_0 c\left\{\frac{1}{1+4\left(\dfrac{1}{kr}\right)^4} - i\frac{\dfrac{1}{kr}+2\left(\dfrac{1}{kr}\right)^3}{1+4\left(\dfrac{1}{kr}\right)^4}\right\}, \end{aligned} \tag{2.30}$$

which states that the impedance is independent of θ although the pressure and velocity magnitude depend on cos θ. This is a rather surprising result. Note that the maximum pressure and velocity occur in the direction in which the sphere vibrates ($\theta = 0°$) (Figure 2.5). On the other hand, the pressure and velocity are both zero along the line that is perpendicular to the direction of vibration, which is quite different from the case of the breathing sphere. However, as we can see from Figure 2.6, the sound waves in the farfield are somewhat similar to those in the case of the breathing sphere; the waves become plane waves in the farfield.

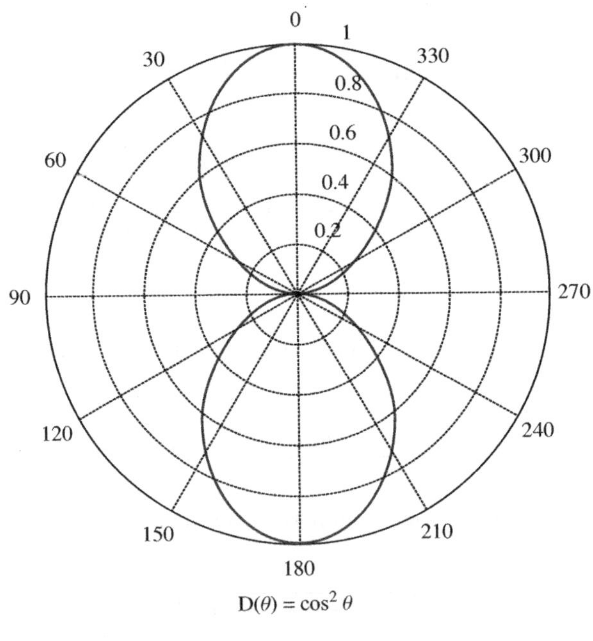

Figure 2.5 The directivity factor of a trembling sphere; the circumferential angle is expressed in degrees and the radial distance is a nondimensional arbitrary unit

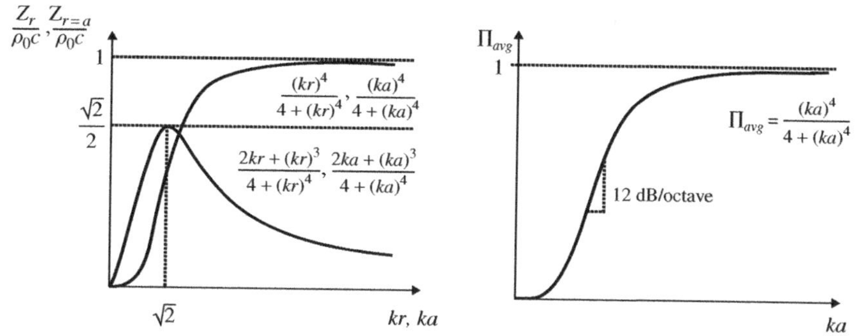

(a) The acoustic impedance(Z_r) and the radiation impedance($Z_{r=a}$)

(b) The radiation power (Π_{avg})

Figure 2.6 Impedances of the trembling sphere: (a) the acoustic impedance and the radiation impedance and (b) the radiation power. Note that the radiation impedance and the radiation power are proportional to $(kr)^4$ and $(ka)^4$, respectively. The corresponding characteristics of the breathing sphere are proportional to $(kr)^2$ and $(ka)^2$. The radiation of the trembling sphere therefore depends on the viewing location and the size of the diameter relative to the wavelength of interest

In relation to our investigation in the breathing sphere case, let us look at the radiation characteristics of the trembling sphere, in particular, the radiation impedance. From Equation (2.30), we obtain the impedance as

$$Z_{r=a} = \rho_0 c \left\{ \frac{(ka)^4}{4+(ka)^4} - i\frac{2(ka)+(ka)^3}{4+(ka)^4} \right\}$$
$$= \rho_0 c \left\{ \frac{1}{1+4\left(\frac{1}{ka}\right)^4} - i\frac{\frac{1}{ka}+2\left(\frac{1}{ka}\right)^3}{1+4\left(\frac{1}{ka}\right)^4} \right\}. \quad (2.31)$$

Equations (2.12) and (2.31) show that the radiation of the breathing sphere and trembling sphere depend on $(ka)^2$ and $(ka)^4$, respectively. If ka is small (i.e., if the radius of the sphere is small compared to the wavelength of interest), then the reactive term which is the imaginary part of the impedance dominates the radiation characteristics. On the other hand, when ka is large, then the resistive term governs the impedance and the trembling sphere effectively radiates sound waves.

We can also obtain the expression for the normalized radiation power (Π_{avg}) as

$$\Pi_{avg} = \frac{(ka)^4}{4+(ka)^4}\cos^2\theta = \frac{1}{1+4\left(\frac{1}{ka}\right)^4}\cos^2\theta. \quad (2.32)$$

Figure (2.6b) illustrates the behavior of the radiation power. From this figure, we can readily see that the radiation power rapidly decreases as ka becomes small; it decreases by 12 dB/octave.[6]

[6] It is 6 dB/octave in the case of the breathing sphere.

It should be noted that the pressure and velocity depend on θ. When $\theta = 90°$ or $270°$, the radiation pressure is zero. On the other hand, the radiated pressures reach their maxima when $\theta = 0°$ or $180°$. To effectively express this angle dependency of the radiation, we define the directivity factor as

$$D = \frac{I}{I_{\text{sphere}}}, \tag{2.33}$$

where I_{sphere} is the intensity radiated from the breathing sphere with radiation power equivalent to that of the radiator of interest. I denotes the intensity of the radiator whose directivity we wish to characterize. Figure 2.5 illustrates the directivity factor of the trembling sphere. As indicated by Equations (2.32) and (2.33), the radiated sound from a trembling sphere (e.g., sound radiation from a volume source such as a conventional audio speaker or percussion musical instrument) tends to have more directional dependency as the wavelength becomes smaller than its radiator size. This is one reason why we normally need two speakers separated by a significant distance, effectively generating a stereo sound.

To summarize, we have studied the basic units of radiators, that is, the breathing sphere and the trembling sphere. The characteristics of the radiators are expressed in terms of sound pressure, velocity, intensity, acoustic impedance, radiation impedance, and radiation power. For the breathing sphere, the sound pressure was found to be inversely proportional to the observation distance from the origin, and the power is inversely proportional to the square of the distance. The expression ka, which describes the size of the wavelength of interest relative to the radiator size, determines the radiation characteristics. As ka becomes larger, the radiated sound becomes planar. Note that for relatively small values of ka, the radiated power increases by 6 dB when ka doubles in size.

For a trembling sphere, the radiated sound depends strongly on the axis of propagation; in other words, it has very strong directivity. For relatively small values of ka, the radiated power increases by 12 dB when ka doubles in size. In this case, it is more rapid than in the breathing sphere.[7] We anticipate that these two radiators can be used to construct any type of radiator by the principle of superposition. In other words, we can draw any radiated sound field in space using these two basic paint brushes. Their characteristics are understood by using their impedances.

We can also anticipate that any visualized sound can be considered as, at least to the first-order approximation, the superposition of the breathing as well as the trembling spheres. How these two fundamental participate in the visualized object has to be analyzed based on the aforementioned understanding.

2.3 Radiation from a Baffled Piston

We have seen that the vibration of fluid particles induces sound propagation in space. The propagation strongly depends on the characteristic impedance of the medium. However, radiation is not only governed by the medium impedance but also by the relation between the fluid particle velocity and the pressure on the surface of the radiator; this is defined to be the radiation impedance. Section 2.2 introduces two basic radiators that feature all the necessary physics to understand radiation.

There are basically two ways to make sound. As the Kirchhoff–Helmholtz integral equation (Equation (1.121)) states, sound can be generated by the fluctuation of fluid particles which are attached to the surface of a vibrating structure, or, if there is a certain pressure fluctuation, then sound propagation will be induced from the pressure changes. The former is generally called

[7] The radiation power decreases more rapidly compared to that of the breathing sphere as ka becomes smaller. The breathing sphere is therefore a better radiator than the trembling sphere for a given velocity u.

"velocity source or velocity sound source" and the latter is called "pressure source or pressure sound source."[8]

Figure 2.7 illustrates the boundary condition for the integral equation (Equation (1.121)). We can rewrite the integral equation that emphasizes the individual contribution of pressure and velocity sources as

$$p(\mathbf{r}) = \int_{S_p} (p(\mathbf{r}_0)\nabla_0 g(\mathbf{r}|\mathbf{r}_0) - g(\mathbf{r}|\mathbf{r}_0)\nabla_0 p(\mathbf{r}_0)) \cdot \mathbf{n} dS$$
$$+ \int_{S_u} (p(\mathbf{r}_0)\nabla_0 g(\mathbf{r}|\mathbf{r}_0) - g(\mathbf{r}|\mathbf{r}_0)\nabla_0 p(\mathbf{r}_0)) \cdot \mathbf{n} dS. \quad (2.34)$$

where the complex pressure amplitude $(p(\mathbf{r}_0))$ and the Green's function $(g(\mathbf{r}|\mathbf{r}_0))$ are function of frequency ω, and the frequency dependency is omitted here for convenience.

Equation (2.34) conveys that the sound pressure at an arbitrary location is what is propagated from the location where it is produced by the pressure and velocity sources. To understand how we can use this equation, consider the sound radiation from a baffled piston as illustrated in Figure 2.8. Note that we only have the second integral of Equation (2.34) in this case. If we apply Equation (2.34) to this specific case by employing the surface of the integral $(S_{0\pm})$ (Figure 2.8), then we obtain[9]

$$p(\mathbf{r}) = -2i\rho_0\omega \int_{S_{0+}} \frac{e^{ikR}}{4\pi R} u_n dS_0. \quad (2.35)$$

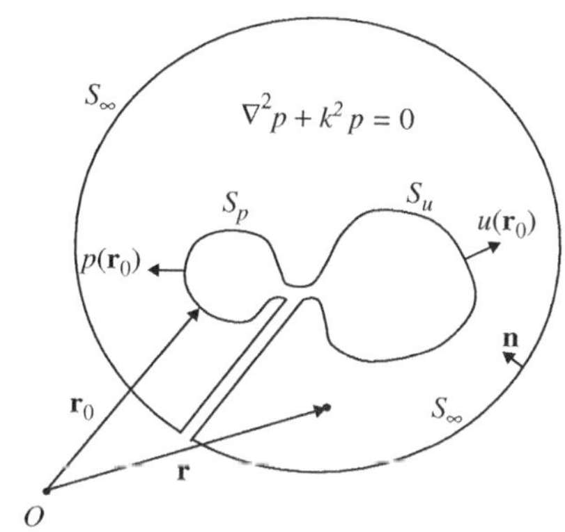

Figure 2.7 Volume integral of the Kirchhoff–Helmholtz integral equation. (S_p and S_u express the boundary surface for the pressure p and velocity u, S_∞ denotes a surface that is infinitely far from the origin, $\mathbf{n}$ is the surface normal vector, and $\mathbf{r}$ and $\mathbf{r}_0$ indicate the observation position and the boundary position vector, respectively.)

[8] These two methods of sound generation are usually used in musical instruments. For example, a flute uses the pressure fluctuation. On the other hand, the violin and cello use the vibration of the structure that is induced by the excitation of the bow on the string.

[9] The surface integral S_0 is composed of S_0^+ and S_0^-. The velocity distributions on each surface are equal in magnitude but different in direction. The boundary condition of the baffle, which requires the velocities normal to the surface to be 0, is therefore automatically satisfied. Green's functions which are employed satisfy the aforementioned boundary conditions and the linear acoustic wave equation.

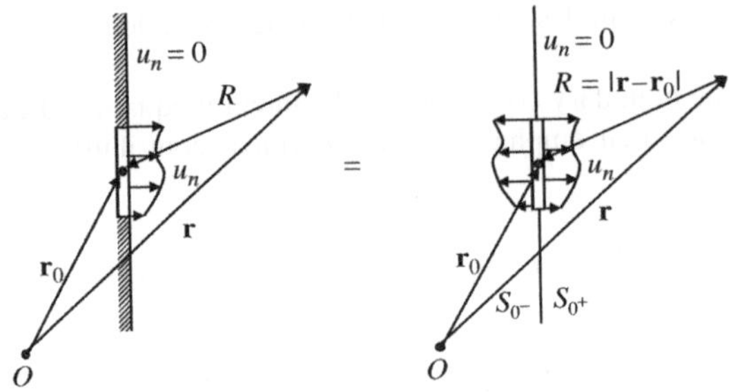

Figure 2.8 Surface integral to calculate the sound radiation from a baffled piston and nomenclature of the coordinate

This result can be obtained by realizing that $\partial g/\partial n_0 = \mathbf{n} \cdot \nabla_0 g$ on S_{0+} and $\partial g/\partial n_0$ on S_{0-} (Figure 2.8) have opposite signs, but the pressure on S_{0+} and S_{0-} is equal. Therefore, the first term of the second integral in Equation (2.34) disappears, and we only have the second term. Using free-space Green's function in Equation (2.34), we can derive Equation (2.35) (Section 2.7.3.4).

Note that Equation (2.35) states that the radiated sound pressure is induced by the vibrating surface velocity. The waves generated by the vibrating surface propagate like a monopole in space. The magnitude of vibration is simply a scale factor of the sound pressure that is propagated from the surface to the observation position of interest. It can be demonstrated by dividing the integral surface into small elements. We can then see that the velocity of each element (u_n) causes fluid particles to vibrate and that the disturbances then propagate to the observation position ($\mathbf{r}$) as the monopole propagator. Equation (2.35) can also be regarded as an expression of Huygens principle in an integral form. Note that the denominator of the integrand takes the absolute distance. This is because the propagator, which is monopole, depends only on the distance. This is one of the characteristics of a monopole source.

We start with the case where u_n is constant. In other words, the piston is a rigid vibrator (Figure 2.9). By using Equation (2.35), the pressure at an arbitrary position (z) can be expressed by

$$p(z) = -2i\rho_0\omega \int_0^a \frac{e^{ik\sqrt{z^2+\zeta^2}}}{4\pi\sqrt{z^2+\zeta^2}} u_n 2\pi\zeta\, d\zeta. \tag{2.36}$$

This equation can be simplified as

$$p(z) = \rho_0 c u_n (e^{ikz} - e^{ikR_a}), \tag{2.37}$$

where $R_a = \sqrt{z^2 + a^2}$ is the distance from the edge of a circular piston to z (detailed derivation is given in Section 2.7.1). This result has significant physical implications. There are two distinct contributions in the z direction. The first (e^{ikz}) is the pressure coming from the center of the piston, and the second term (e^{ikR_a}) is from the piston's rim. The coefficient ($\rho_0 c u_n$) is the pressure of the plane wave that is generated by the piston's motion. Note that these two pressure waves interfere with each other. The interference sometimes mutually cancels the waves or reinforces them. In the farfield (i.e., when z is large compared to the wavelength of interest), the cancelation is more likely to dominate the result. If the wavelength of interest is fairly large compared to the piston's diameter ($a/\lambda \ll 1$), then perfect cancelation occurs independently of the observation position.

If z is significantly large compared to the radius of the piston, then Equation (2.37) becomes[10]

$$p(z) = -\rho_0 c u_n \frac{ika}{2}\frac{a}{z} e^{ikz}. \tag{2.38}$$

[10] For a detailed derivation, see Section 2.7.1.

This essentially states that the farfield radiation of the baffled piston is similar to monopole radiation. This is a general observation which can be made in most cases. The nearfield effect ceases rapidly and only monopole behavior persists in the farfield (Section 2.7.1). Note that Equation (2.38) describes the sound pressure along the center line of the piston (z axis). Figure 2.10 depicts the coordinates and nomenclature that we use to predict the radiation from the vibrating piston to an arbitrary position, including the z axis. The sound pressure ($p(r,\theta)$) on the $x-z$ plane is given as[11]

$$p(r,\theta) = -ik\rho_0 c u_n \frac{e^{ikr}}{r} a^2 \frac{J_1(ka\sin\theta)}{ka\sin\theta}, \tag{2.39}$$

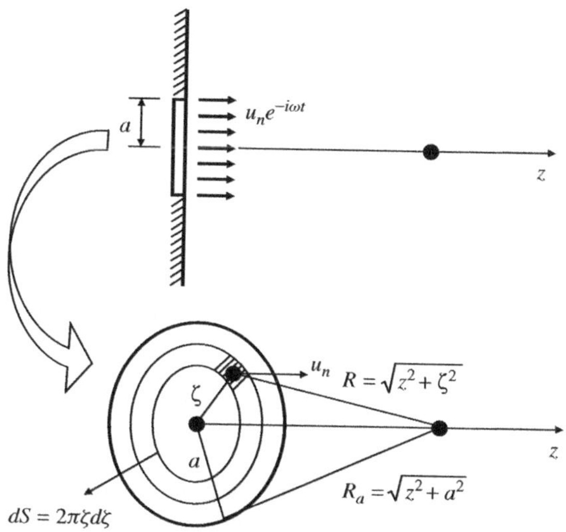

Figure 2.9 The radiated sound field from an infinitely baffled circular piston

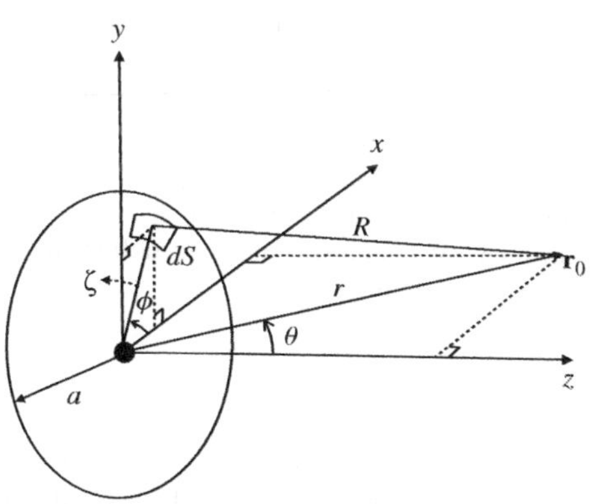

Figure 2.10 Coordinate set-up and variables for obtaining radiated sound field on $x-z$ plane from an infinitely baffled circular piston

[11] Section 2.7.2 includes the detailed derivation.

where J_1 is a Bessel function of the first kind. It is interesting to look at the sound pressure at $\theta = 0°$, that is,

$$p(r, 0) = -i\rho_0 c \frac{ka^2}{r} e^{ikr} \frac{u_n}{2}. \tag{2.40}$$

This is identical to Equation (2.38), meaning that Equation (2.39) is a general expression of the radiation sound pressure on the $x - z$ plane. Note that Equations (2.39) and (2.40) do not exhibit any dependency with respect to ϕ (Figure 2.10). This is a simple consequence of the symmetry property of radiation: we assumed that the velocity distribution on the piston is uniform (u_n). The sound pressure on the $y - z$ plane, therefore, will be the same as that on the $x - z$ plane because the piston's vibratory motion is symmetrical with respect to the x and y axes. Using Equations (2.39) and (2.40), we can obtain an expression which provides further significant physical insight, that is,

$$p(r, \theta) = p(r, 0) \cdot 2 \cdot \frac{J_1(ka \sin\theta)}{ka \sin\theta}. \tag{2.41}$$

Also, note that the mean intensity is

$$I_{avg} = \frac{|p|^2}{2\rho_0 c}. \tag{2.42}$$

Therefore, the intensity ratio between the time-averaged intensity at an arbitrary position with respect to that of the axis attached to the center of the piston can be expressed as

$$\frac{I_{avg}(r, \theta)}{I_{avg}(r, 0)} = \left[2\frac{J_1(ka \sin\theta)}{ka \sin\theta}\right]^2 \tag{2.43}$$

by using Equations (2.41) and (2.42).

Figure 2.11 depicts (Equation (2.43)) in terms of ka. Equation (2.43) is often called the directivity or directivity index. Other names such as spreading and spreading index are also widely used. From Figure 2.11, we can see that a key factor governing the radiation characteristics is ka. This basically measures how large the radiator is compared to the wavelength of interest. Equation (2.43) states that the angular dependency of the radiation significantly decreases as the radiator size decreases relative to the wavelength, or when we have lower frequency radiation. On the other hand, when ka becomes larger, or the frequency becomes higher, the radiation strongly depends on the angle. In other words, the directivity tends to be angle dependent. As illustrated in Figure 2.11, there is a silent region. In other words, there are some angles where we cannot hear sound coming from the piston.

We now examine the radiation impedance of the circular baffled piston, which shows the relation between pressure and particle velocity on the surface of the piston. Note that the radiation impedance of the plate varies with the position of the plate. The velocity is constant (u_n) on the surface, but the resulting pressure is not uniform. We therefore need to define the radiation impedance of this case. One possible definition can be written as

$$Z_r = \frac{p_{avg}}{u_n} = \frac{F/S}{u_n}, \tag{2.44}$$

where F is force acting on the source surface (S) and p_{avg} is the average pressure on the surface. To obtain F, we need to calculate the pressure on the surface of the piston ($p_S(\varsigma', \phi')$). This can be regarded as the sum of the pressure induced by the other area over the entire piston surface, that is,

$$p_S(\varsigma', \phi') = \frac{ik\rho_0 c u_n}{2\pi} \int_S \frac{e^{ikR}}{R} dS = \frac{ik\rho_0 c u_n}{2\pi} \int_0^{2\pi} \int_0^a \frac{e^{ikR}}{R} \varsigma \, d\varsigma \, d\phi, \tag{2.45}$$

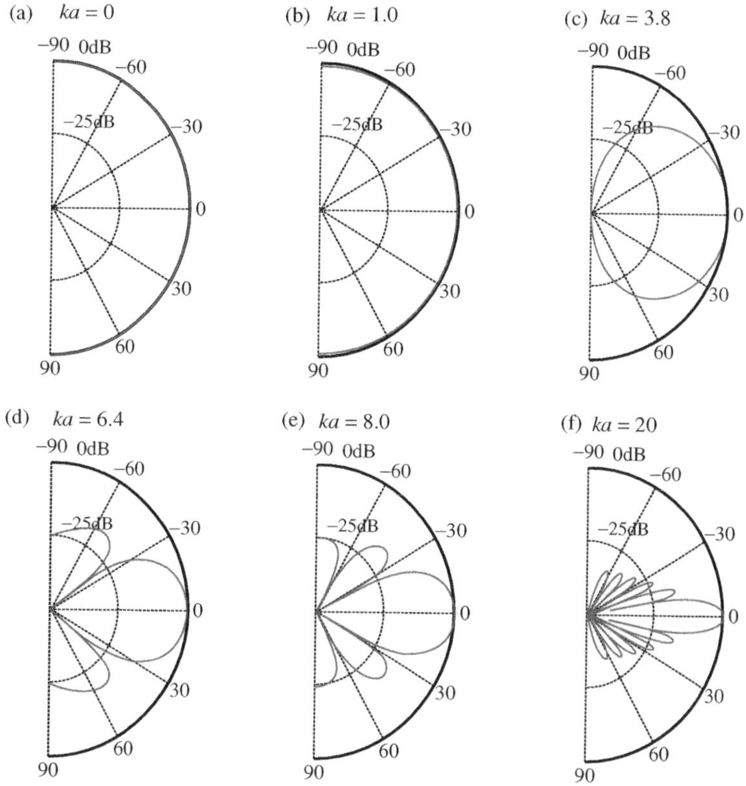

$$DI = \left[2 \frac{J_1(ka \sin \theta)}{ka \sin \theta} \right]^2$$

Figure 2.11 (a–f) The directivity or directivity index of a baffled circular piston

where Figure 2.12a illustrates the variables for the integration. The force acting on the piston surface can be given by

$$\begin{aligned} F &= \int_S p_S dS' = \int_0^{2\pi} \int_0^a p_S(\zeta', \phi')\zeta' d\zeta' d\phi' \\ &= \frac{ik\rho_0 c u_n}{2\pi} \int_0^{2\pi} \int_0^a \left[\int_0^{2\pi} \int_0^a \frac{e^{ikR}}{R} \zeta d\zeta d\phi \right] \zeta' d\zeta' d\phi'. \end{aligned} \tag{2.46}$$

The integration of Equation (2.46) can be written as:

$$\begin{aligned} \int_0^{2\pi} \int_0^a \left[\int_0^{2\pi} \int_0^a \frac{e^{ikR}}{R} \zeta d\zeta d\phi \right] \zeta' d\zeta' d\phi' &= 2 \int_0^{2\pi} \int_0^a \left[\int_{-\pi/2}^{\pi/2} \int_0^{2\zeta' \cos\varphi} \frac{e^{ikR}}{R} R dR d\varphi \right] \zeta' d\zeta' d\phi' \\ &= \frac{2\pi}{ik} \int_0^{2\pi} \int_0^a [1 - J_0(2k\zeta') - iH_0(2k\zeta')]\zeta' d\zeta' d\phi', \end{aligned} \tag{2.47}$$

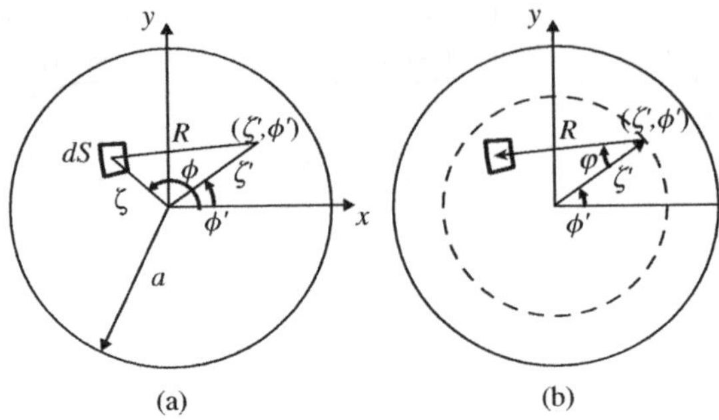

Figure 2.12 The variables on the surface of the disk for integrating with respect to (a) ζ and ϕ and (b) R and φ

where $J_0(2k\zeta)$ is the first kind of Bessel function of zero order, and $H_0(2k\zeta)$ is the zero-order Struve function. Figure 2.12b illustrates the variables of the disk used to calculate the integration.

The force acting on the piston surface can then be given by

$$
\begin{aligned}
F &= \frac{ik\rho_0 c u_n}{2\pi}\frac{2\pi}{ik}\int_0^{2\pi}\int_0^{a}[1 - J_0(2k\zeta') - iH_0(2k\zeta')]\zeta' d\zeta' d\phi' \\
&= \rho_0 c u_n \int_0^{2\pi}\int_0^{a}[1 - J_0(2k\zeta') - iH_0(2k\zeta')]\zeta' d\zeta' d\phi' \\
&= \rho_0 c u_n \pi a^2\left[1 - \frac{J_1(2ka)}{ka} - i\frac{H_1(2ka)}{ka}\right],
\end{aligned}
\tag{2.48}
$$

where $J_1(2ka)$ is the first kind of Bessel function of the first order, and $H_1(2ka)$ is the Struve function of the first order. p_{avg} is then defined as

$$
p_{avg} = \rho_0 c u_n\left[1 - \frac{J_1(2ka)}{ka} - i\frac{H_1(2ka)}{ka}\right], \tag{2.49}
$$

and the radiation impedance of a baffled circular piston with radius a can be given by

$$
Z_r = \rho_0 c\left[1 - \frac{J_1(2ka)}{ka} - i\frac{H_1(2ka)}{ka}\right]. \tag{2.50}
$$

Figure 2.13 shows the normalized radiation impedance ($Z_r/\rho_0 c$) with respect to ka. As ka becomes large (i.e., radius becomes larger than wavelength), the resistance term approaches the medium impedance ($\rho_0 c$). On the other hand, the reactance is quickly diminished. It shows that the baffled piston radiates very effectively, as if generating a plane wave, when the wavelength of interest is larger than the radius a.

To summarize this section, we have studied the radiation characteristics of the sound generated by the vibration of a baffled piston. These are essentially governed by the size of the radiator and the distance from which the sound pressure is observed compared to the wavelength. The size and distance of the observation position have to be scaled by the wavelength, and the absolute scale does not possess any significant meaning with respect to the radiation characteristics. We also found that the directivity index is a good measure to examine the angle dependency of the radiation. The radiation tends to be of monopole type when we have a smaller radiator compared

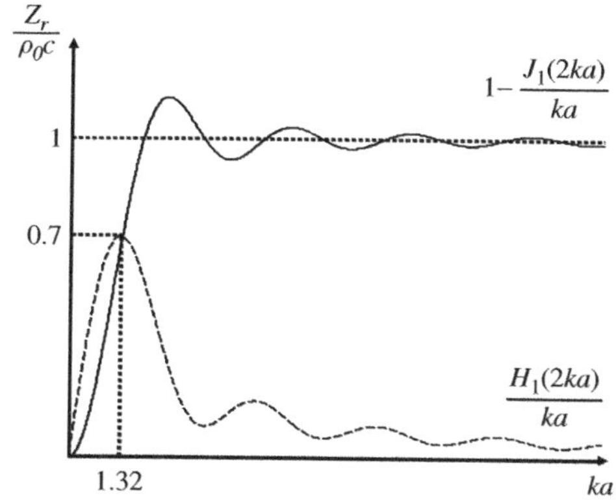

Figure 2.13 Radiation impedance, which is normalized to $\rho_0 c$, of the circular baffled piston with respect to ka. The solid line represents the resistance term, and the dashed line the reactance

to the wavelength. On the other hand, as the radiator increases in size compared to the wavelength, there will be more interference. As a result, there will be a very rapid angle fluctuation in the directivity index.

If the size of a baffle is finite, we cannot expect similar results. However, if for example, we produce a wave using a speaker unit that is installed in a box, then radiation characteristics as described in this section may or may not be observed depending on the size of the box compared to the wavelength. Overall, examination of the characteristics of a baffled piston will provide basic understanding of sound radiation. It is obvious that if we want to determine precisely how sound is radiated from an arbitrarily vibrating body or surface, it is necessary to use the integral equations (addressed in Section 2.7) or conduct experiments.

2.4 Radiation from a Finite Vibrating Plate

It may be possible to apply what we learned regarding the radiation from a baffled piston to the radiation from a vibrating plate. This idea stems from the realization that the vibrating plate can be modeled as numerous vibrating pistons, as illustrated in Figure 2.14.[12]

Figure 2.14a depicts how the vibration of a plate can be considered to be made by the n modes of the plate. The radiation due to the plate vibration can therefore be considered to be composed of the radiation of n modes of vibration. The radiation due to each mode of vibration can be superimposed by the vibration of many pistons, as illustrated in Figure 2.14b. We can assume that each mode of vibration can have equivalent pistons.[13] The piston's amplitude of vibration has to have the same volume velocity as the corresponding mode of vibration.

Equation (2.35), the Rayleigh integral equation, can be applied to each equivalent piston. The superposition concept can be used, as illustrated in Figure 2.14. The radiated pressure field from the vibrating plate can be readily obtained. To understand the radiation from the vibrating plate, we look at basic radiations patterns such as those illustrated in Figure 2.15. The radiated fields are obtained using the Rayleigh integral equation. The radiator's typical dimensions are larger than

[12] We can assume that the plate vibrates with sufficiently small amplitude so that the radiated pressure obeys the linear acoustic wave equation. Therefore, the principle of superposition holds. Figure 2.14, in fact, shows how we can apply the principle to predict the radiated sound field.

[13] This concept essentially motivates us to establish the BEM.

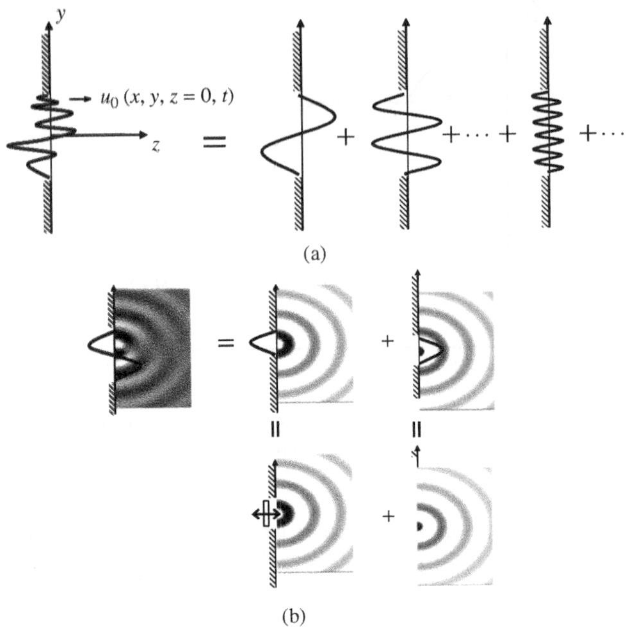

Figure 2.14 (a) Radiation from a finite plate and (b) its possible modeling

the wavelengths that are generated; the radiation efficiencies are therefore fairly good for all three cases. It should be noted that the sound pressures at the distance from the radiators (z_l) become negligible when the distance is long compared to the wavelength, except for case Figure 2.15a. This is because the farfield sound propagation tends to a plane wave, resulting in a perfect cancelation for the cases of Figure 2.15b,c. Note that the observation positions of these cases are exactly on the axis of symmetry, therefore the waves have a phase difference of 180°. This kind of cancelation becomes more and more significant when we have higher order modes.

To look more specifically at the radiation from a plate, we describe the waves in space using rectangular coordinates for convenience. Let us also assume that the wave is harmonic in space and time, without loss of generality. The acoustic wave can now be written as

$$p(x, y, z, t) = Pe^{i\mathbf{k}\cdot\mathbf{r}}e^{-i\omega t} = Pe^{i(k_x x + k_y y + k_z z)}e^{-i\omega t}. \tag{2.51}$$

Equation (2.51) must satisfy the linear acoustic wave equation; the following equality must therefore hold, that is,

$$k_x^2 + k_y^2 + k_z^2 = k^2, \tag{2.52}$$

where $k = \omega/c$, which is a dispersion relation that relates spatial and time domain fluctuation. If we have a vibrating body only on the (x, y) plane which vibrates as illustrated in Figures 2.14 and 2.15, then the wavenumbers in the x and y directions, (k_x and k_y) can be written as

$$k_x = \frac{m\pi}{L_x} \tag{2.53}$$

$$k_y = \frac{n\pi}{L_y}, \tag{2.54}$$

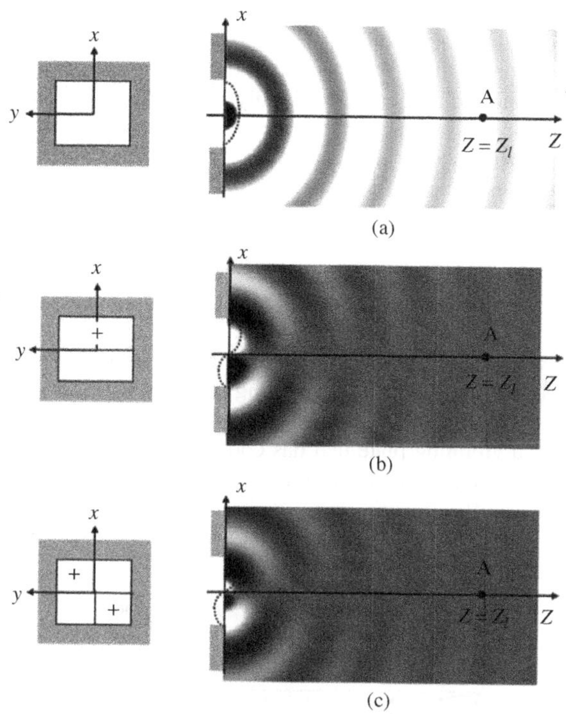

Figure 2.15 Basic examples of plate vibration and radiation (A indicates the observation position, and can be expressed as $(0, 0, z_l)$ in the Cartesian coordinate)

where L_x and L_y are the length of the plate in the x and y directions, respectively, and m and n are integers. These relationships can also be obtained from the boundary conditions of the plates. Figure 2.16a illustrates what is implied by these expressions. Figure 2.16a, Equations (2.52)–(2.54) tell us that the wavenumber in the z direction (k_z), which describes how the wave in the z direction propagates, could be real or imaginary. This depends on whether the free-space wavenumber (k) is smaller or greater than the wavenumber in the (x, y) plane, which is $\sqrt{k_x^2 + k_y^2}$. The former case can be written in terms of the inequality between the wavenumbers, that is,

$$k_x^2 + k_y^2 < k^2. \tag{2.55}$$

In this case, (k_x, k_y) is located inside a radiation circle whose radius is the free-space wavenumber k and its equation is $k^2 = k_x^2 + k_y^2 + k_z^2$. The wavenumber in the z direction is always positive and real, as indicated by Equations (2.52) and (2.55). This means that a wave in the z direction continuously changes its phase as it propagates. In other words, the propagator can be mathematically written as $e^{ik_z z}$. On the other hand, the latter case satisfies the inequality:

$$k_x^2 + k_y^2 > k^2 \tag{2.56}$$

In this case, k_z has to be imaginary and the wave therefore decays exponentially. Note that (k_x, k_y) is located at the outside of the radiation circle (Figure 2.16a).[14] This figure also illustrates

[14] The radiation circle is normally defined in this context. For example, the evanescent wave and propagating wave are decided by the region in which the wavenumber is located.

that the wave is less likely to propagate in the z direction with a larger k_x or k_y, in other words, as the wavelength becomes increasingly smaller. This is simply because the wavenumbers in the x and y directions approach the region far away from the radiation circle as the wavelength becomes smaller. Therefore, propagation in the z direction is not highly plausible. To summarize, the development of either exponentially decaying propagation (evanescent wave) in the z direction or a continuous phase changing wave is determined by the location (k_x, k_y) in reference to the radiation circle.

Figure 2.16 also illustrates the radiation characteristics of the plate when it has higher order modes in either the x or y directions. If $k_x = 0$ but k_y is getting larger, then the vibration mode tends to have more oscillation in the y direction as illustrated in Figure 2.16b. This kind of mode is often referred to as edge mode. We can easily see that the radiation due to this edge mode will be far less effective than that of the baffled piston because of possible wave cancelation, as illustrated in Figure 2.15.

Corner modes occur when the vibrations and radiations with respect to x and y are equally likely to participate.[15] The radiation impedance of a vibrating structure that has a finite area (for example, baffled piston or plate, etc.) should be considered by average concept over the area as explained in Section 2.3. In particular, a vibrating plate that has corner and edge modes must not only consider average sound pressure on the plate but also averaged velocity on the area. For instance, the normal

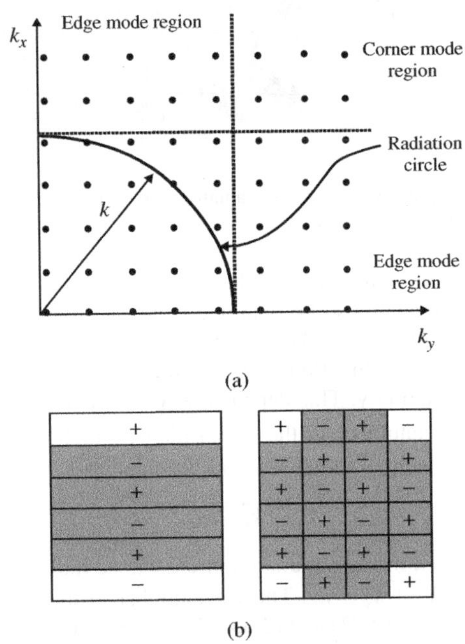

Figure 2.16 Radiation from a finite plate (corner mode and edge mode): (a) corner modes, edge modes, and radiation circle in wavenumber domain and (b) corner mode and edge mode (left: the edge mode dominates the radiation, right: the corner mode dominates the radiation), (+, − indicate that there is a 180° phase difference)

[15] These characteristics may or may not be good depending on how we want to use them. For example, if we want to use these characteristics to reduce the sound or noise, then we would like to have more corner or edge modes. On the other hand, to create a good radiator, we want to avoid these modes.

velocity distribution on a baffled plate can be expressed as

$$u_n(x_0, y_0) = U \sin(k_x x_0) \sin(k_y y_0), \quad \begin{cases} 0 \le x_0 \le L_x \\ 0 \le y_0 \le L_y \end{cases} \tag{2.57}$$

where U is the velocity magnitude on the plate, the subscript 0 indicates source coordinate. Sound pressure at an arbitrary observation point in the positive z direction can be determined by Rayleigh integral equation of Equation (2.35). Sound pressure by the vibration (Equation (2.57)) can therefore be written as

$$p(x, y, z) = -\frac{ik\rho_0 cU}{2\pi} \int_0^{L_y} \int_0^{L_x} \left\{ \sin\left(k_x x_0\right) \sin(k_y y_0) \frac{e^{ikR}}{R} \right\} dx_0 dy_0, \tag{2.58}$$

where R represents the distance between an observation point and an infinitesimal area $(dx_0 dy_0)$ on the plate. The radiation impedance of a vibrating baffled plate can be obtained according to the definition of Equation (2.50) in Section 2.3. Note, however, that the velocity in the denominator also has to be averaged over the area of the plate because the velocity at a point on the plate cannot represent the entire velocity distribution.

We can also extend the concept of studying sound propagation by using wavenumbers into a general three-dimensional case. Acoustic holography is essentially based on this concept. In acoustic holography, the spatial distribution of a sound wave is expressed in terms of the wavenumber domain of interest, and then propagated to a plane that is not measured by using a propagator. For example, we transform the sound wave field in the (x, y) plane to the (k_x, k_y) plane, and then use the k_z propagator. Figure 2.17 illustrates the basic procedures of acoustic holography.

To summarize, the radiation from a plate vibration depends very strongly on the vibration characteristics of the plate. It is understood that the radiation from a plate is effective when the size of the plate is much greater than the wavelength that we wish to generate. On the other hand, if the wavelength is much smaller than the size of the plate, then an edge or corner mode will arise. The associated radiation efficiency tends to be smaller as there are more modes of vibration. This result can be visualized if we consider the possible motions of fluid particles on a plate.

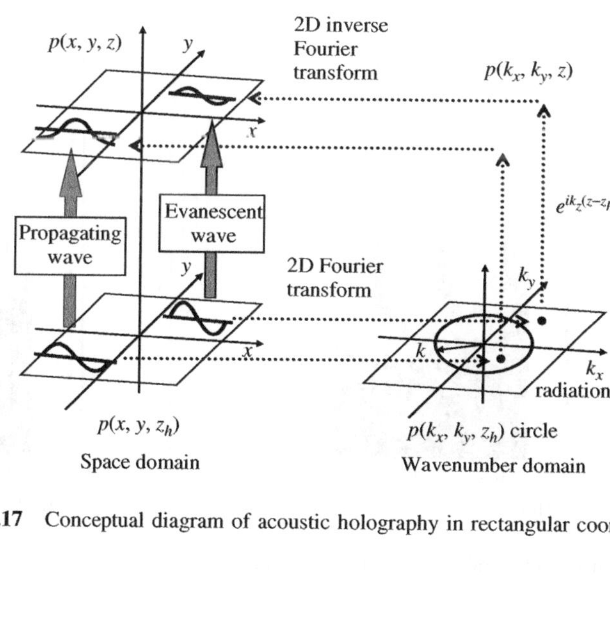

Figure 2.17 Conceptual diagram of acoustic holography in rectangular coordinates [2]

2.5 Diffraction and Scattering

Figure 2.18 illustrates typical diffraction and scattering phenomena. We can clearly see from this figure that the diffraction tends to be stronger as the wavelength becomes larger. In other words, we have more diffraction at the back of the wall as the wavelength increases. Strictly speaking, as the wall's height decreases compared to the wavelength there will be more diffraction.

Diffraction is generally used to describe the physical circumstances under which we can hear sound but cannot see the sound source. As illustrated in Figure 2.18, these circumstances typically occur when a sound wave meets a discontinuity that is open in half space. The sound waves arriving at the lower part of the barrier are more likely to be reflected whereas, at the upper part of the barrier, the waves are mostly scattered. The reflected and incident waves are mostly dominant in the left part of the barrier, and the scattered waves dominate the remainder of the region. We often call this phenomenon diffraction. However, we also refer to such waves as "scattering." Scattering describes waves that are induced due to an abrupt impedance change in space, especially when the waves spread out in space. Notably, reflection and scattering are due to the presence of an impedance discontinuity, and the type of acoustic wave will dictate whether reflection or scattering occurs.

Section 2.7.3 mathematically describes how diffraction depends on the wavelength, frequency, and observation position with respect to wavelength. As we noted in the earlier part of this chapter, radiation, diffraction, and scattering can be intuitively understood in a unified way. All of these phenomena are consequences of the impedance mismatch in space and frequency. Depending on the type of impedance mismatch, the sound waves that can occur are diffraction, scattering, refraction, or reflection. Note that when an incident wave is oblique on the flat surface of a discontinuity, the transmitted wave normally has a different angle of transmission; we call this phenomenon "refraction." Diffraction can also be regarded as a phenomenon that can occur when an incident wave meets an impedance mismatch in space. The impedance mismatch in space produces not only reflection but also scattering at a relatively sharp edge or edges, as illustrated in Figure 2.18. In other words, we can regard the impedance mismatch as a predictor of how much sound pressure will be propagated because it always creates pressure and velocity fluctuation. Thus, understanding of the diffraction problem is possible by finding the solution in a half space for the case of a

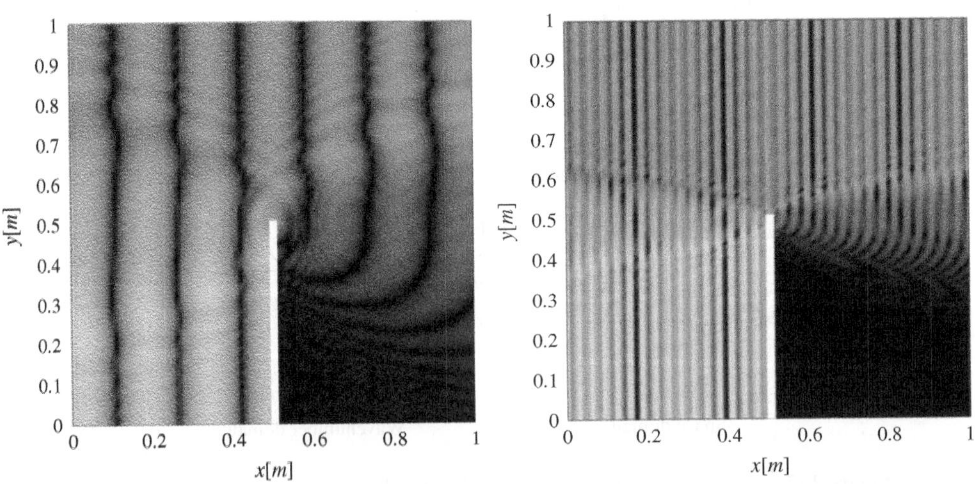

Figure 2.18 Diffraction around a straight barrier (note that the diffraction strongly depends on the wavelength)

piston excitation on an infinite baffle or finite baffle (Section 2.7.3 mathematically addresses the diffraction and radiation problems).

We begin by looking at a scattering problem in terms of a unified concept that regards diffraction, scattering, and other phenomenon as spatial- and frequency-dependent impedance mismatch problems. Beginning with the simplest case which considers all the necessary fundamentals, suppose that we have a plane wave impinging on an arbitrary scatterer, as illustrated in Figure 2.19.

The complex amplitude in space of the incident wave is denoted p_i and the complex amplitude of the scattering wave is denoted p_{sc}. The total sound pressure can then be expressed as

$$p_t = p_i + p_{sc}. \tag{2.59}$$

The total pressure has to satisfy the following boundary condition, that is,

$$\nabla p_t \cdot \mathbf{n} = 0. \tag{2.60}$$

We assumed that the boundary is acoustically rigid and $\mathbf{n}$ is a unit normal vector on the surface (Figure 2.19). Let p_i propagate in the direction of the wavenumber vector $\mathbf{k}$. The incident wave at the position $\mathbf{r}$ can then be written as

$$p_i = Be^{i\mathbf{k}\cdot\mathbf{r}}, \tag{2.61}$$

where B is the complex amplitude of the incident wave (Figure 2.19). Equations (2.59)–(2.61) lead us to a relation between the scattered wave and the incident wave, that is,

$$\nabla p_{sc} \cdot \mathbf{n} = -iB\mathbf{k} \cdot \mathbf{n}e^{i\mathbf{k}\cdot\mathbf{r}}. \tag{2.62}$$

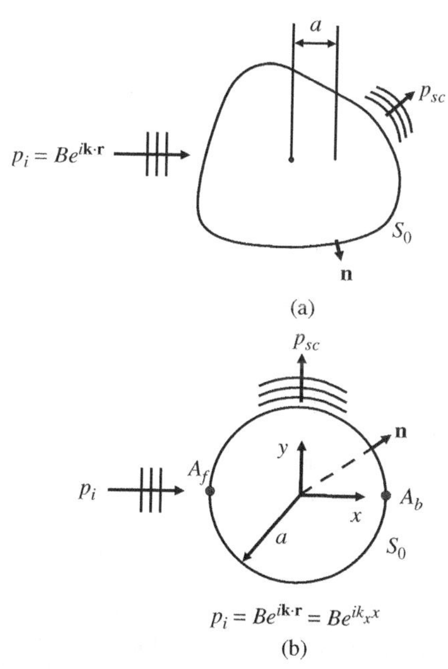

Figure 2.19 Scattering by (a) arbitrary scatterer and by (b) a sphere. (p_i and p_{sc} represent the complex amplitude of incident and scattered waves; S_0 and $\mathbf{n}$ represent the surface of the scatterer and a unit vector normal to the surface; and a is the characteristic length of the scatterer.)

Equation (2.62) essentially states that fluid particles on the scatterer have to move as if they are attached to it.[16] If we rewrite Equation (2.62) in terms of the velocity of the scatterer by using the linearized Euler equation, then we obtain

$$\nabla p_{sc} = \rho_0(i\omega)\mathbf{u}_{sc}, \tag{2.63}$$

where $\mathbf{u}_{sc}$ is the magnitude of the scattering velocity in vector form.

By starting with the simplest case, we consider the scattering of a rigid sphere to explore what is meant by Equations (2.62) and (2.63). Rewriting Equation (2.62) using the coordinate of Figure 2.19b, we obtain

$$\frac{\partial p_{sc}}{\partial n} = -ikB\cos\theta e^{i\mathbf{k}\cdot\mathbf{r}}, \tag{2.64}$$

where θ represents the angle between the propagation vector $\mathbf{k}$ and the normal vector $\mathbf{n}$. Equations (2.63) and (2.64) allows us to find the velocity in the normal direction, which is $\mathbf{u}_{sc}\cdot\mathbf{n}$, that is,

$$\mathbf{u}_{sc}\cdot\mathbf{n} = -ikB\cos\theta\left(\frac{1}{ik\rho_0 c}\right)e^{i\mathbf{k}\cdot\mathbf{r}}\Big]_{on\ S_0}, \tag{2.65}$$

where S_0 is the surface of the sphere (Figure 2.19b). Since the radius of the sphere is a, we can rewrite Equation (2.65) as

$$\mathbf{u}_{sc}\cdot\mathbf{n} = -\frac{B}{\rho_0 c}\cos\theta e^{ika\cos\theta}. \tag{2.66}$$

The normal velocity on the rigid sphere is a function of θ as shown in Figure 2.20, which shows that the real part of the directional component is $\cos\theta e^{ika\cos\theta}$ with respect to ka.

If the size of the scatterer is substantially smaller than the wavelength of interest (i.e., if $ka \ll 1$), then Equation (2.66) can be approximately written as follows by expanding it in a Taylor series and taking only the first term:

$$\mathbf{u}_{sc}\cdot\mathbf{n} \cong -\frac{B}{\rho_0 c}\cos\theta(1 + ika\cos\theta). \tag{2.67}$$

This means that the scattering field is essentially induced by the radiation of the sphere, which vibrates with a velocity as described by Equation (2.67). Note that Equation (2.67) is composed of two distinctly different terms. The scattered field results from two different types of vibration. The first term is mainly governed by cos θ. Therefore, the vibration is an asymmetric velocity distribution with respect to the angles of 90° and 270°.

When $\theta = 0°$, the angle between the wavenumber vector and the surface normal of the sphere is 0° (indicated by the location of A_b on Figure 2.19b). The first term of Equation (2.67) is defined by $-B/\rho_0 c$ and the second by $-ikaB/\rho_0 c$. On the other hand, when $\theta = 180°$ (which corresponds to the position A_f; see Figure 2.19b), the first term of Equation (2.67) will be $B/\rho_0 c$ and the second becomes $-ikaB/\rho_0 c$. The first term states that we have a velocity of $B/\rho_0 c$ at A_f in the direction of the surface normal. Velocity has the same magnitude but exactly the opposite direction, that is, the negative surface normal direction at A_b. This simply means that the sphere oscillates in the same manner as a trembling sphere. The second term, on the other hand, has i and $-i$ phase difference at A_b and A_f. This means that the sphere vibrates as if it were a breathing sphere. Lastly, it is also valuable to note that the second term becomes smaller as ka decreases. In other words, if we have a sphere with diameter smaller than the wavelength of interest, then the second term contributes less than the first term. This means, rather surprisingly, that the trembling sphere effect become stronger as the size of the scatterer becomes smaller.

[16] Note that the pressure gradient induces the motion of fluid particles as described by the linearized Euler equation.

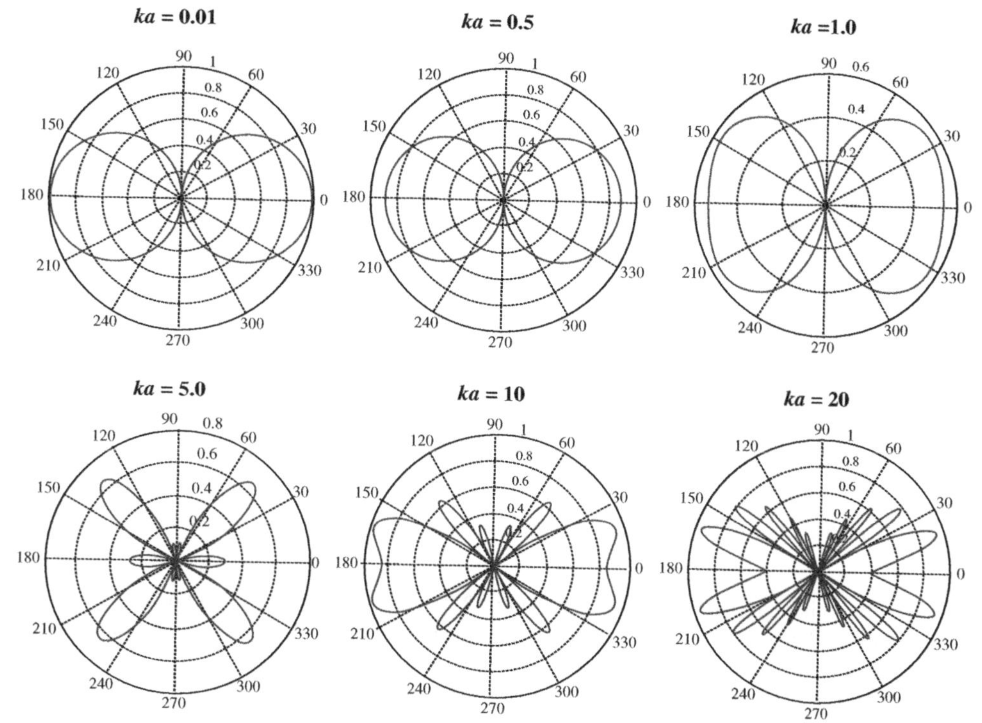

Figure 2.20 Directional component ($\cos\theta e^{ika\cos\theta}$) of the normal velocity ($\mathbf{u}_{sc}\cdot\mathbf{n}$) on the rigid sphere with respect to *ka*

We can generalize this relation by using Equation (2.62) and the linearized Euler equation, that is,

$$\mathbf{u}_{sc}\cdot\mathbf{n} = -\frac{B}{\rho_0 c}e^{i\mathbf{k}\cdot\mathbf{r}_0}\mathbf{e}_k\cdot\mathbf{n}e^{-i\omega t}, \tag{2.68}$$

where $\mathbf{e}_k = \mathbf{k}/k$ is the unit propagation vector and $\mathbf{r}_0$ is the position vector, which indicates the surface of the scatterer. If the wavelength is much larger than the characteristic length of the scatterer, then we can linearize Equation (2.68) such that it becomes

$$\mathbf{u}_{sc}\cdot\mathbf{n} \cong -\left\{\frac{B}{\rho_0 c}\mathbf{e}_k\cdot\mathbf{n} + i\frac{B}{\rho_0 c}\left(\mathbf{k}\cdot\mathbf{r}_0\right)\mathbf{e}_k\cdot\mathbf{n}\right\}e^{-i\omega t}. \tag{2.69}$$

This is a general form of Equation (2.67). The radiated sound field due to this velocity distribution can be readily obtained by using the Rayleigh integral equation. Section 2.7.4 presents details of the associated integration. We attempted to understand Equation (2.67) by examining the contribution of its two terms. Similarly, Equation (2.69) can be conceptually envisaged. The first term is real and the second is imaginary. The vector expressions describe how the scatterer, in general, vibrates.

We have seen that scattering is mainly dominated by the ratio of the size of the scatterer to the wavelength of interest (*ka*) as well as the scattering direction, that is, the angle θ in Equation (2.67). The first type dependency is in line with our expectations because the scattering is induced by the radiation and the radiation strongly depends on the scale factor *ka*, as we already investigated using breathing and trembling spheres. However, the angle dependency has not been well investigated, and

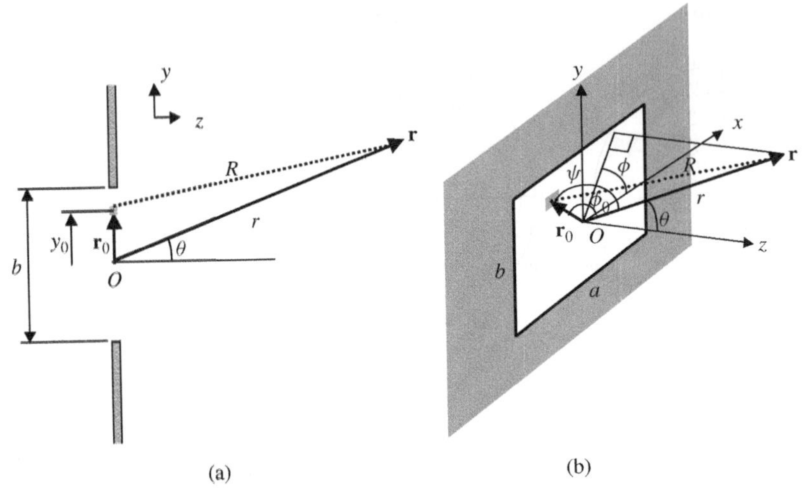

Figure 2.21 (a) Two-dimensional and (b) three-dimensional rectangular slit with corresponding nomenclature

is presumably a characteristic of scattering. Figure 2.21 exhibits two fundamental scatterers which can demonstrate the scatterers' geometrical effect on the radiation. The first is a two-dimensional case and the second is its extension to three dimensions.

The radiated sound pressure for the two-dimensional slit can be found as

$$p(r,\theta) = \sqrt{2}(1-i)\rho_0 c b u_0 \sqrt{\frac{k}{\pi r}} e^{ikr} \frac{\sin((kb/2)\sin\theta)}{(kb/2)\sin\theta} \tag{2.70}$$

(see Section 2.7.3.5 for details). This result says that the scattered sound propagates with respect to its radial direction (Figure 2.21) as if it is a plane wave (the e^{ikr} term of Equation (2.70)). However, it decays with $\sqrt{1/r}$, which is a typical characteristic of a two-dimensional free-field. Note also that the real and imaginary parts contribute equally. The most distinctive characteristics, however, are described in the last term of Equation (2.70), which depends mainly on the size of the slit b relative to the wavelength λ, which is kb, and the angle θ. The last term is a sinc function. It implies that the scattered field becomes more directional for smaller kb and θ, as intuitively seen.

These phenomena are maintained even for a rectangular slit case. The scattered field of the rectangular slit (Figure 2.21) can be obtained as

$$p(r,\theta,\phi) = -i\frac{k\rho_0 c}{2\pi} ab u_0 \frac{e^{ikr}}{r} \frac{\sin((ka/2)\sin\theta\cos\phi)}{(ka/2)\sin\theta\cos\phi} \times \frac{\sin((kb/2)\sin\theta\cos\phi)}{(kb/2)\sin\theta\cos\phi}. \tag{2.71}$$

(see Section 2.7.3.5 for the detailed derivation.) u_0 is the velocity in the z direction (Figure 2.21). This result is quite similar to Equation (2.70) which describes how the scattered field decays as a monopole as r increases, but is proportional to the magnitude of the velocity flux through the slit (abu_0). The directivity that is expressed in the last term is a function of the relative geometry of the slit to the wavelength, that is, ka and kb.

We can also consider that Equations (2.70) and (2.71) describing a diffraction field due to the slits or scatterers. It can also be argued that the diffractions in these cases strongly depend on the nondimensional scale factors (ka and kb) and observation angle θ.

The next step is to investigate diffractions that are induced by more general cases. Any general diffraction problem can be tackled using numerical analysis: for example, Boundary Element Method (BEM). However, it will be interesting to investigate the general diffraction problem by starting with a very simple case (Figure 2.22) then expanding to more general case as illustrated in Figure 2.23.

To understand the diffraction phenomenon, we begin by studying a typical example: diffraction phenomenon by a sound barrier as shown in Figure 2.22, for example. The closed form solution that describes diffraction due to a semi-infinite barrier (Figure 2.22a), the simplest case, can be obtained,

$$p(y,z) = \frac{e^{i(kr+\pi/4)}}{\sqrt{8\pi kr}} D(\theta,\varphi) + \begin{cases} 0, & \theta < \varphi \\ e^{i(k\sin\varphi)y + i(k\cos\varphi)z}, & \theta > \varphi \end{cases}, \tag{2.72}$$

where

$$D(\theta,\varphi) = \frac{1}{\cos(\theta/2+\varphi/2)} - \frac{1}{\sin(\theta/2-\varphi/2)}. \tag{2.73}$$

Equation (2.72) states that the diffraction depends upon the angles θ and φ. When $\theta < \varphi$, we cannot see the sound source. In other words, we are in the shadow zone and the diffraction is dominated by the scattered field induced at the edge of the wall. On the other hand, when $\theta > \varphi$, we can see the sound source on the positive z axis (Figure 2.22). The diffraction field is composed of two parts: the scattered field from the edge of the wall and the direct sound field. In this regard, we can say that the diffraction is the result of the edge scattering. In other words, we hear sound coming from the sound sources. This means that the edge condition strongly affects the diffraction in the shadow zone.

The analytic solution provides many of the elements of the physics associated with the diffraction of a barrier. It also serves as a source for understanding diffraction. However, it is also true that it entails too much information to design the barrier. In this respect, Figures 2.24 and 2.25 effectively provide the practical parameters that are associated with the diffraction of a barrier. In this respect, we first consider the simplest two-dimensional diffraction problem, illustrated in Figure 2.24. First, let us investigate the possible geometrical scales that have to be considered in the diffraction. These are the direct distance d between the source S and the receiver R, unless there is a barrier, and the distance $(L_A + L_B)$, which is the shortest distance that the sound from the source can travel to the receiver. It would also be rational to scale these geometrical distances with respect to the

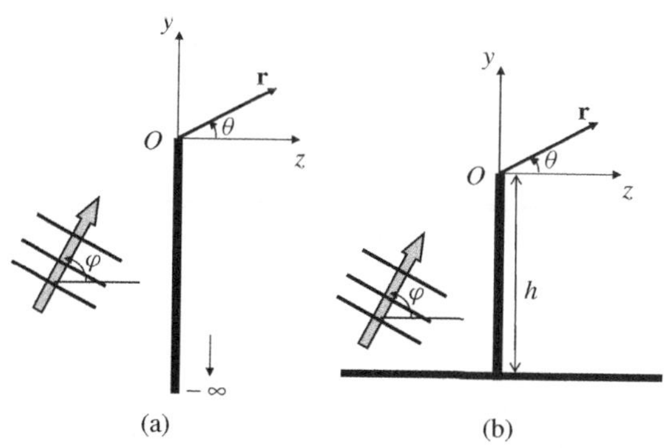

Figure 2.22 Two-dimensional diffraction problem for a plane wave source (it is assumed that the wavelength is much larger than the thickness of the wall, and the wall is acoustically rigid): (a) semi-infinite and (b) finite barrier case

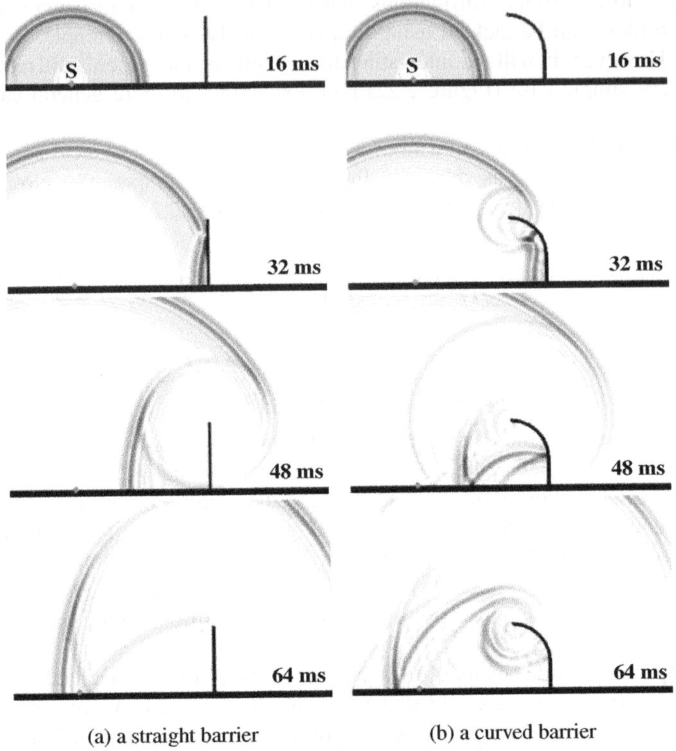

Figure 2.23 Diffraction of barriers when we have a monopole source, obtained using FDTD (finite difference time domain): (a) by a straight barrier and (b) by a curved barrier with respect to time (S denotes a monopole source on the ground). (Courtesy of H. Tachibana (University of Tokyo).)

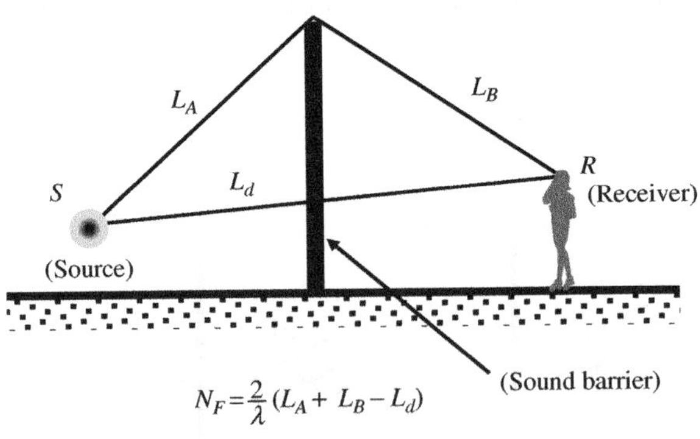

Figure 2.24 The sound barrier and associated nomenclature, where N_F is the Fresnel number, S represents the source position, and R is the receiver's location. We also assume that the wall thickness is small relative to the wavelength, and acoustically hard

wavelength of interest. This quickly becomes clear if we imagine what the wave would look like when we have a wavelength that is much larger than the distance $L_A + L_B$. As the wavelength becomes increasingly larger, the barrier's height would lose its presence in terms of diffraction. On the other hand, if the wavelength is much smaller than $L_A + L_B$, then the listener would perceive that the sound comes from the edge of the barrier.

Considering a triangle of sides L_A, L_B, and L_d leads us to deduce the relative distance between $L_A + L_B$ and L_d with regard to the wavelength. This parameter, which is needed to view the diffraction problem, is the Fresnel number (N_F) and is defined

$$N_F = \frac{L_A + L_B - L_d}{\lambda/2}. \tag{2.74}$$

The transmission loss of a barrier (TL_{br}) is generally expressed as[17]

$$TL_{br} = 20\log_{10}\left(\frac{\sqrt{2\pi N_F}}{\tanh\sqrt{2\pi N_F}}\right) + 5\text{dB} \quad (N_F > 0). \tag{2.75}$$

Figure 2.25 depicts Equation (2.75) in terms of a log-log diagram. This provides a very practical means to design a barrier. For example, once we know the desired sound attenuation by using a barrier, then the graph provides us with the corresponding Fresnel number (N_F). We can then estimate the required height of the barrier from the Fresnel number. We could also regard the Fresnel number as a means to represent the effect of the barrier or, more generally, a spatially distributed impedance discontinuity, on diffraction. When sound meets this discontinuity, it is reflected and scattered, depending on the relative size compared to that of the geometry of impedance mismatch. Fresnel number determines or represents a complicated sound wave propagation associated diffraction; it is surprising that one number can express everything about the diffraction.

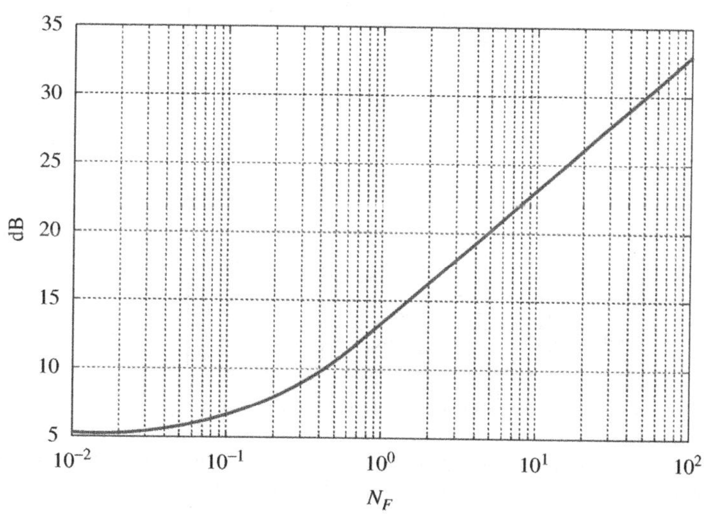

Figure 2.25 The sound attenuation due to a sound barrier (N_F is the Fresnel number)

[17] The total transmission loss (TL_b) can be written as $TL_b = TL + TL_{br}$, where TL is the transmission loss when there is no sound barrier.

A similar phenomenon to that found with diffraction can also be found for refraction. Refraction is generally due to a change in the media's characteristic impedance. For example, as illustrated in Figure 2.26, we can observe the refraction of sound due to the inhomogeneous characteristics of the media because of varying temperature. If we have a gradual increase in temperature from the ground surface which, for example, could occur during the night, then the sound will propagate toward the ground surface (Figure 2.26).

Note also that there is a region where we cannot hear the sound, which is often called the "shadow zone."[18] This is due to a gradual increase or decrease of the characteristic impedance from the ground surface.[19] To understand this in more depth, consider the case of multilayer media as illustrated in Figure 2.27. As the propagation vectors depict, the propagation becomes stiffer in the x direction as we have greater propagation speed in the y direction. In reality, within the sea and atmosphere, it is possible to have more complicated media characteristics. As a result, some sound can propagate distances of more than 10 or 10 000 km. Impedance change in space brings refraction.

To summarize this section, the scattered field due to the presence of a sphere in an incident sound field is composed of the part that is induced by a breathing sphere and that made by a trembling sphere. The magnitude is proportional to the magnitude of the incident wave (B). The contribution of the breathing sphere becomes smaller as the wavelength becomes shorter than the size of the scatterer or if ka becomes smaller, when its effect on the scattered field is linearly reduced.[20] For scatterers that have general geometries and impedance distributions in space, however, similar scattered fields are produced (as described in Section 2.7.3).

Diffraction occurs when the scatterer is large compared to the wavelength of interest. In this case, waves are reflected and scattered as they experience continuous distribution of impedance

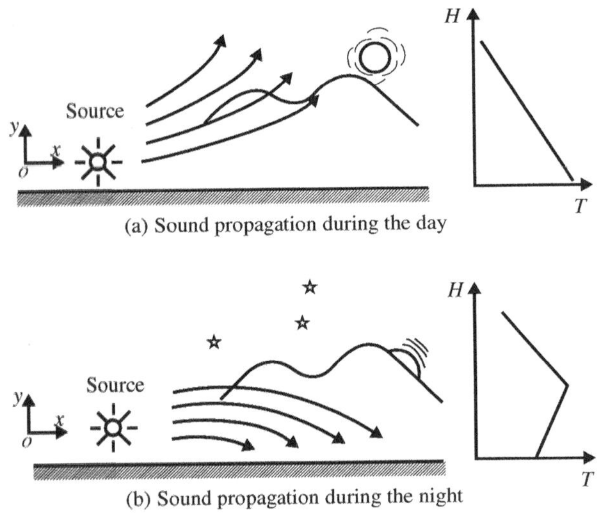

Figure 2.26 Refraction due to the media's impedance change (T is temperature and H is height): (a) sound propagation during the day and (b) sound propagation during the night

[18] We often refer to this region as the "acoustical shadow zone."

[19] The characteristic impedance increases as we rise from ground level for Figure 2.26a, and decreases for the case of Figure 2.26b. However, the speed of sound propagation decreases for the former, and increases for the latter.

[20] The second term in Equation (2.67) has $-i$. This means that the radiation due to the trembling sphere is due to the acceleration of the fluid particle on the surface.

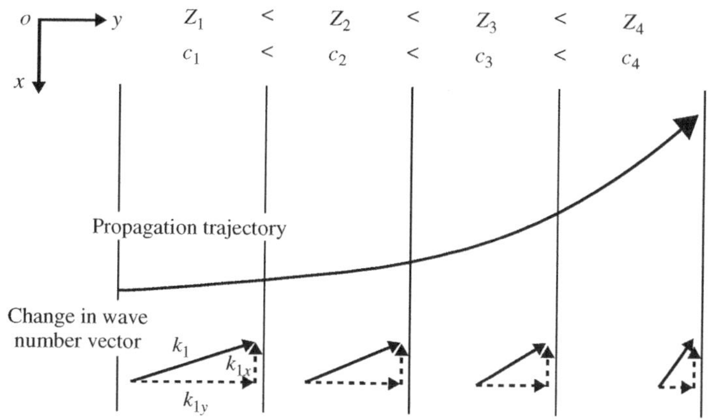

Figure 2.27 Sound propagation in a medium where the characteristic impedance changes smoothly

mismatch in space. It is interesting to note that diffraction in a shadow zone is dominated by the scattering at the edge of barrier: we hear sound from the distributed sound sources on the edge.

Refraction is a phenomena induced by impedance changes in space, in particular when there is a change in impedance perpendicular to the direction of sound propagation.

2.6 Chapter Summary

It is plausible that radiation, scattering, diffraction, and refraction can be understood in a unified concept. Depending on the type of impedance mismatch and the wavelength of interest, the radiated sound will be scattered, diffracted, or refracted in space.

We started to explore the nature of radiation, scattering, diffraction, and refraction, as well as their implicit relationship by looking at the radiation of the breathing sphere and the trembling sphere. These two radiators are basic units that can create any radiator or any radiation field by their linear combinations. The radiations are mainly dominated by the relative size (ka) of the radiator compared to the wavelength, and the radiated sound field is mainly governed by the relative distance (kr) from the radiator compared to the wavelength of interest. If ka is very small, the radiation power of the breathing sphere increases by 6 dB as ka increases twice. For the case of the trembling sphere, the increasing rate is 12 dB.

The scattered sound field was investigated in the cases where a wave is scattered by a rigid sphere and passes through a two-dimensional slit. In these cases, the scattered field strongly depends on the angle of the incident wave, the boundary condition of a scatterer, and the ratio of the wavelength of interest relative to the size of a scatterer (ka or kb).

Diffraction was also studied using a typical two-dimensional example. As we saw for the scattering of a slit, diffraction also depends on the angle of the incident wave and the angle from the edge. In the region where we can see the source, the observed sound field is the sum of the incident and the scattered sound fields. However, in the region where the sound source is invisible, the diffraction from the edge of the barrier is dominant. The Fresnel number is widely accepted as a practical means to design a barrier. This parameter requires the difference between the distance that the wave traveled passing the edge of the barrier and the straight distance between the receiver and the source, normalized by the half wavelength. The number, which is compact and less complex compared to how diffraction wave behave in space, is a guideline in the design of a sound barrier.

2.7 Essentials of Radiation, Scattering, and Diffraction

2.7.1 Radiated Sound Field from an Infinitely Baffled Circular Piston

The sound pressure, radiated by an infinitely balled circular piston that has a radius a and vibrates with a velocity u_n, can be written as

$$p(z) = -\frac{ik}{2\pi}\rho_0 c \int_0^a \frac{e^{ik\sqrt{z^2+\zeta^2}}}{\sqrt{z^2+\zeta^2}} u_n 2\pi\zeta d\zeta. \tag{2.76}$$

Figure 2.9 depicts the associated notations and their definitions. $p(z)$ is the sound pressure along the line of symmetry of the circular piston. If we define the variables $z^2+\zeta^2=\alpha^2$ and $2\zeta d\zeta = 2\alpha d\alpha$ for a fixed z, Equation (2.76) can be rewritten as

$$\begin{aligned} p(z) &= -\frac{ik}{2\pi}\rho_0 c u_n 2\pi \int_z^{R_a} e^{ik\alpha} d\alpha \\ &= \rho_0 c u_n (e^{ikz} - e^{ikR_a}), \end{aligned} \tag{2.77}$$

where $R_a = \sqrt{z^2+a^2}$ is the distance from the edge of a circular piston to z.

Let us examine what happens when z becomes very large. It is obvious that the difference between R_a and z is negligible when the observation point becomes farther from the origin. In this case, the pressure at z is expressed as

$$\begin{aligned} p(z) &= -\rho_0 c u_n \left\{ e^{ik(R_M+\delta)} - e^{ik(R_M-\delta)} \right\} \\ &= -\rho_0 c u_n e^{ikR_M} 2i \sin k\delta, \end{aligned} \tag{2.78}$$

where R_M and δ represent long and short distances. These are defined as

$$R_M = \frac{R_a + z}{2}, \tag{2.79}$$

$$\delta = \frac{R_a - z}{2}. \tag{2.80}$$

Equations (2.79) and (2.80) can be approximated as

$$\begin{aligned} \delta &= \frac{1}{2}(\sqrt{z^2+a^2} - z) \\ &= \frac{1}{2} z \left(\sqrt{1 + \left(\frac{a}{z}\right)^2} - 1 \right) \\ &= \frac{1}{2} z \left(1 + \frac{1}{2}\left(\frac{a}{z}\right)^2 + \cdots - 1 \right) \\ &\cong \frac{1}{4}\frac{a^2}{z} \end{aligned} \tag{2.81}$$

and

$$R_M \cong z. \tag{2.82}$$

Pressure $p(z)$ is then defined by

$$p(z) = -\rho_0 c u_n \frac{ika}{2}\frac{a}{z}e^{ikz}. \tag{2.83}$$

In conclusion, the sound pressure in the farfield is inversely proportional to the distance from the source, as for the sound field generated by a monopole source.

2.7.2 *Sound Field at an Arbitrary Position Radiated by an Infinitely Baffled Circular Piston*

From Figure 2.10, R can be defined

$$R^2 = r^2 + \varsigma^2 - 2r\varsigma \sin\theta\cos\phi = r^2\left[1 - 2\frac{\varsigma}{r}\sin\theta\cos\phi + \frac{\varsigma^2}{r^2}\right] \tag{2.84}$$

which can be written as

$$R = r\left(1 - 2\frac{\varsigma}{r}\sin\theta\cos\phi + \frac{\varsigma^2}{r^2}\right)^{\frac{1}{2}}. \tag{2.85}$$

By using the binomial expansion, this can be approximated as

$$R \cong r\left(1 - \frac{\varsigma}{r}\sin\theta\cos\phi + \frac{\varsigma^2}{r^2}\left(\frac{1}{2} - \frac{1}{2}\sin^2\theta\cos^2\phi\right)\right). \tag{2.86}$$

Therefore, kR can be expressed as

$$kR \cong kr - k\varsigma\sin\theta\cos\phi + \frac{k\varsigma^2}{2r}(1 - \sin^2\phi\cos^2\theta). \tag{2.87}$$

Consider the case of $ka \ll \frac{r}{a}$, that is, the observation point is at a great distance and the wavelength is very long compared to the radius of the circular piston. In this case, the third term in the right-hand side of Equation (2.87) is negligible compared to the other terms. We therefore have

$$\frac{e^{ikR}}{R} \cong \frac{e^{ikr}}{r}e^{-ik\varsigma\sin\theta\cos\phi}. \tag{2.88}$$

By using the Rayleigh integral equation, the pressure at $\mathbf{r}_0$ (Figure 2.10) is expressed as

$$\begin{aligned} p(r,\theta) &= \frac{e^{ikr}}{r}\frac{-ik\rho_0 c}{2\pi}u_n \int_0^a\int_0^{2\pi} e^{-ik\varsigma\sin\theta\cos\phi}\varsigma\, d\varsigma\, d\phi \\ &= \frac{-ik\rho_0 c}{2\pi}\frac{e^{ikr}}{r}u_n \int_0^a \left[4\int_0^{\pi/2}\cos\left(k\varsigma\sin\theta\cos\phi\right)d\phi\right]\varsigma\, d\varsigma. \end{aligned} \tag{2.89}$$

According to the characteristics of the Bessel function, which is

$$J_0(z) = \frac{2}{\pi}\int_0^{\frac{\pi}{2}}\cos(z\cos\alpha)d\alpha, \tag{2.90}$$

the sound pressure (Equation (2.89)) can be rewritten as

$$
\begin{aligned}
p(r,\theta) &= \frac{-ik\rho_0 c}{2\pi}\frac{e^{ikr}}{r}u_n\int_0^a \frac{\pi}{2}\cdot 4J_0(k\zeta\sin\theta)\zeta d\zeta \\
&= -ik\rho_0 c u_n \frac{e^{ikr}}{r}\int_0^a J_0(k\zeta\sin\theta)\zeta d\zeta \\
&= -ik\rho_0 c u_n \frac{e^{ikr}}{r}\cdot a^2\frac{J_1(ka\sin\theta)}{ka\sin\theta}.
\end{aligned}
\tag{2.91}
$$

The sound pressure (Equation (2.91)) on the z axis is the same as that of Equation (2.83) because when θ approaches 0, $\frac{J_1(ka\sin\theta)}{ka\sin\theta}$ tends to 1/2.

2.7.3 *Understanding Radiation, Scattering, and Diffraction Using the Kirchhoff–Helmholtz Integral Equation*

Radiation, diffraction, and scattering can be understood as various forms of sound fields that are produced by spatially distributed impedance mismatch. Depending on the impedance distribution in space, the resulting sound field can vary widely. It is well known that the sound fields that are induced by the spatial impedance distribution can be predicted by various numerical or theoretical approaches. In this chapter, analytical solutions of particular significance, which allow us to understand associated fundamental physical meanings, are introduced. We first define and establish notation and summarize basic equations.

2.7.3.1 Symbols and Basic Equations

- Expression of a time harmonic function: $e^{-i\omega t}$.
- Position of prediction: $\mathbf{r}$
  - in the orthogonal coordinate systems,
  - one-dimensional coordinate system: (x),
  - two-dimensional coordinate system (x, y) or (r, θ) for polar coordinate system,
  - three-dimensional coordinate system (x, y, z) for Cartesian coordinate, (r, φ, z) for cylindrical coordinates, or (r, θ, ϕ) for spherical coordinate system,
- Position of sound source: $\mathbf{r}_0$
  - in the orthogonal coordinate systems,
  - one-dimensional coordinates: (x_0),
  - two-dimensional coordinates: (x_0, y_0), (r_0, θ_0),
  - three-dimensional coordinates: (x_0, y_0, z_0), (r_0, φ_0, z_0), (r_0, θ_0, ϕ_0),
- The distance between the sound source and prediction point: R,
- Complex amplitude of sound pressure: $p(\mathbf{r})$,
- Complex amplitude of particle velocity: $u_n(\mathbf{r})$, n denotes the direction,
- Relationship between sound pressure and velocity: linearized Euler equation

$$
-\nabla p = \rho_0\frac{\partial \boldsymbol{u}}{\partial t}
\tag{2.92}
$$

Green's function: $g(\mathbf{r}|\mathbf{r}_0)$

- Green's law:

$$\oint_C \mathbf{F}\cdot\mathbf{n}dl = \int_S \nabla\cdot\mathbf{F}dS \tag{2.93}$$

(Line integral calculated for counter-clockwise direction. C represents a contour formed by a certain line and S denotes the surface. In addition, $\mathbf{n}$ represents the unit normal vector pointing toward outside the domain wrapped by the contour.)

- Gauss's theorem:

$$\int_S \mathbf{F}\cdot\mathbf{n}dS = \int_V \nabla\cdot\mathbf{F}dV \tag{2.94}$$

(V is volume of the region of interest.)

2.7.3.2 One-Dimensional Kirchhoff–Helmholtz Integral Equation

In order to understand the Kirchhoff–Helmholtz integral equation, we shall first investigate it for a one-dimensional case.

The one-dimensional acoustic governing equation can be expressed as a Helmholtz integral equation in the frequency domain, that is,

$$\frac{d^2p(x)}{dx^2} + k^2p(x) = 0, \tag{2.95}$$

where p is the complex sound pressure and k represents the wavenumber. x denotes a position in the defined domain $a \le x \le b$ (see Figure 2.28).

Consider Green's function g which satisfies

$$\frac{d^2g(x|x_0)}{dx^2} + k^2g(x|x_0) = -\delta(x-x_0), \tag{2.96}$$

where $\delta(x-x_0)$ is a one-dimensional delta function. By multiplying Equation (2.95) by g and Equation (2.96) by p and subtracting the latter from the former, we arrive at

$$g\left(\frac{d^2}{dx^2}+k^2\right)p - p\left(\frac{d^2}{dx^2}+k^2\right)g = p\cdot\delta(x-x_0). \tag{2.97}$$

The left-hand side of Equation (2.97) is reduced to

$$-\frac{d}{dx}\left(p\frac{dg}{dx} - g\frac{dp}{dx}\right) = p\cdot\delta(x-x_0). \tag{2.98}$$

Integrating Equation (2.98) for x over the domain between a and b, we have

$$\int_a^b p(x)\cdot\delta(x-x_0)dx = -\int_a^b \frac{d}{dx}\left[p(x)\frac{dg(x|x_0)}{dx} - g(x|x_0)\frac{dp(x)}{dx}\right]dx \tag{2.99}$$

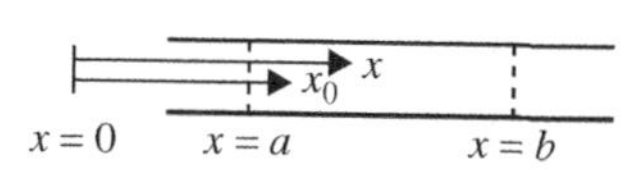

Figure 2.28 One-dimensional coordinate system with free-field boundary conditions

or

$$p(x_0) = \left[p(x) \frac{dg(x|x_0)}{dx} - g(x|x_0) \frac{dp(x)}{dx} \right]_{x=b}^{x=a}. \tag{2.100}$$

The reciprocity of the Green's function, $g(x|x_0) = g(x_0|x)$, is applied to Equation (2.100) and x is replaced with x_0, we obtain

$$p(x) = \left[p(x_0) \frac{dg(x|x_0)}{dx_0} - g(x|x_0) \frac{dp(x_0)}{dx_0} \right]_{x_0=b}^{x_0=a}. \tag{2.101}$$

This is the one-dimensional Kirchhoff–Helmholtz integral equation. Equation (2.101) means that if the sound pressure amplitude and the velocity amplitude at $x_0 = a, b$ are given, then sound pressure at any position x can also be obtained. In other words, the velocity and sound pressure at a and b propagate to x using Green's function as the propagator. Note that any function that satisfies Equation (2.96) can be used as a Green's function. The most commonly selected is a free-field Green's function, that is,

$$g_f(x|x_0) = -\frac{e^{ikR}}{2ik}, \quad R = |x - x_0|, \tag{2.102}$$

where the subscript f denotes a free-field. This Green's function can be obtained from Equation (2.96). Because it is a one-dimensional free-field, the solution of Equation (2.96) can be written as

$$g_f(x|x_0) = \begin{cases} Ae^{ik(x-x_0)}, & x > x_0 \\ Be^{-ik(x-x_0)}, & x < x_0. \end{cases} \tag{2.103}$$

Since the sound pressure must be the same at $x = x_0$, A and B are the same. If Equation (2.96) is integrated from $x = x_0 - \varepsilon$ to $x = x_0 + \varepsilon$, we obtain

$$\left. \frac{dg_f}{dx} \right|_{x=x_0-\varepsilon}^{x=x_0+\varepsilon} = -1. \tag{2.104}$$

This means that, while the sound pressure is continuous, the rate change of spatial distribution of pressure is not continuous and the difference is -1. Substituting Equation (2.103) into Equation (2.104), we have

$$Aik + Aik = -1. \tag{2.105}$$

Hence, Equation (2.103) can be rewritten over the entire domain as

$$g_f(x|x_0) = -\frac{e^{ikR}}{2ik},$$
$$R = |x - x_0|. \tag{2.106}$$

Note also that the Green's function which satisfies the Neumann boundary condition (see Figure (2.29)) in a semi-infinite space must have zero velocity at $x = 0$. The function can be obtained by superimposing the free-field Green's function (Equation (2.102)), which can be expressed as

$$\begin{aligned} g_N(x|x_0) &= g_f(x|x_0) + g_f(x|-x_0) \\ &= -\frac{e^{ikR}}{2ik} - \frac{e^{ikR'}}{2ik}, \end{aligned} \tag{2.107}$$

where $R = |x - x_0|$, $R' = |x + x_0|$ and the subscript N denotes Neumann's boundary condition.

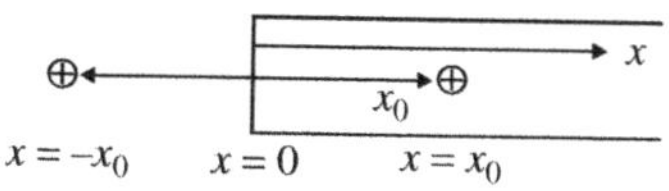

Figure 2.29 Coordinate system to induce one-dimensional Green's function which satisfies Neumann's boundary condition

When $x_0 \to 0$, Equation (2.107) becomes

$$g_N(x|x_0) = \lim_{x_0 \to 0} g_N(x|x_0) = -\frac{e^{ikx}}{ik}. \tag{2.108}$$

In this case, $\left.\frac{dg_N}{dx_0}\right]_{x_0=0} = 0.$

For a one-dimensional plane wave satisfying the Sommerfeld radiation condition, Equation (2.108) is negligible at $x = \infty$. By using the linearized Euler equation and Equation (2.108), we have

$$p(x) = \rho_0 c e^{ikx} u(0), \tag{2.109}$$

which is a one-dimensional Rayleigh integral equation.

We now consider a Green's function that satisfies the Dirichlet boundary condition in a semi-infinite space. A Green's function whose value becomes 0 at $x = 0$ can be written as

$$\begin{aligned} g_D(x|x_0) &= g_f(x|x_0) - g_f(x|-x_0) \\ &= -\frac{e^{ikR}}{2ik} + \frac{e^{ikR'}}{2ik} \end{aligned} \tag{2.110}$$

where $R = |x - x_0|$, $R' = |x + x_0|$, and the subscript D denotes the Dirichlet boundary condition (see Figure 2.30). If $x_0 \to 0$ in Equation (2.110),

$$\left.\frac{dg_D(x|x_0)}{dx_0}\right|_{x_0=0} = \lim_{x_0 \to 0} \frac{dg_D(x|x_0)}{dx_0} = e^{ikx}. \tag{2.111}$$

Because $g_D = 0$ at $x = 0$, and using Equation (2.111), Equation (2.101) can be reduced to

$$p(x) = e^{ikx} p(0). \tag{2.112}$$

We assumed that the boundary value at $x = \infty$ is negligible.

We now expand the theory to two-dimensional cases.

2.7.3.3 Two-Dimensional Kirchhoff–Helmholtz Integral Equation

In the two-dimensional case, the governing equation can be written as

$$\nabla^2 p(\mathbf{r}) + k^2 p(\mathbf{r}) = 0, \tag{2.113}$$

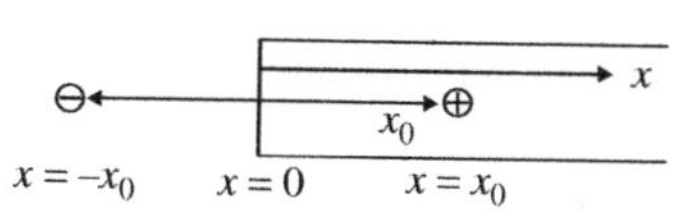

Figure 2.30 Coordinate system to induce one-dimensional Green's function which satisfies Dirichlet boundary condition

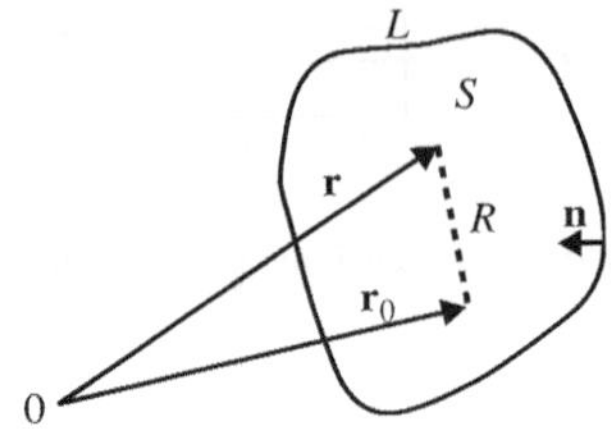

Figure 2.31 Coordinate system used for a two-dimensional Kirchhoff–Helmholtz integral equation

where $\mathbf{r}$ represents all points other than $\mathbf{r}_0$ (see Figure 2.31). For $\mathbf{r} = \mathbf{r}_0$, it satisfies

$$\nabla^2 g(\mathbf{r}|\mathbf{r}_0) + k^2 g(\mathbf{r}|\mathbf{r}_0) = -\delta(\mathbf{r} - \mathbf{r}_0). \tag{2.114}$$

From Equations (2.113) and (2.114), we have

$$g(\nabla^2 + k^2)p - p(\nabla^2 + k^2)g = p(\mathbf{r})\delta(\mathbf{r} - \mathbf{r}_0). \tag{2.115}$$

Rewriting the left-hand side, we have

$$\nabla \cdot (g\nabla p - p\nabla g) = p(\mathbf{r})\delta(\mathbf{r} - \mathbf{r}_0). \tag{2.116}$$

Integrating over the surface of interest S and applying Green's law in Equation (2.114) in Section 2.7.3.1,

$$\int_S p(\mathbf{r})\delta(\mathbf{r} - \mathbf{r}_0)dS = \int_S \nabla \cdot [g(\mathbf{r}|\mathbf{r}_0)\nabla p(\mathbf{r}) - p(\mathbf{r})\nabla g(\mathbf{r}|\mathbf{r}_0)]\,dS \tag{2.117}$$

or

$$p(\mathbf{r}_0) = \oint_L [p(\mathbf{r})\nabla g(\mathbf{r}|\mathbf{r}_0) - g(\mathbf{r}|\mathbf{r}_0)\nabla p(\mathbf{r})] \cdot \mathbf{n}dL, \tag{2.118}$$

where $\mathbf{n}$ is a unit normal vector from the boundary to the region of interest, as depicted in Figure 2.31. By using the reciprocity of Green's function, $g(\mathbf{r}|\mathbf{r}_0) = g(\mathbf{r}_0|\mathbf{r})$. By replacing $\mathbf{r}$ with $\mathbf{r}_0$, we obtain the two-dimensional Kirchhoff–Helmholtz integral equation as

$$p(\mathbf{r}) = \oint_L [p(\mathbf{r}_0)\nabla_0 g(\mathbf{r}|\mathbf{r}_0) - g(\mathbf{r}|\mathbf{r}_0)\nabla_0 p(\mathbf{r}_0)] \cdot \mathbf{n}dL, \tag{2.119}$$

where $\oint_L$ denotes the line integral.

The two-dimensional free-field Green's function is expressed by a Hankel function of the first kind as

$$g_f(\mathbf{r}|\mathbf{r}_0) = \frac{i}{4}H_0^{(1)}(kR), \quad R = |\mathbf{r} - \mathbf{r}_0|. \tag{2.120}$$

In the farfield ($kR \to \infty$), it becomes approximately

$$g_f(\mathbf{r}|\mathbf{r}_0) \cong \frac{1}{\sqrt{2}}(1 + i)\frac{e^{ikR}}{\sqrt{\pi kR}}. \tag{2.121}$$

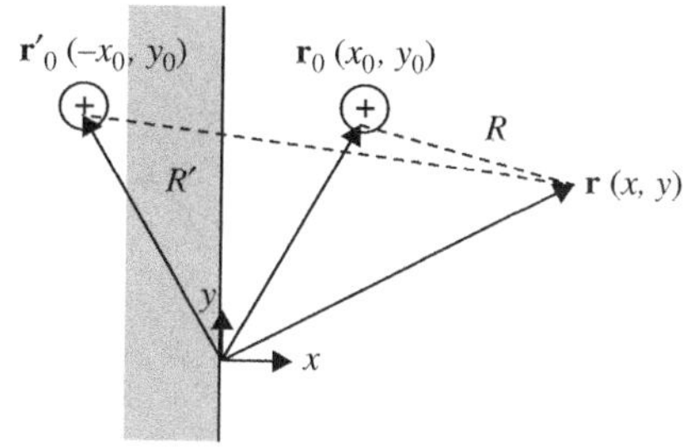

Figure 2.32 Coordinate system used for a two-dimensional Green's function which satisfies the Neumann boundary condition

First, the Green's function that satisfies Neumann boundary condition is similar to the one-dimensional case (see Figure 2.32), that is,

$$\begin{aligned} g_N(x, y|x_0, y_0) &= g_f(x, y|x_0, y_0) + g_f(x, y| - x_0, y_0) \\ &= \frac{i}{4}H_0^{(1)}(kR) + \frac{i}{4}H_0^{(1)}(kR') \end{aligned} \tag{2.122}$$

where $R = \sqrt{(x - x_0)^2 + (y - y_0)^2}$ and $R' = \sqrt{(x + x_0)^2 + (y - y_0)^2}$. As $x_0 \to 0$, Equation (2.122) can be rewritten as

$$\begin{aligned} g_N(x, y|x_0, y_0) &= \lim_{x_0 \to 0} g_N(x, y|x_0, y_0) \\ &= \frac{i}{2}H_0^{(1)}(kR), \end{aligned} \tag{2.123}$$

where $R = \sqrt{x^2 + (y - y_0)^2}$. Note that $\frac{dg_N}{dx_0} = 0$, and the boundary value at infinity can be neglected. Using Equations (2.118), (2.123), and (2.113) from Section 2.7.3.1, Equation (2.119) becomes

$$p(x, y) = -\frac{k\rho_0 c}{2}\int_{-\infty}^{\infty} H_0^{(1)}(kR)u_x(0, y_0)dy_0. \tag{2.124}$$

This equation is a two-dimensional Rayleigh integral equation. By using Equation (2.121), Equation (2.124) can be approximated in the farfield as

$$p(x, y) \cong \sqrt{2}(1 - i)\rho_0 c\sqrt{\frac{k}{\pi}}\int_{-\infty}^{\infty} \frac{e^{ikR}}{\sqrt{R}}u_x(0, y_0)dy_0. \tag{2.125}$$

We now find the Green's function that satisfies the Dirichlet boundary condition in a semi-infinite space (see Figure 2.33). By using a two-dimensional Green's function as used in the one-dimensional case, it can be written as

$$\begin{aligned} g_D(x, y|x_0, y_0) &= g_f(x, y|x_0, y_0) - g_f(x, y| - x_0, y_0) \\ &= \frac{i}{4}H_0^{(1)}(kR) - \frac{i}{4}H_0^{(1)}(kR'), \end{aligned} \tag{2.126}$$

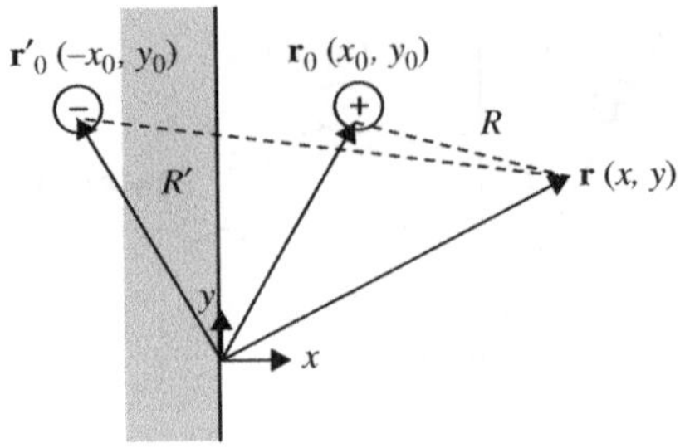

Figure 2.33 Coordinate system used for a two-dimensional Green's function which satisfies the Dirichlet boundary condition

where $R = \sqrt{(x - x_0)^2 + (y - y_0)^2}$ and $R' = \sqrt{(x + x_0)^2 + (y - y_0)^2}$. As $x_0 \to 0$,

$$\left.\frac{\partial g_D\left(x, y|x_0, y_0\right)}{\partial n}\right|_{x_0=0} = \left.\frac{\partial g_D\left(x, y|x_0, y_0\right)}{\partial x_0}\right|_{x_0=0}$$
$$= \lim_{x_0 \to 0} \frac{\partial g_D(x, y|x_0, y_0)}{\partial x_0} = \frac{i}{2}\frac{kx}{R} H_1^{(1)}(kR), \tag{2.127}$$

where $R = \sqrt{x^2 + (y - y_0)^2}$. Because $g_D = 0$, Equation (2.119) can be rewritten (by using Equation (2.127)) as

$$p(x, y) = \frac{i}{2} kx \int_{-\infty}^{\infty} \frac{H_1^{(1)}(kR)}{R} p(0, y_0) dy_0. \tag{2.128}$$

Again, it is assumed that the boundary value at infinity is negligible, just as in the one-dimensional case.

We now consider the three-dimensional case.

2.7.3.4 Three-Dimensional Kirchhoff–Helmholtz Integral Equation

The sound pressure at an arbitrary position $\mathbf{r}(\mathbf{r} \neq \mathbf{r}_0)$ must satisfy

$$\nabla^2 p(\mathbf{r}) + k^2 p(\mathbf{r}) = 0. \tag{2.129}$$

At $r = \mathbf{r}_0$, the Green's function must satisfy the Helmholtz equation, that is,

$$\nabla^2 g(\mathbf{r}|\mathbf{r}_0) + k^2 g(\mathbf{r}|\mathbf{r}_0) = -\delta(\mathbf{r} - \mathbf{r}_0). \tag{2.130}$$

By multiplying by g and Equation (2.130) by p, and by subtracting the latter from the former, we have

$$g(\nabla^2 + k^2)p - p(\nabla^2 + k^2)g = p(\mathbf{r})\delta(\mathbf{r} - \mathbf{r}_0). \tag{2.131}$$

Rewriting the left-hand side of Equation (2.131), we obtain

$$p(\mathbf{r})\delta(\mathbf{r} - \mathbf{r}_0) = \nabla \cdot (g\nabla p - p\nabla g). \tag{2.132}$$

Integrating over the volume of interest V yields

$$\int_V p(\mathbf{r})\delta(\mathbf{r}-\mathbf{r}_0)dV = \int_V \nabla\cdot[g(\mathbf{r}|\mathbf{r}_0)\nabla p(\mathbf{r}) - p(\mathbf{r})\nabla g(\mathbf{r}|\mathbf{r}_0)]dV. \tag{2.133}$$

Note that the volume excludes all sound sources. In other words, it is a source-free volume. By using Gauss's theorem to make the right-hand side of Equation (2.133) a surface-integral form, it can be rewritten as

$$p(\mathbf{r}_0) = \int_S [p(\mathbf{r})\nabla g(\mathbf{r}|\mathbf{r}_0) - g(\mathbf{r}|\mathbf{r}_0)\nabla p(\mathbf{r})]\cdot\mathbf{n}dS, \tag{2.134}$$

where $\mathbf{n}$ is a unit normal vector from the boundary to the region of interest. By using Green's function's reciprocity, $g(\mathbf{r}|\mathbf{r}_0) = g(\mathbf{r}_0|\mathbf{r})$. By replacing $\mathbf{r}$ by $\mathbf{r}_0$, the three-dimensional Kirchhoff–Helmholtz integral equation can be obtained as

$$p(\mathbf{r}) = \int_S [p(\mathbf{r}_0)\nabla_0 g(\mathbf{r}|\mathbf{r}_0) - g(\mathbf{r}|\mathbf{r}_0)\nabla_0 p(\mathbf{r}_0)]\cdot\mathbf{n}dS. \tag{2.135}$$

The three-dimensional free-field Green's function is

$$g_f(\mathbf{r}|\mathbf{r}_0) = \frac{e^{ikR}}{4\pi R}, \quad R = |\mathbf{r}-\mathbf{r}_0|. \tag{2.136}$$

The Green's function (Figure 2.34) that satisfies the Neumann boundary condition in a semi-infinite space is

$$\begin{aligned} g_N(x, y, z|x_0, y_0, z_0) &= g_f(x, y, z|x_0, y_0, z_0) + g_f(x, y, z|x_0, y_0, -z_0) \\ &= \frac{e^{ikR}}{4\pi R} + \frac{e^{ikR'}}{4\pi R'}, \end{aligned} \tag{2.137}$$

where

$$R = \sqrt{(x-x_0)^2 + (y-y_0)^2 + (z-z_0)^2}$$

and

$$R' = \sqrt{(x-x_0)^2 + (y-y_0)^2 + (z+z_0)^2}.$$

As $z_0 \to 0$, Equation (2.137) becomes

$$g_N(x, y, z|x_0, y_0, z_0) = \lim_{z_0\to 0} g_f(x, y, z|x_0, y_0, z_0) = \frac{e^{ikR}}{2\pi R}, \tag{2.138}$$

where $R = \sqrt{(x-x_0)^2 + (y-y_0)^2 + z^2}$. In this case, $\frac{\partial g_N}{\partial z_0} = 0$ at the boundary. Using Equation (2.92) (Section 2.7.3.1) and Equation (2.138), we then have

$$p(x, y, z) = -2i\rho_0\omega \int_{-\infty}^{\infty}\int_{-\infty}^{\infty} \frac{e^{ikR}}{4\pi R} u_z(x_0, y_0, 0)dx_0 dy_0, \tag{2.139}$$

which is the *Rayleigh's first integral equation* in terms of Cartesian coordinates.

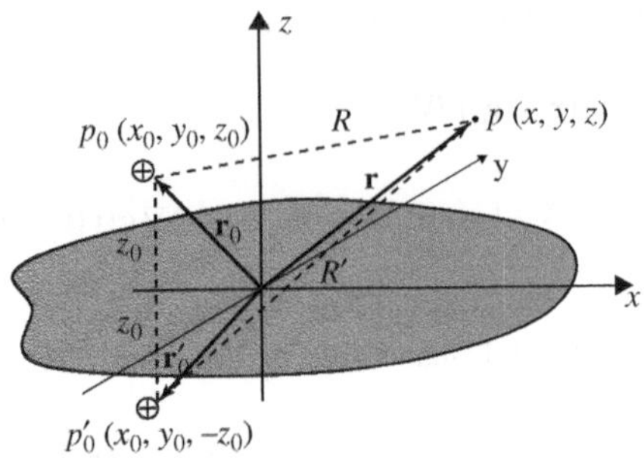

Figure 2.34 Coordinate system used for a three-dimensional Green's function which satisfies the Neumann boundary condition

We now find the Green's function that satisfies the Dirichlet boundary condition. By using the free-field Green's function as for the cases of one and two dimensions, it can be written (see Figure 2.35):

$$g_D(x, y, z|x_0, y_0, z_0) = g_f(x, y, z|x_0, y_0, z_0) - g_f(x, y, z|x_0, y_0, -z_0)$$
$$= \frac{e^{ikR}}{4\pi R} - \frac{e^{ikR'}}{4\pi R'} \tag{2.140}$$

where

$$R = \sqrt{(x - x_0)^2 + (y - y_0)^2 + (z - z_0)^2},$$

and

$$R' = \sqrt{(x - x_0)^2 + (y - y_0)^2 + (z + z_0)^2}.$$

As $z_0 \to 0$,

$$\left.\frac{\partial g_D(x, y, z|x_0, y_0, z_0)}{\partial n_0}\right|_{z_0=0} = \left.\frac{\partial g_D(x, y, z|x_0, y_0, z_0)}{\partial z_0}\right|_{z_0=0}$$
$$= \lim_{z_0 \to 0} \frac{\partial g_D(x, y, z|x_0, y_0, z_0)}{\partial z_0}$$
$$= \frac{1}{2\pi}\frac{z}{R^3}(1 - ikR)e^{ikR}, \tag{2.141}$$

where $R = \sqrt{(x - x_0)^2 + (y - y_0)^2 + z^2}$. Since $g_D = 0$ at the boundary, by using Equation (2.141) we can predict the pressure at any position in the source-free volume by

$$p(x, y, z) = \frac{z}{2\pi}\int_{-\infty}^{\infty}\int_{-\infty}^{\infty} \frac{(1 - ikR)}{R^3} e^{ikR}\, p(x_0, y_0, 0)\,dx_0 dy_0. \tag{2.142}$$

2.7.3.5 Diffraction by a Slit

We have obtained integral functions that can be used to predict sound fields in all possible dimensions. This allows us to explore how sound propagates and changes under various spatially distributed impedance boundary conditions or impedance mismatch.

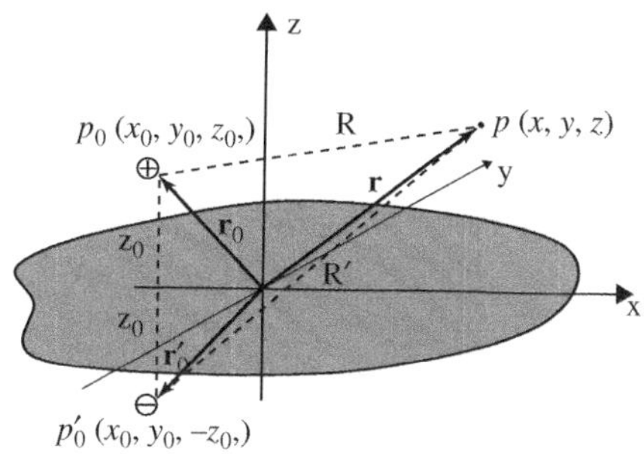

Figure 2.35 Coordinate system used for a three-dimensional Green's function which satisfies the Dirichlet boundary condition

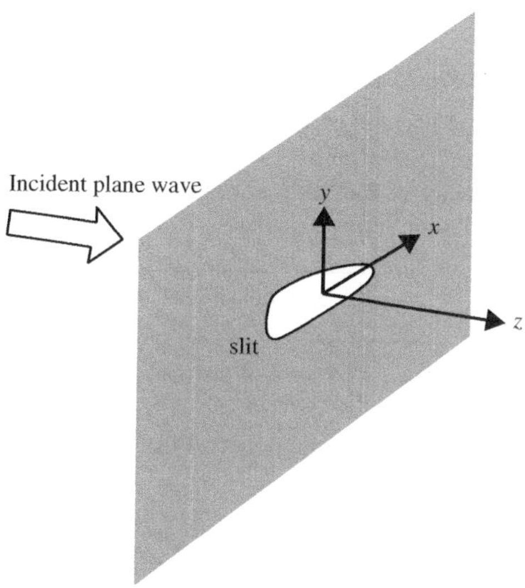

Figure 2.36 Infinite baffle with a slit

We first consider diffraction when there is a slit in an infinite baffle, as shown in Figure 2.36. Assume that the baffle is rigid and the incident wave is a plane wave that arrives at the plate surface (baffle) in the normal direction. Assume that the plane wave's velocity at the slit is uniform over the surface, which is u_0.

Two-Dimensional Slit

If the width of the slit is b and it is placed from $-\infty$ to ∞ in the x direction, it becomes a two-dimensional diffraction problem as illustrated in Figure 2.37. In this case, applying the Rayleigh integral equation (Equation (2.125) of Section 2.7.3.3) to the farfield, we can write

$$p(r, \theta) \cong \sqrt{2}(1 - i)\rho_0 c u_0 \sqrt{\frac{k}{\pi}} \int_{-b/2}^{b/2} \frac{e^{ikR}}{\sqrt{R}} dy_0, \tag{2.143}$$

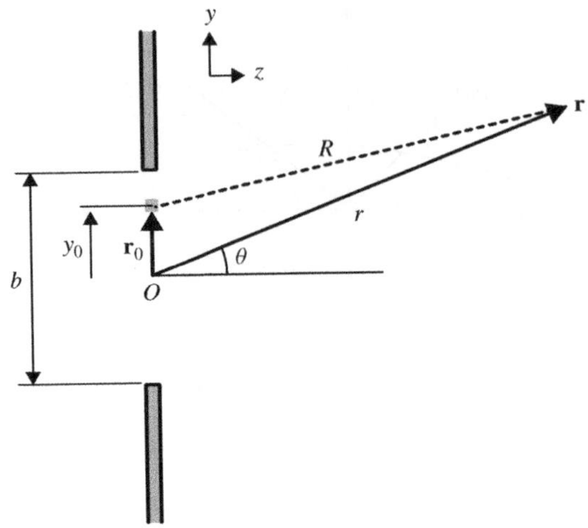

Figure 2.37 Coordinate system of two-dimensional slit

where u_0 is the impinging velocity at the slit's entrance. Note that R (Equation (2.143)) in the farfield can be written using a Taylor series expansion as

$$R = \sqrt{r^2 + y_0^2 - 2ry_0\cos(\pi/2 - \theta)}$$
$$= r\sqrt{1 + \left(\frac{y_0}{r}\right)^2 - 2\frac{y_0}{r}\sin\theta} \cong r - y_0\sin\theta. \tag{2.144}$$

If Equation (2.144) is substituted into Equation (2.143), the sound pressure at the farfield is expressed as

$$p(r,\theta) \cong \sqrt{2}(1-i)\rho_0 c u_0 \sqrt{\frac{k}{\pi}} \int_{-b/2}^{b/2} \frac{e^{ik(r - y_0\sin\theta)}}{\sqrt{r - y_0\sin\theta}} dy_0$$
$$\cong \sqrt{2}(1-i)\rho_0 c u_0 \sqrt{\frac{k}{\pi r}} e^{ikr} \int_{-b/2}^{b/2} e^{-iky_0\sin\theta} dy_0$$
$$\cong \sqrt{2}(1-i)\rho_0 c u_0 \sqrt{\frac{k}{\pi r}} e^{ikr} \frac{e^{-ik(b/2)\sin\theta} - e^{ik(b/2)\sin\theta}}{-ik\sin\theta}$$
$$\cong \sqrt{2}(1-i)\rho_0 c u_0 \sqrt{\frac{k}{\pi r}} e^{ikr} \frac{-2i\sin((kb/2)\sin\theta)}{-ik\sin\theta}$$
$$\cong \sqrt{2}(1-i)\rho_0 c b u_0 \sqrt{\frac{k}{\pi r}} e^{ikr} \frac{\sin((kb/2)\sin\theta)}{(kb/2)\sin\theta}. \tag{2.145}$$

Equation (2.145) states that directivity pattern of $|p(r,\theta)|^2$ can be expressed as

$$|p(\theta)|^2 = \left|\frac{\sin((kb/2)\sin\theta)}{(kb/2)\sin\theta}\right|^2. \tag{2.146}$$

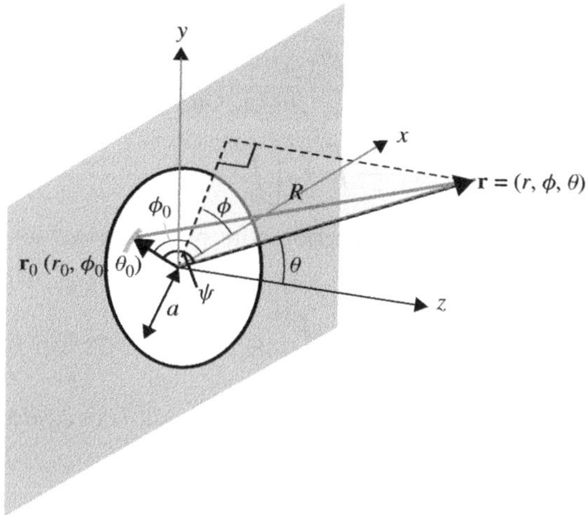

Figure 2.38 Coordinate system of round slit

This is a sinc function with regard to kb and $\sin\theta$. The smaller the slit's height (b) relative to the wave, the stronger is the directivity of the field that is formed. The greater the value of kb or θ, less diffraction by the slit is expected. This can also be understood as a typical scattering phenomenon by a two-dimensional slit.

Diffraction by a Round Slit

By applying the Rayleigh integral equation to the circular slit, as shown in Figure 2.38, the sound pressure can be expressed as

$$p(r, \theta, \phi) \cong -\frac{ik\rho_0 c}{2\pi} u_0 \int_0^a \int_0^{2\pi} \frac{e^{ikR}}{R} r_0 d\phi_0 dr_0. \tag{2.147}$$

The farfield approximation of R (Equation (2.147)) is

$$\begin{aligned} R &= \sqrt{r^2 + r_0^2 - 2rr_0\cos\psi} \\ &= r\sqrt{1 + \left(\frac{r_0}{r}\right)^2 - 2\frac{r_0}{r}\cos\psi} \cong r - r_0\cos\psi. \end{aligned} \tag{2.148}$$

In Equation (2.148), $\cos\psi$ can be obtained by the inner products of unit vectors in the $\mathbf{r}$ and $\mathbf{r}_0$ directions ($\mathbf{e}_r$ and $\mathbf{e}_{r0}$). Note also that, due to symmetry, $p(r, \theta, \phi)$ does not change its value relative to ϕ. Therefore, when $\phi = 0$, $\cos\psi$ is written as

$$\begin{aligned} \cos\psi &= \mathbf{e}_r \cdot \mathbf{e}_{r0} \\ &= (\sin\theta\mathbf{e}_x + \cos\theta\mathbf{e}_z) \cdot (\cos\phi_0\mathbf{e}_x + \sin\phi_0\mathbf{e}_y) \\ &= \sin\theta \cdot \cos\phi_0. \end{aligned} \tag{2.149}$$

Equation (2.148) can therefore be rewritten as

$$R \cong r - r_0 \sin\theta\cos\phi_0. \tag{2.150}$$

Substituting Equation (2.150) into Equation (2.147), we have

$$p(r,\theta,\phi) \cong -\frac{ik\rho_0 c}{2\pi}u_0\int_0^a\int_0^{2\pi}\frac{e^{ik(r-r_0\sin\theta\cos\phi_0)}}{r-r_0\sin\theta\cos\phi_0}r_0 d\phi_0 dr_0$$

$$\cong -\frac{ik\rho_0 c}{2\pi}u_0\frac{e^{ikr}}{r}\int_0^a r_0\left[\int_0^{2\pi}e^{-ikr_0\sin\theta\cos\phi_0}d\phi_0\right]dr_0. \tag{2.151}$$

Note that the imaginary part is an odd function with respect to ϕ_0; its integral is therefore 0. The integral of Equation (2.151) is reduced to

$$\int_0^{2\pi}e^{-ikr_0\sin\theta\cos\phi_0}d\phi_0 = \int_0^{2\pi}\cos(kr_0\sin\theta\cos\phi_0)d\phi_0$$

$$= 2\int_0^{\pi}\cos(kr_0\sin\theta\cos\phi_0)d\phi_0. \tag{2.152}$$

Noting that one of the characteristics of a Bessel function is

$$J_{2n}(u) = \frac{(-1)^n}{\pi}\int_0^{\pi}\cos 2n\phi_0\cdot\cos(u\cos\phi_0)d\phi_0, \tag{2.153}$$

Equation (2.152) can be rewritten as

$$\int_0^{2\pi}e^{-ikr_0\sin\theta\cos\phi_0}d\phi_0 = 2\pi J_0(kr_0\sin\theta). \tag{2.154}$$

Equation (2.151) then becomes

$$p(r,\theta,\phi) \cong -ik\rho_0 c u_0\frac{e^{ikr}}{r}\int_0^a r_0 J_0(kr_0\sin\theta)dr_0. \tag{2.155}$$

By changing the variable using $\zeta = kr_0\sin\theta$, Equation (2.155) can be rewritten as

$$\int_0^a r_0 J_0(kr_0\sin\theta)dr_0 = \frac{1}{(k\sin\theta)^2}\int_0^{ka\sin\theta}\zeta J_0(\zeta)d\zeta. \tag{2.156}$$

Using another characteristic of the Bessel function, which is

$$\zeta J_n'(\zeta) = \zeta J_{n-1}(\zeta) - \zeta J_n(\zeta), \tag{2.157}$$

Equation (2.156) can be simplified by integration by parts as

$$\frac{1}{(k\sin\theta)^2}\int_0^{ka\sin\theta}\zeta J_0(\zeta)d\zeta$$

$$= \frac{1}{(k\sin\theta)^2}\left[\int_0^{ka\sin\theta}\zeta J_1'(\zeta)\,d\zeta + \int_0^{ka\sin\theta}J_1(\zeta)d\zeta\right]$$

$$= \frac{1}{(k\sin\theta)^2}\left[\zeta J_1(\zeta)\,|_0^{ka\sin\theta} - \int_0^{ka\sin\theta}J_1(\zeta)d\zeta + \int_0^{ka\sin\theta}J_1(\zeta)d\zeta\right]$$

$$= \frac{ka\sin\theta}{(k\sin\theta)^2}J_1(ka\sin\theta)$$

$$= \frac{a^2 J_1(ka\sin\theta)}{ka\sin\theta}. \tag{2.158}$$

If this is substituted into Equation (2.155), the sound pressure in the farfield is ultimately expressed as

$$p(r,\theta,\phi) \cong -ik\rho_0 ca^2 u_0 \frac{e^{ikr}}{r} \frac{J_1(ka\sin\theta)}{ka\sin\theta}. \tag{2.159}$$

From Equation (2.159), the directivity pattern of $|p(r,\theta,\phi)|^2$ is

$$|p(\theta,\phi)|^2 = \left|\frac{J_1(ka\sin\theta)}{ka\sin\theta}\right|^2. \tag{2.160}$$

This is a function of ka and $\sin\theta$, and behaves similarly to that predicted for the two-dimensional slit.

Diffraction by a Rectangular Slit

By applying the Rayleigh integral equation (Figure 2.39 shows the corresponding coordinate), the sound pressure is expressed as

$$p(r,\theta,\phi) = -\frac{ik\rho_0 c}{2\pi} u_0 \int_{-b/2}^{b/2}\int_{-a/2}^{a/2} \frac{e^{ikR}}{R} dx_0 dy_0. \tag{2.161}$$

In Equation (2.161), R in farfield is expressed as

$$\begin{aligned} R &= \sqrt{r^2 + r_0^2 - 2rr_0\cos\psi} \\ &= r\sqrt{1 + \left(\frac{r_0}{r}\right)^2 - 2\frac{r_0}{r}\cos\psi} \cong r - r_0\cos\psi. \end{aligned} \tag{2.162}$$

In Equation (2.162), as noted before, we can obtain $\cos\psi$ by the inner products of the unit factors in the $\mathbf{r}$ and $\mathbf{r}_0$ directions ($\mathbf{e}_r$ and $\mathbf{e}_{r0}$):

$$\begin{aligned} \cos\psi &= \mathbf{e}_r \cdot \mathbf{e}_{r_0} \\ &= (\sin\theta\cos\phi\,\mathbf{e}_x + \sin\theta\sin\phi\,\mathbf{e}_y + \cos\theta\,\mathbf{e}_z) \cdot \left(\frac{x_0}{r_0}\mathbf{e}_x + \frac{y_0}{r_0}\mathbf{e}_y\right) \\ &= \frac{x_0}{r_0}\sin\theta\cos\phi + \frac{y_0}{r_0}\sin\theta\sin\phi. \end{aligned} \tag{2.163}$$

Equation (2.162) can therefore be rewritten as

$$R \cong r - x_0\sin\theta\cos\phi - y_0\sin\theta\sin\phi. \tag{2.164}$$

When Equation (2.164) is substituted into Equation (2.161), $p(r,\theta,\phi)$ can be written approximately as

$$\begin{aligned} &p(r,\theta,\phi) \\ &\cong -\frac{ik\rho_0 c}{2\pi} u_0 \int_{-b/2}^{b/2}\int_{-a/2}^{a/2} \frac{e^{ik(r - x_0\sin\theta\cos\phi - y_0\sin\theta\sin\phi)}}{r - x_0\sin\theta\cos\phi - y_0\sin\theta\sin\phi} dx_0 dy_0 \\ &\cong -\frac{ik\rho_0 c}{2\pi} u_0 \frac{e^{ikr}}{r} \int_{-b/2}^{b/2} e^{-iky_0\sin\theta\sin\phi} \left[\int_{-a/2}^{a/2} e^{-ikx_0\sin\theta\cos\phi} dx_0\right] dy_0. \end{aligned} \tag{2.165}$$

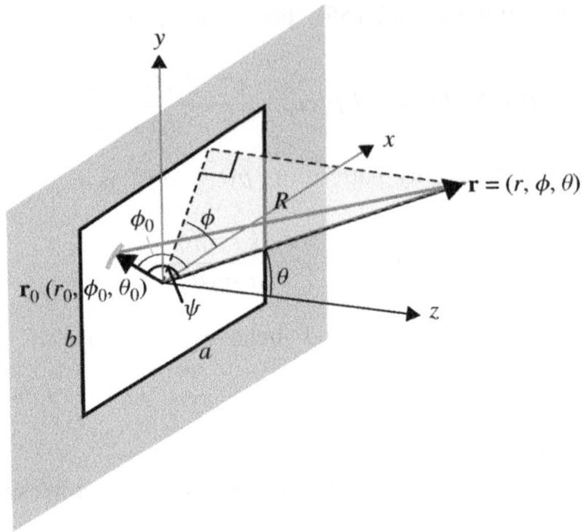

Figure 2.39 Coordinate system of rectangular slit

The first integral of Equation (2.165) can be written as

$$\int_{-a/2}^{a/2} e^{-ikx_0 \sin\theta\cos\phi} dx_0 = \frac{e^{i(ka/2)\sin\theta\cos\phi} - e^{-i(ka/2)\sin\theta\cos\phi}}{ik\sin\theta\cos\phi}$$

$$= a\frac{\sin((ka/2)\sin\theta\cos\phi)}{(ka/2)\sin\theta\cos\phi}. \quad (2.166)$$

Substituting Equation (2.166) into Equation (2.165) and solving the second integral provides the sound pressure level in the farfield as

$$p(r,\theta,\phi) = -\frac{ik\rho_0 c}{2\pi}abu_0\frac{e^{ikr}}{r}\frac{\sin((ka/2)\sin\theta\cos\phi)\sin((kb/2)\sin\theta\cos\phi)}{(ka/2)\sin\theta\cos\phi(kb/2)\sin\theta\cos\phi}. \quad (2.167)$$

Therefore, the directivity pattern of $|p(r,\theta,\phi)|^2$ can be expressed as

$$|p(\theta,\phi)|^2 = \left|\frac{\sin((ka/2)\sin\theta\cos\phi)\sin((kb/2)\sin\theta\cos\phi)}{(ka/2)\sin\theta\cos\phi(kb/2)\sin\theta\cos\phi}\right|^2, \quad (2.168)$$

where the maximum value is 1. Again note that the directivity strongly depends on the normalized scale factors (*ka, kb*), which describe the size of the rectangular slit relative to the wavelength of interest.

2.7.4 *Scattered Sound Field Using the Rayleigh Integral Equation*

From Section 2.5, the scattered field is derived as,

$$p_{sc} = \frac{ik\rho_0 c}{2\pi}\int_{S_0}\frac{e^{ikR}}{R}\left\{\frac{B}{\rho_0 c}\mathbf{e}_k\cdot\mathbf{n} + i\frac{B}{\rho_0 c}(\mathbf{k}\cdot\mathbf{r}_0)\,\mathbf{e}_k\cdot\mathbf{n}\right\}dS \quad (2.169)$$

The first term can be rewritten as

$$\text{first term} = \frac{ik\rho_0 c}{2\pi}\int_{S_0}\frac{e^{ikR}}{R}\left(\frac{B}{\rho_0 c}\right)\mathbf{e}_k\cdot\mathbf{n}\,dS$$
$$= \frac{ikB}{2\pi}\int_{S_0}\frac{e^{ikR}}{R}\mathbf{e}_k\cdot\mathbf{n}\,dS = \frac{ikB}{2\pi}\int_{V_0}\nabla\cdot\frac{e^{ikR}}{R}\mathbf{e}_k\,dV, \tag{2.170}$$

and the second term can be rearranged as

$$\text{second term} = \frac{ik\rho_0 c}{2\pi}\int_{S_0} i\frac{B}{\rho_0 c}\cdot\frac{e^{ikR}}{R}(\mathbf{k}\cdot\mathbf{r}_0)(\mathbf{e}_k\cdot\mathbf{n})\,dS$$
$$= \frac{k^2 B}{2\pi}\int_{S_0}\frac{e^{ikR}}{R}(\mathbf{e}_k\cdot\mathbf{r}_0)(\mathbf{e}_k\cdot\mathbf{n})\,dS. \tag{2.171}$$

In a farfield,

$$\text{second term} = \frac{k^2 B}{2\pi}\cdot\frac{e^{ikR}}{R}\int_{S_0}(\mathbf{e}_k\cdot\mathbf{r}_0)(\mathbf{e}_k\cdot\mathbf{n})\,dS. \tag{2.172}$$

If Gauss's theorem is applied, then

$$\text{second term} = \frac{k^2 B}{2\pi}\cdot\frac{e^{ikR}}{R}\int_{V_0}\nabla\cdot(\mathbf{e}_k\cdot\mathbf{r}_0)\mathbf{e}_k\,dV$$
$$= \frac{k^2 B}{2\pi}\cdot\frac{e^{ikR}}{R}V_0 = \frac{k^2 B}{2\pi R}V_0 e^{ikR} \tag{2.173}$$

where V_0 is the volume of the scatterer. Therefore, Equation (2.169) can be rewritten as

$$p_{sc} = \frac{ikB}{2\pi}\int_{V_0}\nabla\cdot\frac{e^{ikR}}{R}\mathbf{e}_k\,dV + \frac{k^2 B}{2\pi R}V_0 e^{ikR}. \tag{2.174}$$

This means that the scattered sound field can be divided into two different fields: that constructed by a monopole sound source (proportional to the volume of the scatterer, second term) and that due to a dipole source contribution (first term).

References

1. Kim, Y.H. (2010) Radiation, scattering, and diffraction, in *Sound Propagation: An Impedance Based Approach*, John Wiley & Sons (Singapore) Pte. Ltd, pp. 177–271.
2. Kim, Y.-H. (2004) Can we hear the shape of a noise source? The 18th International Congress on Acoustics (ICA 2004), Plenary Lecture, Kyoto International Conference Hall, April 4–9, 2004.

The first term can be rewritten as

[illegible]

[illegible]

[illegible]

Part II

Sound Visualization

Since the time that Louis Bertrand Castel attempted to relate sound to color (Figure II.1), many attempts have been made to visually represent sound in space. The motivations for such endeavors vary, as in his case, from artistic to scientific or engineering purposes. The scientific and engineering related motivations are mostly based on the fact that we obtain extremely rich information from vision in contrast to that obtained from hearing. In this respect, development of acoustic cameras or videos seems to be the right research direction. However, as can be easily understood, in mathematical terms, this is a typical inverse problem, simply because this is an attempt to get more information than is measured or is available. As M. Kac [1] noted, in the study of the relation between sound generation by membrane vibration and its reception in space, inverse problems can be regarded as "the attempts to obtain what is not available based on what is [2]." One can also find many different descriptions of the meaning of "inverse problem" in Refs. [3–9]. Normally, in an inverse problem, the available data are often insufficient to predict or describe quantities that are not measured. This circumstance is commonly referred to as an "ill-posed problem" in literature (for example, see Refs. [1, 3–9]).

The ill-posed problem for the one-dimensional case is the simplest form of this problem and is depicted in Figure II.2. It can be easily seen that the prediction of $p(x_1, t)$ from $p(x_0, t)$ depends entirely on the propagation or prediction model. Figure II.3 well demonstrates that this inverse problem generally depends on "the selected basis function, or the prediction model for the aforementioned one-dimensional case." The prediction completely depends on how well the basis function mimics the entire data set, instead of only the measured values. This implies that when we attempt to see the shape of noise/sound sources, how well we see the shape of the noise source completely depends on this basis function. This is simply because we predict what is not available by using the selected function.

Therefore, one of the common ways to classify methods that are used in noise/sound source identification is to investigate them with regard to the type of basis functions that they employ. One class can be regarded as the "nonparametric method" that uses the basis functions that do not model the signal. Orthogonal functions fall into this category. A typical method of this class is the Fourier transform. Acoustic holography, in fact, uses this type of function, whereby it maps the sound field of interest with regard to every frequency and corresponding harmonic functions. Chapter 3 addresses how sound can be visualized by using acoustic holography. Figure II.4 is a representative example showing what we can visualize using acoustic holography.

Sound Visualization and Manipulation, First Edition. Yang-Hann Kim and Jung-Woo Choi.

THE COLOR-ORGAN

Figure II.1 Color organ: Louis-Bertrand Castel, a Jesuit priest and mathematician designed and constructed a model of the first color organ, a stringed musical instrument combined with moving transparent and colored tapes. It related seven colors to the seven complete notes of the occidental musical scale [10]

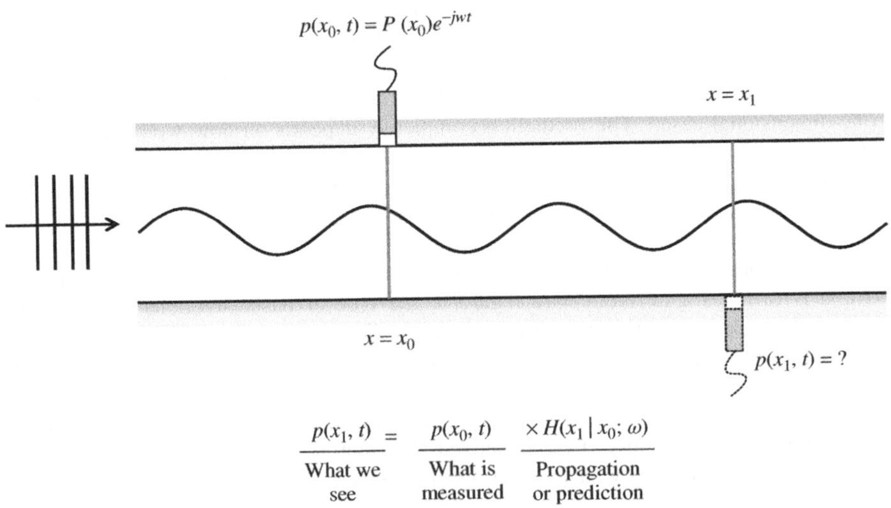

Figure II.2 One-dimensional case of sound visualization

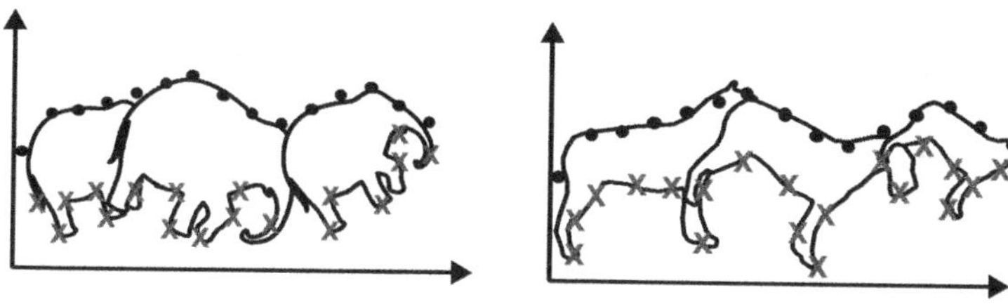

Figure II.3 The inverse problem and basis function (•: measured data, ×: predicted data). Are the measured data the parts of elephants or greyhounds? (This illustration is a modified version of the figure [11].)

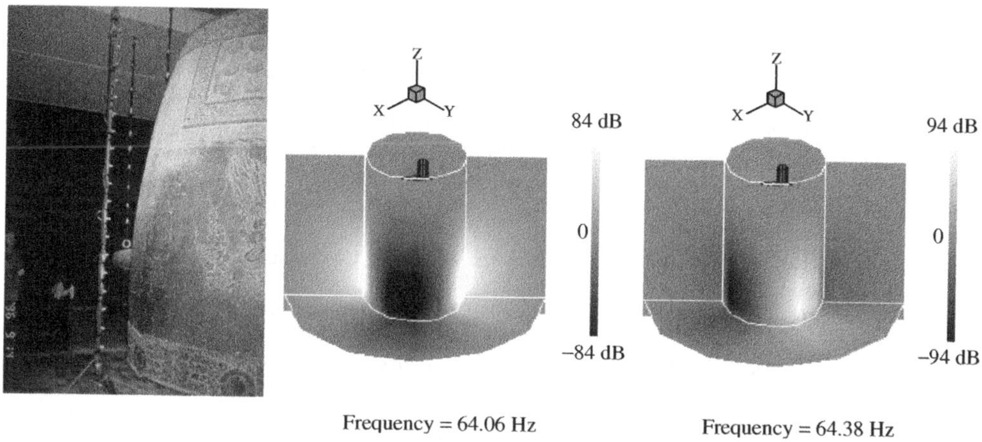

Figure II.4 Sound visualization of ancient Korean bell using near-field acoustic holography. (King Song-Dok Bell, experiment in 1996, NOVIC, KAIST.)

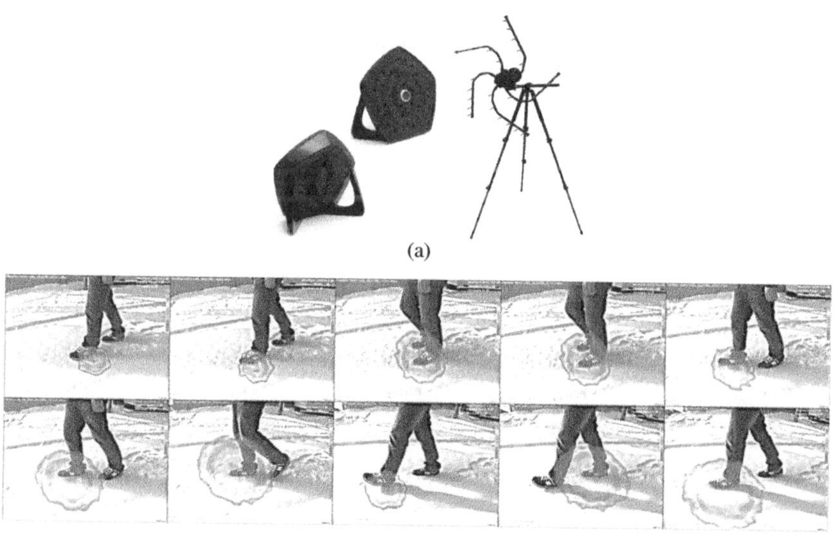

Figure II.5 Real-time sound visualization by beamforming. (a) Microphone arrays for beamforming and (b) visualization of the sound field generated by stepping on snow. (Photograph courtesy of SM Instrument, Korea.)

Another typical method is often called the "parametric method" (Chapter 4). Its name comes from the fact that in this method, certain parameters are used to model the signal. The basis function can be the "elephant" or "greyhound," as shown in Figure II.3. A typical method of this class is the "beamforming method." Different types of basis functions can be used in the methods; the selection of the basis function depends entirely on the sound field that the basis function tries to map (for example, see Refs. [12, 13]). As illustrated in Figure II.3, we can select the "elephant" or other shapes, depending on what we want to predict! The basis function maps the signal by changing its parameter; in the case of plane wave beamforming, this parameter is an angle. Figure II.5 illustrates typical microphone arrays for beamforming and beamforming images visualized in real-time.

References

1. Kac, M. (1966) Can one hear the shape of the drum? *American Mathematical Monthly*, **73**(4), 1–23.
2. Kim, Y.-H. (2004) Can we hear the shape of a noise source? The 18th International Congress on Acoustics (ICA 2004), Kyoto International Conference Hall, Japan, April 4 to 9.
3. Herman, G. T., Tuy, H. K., Langenberg, K. J. (1987) *Basic Methods of Tomography and Inverse Problems*, Malvern Physics Series, Philadelphia, PA.
4. Bui, H. D. (1994) *Inverse Problems in the Mechanics of Materials: An Introduction*, CRC Press, Boca Raton, FL.
5. Kurpisz, K. and Nowak, A. J. (1995) *Inverse Thermal Problems*, Computational Mechanics Publications, Southampton, Boston, MA.
6. Kirsch, A. (1996) *An Introduction to the Mathematical Theory of Inverse Problems*, Springer-Verlag, New York.
7. Bertero, M. and Boccacci, P. (1998) *Introduction to Inverse Problems in Imaging*, IOP Publishing, Bristol, Philadelphia, PA.
8. Isakov, V. (1998), *Inverse Problems for Partial Differential Equations*, Springer-Verlag, New York.
9. Ghosh Roy, D. N. and Couchman, L. S. (2002) *Inverse Problems and Inverse Scattering of Plane Waves*, Academic Press, New York.
10. Peacock, K. (1988) Instruments to perform color-music: two centuries of technological experimentation. *Leonardo*, **21**(4), 397–406.
11. Weber, R.L. *Science with a Smile*, Institute of Physics Publishing, 1992, pp. 111–112.
12. Pillai, S. U. (1989) *Array Signal Processing*, Springer-Verlag, New York.
13. Johnson, D. H. and Dudgeon, D. E. (1993) *Array Signal Processing Concepts and Techniques*, Prentice-Hall, Englewood Cliffs, NJ.

3

Acoustic Holography

3.1 Introduction

One of the subtle problems that make noise control difficult for engineers is "invisibility" of noise or sound. A visual image of noise often helps to determine an appropriate means for noise control. There have been many attempts to fulfill this rather challenging objective. Theoretical (or numerical) means for visualizing the sound field have been attempted, and as a result, a great deal of progress has been made. However, most of the numerical methods are not quite ready for practical applications to noise control problems. In the meantime, rapid progress with instrumentation has made it possible to use multiple microphones and fast signal processing systems. Although these systems are not perfect, they are useful. A state of the art system has recently become available, but it still has many problematic issues; for example, how can one implement the visualized noise field. The constructed noise or sound picture always consists of bias and random errors, and consequently, it is often difficult to determine the origin of the noise and the spatial distribution of the noise field. Section 3.2 of this chapter introduces a brief history, which is associated with "sound visualization," acoustic source identification methods and what has been accomplished with a line or surface array. Section 3.3 introduces difficulties and recent studies, including de-Dopplerization and de-reverberation methods, both essential for visualizing a moving noise source, such as occurs for cars or trains. This section also addresses what produces ambiguity in realizing real sound sources in a room or closed space. Another major issue associated with sound/noise visualization is whether or not we can distinguish between mutual dependences of noise in space (Section 3.3.5); for example, we are asked to answer the question, "Can we see two birds singing or one bird with two beaks?"

3.2 The Methodology of Acoustic Source Identification

The famous article written by M. Kac [1], "Can one hear the shape of the drum?" clearly addresses what the inverse problem essentially means. The inverse problem can be regarded as an attempt to obtain what is not available using what is available. The example above demonstrates the relation between sound generation by membrane vibration and its reception in space. One can find many other examples of inverse problems [2–8]. Often, in the inverse problem, it is hard to predict or describe what data are not measured because the available data are not sufficient. This circumstance is commonly referred to as an "ill-posed problem" in the literature (for example, see [1–19]). Figure 3.1 demonstrates what might happen in practice; the prediction depends on how well the

---

Sound Visualization and Manipulation, First Edition. Yang-Hann Kim and Jung-Woo Choi.

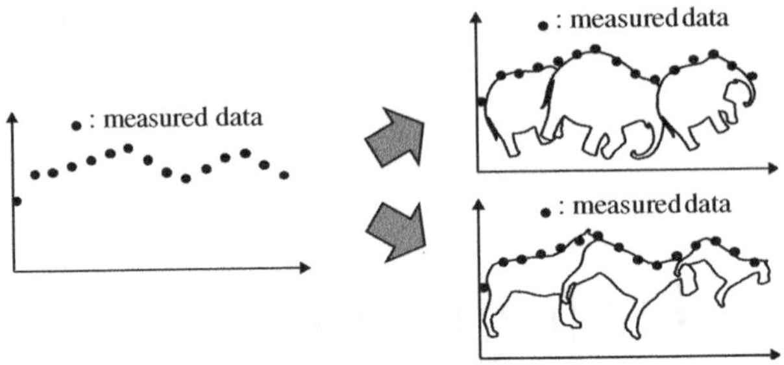

Figure 3.1 The inverse problem and basis function. Are the measured data the parts of elephants or greyhounds?

basis function (the elephants or dogs in Figure 3.1) mimics what happens in reality. When we try to see the shape of noise/sound sources, how well we see the shape of a noise source completely depends on this basis function. This is simply because we predict what is not available by using the selected basis function.

To classify the methods which are used in noise/sound source identification, it can be said that examining the types of basic functions which are employed is one of the commonly used ways. In this respect, one type is employed in the "nonparametric method", using basis functions which do not model the signal. In other words, this kind of method does not model the basis function for mapping the sound field that is not measured. All orthogonal functions fall into this category. One of the typical methods of this kind uses the Fourier transform. Acoustic holography, uses this type of basis function; it maps the sound field of interest with regard to every frequency that is measured; it sees the sound field in the frequency domain. In fact, the ideas of acoustic holography originated from optics [20–29]. Acoustic holography was simply extended or modified from the

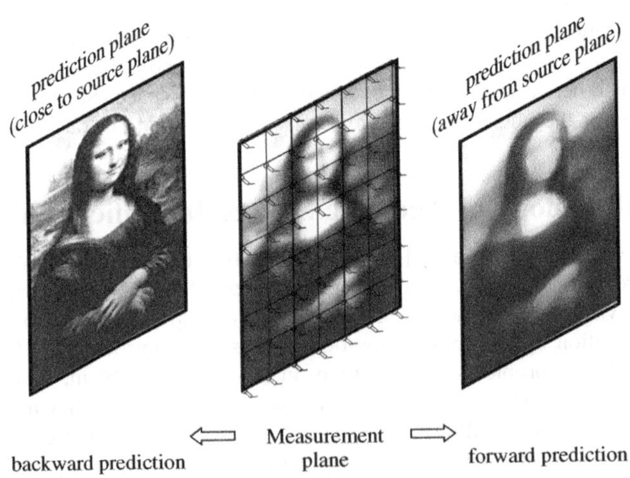

Figure 3.2 Illustration of acoustic holography. (The nearfield acoustic holography measures evanescent wave on the measurement plane. The measurement plane is always finite: finite aperture. Therefore, we get only limited data.)

basic idea of optical holography. Nearfield acoustic holography [30, 31] has been recognized as a very useful means to predict the true appearance of the source (Figure 3.2) (the nearfield effect on the resolution was first introduced in the field of microwave [32]). The highlight of this method is to include or measure the exponentially decaying waves as they propagate from the sound source so that the method can completely reconstruct the sources.

Another type is often associated with the "parametric method." The name "parametric method" derives from the fact that such methods model the signal using certain parameters. In other words, options to choose the basis function can be given to us depending upon the prediction of the sound source. A typical method of this kind is the so called "beamforming method." Different types of basis functions can be chosen for this method; it entirely depends on the sound field that the basis function tries to map [33, 34]. We can select either the elephants or dogs (others are also possible) in Figure 3.1, depending on what we want to predict! This type of mapping gives us information about the source location. As illustrated in Figure 3.3, the basis function maps the signal by changing its parameter; in the case of the plane wave beamforming, the incident angle of the plane wave can be regarded as a parameter.

The main issues, which have been discussed for this kind of mapping method, are directly related to the structure of the correlation matrix that comes from the measured acoustic pressure vector and its complex conjugate (see Figure 3.3 for the details). In this method, each scan vector is multiplied by changing its parameter; it is the angle of arrival for the plane wave of Figure 3.3, and the correlation matrix is given as illustrated in Figure 3.3. The scan vector is a basis function in this case. As one can see immediately, this problem is directly related to the structure of the correlation matrix and the basis function used. The signal-to-noise ratio (SNR) of the measured correlation matrix determines the effectiveness of the estimation. There have been many attempts to improve the estimator's performance with regard to the SNR [34, 35]. These methods have been mainly developed for application in radar and the underwater community [36]. This technique has also

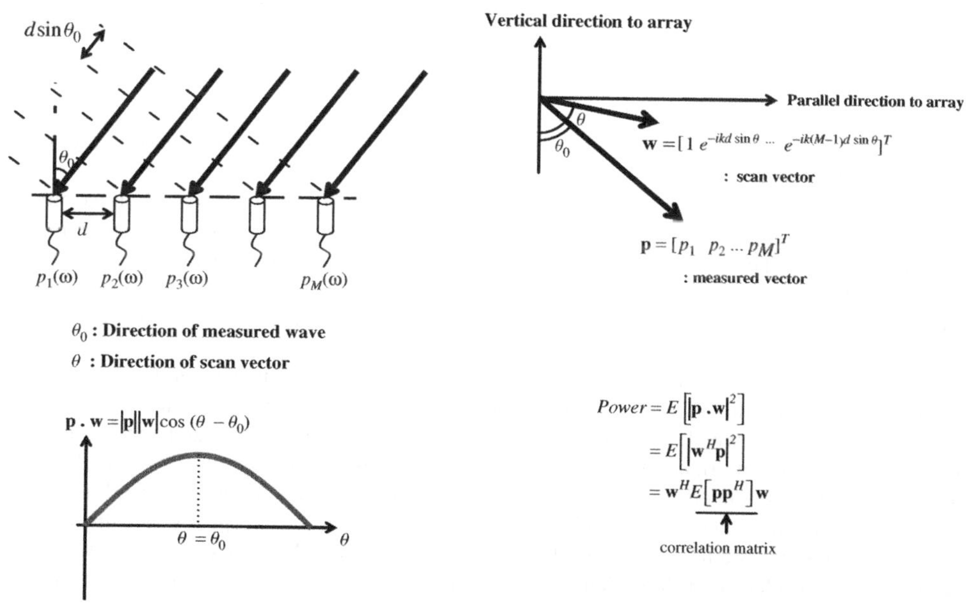

Figure 3.3 The beamforming method

been applied to a noise source location finding problem. High speed train noise source estimation [37–39] is one of the examples. Various shapes of arrays have been tried to improve the spatial resolution [40–42]. But it is obvious that these methods have no correlation with seeing the shape of sound or noise source; they only provide us its location. Therefore, we will not discuss the beamforming method in this chapter. In the next section, the problems that we have discussed will be defined.

3.3 Acoustic Holography: Measurement, Prediction, and Analysis

3.3.1 Introduction and Problem Definitions

Acoustic holography consists of three components. One component is measurement, which consists of measuring the sound pressure on the hologram plane. The second is predicting the acoustic variables, including the velocity distribution, on the plane of interest. The last part involves analyzing the holographic reconstruction. The last part has not been recognized as being as important as other parts in the past years. However, this part gives us what the sound picture really means: Visualization.

The issues associated with measurement are all related to the measurement configuration that constitutes the hologram; we measure the sound pressure at discrete measurement positions and the measurement area is finite (finite aperture) as illustrated in Figure 3.2. References [43–51] explain the necessary steps to avoid spatial aliasing, wrap around error, and the effect of including evanescent waves on the resolution (Nearfield acoustic holography). If sensors are incorrectly located on the hologram surface, this would result in errors in the predicted results. Similar errors can be produced when there is a magnitude and phase mismatch between sensors. This is well summarized in [52]. There have been many attempts to reduce the aperture effect. One method is to extrapolate the pressure data based on what is measured [49, 51]. Another method actually allows the measuring of sound pressure in sequence and interpreting the measured sound pressures with respect to reference signals, assuming that the measured sound pressure field is stationary during the measurement and the numbers of independent sources are smaller than the number of reference microphones [53–61]. Another method actually allows scanning or moving the microphone array, therefore, extending the aperture size as much as one can [62–65]. This also allows one to measure the sound pressure generated by moving sound sources, such as in the case of a vehicle's exterior noise.

The prediction problem is rather well defined and relatively straightforward. Basically, the solution of the acoustic wave equation usually results in the sound pressure distribution on the measurement plane. The prediction can be attempted by using a Green's function; an example may be found in the Kirchhoff-Helmholtz (KH) integral equation. It is noteworthy, however, that the prediction depends on the shape of the measurement and prediction surface, and also the presence of sound reflections [53, 66–87].

The acoustic holography analysis problem has been introduced rather recently. As mentioned earlier in this section, this is one of the essential issues connected to the general inverse problem. One basic question is whether what we see and imagine is related to what happens in reality. There are two different sound/noise sources: one is what is really radiating the sound, and the other one is what is reflecting the sound. The former is often called "active sound/noise," and the latter is "passive sound/noise." This is an important practical concept for establishing noise control strategies; we want to eliminate the active noise source. Another concern is whether the sources are independently or dependently correlated (Figure 3.25). The concept of an independent and dependent source has to be addressed properly to understand the issues.

3.3.2 Prediction Process

The prediction process is related to how we can predict the unmeasured sound pressure or other acoustic variables based on the measured sound pressure information. The following equation relates the unmeasured and measured pressure.

$$p(\mathbf{r},\omega)=\int_{S_h}\left(\frac{\partial p(\mathbf{r}_h,\omega)}{\partial n_h}g(\mathbf{r}|\mathbf{r}_h;\omega)-p(\mathbf{r}_h,\omega)\frac{\partial g(\mathbf{r}|\mathbf{r}_h;\omega)}{\partial n_h}\right)dS_h \tag{3.1}$$

Equation (3.1) is the KH integral equation of Equation (2.134), where $g(\mathbf{r}|\mathbf{r}_h;\omega)$ has to satisfy the inhomogeneous wave equation and the free-field Green's function is normally well used. The derivative with respect to n_h represents the inner product of the surface normal vector $\mathbf{n}$ and gradient (∇_h) of $p(\mathbf{r}_h,\omega)$ or $g(\mathbf{r}|\mathbf{r}_h;\omega)$. Note that the sign of the integral equation is reversed, as compared to Equation (2.134), because the surface normal is taken to the outside of the volume. This equation essentially says that we can predict sound pressure anywhere if we know the sound pressures and velocities on the boundary (Figure 3.4). However, it is noteworthy that measuring the velocity on the boundary is more difficult than measuring sound pressure. This rather practical difficulty can be solved by introducing a Green's function that satisfies the Dirichlet boundary condition: $g_D(\mathbf{r}|\mathbf{r}_h;\omega)$. Then, Equation (3.1) becomes

$$p(\mathbf{r},\omega)=\int_{S_h}\left(-p(\mathbf{r}_h,\omega)\frac{\partial g_D(\mathbf{r}|\mathbf{r}_h;\omega)}{\partial n_h}\right)dS_h. \tag{3.2}$$

This equation allows us to predict sound pressure on any surface of interest. It is noteworthy that we can choose a Green's function as long as it satisfies the linear inhomogeneous wave equation, or the inhomogeneous Helmholtz equation in the frequency domain. That is,

$$\nabla^2 g(\mathbf{r}\,|\,\mathbf{r}_h;\omega)+k^2 g(\mathbf{r}\,|\,\mathbf{r}_h;\omega)=-\delta(\mathbf{r}-\mathbf{r}_h). \tag{3.3}$$

Therefore, we can select a Green's function in such a way that can eliminate one of the terms in the right-hand side of Equation (3.1); Equation (3.2) is one of the cases.

To see what essentially happens in the prediction process, let us consider Equation (3.2) when the measurement and prediction plane are both planar; the planar acoustic holography assumes that the sound field is free from reflection (Figure 3.5). Then we can write Equation (3.2) as

$$p(\mathbf{r},\omega)=\int_{S_h}p(\mathbf{r}_h,\omega)K_{PP}(\mathbf{r}-\mathbf{r}_h;\omega)dS_h, \tag{3.4}$$

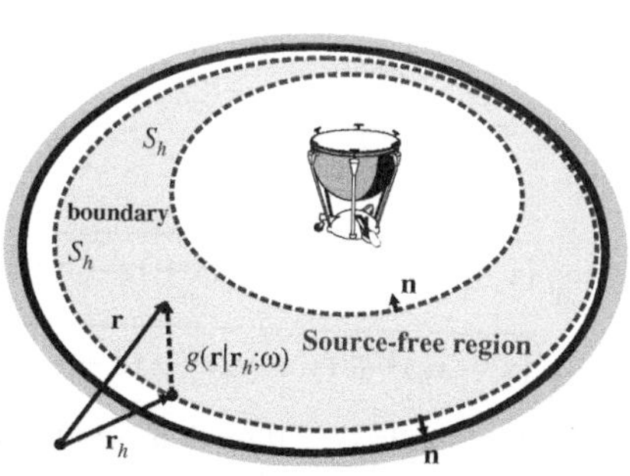

Figure 3.4 Geometry and nomenclature for the Kirchhoff-Helmholtz integral equation (Equation 3.1)

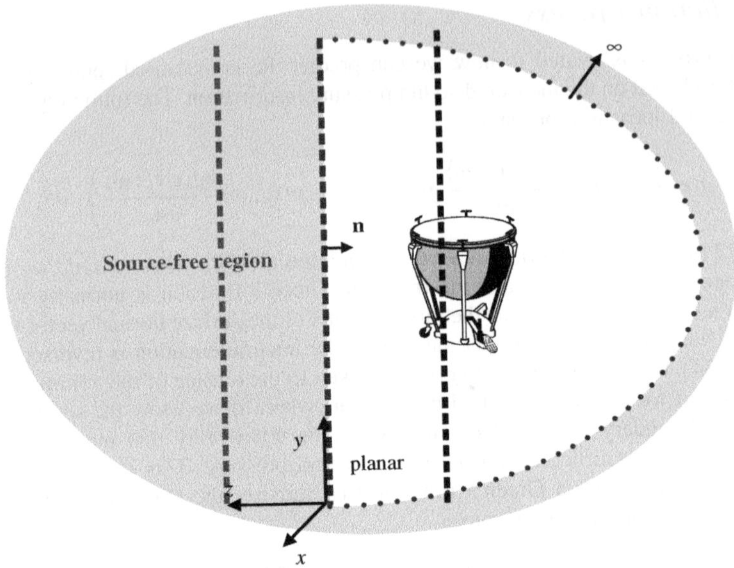

Figure 3.5 Illustration of the planar acoustic holography

where

$$k = \frac{2\pi f}{c}.$$
$$\mathbf{r} = (x, y, z)$$
$$\mathbf{r}_h = (x_h, y_h, z_h). \tag{3.5}$$

The K_{pp} can be readily obtained by using two free-field Green's functions (g_f) that are located at z_h and $-z_h$, so that it satisfies the Dirichlet boundary condition. In rectangular coordinate, the Green's function is written as

$$g_D(\mathbf{r}|\mathbf{r}_h; \omega) = g_f(\mathbf{r}|x_h, y_h, z_h; \omega) - g_f(\mathbf{r}|x_h, y_h, -z_h; \omega)$$
$$= \frac{e^{ikr_1}}{r_1} - \frac{e^{ikr_2}}{r_2} \tag{3.6}$$

where

$$r_1 = \sqrt{(x - x_h)^2 + (y - y_h)^2 + (z - z_h)^2},$$
$$r_2 = \sqrt{(x - x_h)^2 + (y - y_h)^2 + (z + z_h)^2}. \tag{3.7}$$

The normal direction on hologram plane is $-z_h$ direction (see Figure 3.5), therefore the normal derivative on the hologram plane will be calculated [45] as,

$$K_{pp}(\mathbf{r}; \omega) = -\frac{\partial g_D}{\partial n_h}\bigg|_{at\ z_h=0} = \frac{\partial g_D}{\partial z_h}\bigg|_{at\ z_h=0} = \frac{\partial}{\partial z_h}\left(\frac{e^{ikr_1}}{4\pi r_1} - \frac{e^{ikr_2}}{4\pi r_2}\right)\bigg|_{at\ z_h=0}$$

$$= \frac{\partial r_1}{\partial z_h}\frac{\partial}{\partial r_1}\frac{e^{ikr_1}}{4\pi r_1} - \frac{\partial r_2}{\partial z_h}\frac{\partial}{\partial r_2}\frac{e^{ikr_2}}{4\pi r_2}\bigg|_{at\ z_h=0}$$

$$= \frac{z}{2\pi}(1 - ikr)\frac{e^{ikr}}{r^3} \tag{3.8}$$

where

$$r = \sqrt{x^2 + y^2 + z^2}. \tag{3.9}$$

This is a convolution integral, and therefore, we can write this in wavenumber domain as

$$P(k_x, k_y, z, \omega) = P(k_x, k_y, z_h, \omega)\exp[ik_z(z - z_h)], \tag{3.10}$$

where

$$P(k_x, k_y, z, \omega) = \int_{-\infty}^{\infty}\int_{-\infty}^{\infty} p(\mathbf{r}, \omega)e^{-i(k_x x + k_y y)}dxdy,\ k_z = \sqrt{k^2 - k_x^2 - k_y^2}. \tag{3.11}$$

This equation essentially predicts sound pressure with respect to the wavenumber (k_x, k_y). If $k^2 \geq k_x^2 + k_y^2$, the wave in z-direction (k_z) is propagating in space. Otherwise, the z-direction wave decays exponentially; it is an evanescent wave (Figure 3.6).

We have derived the formulas that can predict what we did not measure based on what we measured, by using a Green's function. It is noteworthy that we can get the same results if we use the characteristic solutions of the Helmholtz equation. Section 3.3.3 describes its details.

Equation (3.10) also allows us to predict the sound pressure on the source plane, when $z = z_s$. It is an inverse problem because it predicts pressure distribution on the source plane based on the hologram pressure (Figures 3.5 and 3.6).

Figure 3.7 essentially illustrates how we can process the data for predicting what we do not measure based on what we measure. The prediction process, of course, cannot be perfect, therefore must have inevitable errors. It can be a planar, cylindrical, or spherical, depending on the selected coordinates. If one discretized the surface of arbitrary shape, then the prediction process falls into Boundary Element Method (BEM).

There are four major areas that cause errors in acoustic holography prediction. One is related to the integration of Equation (3.4). This equation has to be implemented on the discretized surface (Figure 3.8). The surface, therefore, has to be spatially sampled according to the selected surface. This spatial sampling can produce a spatial aliasing, depending on the spatial distribution of sound source. It is noteworthy that the information about the sound source: its location, spatial, and temporal characteristics are not available when we use the acoustic holography technique. The sampling

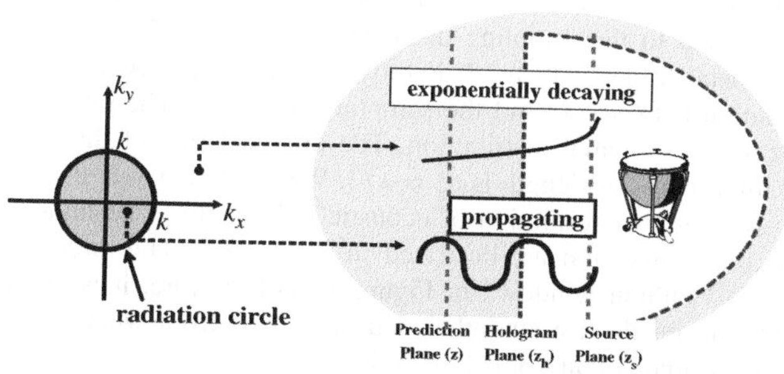

Figure 3.6 Propagating and exponentially decaying waves in acoustic holography

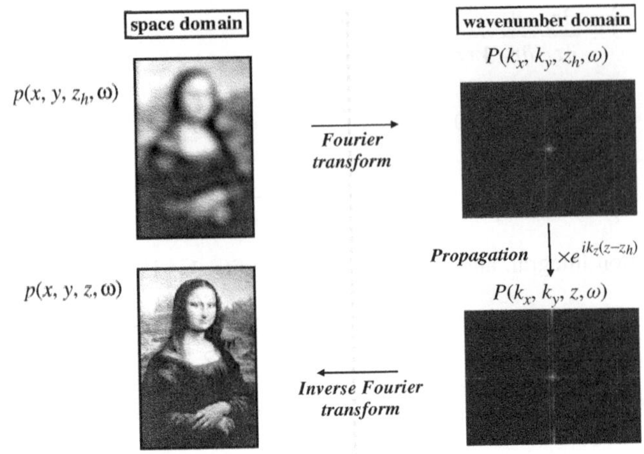

Figure 3.7 The data processing procedure of acoustic holography

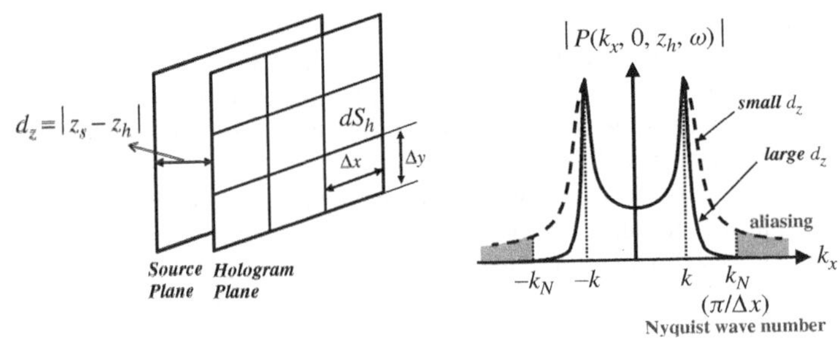

Figure 3.8 Error due to discrete measurement. (Spatial aliasing problem: Microphone spacing determines Nyquist wavenumber. This wavenumber has to be smaller than maximum wavenumber of acoustic pressure distribution on the hologram plane. Microphone spacing, therefore, has to get smaller as the distance between hologram (measurement plane) and the source plane decreases. Rule of thumb is $\Delta x, \Delta y < d_z$.)

wavenumber must be larger than twice the maximum wavenumber of interest in analogous with time-domain sampling theorem. As illustrated in Figure 3.8, the distance between hologram and source plane is also related to the sampling distance Δx and Δy. The closer to the source, the smaller the sampling distance needs to be. It is only qualitative conjecture, and there is, in strict sense, no definite relation between d_z and the sampling distances. The relation, of course, has to be dependent on the sources' spatial distribution. The practical rule of thumb, determining d_z in association with the minimum wavelength is $d_z = \lambda/3$. We must also note that the size of the aperture determines the wavenumber resolution of acoustic holography. The finite aperture inevitably produces very sharp data truncation as illustrated in Figure 3.9. This produces unrealistic high wavenumber noise (see "without window" in Figure 3.9). Therefore, it is often required that we use a window, which can result in a smoother data transition from what is on the measurement plane to what is not measured (Figure 3.9) [49–51].

The spatial Fourier transform in the prediction process (Equation (3.10)) has to be done in the domain for which data is available: finite Fourier transform, therefore produces a ghost hologram

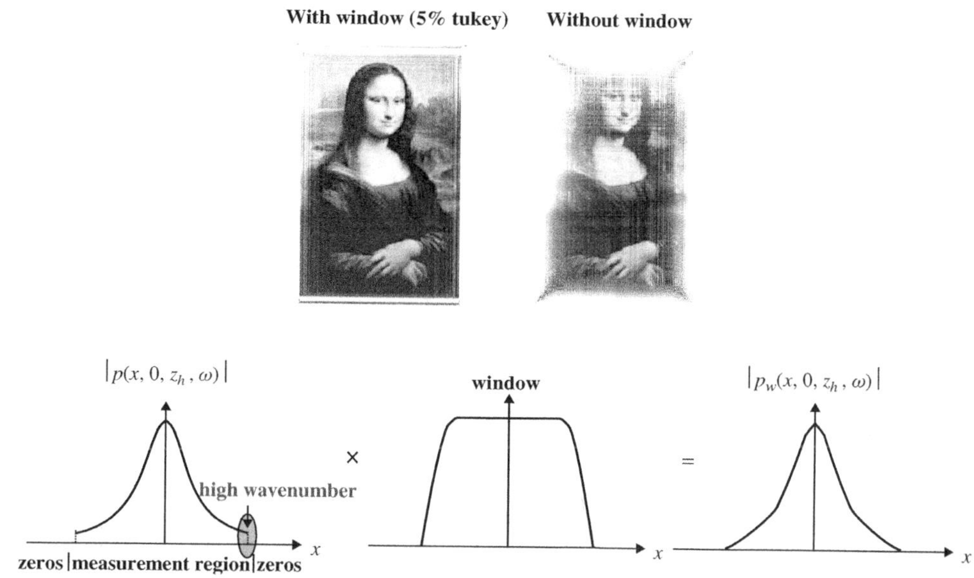

Figure 3.9 Effect of finite aperture. (The rapid change at the aperture edges produces high wavenumber noise.)

as illustrated in Figure 3.10 [45, 49]. This effect can be effectively removed by adding zeros on the hologram data (Figure 3.10). The last thing we have to note is what can happen when we do backward propagation. As we can see in Equation (3.10), when we predict the sound pressure distribution on the plane closer to the sound source ($z < z_h$ (Figure 3.5)) and k_z has an imaginary value (evanescent wave), then the sound pressure distribution of the exponentially decaying part will be unrealistically magnified (Figure 3.11) [46]. Figure 3.12 graphically summarizes the entire processing steps of acoustic holography.

The issues related with the evanescent wave and its measurement is well addressed in the literature: for example, see [88]. The measurement of evanescent waves essentially allows us to

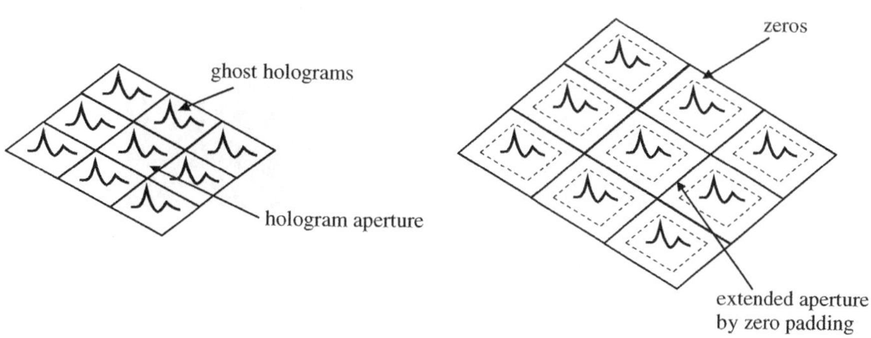

Figure 3.10 Effect of finite spatial Fourier transform on acoustic holography. (The ghost image due to finite Fourier transform, circular convolution can be eliminated by adding zeros.)

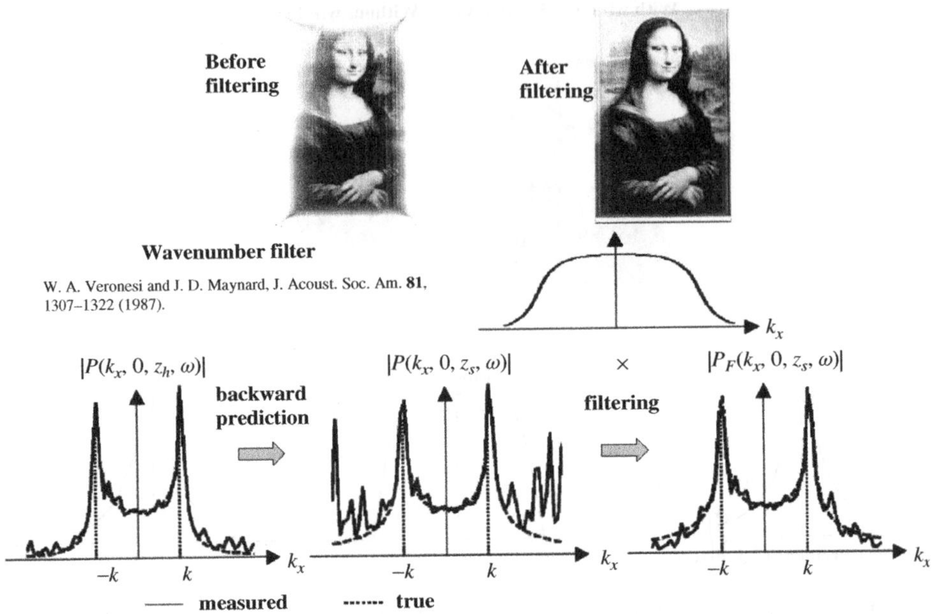

Figure 3.11 Wavenumber filtering in backward prediction. (Evanescent wave components are magnified without filtering.)

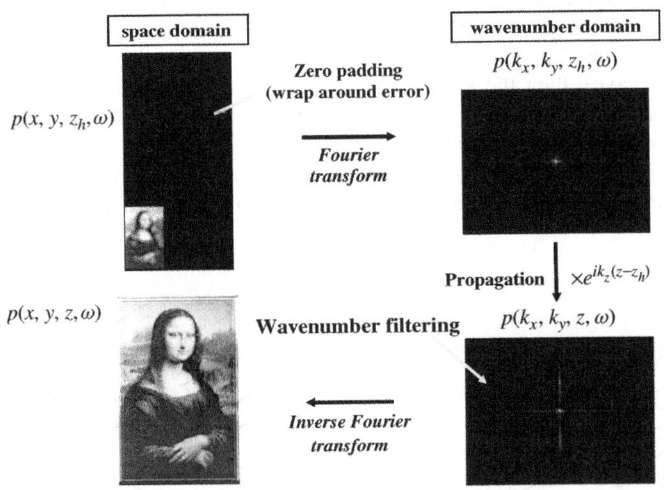

Figure 3.12 Summary of acoustic holography prediction process

have higher resolution than conventional acoustic holography [89–94]. However, it is noteworthy that the evanescent wave component is substantially smaller compared with other propagating components. Therefore, it is easy to produce errors that are associated with sensor or position mismatch [52]; in other words, it is very sensitive to the SNR. The errors due to position and sensor mismatch are of two types. One is bias error and the other is random error. Reference [52] shows that the bias error due to the mismatches is negligible, but the random error is significant in backward prediction. This is related to measurement spacing on a hologram plane (Δ_h), prediction plane (Δ_z), and the distance between the hologram plan and prediction plane (d_z). It is approximately proportional to $24.9(d_z/\Delta_z) + 20\log_{10}(\Delta_h/\Delta_z)$ in a dB scale. The SNR can be amplified when we try to reconstruct the source field; this is one of the typical ill-posed phenomena. Focusing on this problem, there have been many attempts to reduce this effect by using a spatial filter [46, 95–99]. It is often called "regularization" of acoustic holography; Refs. [100–114] discuss this.

Depending on the separable coordinates that we use for acoustic holography, we can construct cylindrical or spherical coordinates; for example, see [53, 67, 69] (Figures 3.13 and 3.14). These methods predict the sound field in exactly the same manner as the planar holography does but with respect to the different coordinates. As we can expect, however, these methods have some advantages. For example, the wrap around error is negligible; in fact, there is no such error in the spherical acoustic holography, and no aperture related error [53]. Very recently, the advantage of using spherical functions has also been introduced [79, 81, 86, 87, 112].

3.3.3 *Mathematical Derivations of Three Acoustic Holography Methods and Their Discrete Forms*

We often use the KH integral equation to explain how we predict what we do not measure based on what we measure. It is noteworthy, however, that the same result can be obtained by using the characteristic solutions of the Helmholtz equation. The following sections address how these can be obtained. Planar, cylindrical, and spherical acoustic holography are derived using characteristic equations in terms of a corresponding coordinate system. The equations for the holography are also expressed in discrete forms that are normally used in the computation.

3.3.3.1 Planar Acoustic Holography

If we see solutions of the Helmholtz equation;

$$\nabla^2 p(\mathbf{r}, \omega) + k^2 p(\mathbf{r}, \omega) = 0 \tag{3.12}$$

in terms of Cartesian coordinate, then we can write them as

$$p(x, y, z) = X(x)Y(y)Z(z) \tag{3.13}$$

where, $k = \frac{\omega}{c} = \frac{2\pi f}{c}$. The frequency dependency ω is omitted for simplicity. We assumed then p can be separable with respect to *X*, *Y*, and *Z*. Equations (3.12) and (3.13) give us the characteristic equation, that is,

$$\psi(x, y, z|k_x, k_y, k_z) = \begin{Bmatrix} e^{ik_x x} \\ e^{-ik_x x} \end{Bmatrix} \begin{Bmatrix} e^{ik_y y} \\ e^{-ik_y y} \end{Bmatrix} \begin{Bmatrix} e^{ik_z z} \\ e^{-ik_z z} \end{Bmatrix} \tag{3.14}$$

where,

$$k^2 = {k_x}^2 + {k_y}^2 + {k_z}^2 \tag{3.15}$$

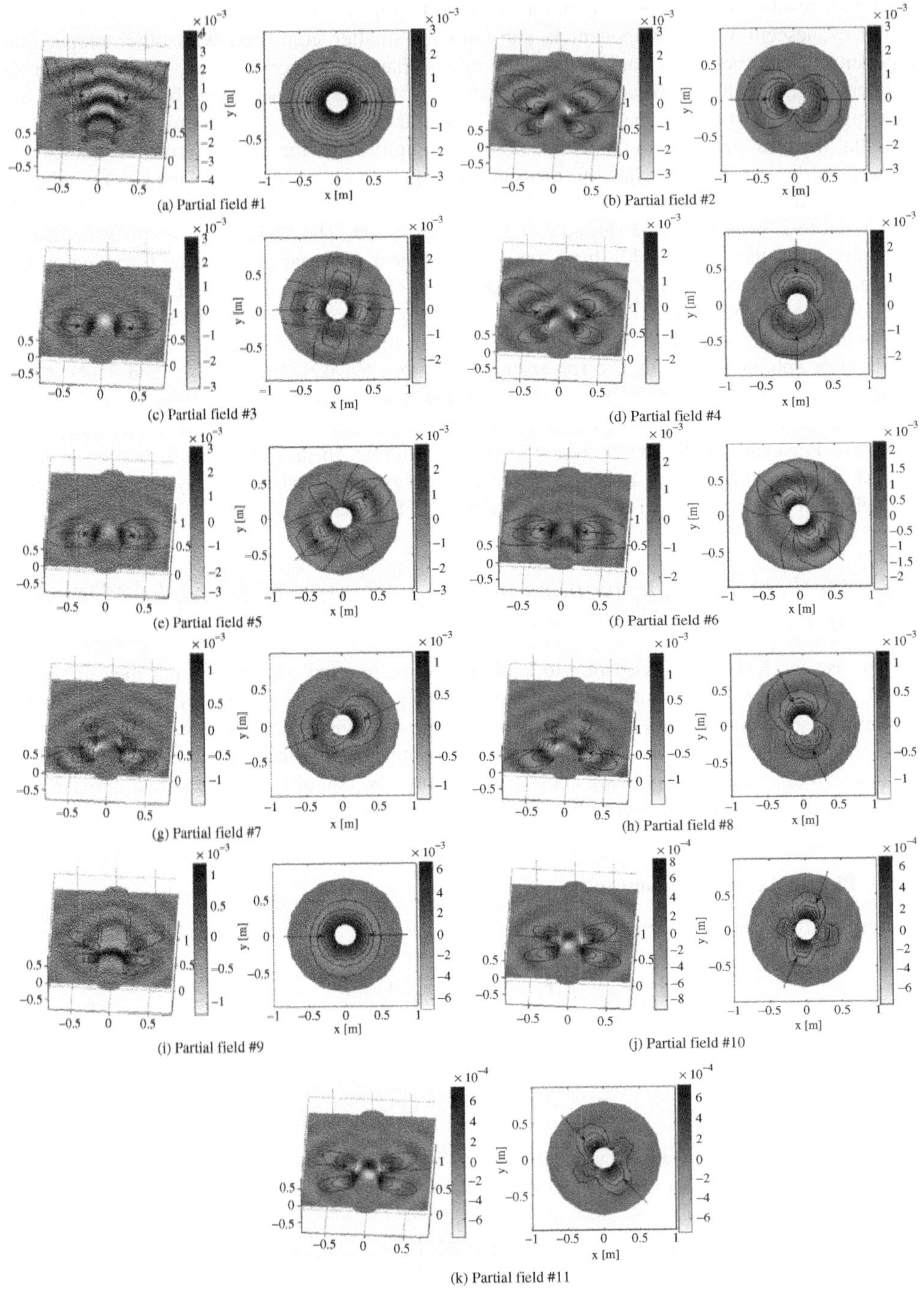

Figure 3.13 Cylindrical holography © 2007, JASA. Reproduced, with permission

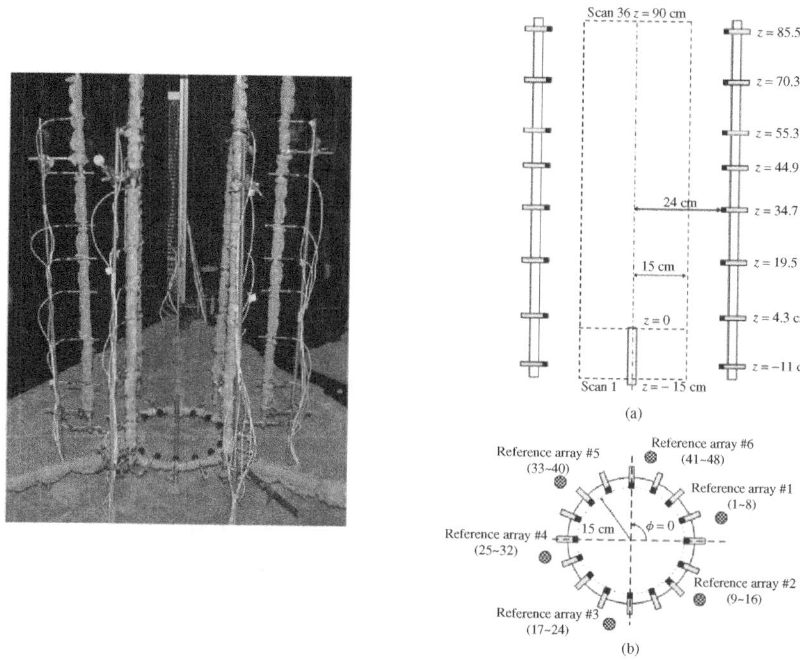

Figure 3.13 (*continued*)

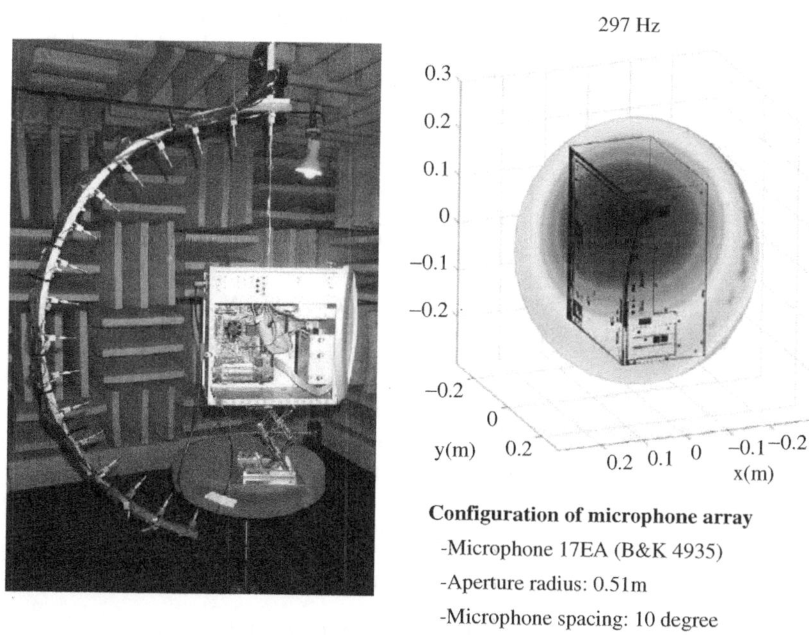

Figure 3.14 Spherical holography

Now we can write

$$p(\mathbf{r}) = \int P(\mathbf{k})\psi(\mathbf{r}|\mathbf{k})d\mathbf{k}, \tag{3.16}$$

where

$$\mathbf{k} = (k_x, k_y, k_z). \tag{3.17}$$

Let's assume that the sound sources are all located at $z < z_s$ and we measure at $z_h > z_s$. Then we can write Equation (3.16) as

$$p(x, y, z) = \frac{1}{(2\pi)^2}\int_{-\infty}^{\infty}\int_{-\infty}^{\infty} P(k_x, k_y, z)e^{i(k_x x + k_y y)}dk_x dk_y. \tag{3.18}$$

It is noteworthy that we can neglect $-ik_z z$ of Equation (3.14), because of the assumption we made ($z < z_s$ and at $z_h > z_s$). The k_x and k_y can be either positive or negative.

In Equation (3.18),

$$k_z = \begin{cases} \sqrt{k^2 - {k_x}^2 - {k_y}^2}, & \text{when} \quad k^2 > {k_x}^2 + {k_y}^2 \\ i\sqrt{{k_x}^2 + {k_y}^2 - k^2}, & \text{when} \quad k^2 < {k_x}^2 + {k_y}^2. \end{cases} \tag{3.19}$$

Figure 3.6 illustrates how these two different k_z values essentially mean. We measure $p(x, y, z_h)$, therefore we have data of sound pressure data on $z = z_h$. Fourier transform of Equation (3.18) leads us

$$P(k_x, k_y, z_h) = \int_{-\infty}^{\infty}\int_{-\infty}^{\infty} p(x, y, z_h)e^{-i(k_x x + k_y y)}dxdy. \tag{3.20}$$

Using Equations (3.10), (3.18) and (3.20), we can always estimate sound pressure on z, which is away from the source. It is noteworthy that Equation (3.20) has to be performed in discrete domain. In other words, we have to use finite aperture of rectangular shape and it is spatially sampled (Figures 3.8 and 3.15a). If the number of measurement points along x and y direction are M and N respectively, and the corresponding sampling distance is Δx and Δy, then Equation (3.20) can be rewritten as

$$P(k_x, k_y, z) = e^{ik_z(z - z_h)}\left[\Delta x \Delta y \sum_{m=0}^{M-1}\sum_{n=0}^{N-1} p\left(x_m, y_n, z_h\right) e^{-ik_x x_m} e^{-ik_y y_n}\right], \tag{3.21}$$

where

$$x_m = \left(m + \frac{1 - M}{2}\right)\Delta x, \; y_n = \left(n + \frac{1 - N}{2}\right)\Delta y. \tag{3.22}$$

3.3.3.2 Cylindrical Acoustic Holography

A solution can also found in cylindrical coordinate, that is,

$$p(r, \varphi, z) = R(r)\Phi(\varphi)Z(z). \tag{3.23}$$

Figure 3.15 shows the coordinate systems. Then, its characteristic solutions are

$$\psi_m(r, \varphi, z|k_r, k_z) = \begin{Bmatrix} {H_m}^{(1)}(k_r r) \\ {H_m}^{(2)}(k_r r) \end{Bmatrix} \begin{Bmatrix} e^{im\varphi} \\ e^{-im\varphi} \end{Bmatrix} \begin{Bmatrix} e^{ik_z z} \\ e^{-ik_z z} \end{Bmatrix}, \tag{3.24}$$

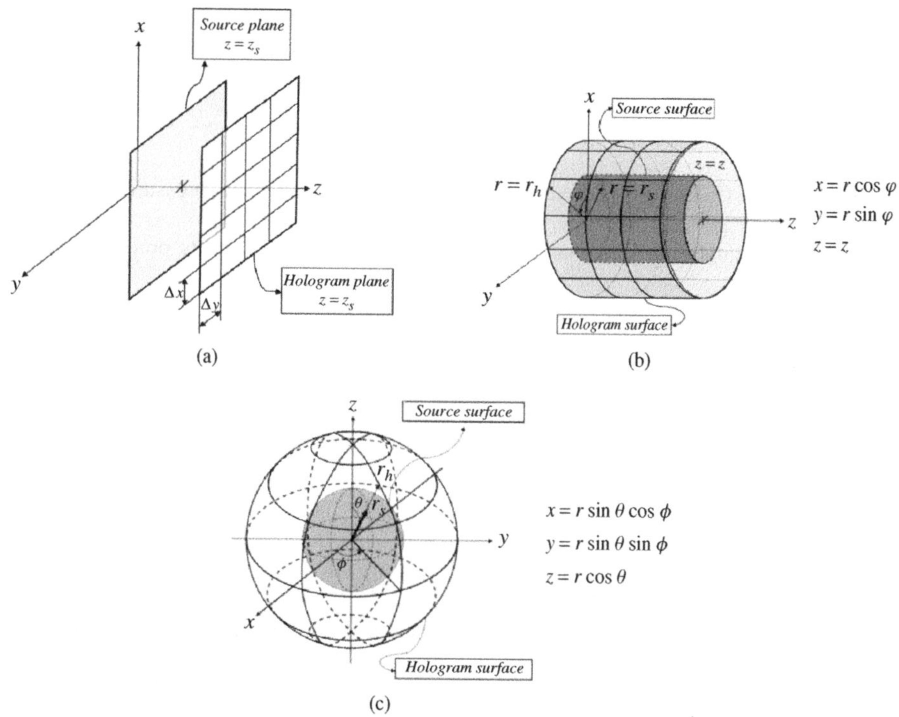

Figure 3.15 Coordinate systems of acoustic holography. (a) Planar acoustic holography. (b) Cylindrical Acoustic holography. (c) Spherical acoustic holography

where

$$k^2 = k_r^{\ 2} + k_z^{\ 2}. \tag{3.25}$$

The $H_m^{(1)}$ and $H_m^{(2)}$ are the first and second kind of cylindrical Hankel function, respectively. The $e^{im\varphi}$ and $e^{-im\varphi}$ express mode shapes in φ -direction.

Using the characteristic function Equation (3.24), we can write a solution of Helmholtz equation with respect to cylindrical coordinate as

$$p(r, \varphi, z) = \frac{1}{2\pi} \sum_{m=-\infty}^{\infty} \int P_m(k_z)\psi_m(r, \varphi, z|k_r, k_z)dk_z. \tag{3.26}$$

Assuming that the sound sources are all located at $r < r_s$ and the hologram surface is situated on the surface; $r = r_h$ and $r_h > r_s$, then there is no wave that is propagating in the negative r-direction, in other words, toward the sources. Then Equation (3.26) can be rewritten as

$$p(r, \varphi, z) = \frac{1}{2\pi} \sum_{m=-\infty}^{\infty} \int_{-\infty}^{\infty} P_m(k_z)e^{im\varphi}e^{ik_z z}H_m^{(1)}(k_r r)dk_z, \tag{3.27}$$

and k_r has to be

$$k_r = \begin{cases} \sqrt{k^2 - k_z^{\ 2}}, & when \quad k^2 \geq k_z^{\ 2} \\ i\sqrt{k_z^{\ 2} - k^2}, & when \quad k^2 < k_z^{\ 2}. \end{cases} \tag{3.28}$$

We measure acoustic pressure at $r = r_h$, therefore $p(r_h, \varphi, z)$ is available. Then $P_m(k_z)$ can be readily obtained.

That is,

$$P_m(k_z) = \frac{1}{2\pi H_m^{(1)}(k_r r_h)} \int_0^{2\pi} \int_{-\infty}^{\infty} p(r_h, \varphi, z) e^{-im\varphi} e^{-ik_z z} dz \, d\varphi. \tag{3.29}$$

Inserting Equation (3.29) into Equation (3.27) provides us acoustic pressure at unmeasured surface at r.

Discretization of Equation (3.29) makes a formula that can be used in its practical calculation. That is,

$$P_m(k_z) = \frac{1}{H_m^{(1)}(k_r r_h)} \frac{1}{L} \Delta z \sum_{l=0}^{L-1} \sum_{n=0}^{N-1} p(r_h, \varphi_l, z_n) e^{-im\varphi_l} e^{-ik_z z_n}, \tag{3.30}$$

where

$$\varphi_l = \frac{(2l+1)\pi}{L}, \quad z_n = \left(n + \frac{1-N}{2}\right) \Delta z, \tag{3.31}$$

L and N are the number of data points in φ and z direction, respectively.

3.3.3.3 Spherical Acoustic Holography

The Helmholtz equation can also be expressed in spherical coordinate (Figure 3.15c). Assuming again that the separation of variable also holds in this case, then we can write

$$p(r, \theta, \phi) = R(r)\Theta(\theta)\Phi(\phi). \tag{3.32}$$

Plugging into Equation (3.12) gives us the characteristic equation

$$\psi_{nm}(r, \theta, \phi) = \begin{Bmatrix} h_n^{(1)}(kr) \\ h_n^{(2)}(kr) \end{Bmatrix} \begin{Bmatrix} P_n^m(\cos\theta) \\ Q_n^m(\cos\theta) \end{Bmatrix} \begin{Bmatrix} e^{im\phi} \\ e^{-im\phi} \end{Bmatrix}. \tag{3.33}$$

The n is nonnegative integer and m can be any integer between $-n$ and n. The $h_n^{(1)}$ and $h_n^{(2)}$ are the spherical Hankel function of the first and second kind. It is also noteworthy that $P_n{}^m$ and Q_n^m are the associated Legendre functions of the first and second kind.

Then we can write the solution of the Helmholtz equation as

$$p(r, \theta, \phi) = \sum_{n=0}^{\infty} \sum_{m=-n}^{n} D_{nm} \psi_{nm}(r, \theta, \phi), \tag{3.34}$$

where D_{nm} are harmonic coefficients of spherical harmonic expansion those include all the information of the pressure field distribution.

Suppose that we have sound sources at $r < r_s$ and the hologram is on the surface; $r = r_h$ and $r_h > r_s$, then, Equation (3.34) can be simplified as

$$p(r, \theta, \phi) = \sum_{n=0}^{\infty} \sum_{m=-n}^{n} D_{nm} h_n^{(1)}(kr) Y_n^m(\theta, \phi), \tag{3.35}$$

where

$$Y_n^m(\theta, \phi) = \sqrt{\frac{(2n+1)}{4\pi} \frac{(n-m)!}{(n+m)!}} P_n{}^m(\cos\theta) e^{im\phi}. \tag{3.36}$$

This is spherical harmonic function. It is noteworthy that we only have the first kind spherical harmonic function because there are waves that only propagate away from the sources. The second kind associated Legendre function was discarded because we will have finite acoustic pressure where $\theta = 0$ *or* π.

Similarly, as we previously stated, the sound pressure data on the hologram is available, therefore we can get D_{nm} of Equation (3.35) by

$$D_{nm} = \frac{1}{h_n^{(1)}(kr_h)} \int_0^{\pi} \int_0^{2\pi} p(r_h, \theta, \phi) Y_n^m(\theta, \phi)^* \sin\theta d\phi d\theta. \tag{3.37}$$

We used the orthogonal property of Y_n^m. The superscript * expresses complex conjugate. Using Equations (3.37) and (3.35), we can estimate acoustic pressure at anywhere away from the sources.

The discrete form of Equation (3.37) can be written as

$$D_{nm} = \frac{1}{h_n^{(1)}(kr_h)} \sum_{l=0}^{L-1} \sum_{q=0}^{Q-1} A_{lq} p(r_h, \theta_l, \phi_q) Y_n^{\ m}(\theta_l, \phi_q) \sin\theta_l, \tag{3.38}$$

where A_{lq} is the solid angle $(\sin\theta_l d\phi_q d\theta_l)$ of each patch positioned at (θ_l, ϕ_q). For regular sampling in θ and ϕ axes, the sampling positions are given by

$$\theta_l = \frac{(2l+1)\pi}{2L}, \quad \phi_q = \frac{(2q+1)\pi}{Q}, \tag{3.39}$$

where L is the number of data points in θ and Q is what is in ϕ direction.

3.3.4 *Measurement*

3.3.4.1 Step-by-Step and Continuous Scanning Methods

To construct a hologram, we commonly measure sound pressure at discrete positions as illustrated in Figure 3.2. However, if the sound generated by the source, and therefore the sound field, can be assumed to be stationary, then we do not have to measure them at the same time. Figure 3.16 illustrates one way to accomplish this measurement.

This method measures the sound pressure field normally a using line array in steps (Figure 3.16a). To understand the issues that are associated with this measurement system for the sake of its simplicity, let us see how we process a signal of frequency ω when there is a single source.

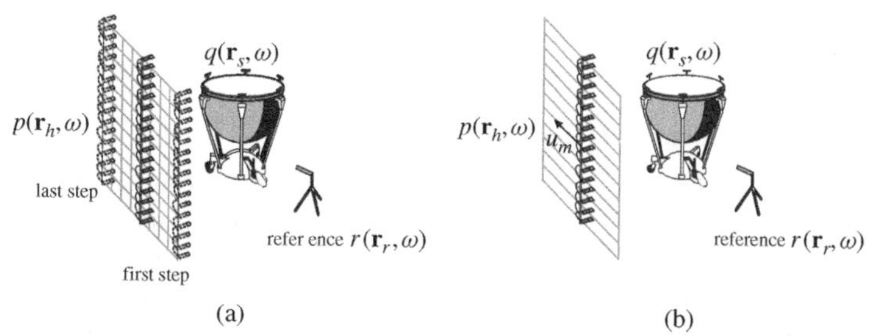

Figure 3.16 Two measurement methods of pressure on the hologram plane. (a) Step-by-step scanning and (b) continuous scanning

The relation between the sound source and sound pressure in the field, or measurement position ($\mathbf{r}_h$) can be written as

$$p(\mathbf{r}_h, \omega) = h(\mathbf{r}_h|\mathbf{r}_s; \omega)q(\mathbf{r}_s, \omega), \tag{3.40}$$

where $q(\mathbf{r}_s, \omega)$ is the source input signal and $h(\mathbf{r}_h|\mathbf{r}_s; \omega)$ is the transfer function between the source input and the measured pressure. This means that if we know the transfer function and input, then we can find the magnitude and phase between the measured positions. Because it is usually not practical to measure the input, we normally use reference signals (Figure 3.16a). By using a reference signal, the pressure can be written as

$$p(\mathbf{r}_h, \omega) = h(\mathbf{r}_h|\mathbf{r}_r; \omega)r(\mathbf{r}_r, \omega), \tag{3.41}$$

where $r(\mathbf{r}_r, \omega)$ is the reference signal. We can obtain $h(\mathbf{r}_h|\mathbf{r}_r; \omega)$ by

$$h(\mathbf{r}_h|\mathbf{r}_r; \omega) = \frac{p(\mathbf{r}_h, \omega)}{r(\mathbf{r}_r, \omega)}. \tag{3.42}$$

The input and reference has the relation of

$$r(\mathbf{r}_r, \omega) = h(\mathbf{r}_r|\mathbf{r}_s; \omega)q(\mathbf{r}_s, \omega) = \frac{h(\mathbf{r}_r|\mathbf{r}_s; \omega)}{h(\mathbf{r}_h|\mathbf{r}_r; \omega)}q(\mathbf{r}_s, \omega), \tag{3.43}$$

where $h(\mathbf{r}_r|\mathbf{r}_s; \omega)$ is the transfer function between the input and reference. As a result, therefore, we can see that Equation (3.41) has the same form as Equation (3.40).

It is noteworthy that Equation (3.40) holds for the case that we have only one sound source and the sound field is stationary random. However, if the number of sound sources is two, then Equation (3.40) has to be

$$p(\mathbf{r}_h, \omega) = h(\mathbf{r}_h|\mathbf{r}_{s,1}; \omega)q(\mathbf{r}_{s,1}, \omega) + h(\mathbf{r}_h|\mathbf{r}_{s,2}; \omega)q(\mathbf{r}_{s,2}, \omega), \tag{3.44}$$

where $q(\mathbf{r}_{s,\kappa}, \omega)$ is the κth input and $h(\mathbf{r}_h|\mathbf{r}_{s,\kappa}; \omega)$ is its transfer function. There are two independent sound fields. This requires, of course, two independent reference signals. It has been well accepted that the number of reference microphones has to be greater than the number of independent sources [56]. However, if this is strictly true, then it means that we have to somehow know the number of sources, and this, in some degree, contradicts what the acoustic holography provides us.

A recent study [61] demonstrates that measured information, where the sources are, and how many independent sources are there, converge to true value as the number of reference microphones increases. This study also shows that sources of high power are likely to be identified even if the number of reference microphones is less than the number of sources. Figure 3.17 shows an example of this method when there are many independent sound fields.

3.3.4.2 Moving Frame Acoustic Holography (MFAH)

On the other hand, one study shows that we can even continuously scan the sound field by using a line array of microphones (Figure 3.16b) [62–65]. This method essentially allows us to extend the aperture size without any limit as long as the sound field is stationary; in fact, Ref. [65] also shows that this method can be used for a slowly varying sound field (quasi-stationary).

This method has to deal with the Doppler shift. For example, let us consider a plane wave in (k_{x0}, k_{y0}, k_{z0}) direction and a pure tone of frequency ω_{h0}. Then the pressure on the hologram plane

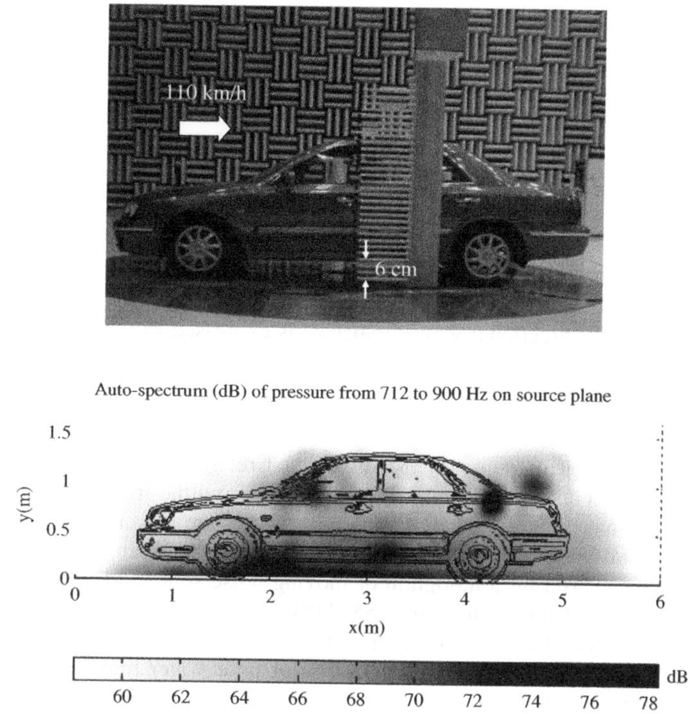

Figure 3.17 Application result of the step-by-step scanning method to the wind noise of a car. This figure is the pressure distribution at 712–900 Hz in a source plane when the flow velocity is 110 km/h. In this experiment, 17 reference microphones are randomly located in the car. © 2004, JASA. Reproduced, with permission, from [116]

can be written as

$$p(\mathbf{r}_h, t) = P_0 \exp[i(k_{x0}x_h + k_{y0}y_h + k_{z0}z_h)] \times \exp(-i\,\omega_{h_0}t), \tag{3.45}$$

where P_0 denotes the complex magnitude of the plane wave. Spatial information of plane wave with respect to x-direction can be represented by wavenumber spectrum, and it can be described as

$$\begin{aligned} P(k_x, y_h, z_h, t) &= \int_{-\infty}^{\infty} p(x_h, y_h, z_h, t)e^{-ik_x x_h}dx_h \\ &= 2\pi P_0 \exp\left[i\left(k_{y0}y_h + k_{z0}z_h\right)\right] \delta(k_x - k_{x0}) \times \exp(-i\,\omega_{h_0}t), \\ &= P(k_{x0}, y_h, z_h)\delta(k_x - k_{x0}) \exp(-i\omega_{h_0}t) \end{aligned} \tag{3.46}$$

where $P(k_{x0}, y_h, z_h) = 2\pi P_0 \exp[i(k_{y0}y_h + k_{z0}z_h)]$ is the wavenumber spectrum of the plane wave at $k_x = k_{x0}$.

If a microphone is moving at the x-velocity of u_m, the measured signal $p_m(x_h, y_h, z_h, t)$ is

$$p_m(x_h, y_h, z_h, t) = p(u_m t, y_h, z_h, t). \tag{3.47}$$

The Fourier transform with respect to time $\mathcal{F}_t$ of Equation (3.47), using Equation (3.45), can be expressed as

$$\begin{aligned}\mathcal{F}_t[p_m(u_m t, y_h, z_h, t)] &= \int_{-\infty}^{\infty} p_m(u_m t, y_h, z_h, t)e^{i\omega t}\,dt \\ &= P_0 \exp[i(k_{y0}y_h + k_{z0}z_h)] \times 2\pi\delta(u_m k_{x0} - \omega_{h0} + \omega). \\ &= P(k_{x0}, y_h, z_h)\delta(u_m k_{x0} - \omega_{h0} + \omega)\end{aligned} \tag{3.48}$$

Equation (3.48) means that the complex amplitude of the plane wave is located at the shifted frequency $\omega_{h0} - u_m k_{x0}$, as shown in Figures 3.18 and 3.19. In general, the relation between the shifted frequency ω and x -directional wavenumber k_x is expressed as

$$k_{x0} = \frac{\omega_{h0} - \omega}{u_m}. \tag{3.49}$$

We can measure the original frequency ω_{h0} by a fixed reference microphone. By the Doppler shift, therefore, we can obtain the wavenumber components from the frequency components of the moving microphone signal. Figure 3.20 illustrates how we get the wavenumber spectrum.

This method essentially uses the relative coordinate between hologram and microphone. Therefore, it can be used for measuring a hologram of moving noise sources (Figure 3.21). This is one of the major contributions of this method [63–65, 118].

A Doppler effect has been handled on the time domain as well as the frequency domain [119]. These methods have been developed for visualization of underwater moving sources [120, 121]. The medium moving induces to not frequency Doppler effects but a wavenumber Doppler effects. References have been developed the wavenumber filter for considering convective fluid [122–124].

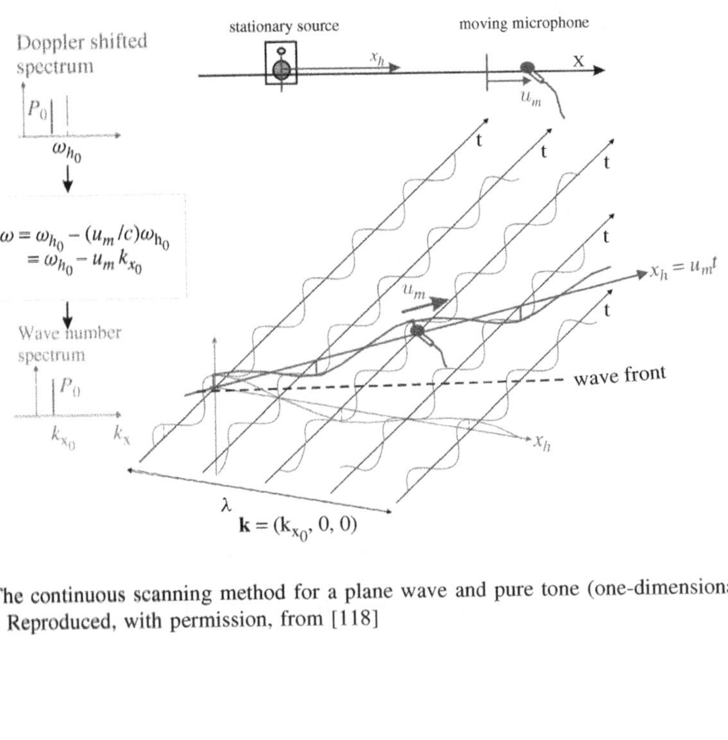

Figure 3.18 The continuous scanning method for a plane wave and pure tone (one-dimensional illustration) © 2001. JASA. Reproduced, with permission, from [118]

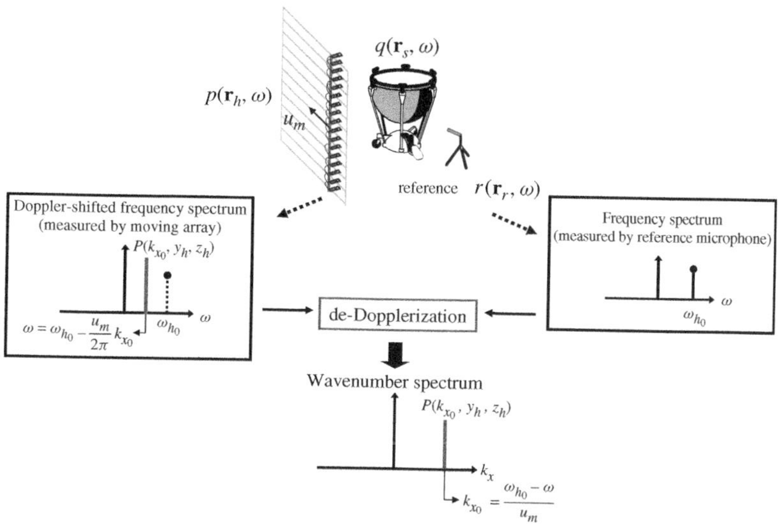

Figure 3.19 de-Dopplerization procedure for line spectrum

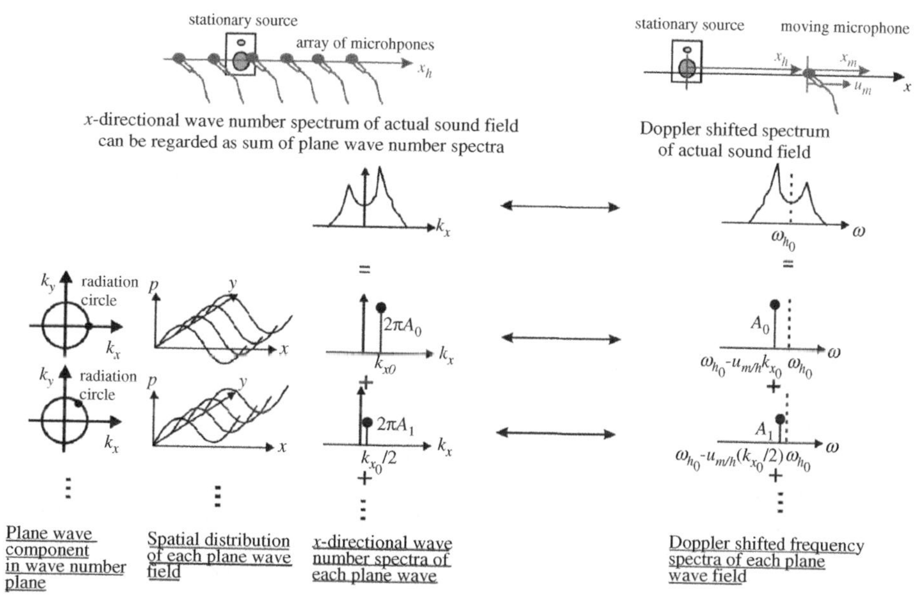

Figure 3.20 The continuous scanning method for more general case (one-dimensional illustration). © 2001. JASA. Reproduced, with permission, from [118]

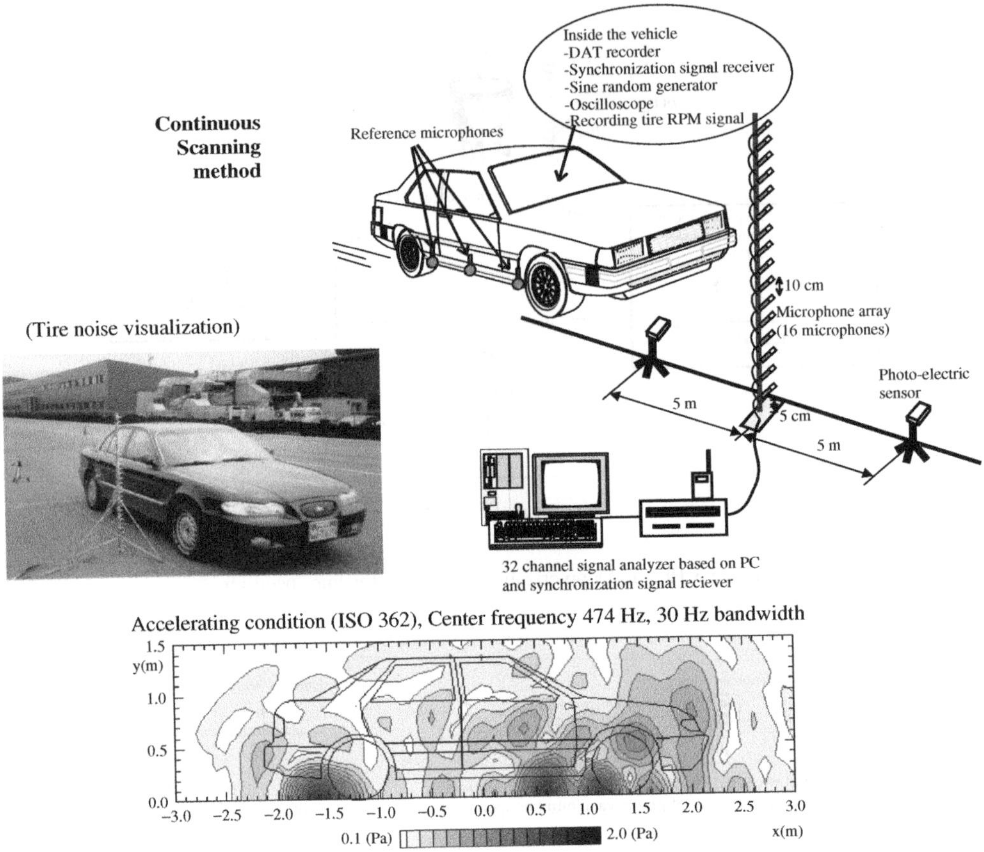

Figure 3.21 Applied result of the continuous scanning method to vehicle pass-by noise. The tire pattern noise distribution (pressure) on the source plane is shown when the vehicle is accelerated from 50 km/h. © 2001. JASA. Reproduced, with permission, from [118]

3.3.5 Analysis of Acoustic Holography

3.3.5.1 Partial Field Decomposition

Once we have a picture of the sound (acoustic holography), then we will immediately have questions about its meaning. What we have is usually a contour plot of sound pressure distribution or vector plot of sound intensity on a plane of our interest. This plot may help us to imagine where the sound source is and how it radiates into space with respect to a frequency of interest. However, in the strict sense, the only thing we can do from the two-dimensional expression of sound pressure or intensity distribution is to guess what was really there. We do not know, precisely, where the sound sources are (Figure 3.22).

It is also noteworthy that there are two types of sound sources. One is what we may call an "active sound source," and the other is a "passive sound source," as we mentioned earlier. The former is the source that radiates sound by itself. The latter is radiating but just reflecting sound. These two different types of sound sources can be distinguished, by eliminating reflected sound [125]. This is directly related to the way of treating the boundary condition on our analysis problem.

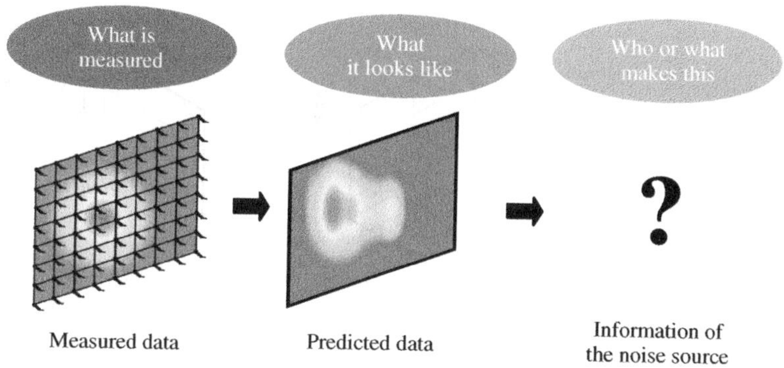

Figure 3.22 Illustration of analysis problem

The boundary condition of a locally reacting surface can be written as [125–127]

$$v(\mathbf{r}_s, \omega) = A(\mathbf{r}_s, \omega) p(\mathbf{r}_s, \omega) + s(\mathbf{r}_s, \omega). \tag{3.50}$$

$v(\mathbf{r}_s, \omega)$ and $p(\mathbf{r}_s, \omega)$ are the velocity and pressure on the wall. $A(\mathbf{r}_s, \omega)$ is the wall admittance and $s(\mathbf{r}_s, \omega)$ is the source strength on the wall. The active sound source is located at a position such that the source strength is not zero. This equation says that we can estimate the source strength if we measure the wall admittance. To do this, it is necessary to first turn off the source or sources, and then measure the wall admittance by putting a known source in the position we desire (Figure 3.23a). The next step is to turn on the sources and get the sound pressure and velocity distribution on the wall, using the admittance information (Figure 3.23b). This provides us what the location of the source and how strong it is (the source power). For example, see Figure 3.24.

Another very important problem is whether or not we can distinguish between independent or dependent sources: "two birds singing" or "one bird with two beaks" (Figure 3.25). This has a rather significant practical application. For example, to effectively control the noise sources, we only need to control independent noise sources. This can be done by using the statistical differences between the signals that are induced by the independent and dependent sound sources.

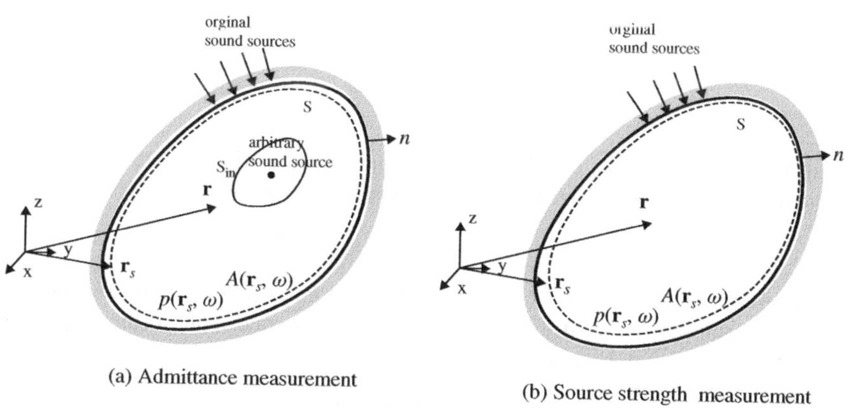

Figure 3.23 Two steps to separate the active and passive source

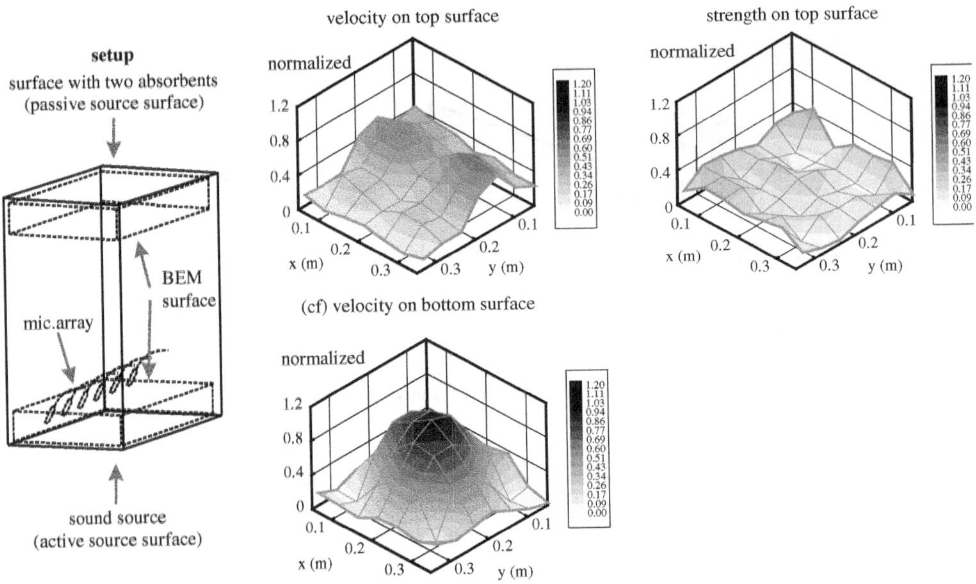

Figure 3.24 Experimental result that separates the active and passive source. The top surface is made by sound absorption material. The speaker on the bottom surface radiates sound. The top surface, which is reflecting sound, is eliminated by the separation

Figure 3.25 Spatially independent or dependent sources

For example, let us consider a two-input/single-output system (Figure 3.26). If the two inputs are independent, the spectrum $S_{pp}(\omega)$ of the output $p(\omega)$ can be expressed as

$$S_{pp}(\omega) = |h_1(\omega)|^2 S_{q1q1}(\omega) + |h_2(\omega)|^2 S_{q2q2}(\omega), \tag{3.51}$$

where $S_{q_\kappa q_\kappa}(\omega)$ is the spectrum of the κ-th input $q_\kappa(\omega)$ and $h_\kappa(\omega)$ is its transfer function. The first and second terms represent the contributions of the first and second input to the output spectrum, respectively. If we can obtain a signal as

$$w_1(\omega) = c_1(\omega) q_1(\omega), \tag{3.52}$$

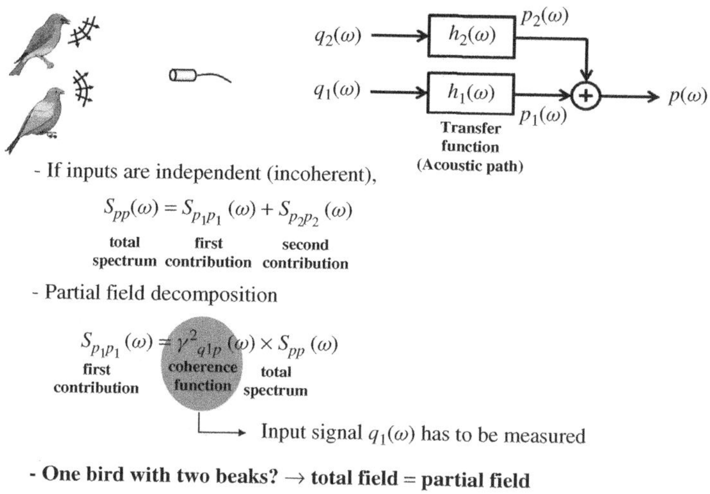

Figure 3.26 Two inputs/single output system and its partial field

then we can estimate the contribution of the first source as [128]

$$S_{p_1p_1}(\omega) = |h_1(\omega)|^2 S_{q_1q_1}(\omega) = \gamma^2_{w1p}(\omega) S_{pp}(\omega), \tag{3.53}$$

where $\gamma^2_{w1p}(\omega)$ is the coherence function between $w_1(\omega)$ and $q_1(\omega)$ (Figure 3.27).

We can simply extend Equation (3.53) to the case of multiple outputs. Acoustic holography is the case (Figure 3.28). The main tenet is how to obtain the signal that satisfies Equation (3.52). We can generally say that by putting sensors closer to the source or sources [56, 57, 129–131]. We may have a better signal that can be used to distinguish between independent or dependent sources. However, it is not well proved and not practical; it is not always easy to put the sensors close to the sources. Very recently, a method that does not require this [132, 133] was developed. Figure 3.29 explains the method's procedures. The first and second steps are the procedures of acoustic holography: measurement and prediction. The third step is to search for the maximum pressure on the source plane. This method assumes that the maximum pressure satisfies Equation (3.52). The fourth step is to estimate the contribution of the first source by using the coherence functions between the maximum pressure and other points like Equation (3.53). The fifth step is to calculate the remaining spectrum by subtracting the first contribution from the output spectrum. The steps are repeated until the contributions of other sources are estimated. For example, see Figures 3.30 and 3.31.

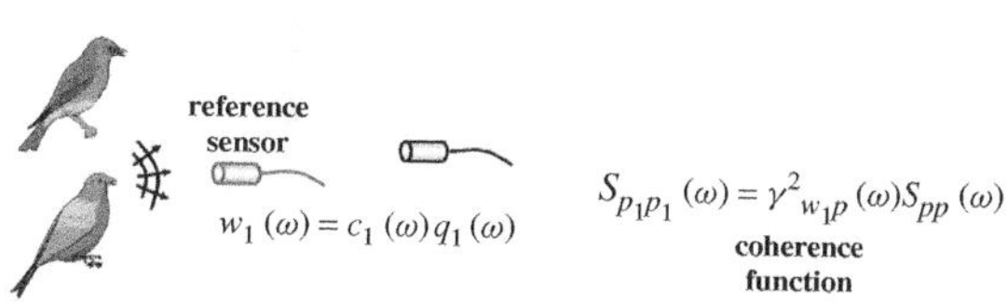

Figure 3.27 Conventional method to get partial field

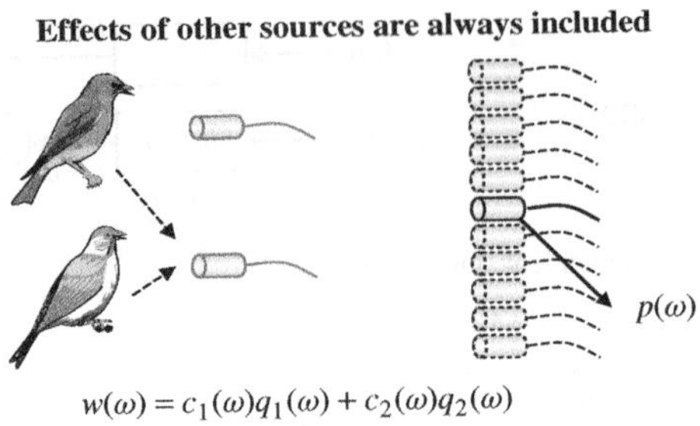

Figure 3.28 Acoustic holography and partial field decomposition

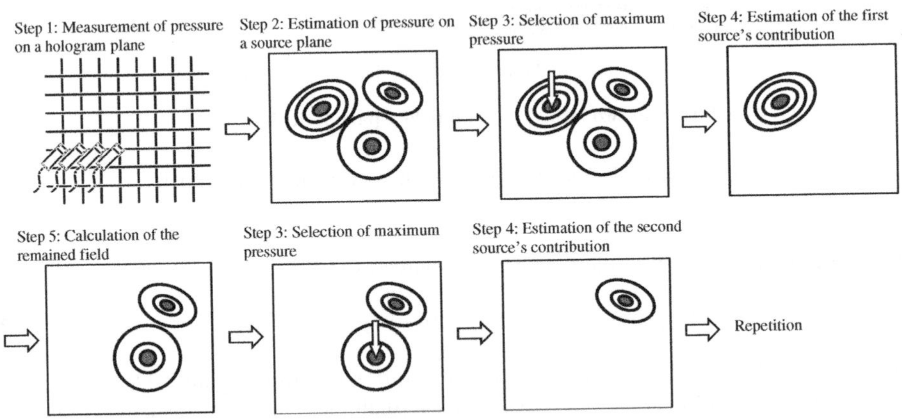

Figure 3.29 Procedure to separate the independent and dependent source

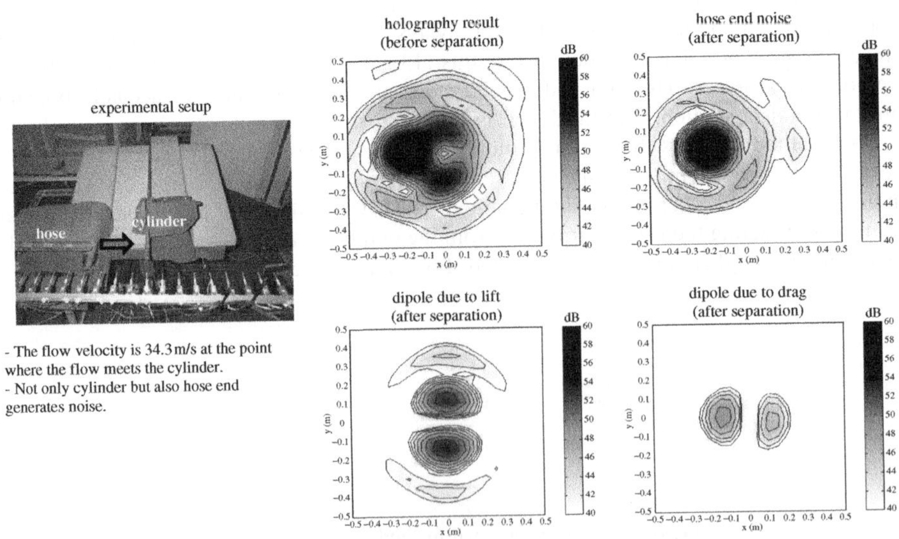

Figure 3.30 The separation method is applied to a vortex experiment. © 2004, JASA. Reproduced, with permission, from [134]

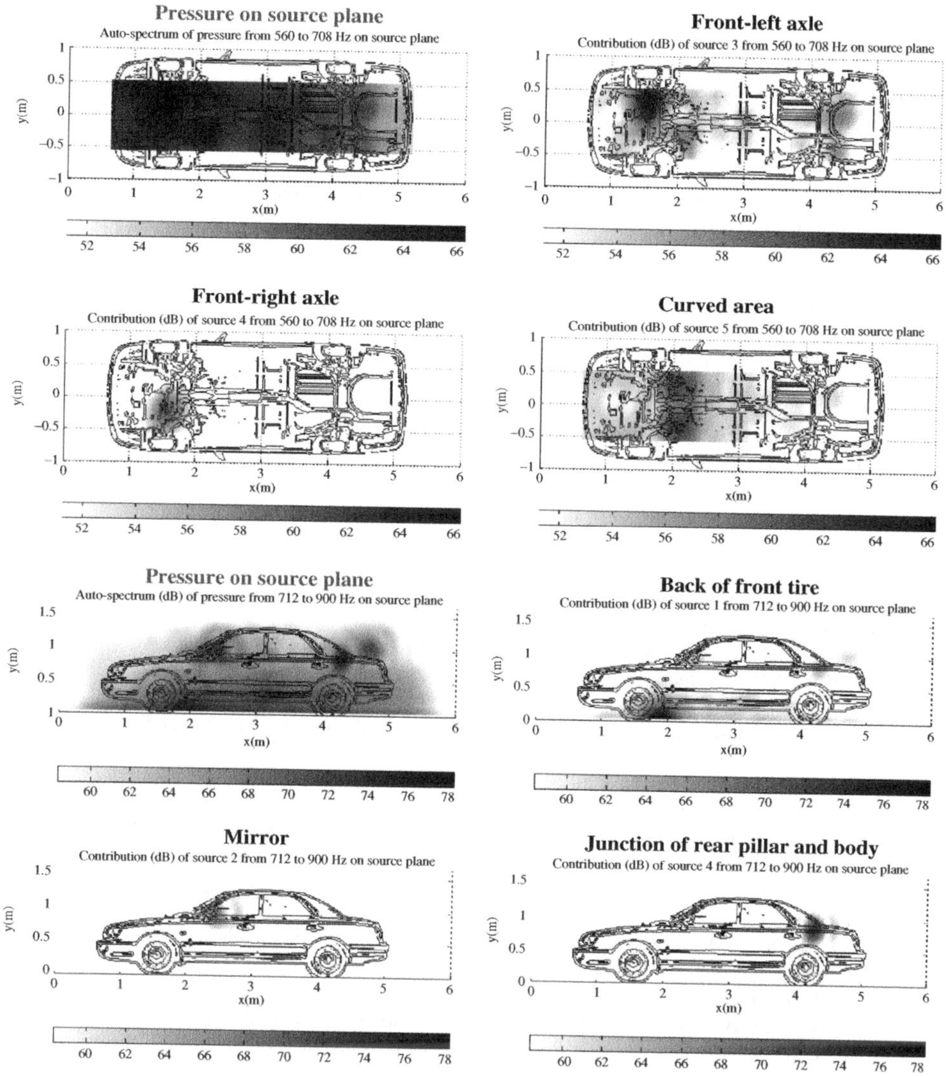

Figure 3.31 Partial field decomposition results of vehicle in wind tunnel. © 2004, JASA. Reproduced, with permission, from [116]

3.4 Summary

As expected, we cannot simply say "Yes," or "No" on the question of whether or not we can see the sound field. However, we now understand that the "analysis" of what we obtained, acoustic holography, has to be properly addressed. In the past, there has not been much attention given to this problem. We now know how better to obtain information from the sound picture. Making a picture is the job of acoustic holography, but the interpretation on the picture belongs to the observer. This chapter reviewed some useful guidelines for more correctly interpreting the sound field in order to derive the correct impression, or the right information from the picture.

References

1. Kac, M. (1966) Can one hear the shape of the drum? *American Mathematical Monthly*, **73**, 1–23.
2. Herman, G.T., Tuy, H.K. and Langenberg, K.J. (1987) *Basic Methods of Tomography and Inverse Problems*, Malvern Physics Series, Adam Hilge, Philadelphia, PA.
3. Bui, H. D. (1994) *Inverse Problems in the Mechanics of Materials: An Introduction*, CRC Press, Boca Raton, FL.
4. Kurpisz, K. and Nowak, A. J. (1995) *Inverse Thermal Problems*, Computational Mechanics Publications, Southampton, Boston, MA.
5. Kirsch, A. (1996) *An Introduction to the Mathematical Theory of Inverse Problems*, Springer-Verlag, New York.
6. Bertero, M. and Boccacci, P. (1998) *Introduction to Inverse Problems in Imaging*, IOP Publishing, Bristol, Philadelphia, PA.
7. Isakov, V. (1998) *Inverse Problems for Partial Differential Equations*, Springer-Verlag, New York.
8. Ghosh Roy, D. N. and Couchman, L. S. (2002) *Inverse Problems and Inverse Scattering of Plane Waves*, Academic Press, New York.
9. Hardamard, J. (1923) *Lectures on Cauchy's Problem in Linear Partial Differential Equations*, Yale University Press, New Haven, CT.
10. Landweber, L. (1951) An iteration formula for Fredholm integral equations of the first kind, *American Journal of Mathemathics*, **73**, 615–624.
11. Cormack, A.M. (1963) Representation of a function by its line integrals, with some radiological applications, *Journal of Applied Physics*, **34**, 2722–2727.
12. Cormack, A.M. (1964) Representation of a function by its line integrals, with some radiological applications II, *Journal of Applied Physics*, **35**, 2908–2913.
13. Tikhonov, A.N. and Arsenin, V.Y. (1977) *Solutions of Ill-Posed Problems*, Winston, Washington, DC.
14. Deans, S.R. (1983) *The Radon Transform and Some of Its Applications*, John Wiley & Sons, Inc., New York.
15. Louis, A.K. (1983) Mathematical problems of computerized tomography. *Proceedings of IEEE*, **71**, 379–389.
16. Protter, M.H. (1987) Can one hear the shape of a drum? Revisited. *Society for Industrial and Applied Mathematics Review*, **29**, 185–197.
17. Chadan, K. and Sabatier, P.C. (1989) *Inverse Problems in Quantum Scattering Theory*, 2nd edn, Springer-Verlag, New York.
18. Louis, A.K. (1992) Medical imaging: state of the art and future development, *Inverse Problems*, **8**, 709–738.
19. Colton, D. and Kress, R. (1998) *Inverse Acoustic and Electromagnetic Scattering Theory*, Chapter 1.19, 2nd edn, Springer-Verlag, New York.
20. (a) Gabor, D. (1948) A new microscopic principle, *Nature*, **161**, 777; (b) which is cited by Hilderbrand, B. P. and Brenden, B. B. (1972) *An Introduction to Acoustical Holography*, Plenum Press, New York.
21. Leith, E. N. and Upatnieks, J. (1962) Reconstructed wavefronts and communication theory, *Journal of the Optical Society of America*, **52**, 1123–1130.
22. Kock, W.E. (1966) Hologram television. *Proceedings of IEEE*, **54**, 331.
23. Tricoles, G. and Rope, E.L. (1967) Reconstructions of visible images from reduced-scale replicas of microwave holograms, *Journal of the Optical Society of America*, **57**, 97–99.
24. Sherman, G.C. (1967) Reconstructed wave forms with large diffraction angles, *Journal of the Optical Society of America*, **57**, 1160–1161.
25. Goodman, J.W. and Lawrence, R.W. (1967) Digital image formation from electronically detected holograms, *Applied Physics Letters*, **11**, 77–79.
26. Aoki, Y. (1967) Microwave holograms and optical reconstruction, *Applied Optics*, **6**, 1943–1946.
27. Porter, R.P. (1970) Diffraction-limited, scalar image formation with holograms of arbitrary shape, *Journal of the Optical Society of America*, **60**, 1051–1059.
28. Wolf, E. (1970) Determination of the amplitude and the phase of scattered fields by holography, *Journal of the Optical Society of America*, **60**, 18–20.
29. Carter, W.H. (1970) Computational reconstruction of scattering objects from holograms, *Journal of the Optical Society of America*, **60**, 306–314.

30. Williams, E.G., Maynard, J.D. and Skudrzyk, E. (1980) Sound source reconstruction using a microphone array, *Journal of the Acoustical Society of America*, **68**, 340–344.
31. Williams, E.G. and Maynard, J.D. (1980) Holographic imaging without the wavelength resolution limit, *Physics Review Letters*, **45**, 554–557.
32. Ash, E.A. and Nichols, G. (1972) Super-resolution aperture scanning microscope, *Nature*, **237**, 510–512.
33. Pillai, S.U. (1989) *Array Signal Processing*, Springer-Verlag, New York.
34. Johnson, D.H. and Dudgeon, D.E. (1993) *Array Signal Processing Concepts and Techniques*, Prentice-Hall, Englewood Cliffs, NJ.
35. Kaveh, M. and Barabell, A.J. (1986) The statistical performance of the MUSIC and the minimum-Norm algorithms in resolving plane waves in noise. *IEEE Transactions on Acoustics, Speech and Signal Processing*, **34**, 331–341.
36. Lasky, M. (1977) Review of undersea acoustics to 1950, *Journal of the Acoustical Society of America*, **61**, 283–297.
37. Barskow, B., King, W.F. and Pfizenmaier, E. (1987) Wheel/rail noise generated by a high-speed train investigated with a line array of microphones, *Journal of Sound and Vibration*, **118**, 99–122.
38. Hald, J. and Christensen, J.J. (2002) A class of optimal broadband phased array geometries designed for easy construction. Proceedings of Inter-Noise 2002.
39. Takano, Y. (2003) Development of visualization system for high-speed noise sources with a microphone array and a visual sensor. Proceedings of Inter-Noise 2003.
40. Elias, G. (1995) Source localization with a two-dimensional focused array: optimal signal processing for a cross-shaped array. Proceedings of Inter-Noise 95, pp. 1175–1178.
41. Nordborg, A., Martens, A., Wedemann, J., and Willenbrink, L. (2001) Wheel/rail noise separation with microphone array measurements. Proceedings of Inter-Noise 2001, pp. 2083–2088.
42. Nordborg, A., Wedemann, J., and Willenbrink, L. (2000) Optimum array microphone configuration. Proceedings of Inter-Noise 2000.
43. Stepanishen, P.R. and Benjamin, K.C. (1982) Forward and backward prediction of acoustic fields using FFT methods, *Journal of the Acoustical Society of America*, **71**, 803–812.
44. Williams, E.G. and Maynard, J.D. (1982) Numerical evaluation of the Rayleigh integral for planar radiators using the FFT, *Journal of the Acoustical Society of America*, **72**, 2020–2030.
45. Maynard, J.D., Williams, E.G. and Lee, Y. (1985) Nearfield acoustic holography (NAH): I. Theory of generalized holography and the development of NAH, *Journal of the Acoustical Society of America*, **78**, 1395–1413.
46. Veronesi, W.A. and Maynard, J.D. (1987) Nearfield acoustic holography (NAH) II. Holographic reconstruction algorithms and computer implementation, *Journal of the Acoustical Society of America*, **81**, 1307–1322.
47. Hayek, S.I. and Luce, T.W. (1988) Aperture effects in planar nearfield acoustical imaging, *Journal of Vibration, Acoustics, Stress and Reliability in Design*, **110**, 91–96.
48. Sarkissian, A., Gaumond, C.F., Williams, E.G. and Houston, B.H. (1993) Reconstruction of the acoustic field over a limited surface area on a vibrating cylinder, *Journal of the Acoustical Society of America*, **93**, 48–54.
49. Hald, J. (1994) Reduction of spatial windowing effects in acoustical holography. Proceedings of Inter-Noise 94, pp. 1887–1890.
50. Kwon, H.-S. and Kim, Y.-H. (1995) Minimization of bias error due to windows in planar acoustic holography using a minimum error window, *Journal of the Acoustical Society of America*, **98**, 2104–2111.
51. Saijou, K. and Yoshikawa, S. (2001) Reduction methods of the reconstruction error for large-scale implementation of nearfield acoustical holography, *Journal of the Acoustical Society of America*, **110**, 2007–2023.
52. Nam, K.-U. and Kim, Y.-H. (1999) Errors due to sensor and position mismatch in planar acoustic holography, *Journal of the Acoustical Society of America*, **106**, 1655–1665.
53. Weinreich, G. and Arnold, E.B. (1980) Method for measuring acoustic radiation fields, *Journal of the Acoustical Society of America*, **68**, 404–411.
54. Williams, E.G., Dardy, H.D. and Fink, R. G. (1985) Nearfield acoustical holography using an underwater, automated scanner, *Journal of the Acoustical Society of America*, **78**, 789–798.
55. Blacodon, D., Candel, S.M. and Elias, G. (1987) Radial extrapolation of wave fields from synthetic measurements of the nearfield, *Journal of the Acoustical Society of America*, **82**, 1060–1072.

56. Hald, J. (1989) STSF – A Unique Technique for Scan-Based Nearfield Acoustic Holography without Restrictions on Coherence. Technical Review No. 1, Brüel & Kjær.
57. Ginn, K.B. and Hald, J. (1989) STSF – Practical Instrumentation and Applications. Technical Review No. 2, Brüel & Kjær.
58. Yoon, S.H. and Nelson, P.A. (2000) A method for the efficient construction of acoustic pressure cross-spectral matrices, *Journal of Sound and Vibration*, **233**, 897–920.
59. Yoon, S.H. and Nelson, P.A. (2000) Estimation of acoustic source strength by inverse methods: part II, experimental investigation of methods for choosing regularization parameters, *Journal of Sound and Vibration*, **233**, 669–705.
60. Kwon, H.-S., Kim, Y.-J. and Bolton, J.S. (2003) Compensation for source nonstationarity in multireference, scan-based nearfield acoustical holography, *Journal of the Acoustical Society of America*, **113**, 360–368.
61. Nam, K.-U. and Kim, Y.-H. (2003) Low coherence acoustic holography. Proceedings of Inter-Noise 2003.
62. Kwon, H.-S. and Kim, Y.-H. (1998) Moving frame technique for planar acoustic holography, *Journal of the Acoustical Society of America*, **103**, 1734–1741.
63. Park, S.-H. and Kim, Y.-H. (1998) An improved moving frame acoustic holography for coherent band-limited noise, *Journal of the Acoustical Society of America*, **104**, 3179–3189.
64. Park, S.-H. and Kim, Y.-H. (2000) Effects of the speed of moving noise sources on the sound visualization by means of moving frame acoustic holography, *Journal of the Acoustical Society of America*, **108**, 2719–2728.
65. Park, S.-H. and Kim, Y.-H. (2001) Visualization of pass-by noise by means of moving frame acoustic holography, *Journal of the Acoustical Society of America*, **110**, 2326–2339.
66. Candel, S.M. and Chassaignon, C. (1984) Radial extrapolation of wave fields by spectral methods, *Journal of the Acoustical Society of America*, **76**, 1823–1828.
67. Williams, E.G., Dardy, H.D. and Washburn, K.B. (1987) Generalized nearfield acoustical holography for cylindrical geometry: theory and experiment, *Journal of the Acoustical Society of America*, **81**, 389–407.
68. Gardner, B.K. and Bernhard, R.J. (1988) A noise source identification technique using an inverse Helmholtz integral equation method, *Journal of Vibration, Acoustics, Stress and Reliability in Design*, **110**, 84–90.
69. Williams, E.G., Houston, B.H. and Bucaro, J.A. (1989) Broadband nearfield acoustical holography for vibrating cylinders, *Journal of the Acoustical Society of America*, **86**, 674–679.
70. Koopmann, G.H., Song, L. and Fahnline, J.B. (1989) A method for computing acoustic fields based on the principle of wave superposition, *Journal of the Acoustical Society of America*, **86**, 2433–2438.
71. Sarkissan, A. (1990) Nearfield acoustic holography for an axisymmetric geometry: a new formulation, *Journal of the Acoustical Society of America*, **88**, 961–966.
72. Tamura, M. (1990) Spatial Fourier transform method of measuring reflection coefficients at oblique incidence. I: theory and numerical examples, *Journal of the Acoustical Society of America*, **88**, 2259–2264.
73. Song, L., Koopmann, G.H. and Fahnline, J.B. (1991) Numerical errors associated with the method of superposition for computing acoustic fields, *Journal of the Acoustical Society of America*, **89**, 2625–2633.
74. Villot, M., Chaveriat, G., and Ronald, J. (1992) Phonoscopy: an acoustical holography technique for plane structures radiating in enclosed spaces, *Journal of the Acoustical Society of America*, **91**, 187–195.
75. Hallman, D.L. and Bolton, J.S. (1993) Multi-reference nearfield acoustical holography in reflective environments. Proceedings of Inter-Noise 93, pp. 1307–1310.
76. Cheng, M.-T., Mann, J.A. III, and Pate, A. (1995) Wave-number domain separation of the incident and scattered sound field in Cartesian and cylindrical coordinates, *Journal of the Acoustical Society of America*, **97**, 2293–2303.
77. Tamura, M., Allard, J.F. and Lafarge, D. (1995) Spatial Fourier-transform method for measuring reflection coefficients at oblique incidence. II. Experimental results, *Journal of the Acoustical Society of America*, **97**, 2255–2262.
78. Wang, Z. and Wu, S.F. (1997) Helmholtz equation – least-squares method for reconstructing the acoustic pressure field, *Journal of the Acoustical Society of America*, **102**, 2020–2032.
79. Wu, S.F. and Yu, J. (1998) Reconstructing interior acoustic pressure fields via Helmholtz equation least-squares method, *Journal of the Acoustical Society of America*, **104**, 2054–2060.
80. Kang, S.-C. and Ih, J.-G. (2000) The use of partially measured source data in nearfield acoustical holography based on the BEM, *Journal of the Acoustical Society of America*, **107**, 2472–2479.

81. Wu, S.F. (2000) On reconstruction of acoustic pressure fields using the Helmholtz equation least squares method, *Journal of the Acoustical Society of America*, **107**, 2511–2522.
82. Rayess, N. and Wu, S.F. (2000) Experimental validation of the HELS method for reconstructing acoustic radiation from a complex vibrating structure, *Journal of the Acoustical Society of America*, **107**, 2955–2964.
83. Zhang, Z., Vlahopoulos, N., Raveendra, S.T., Allen, T. and Zhang, K.Y. (2000) A computational acoustic field reconstruction process based on an indirect boundary element formulation, *Journal of the Acoustical Society of America*, **108**, 2167–2178.
84. Kang, S.-C. and Ih, J.-G. (2000) On the accuracy of nearfield pressure predicted by the acoustic boundary element method, *Journal of Sound and Vibration*, **233**, 353–358.
85. Kang, S.-C. and Ih, J.-G. (2001) Use of nonsingular boundary integral formulation for reducing errors due to nearfield measurements in the boundary element method based nearfield acoustic holography, *Journal of the Acoustical Society of America*, **109**, 1320–1328.
86. Wu, S.F., Rayess, N. and Zhao, X. (2001) Visualization of acoustic radiation from a vibrating bowling ball, *Journal of the Acoustical Society of America*, **109**, 2771–2779.
87. Maynard, J.D. (2003) A new technique combining eigenfunction expansions and boundary elements to solve acoustic radiation problems. Proceedings of Inter-Noise 2003.
88. Williams, E.G. (1999) *Fourier Acoustics: Sound Radiation and Nearfield Acoustical Holography*, Academic Press, London.
89. Thurstone, F.L. (1966) Ultrasound holography and visual reconstruction. *Proceedings of International Symposium on Biomedical Engineering*, **1**, 12–15.
90. Boyer, A.L., Jordan, J.A. Jr., Van Rooy, D.L. *et al.* (1969) Computer reconstruction of images from ultrasonic holograms. Proceedings of 2nd International Symposium on Acoustical Holography.
91. Metilerell, A.F., El-Sum, H.M.A., Dreher, J.J. and Larmore, L. (1967) Introduction to acoustical holography, *Journal of the Acoustical Society of America*, **42**, 733–742.
92. Cram, L.A. and Rossiter, K.O. (1969) Long-wavelength holography and visual reproduction methods. Proceedings of 2nd International Symposium on Acoustical Holography.
93. Graham, T.S. (1969) A new method for studying acoustic radiation using long-wavelength acoustical holography. Proceedings of 2nd International Symposium on Acoustical Holography.
94. Watson, E.E. (1973) Detection of acoustic sources using long-wavelength acoustical holography, *Journal of the Acoustical Society of America*, **54**, 685–691.
95. Fleischer, H. and Axelrad, V. (1986) Restoring an acoustic source from pressure data using Wiener filtering, *Acoustica*, **60**, 172–175.
96. Williams, E.G. (1995) Supersonic acoustic intensity, *Journal of the Acoustical Society of America*, **97**, 121–127.
97. Bai, M.R. (1995) Acoustical source characterization by using recursive Wiener filtering, *Journal of the Acoustical Society of America*, **97**, 2657–2663.
98. Lee, J.C. (1996) Spherical acoustical holography of low-frequency noise sources, *Applied Acoustics*, **48**, 85–95.
99. Carroll, G.P. (1999) The effect of sensor placement errors on cylindrical nearfield acoustic holography, *Journal of the Acoustical Society of America*, **105**, 2269–2276.
100. Veronesi, W.A. and Maynard, J.D. (1989) Digital holographic reconstruction of sources with arbitrarily shaped surfaces, *Journal of the Acoustical Society of America*, **85**, 588–598.
101. Borgiotti, G.V., Sarkissan, A., Williams, E.G., and Schuetz, L. (1990) Conformal generalized nearfield acoustic holography for axisymmetric geometries, *Journal of the Acoustical Society of America*, **88**, 199–209.
102. Photiadis, D.M. (1990) The relationship of singular value decomposition to wave-vector filtering in sound radiation problems, *Journal of the Acoustical Society of America*, **88**, 1152–1159.
103. Kim, G.-T. and Lee, B.-H. (1990) 3-D sound source reconstruction and field reprediction using the Helmholtz integral equation, *Journal of Sound and Vibration*, **136**, 245–261.
104. Sarkissan, A. (1991) Acoustic radiation from finite structures, *Journal of the Acoustical Society of America*, **90**, 574–578.
105. Fahnline, J.B. and Koopmann, G.H. (1991) A numerical solution for the general radiation problem based on the combined methods of superposition and singular value decomposition, *Journal of the Acoustical Society of America*, **90**, 2808–2819.

106. Bai, M.R. (1992) Application of BEM (boundary element method)-based acoustic holography to radiation analysis of sound sources with arbitrarily shaped geometries, *Journal of the Acoustical Society of America*, **92**, 533–549.
107. Borgiotti, G.V. and Rosen, E.M. (1992) The determination of the farfield of an acoustic radiator from sparse measurement samples in the nearfield, *Journal of the Acoustical Society of America*, **92**, 807–818.
108. Kim, B.-K. and Ih, J.-G. (1996) On the reconstruction of the vibro-acoustic field over the surface enclosing an interior space using the boundary element method, *Journal of the Acoustical Society of America*, **100**, 3003–3016.
109. Kim, B.-K. and Ih, J.-G. (2000) Design of an optimal wave-vector filter for enhancing the resolution of reconstructed source field by nearfield acoustical holography (NAH), *Journal of the Acoustical Society of America*, **107**, 3289–3297.
110. Nelson, P.A. and Yoon, S.H. (2000) Estimation of acoustic source strength by inverse methods: Part I, conditioning of the inverse problem, *Journal of Sound and Vibration*, **233**, 643–668.
111. Williams, E.G. (2001) Regularization methods for nearfield acoustical holography, *Journal of the Acoustical Society of America*, **110**, 1976–1988.
112. Wu, S.F. and Zhao, X. (2002) Combined Helmholtz equation – least squares method for reconstructing acoustic radiation from arbitrarily shaped objects, *Journal of the Acoustical Society of America*, **112**, 179–188.
113. Schuhmacher, A., Hald, J., Rasmussen, K.B. and Hansen, P.C. (2003) Sound source reconstruction using inverse boundary element calculations, *Journal of the Acoustical Society of America*, **113**, 114–127.
114. Kim, Y. and Nelson, P.A. (2003) Spatial resolution limits for the reconstruction of acoustic source strength by inverse methods, *Journal of Sound and Vibration*, **265**, 583–608.
115. Lee, M. and Bolton, J.S. (2007) Source characterization of a subsonic jet by using nearfield acoustical holography, *Journal of the Acoustical Society of America*, **121**, 967–977.
116. Nam, K.-U. (2004) Application of acoustic holography to the sound field of low coherence. PhD thesis. KAIST, Daejeon, Republic of Korea.
117. Park, S.-H. (2001) Visualization of noise generated by moving band-limited noise sources by means of an improved moving frame acoustic holograph. PhD thesis. KAIST, Daejeon, Republic of Korea.
118. Kwon, H.-S. and Bolton, J.S. (1998) Partial field decomposition in nearfield acoustical holography by the use of singular value decomposition and partial coherence procedures. Proceedings of Noise-Con 98, pp. 649–654.
119. Yang, D., Wang, Z., Li, B., Luo, Y. and Lian, X. (2011) Quantitative measurement of pass-by noise radiated by vehicles running at high speeds, *Journal of Sound and Vibration*, **330**, 1352–1364.
120. Yang, D., Guo, X., Shi, S. and Ma, J. (2011) The study of joint acoustic holography algorithms based on continuous scanning, *International Journal of Intelligent Systems and Applications*, **3**, 9–16.
121. Bo, H. (2010) Underwater moving noise sources identification based on acoustic holography. Proceedings of International Conference on Intelligent Computing and Intelligent Systems, Vol. 3, pp. 702–705.
122. Ruhala, R.J. and Swanson, D.C. (2002) Planar nearfield acoustical holography in a moving medium, *Journal of the Acoustical Society of America*, **113**, 420–429.
123. Kim, Y.-J. and Niu, Y. (2012) Improved statistically optimal nearfield acoustical holography in subsonically moving fluid medium, *Journal of Sound and Vibration*, **331**, 3945–3960.
124. Kwon, H.-S., Niu, Y. and Kim, Y.-J. (2010) Planar nearfield acoustical holography in moving fluid medium at subsonic and uniform velocity, *Journal of the Acoustical Society of America*, **128**, 1823–1832.
125. Kim, Y.-K. and Kim, Y.-H. (1999) Holographic reconstruction of active sources and surface admittance in an enclosure, *Journal of the Acoustical Society of America*, **105**, 2377–2383.
126. Morse, P.M. and Feshbach, H. (1992) *Methods of Theoretical Physics, Part I*, McGraw-Hill, New York.
127. Beranek, L.L. and Ver, I.L. (1992) *Noise and Vibration Control Engineering*, John Wiley & Sons, Inc., New York.
128. Bendat, J.S. and Piersol, A.G. (1986) *Random Data: Analysis and Measurement Procedures*, 2nd edn, John Wiley & Sons, Inc., New York.
129. Hallman, D. and Bolton, J.S. (1992) Multi-reference nearfield acoustical holography. Proceedings of Inter-Noise 92, pp. 1165–1170.
130. Ruhala, R.J. and Burroughs, C.B. (1998) Separation of leading edge, trailing edge, and sidewall noise sources from rolling tires. Proceedings of Noise-Con 98, pp. 109–114.

131. Tomlinson, M.A. (1999) Partial source discrimination in nearfield acoustic holography, *Applied Acoustics*, **57**, 243–261.
132. Nam, K.-U. and Kim, Y.-H. (2001) Visualization of multiple incoherent sources by the backward prediction of nearfield acoustic holography, *Journal of the Acoustical Society of America*, **109**, 1808–1816.
133. Nam, K.-U., Kim, Y.-H., Choi, Y.-C. *et al.* (2002) Visualization of speaker, vortex shedding, engine, and wind noise of a car by partial field decomposition. Proceedings of Inter-Noise 2002.
134. Nam, K.-U. and Kim, Y.-H. (2004) A partial field decomposition algorithm and its examples for nearfield acoustic holography, *Journal of the Acoustical Society of America*, **116**, 172–185.

4

Beamforming

4.1 Introduction

The pressure field measured by a microphone array contains information relevant to the wavefront that is propagated in space and time. The spatiotemporal information of a sound field can be compactly represented by a few parameters, such as the locations of sound sources, strengths of the sources, or shapes of the wavefronts. The parametric approach searches for these essential parameters in the given measurement data. The parametric label is derived from the fact that these kinds of methods model the measured signal using certain parameters.

The beamforming technique is a representative parametric approach to visualize the sound field. Conventional beamforming attempts to analyze the measured wavefronts by introducing a scan vector, which extracts important parameters related to the source location. This technique visualizes sound by changing a parameter – for example, the direction of the sound source – in the assumed domain.

The performance of the beamforming technique is directly related to the accuracy and robustness of the parameter estimation. As will be explained throughout this chapter, accuracy refers to the ability to estimate the position of a sound source with minimal error and is also connected to the ability to discriminate sources closely positioned in space. Robustness is a measure of consistency of parameter estimation in the presence of noise. The first part of this chapter deals with a method for designing a scan vector that can satisfy both requirements. The scan vector can be designed from either the sound wave propagation model or an analysis of the measured data. As the former does not depend on measured data, it can be considered to be a data-independent approach. The well-known delay-and-sum (DAS) beamformer falls into this category. The latter approach examines the structure of measured data to obtain an enhanced resolution. To analyze the measurement data, a correlation matrix representing the spatial relationship between microphone signals can provide a significant amount of useful information. The signal-to-noise ratio (SNR) of the measured correlation matrix determines the robustness of the estimation. Many representative techniques for improving the estimator performance in regard to the SNR are introduced.

The second half of this chapter is devoted to practical examples exploiting the unique characteristics of the sound sources and measurement geometry. The localization of impulsive sound sources, moving sources such as automobiles and high-speed trains, and post-processing techniques based on the deconvolution of an acoustic image are addressed.

Sound Visualization and Manipulation, First Edition. Yang-Hann Kim and Jung-Woo Choi.

4.2 Problem Statement

When sound propagated from an arbitrary source is measured by an array of sensors, the output signals may be aligned with specific time delays. These time delays are related to the distance between the position of a source and the position of the sensors (Figure 4.1). If the delays in sensor signals are compensated exactly, then compensated signals reinforce each other such that their summation can be maximized. Conventional beamforming methods estimate the angle of arrival for a plane wave or the location of a sound source for a spherical wave by finding the location or angle for which the sum of the signals is maximized.

To explain the concept in more detail, consider a pure-tone sound field of frequency ω that is measured by M microphones linearly arranged in space. The measured complex pressure field can be expressed in vector form as

$$\mathbf{p}(\omega) = [p_1(\omega) \quad p_2(\omega) \quad \cdots \quad p_M(\omega)]^T, \tag{4.1}$$

where the superscript T represents the transpose of a vector or matrix.

This field contains information about the wavefront propagated from a single sound source or a set of sources. The beamforming method assumes that the wavefront can be expressed by a combination of parameters. A simple way to extract a certain parameter α from the given measurement data is to utilize a scan vector that depends on α (Figure 4.2):

$$\mathbf{w}(\alpha, \omega) = [w_1(\alpha, \omega) \; w_2(\alpha, \omega) \; \cdots \; w_M(\alpha, \omega)]^T. \tag{4.2}$$

The scan vector is designed such that the assumed wavefront can be generated by a modeled source at a specific location. The parameter α can be any variable related to the location of a modeled sound source. For example, if we want to know the angle of arrival (θ) of a plane wave source, then such a scan vector could be

$$w_m(\theta, \omega) = \frac{1}{\overline{w}} e^{-i\,k(m-1)\,d\sin\theta} = \frac{1}{M} e^{-i\omega\Delta\tau(m-1)\sin\theta}, \tag{4.3}$$

where m is the index of each microphone ($m = 1, \cdots, M$). Here, $\Delta\tau = d/c$ indicates the time elapsed for a sound wave to travel one microphone spacing d. Therefore, the scan vector represents a time delay operator that delays each microphone signal by a different amount ($\tau = (m-1)\Delta\tau \sin\theta$). Dividing by the total number of microphones ($\overline{w} = M$) normalizes the scan vector such that the beamforming result is invariant with respect to M.

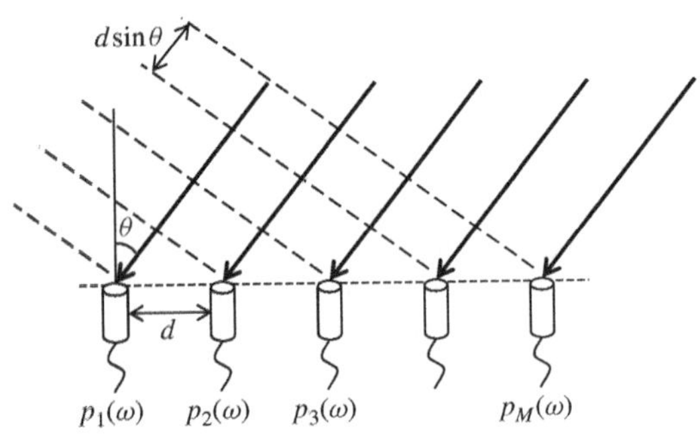

Figure 4.1 Plane wave propagation model

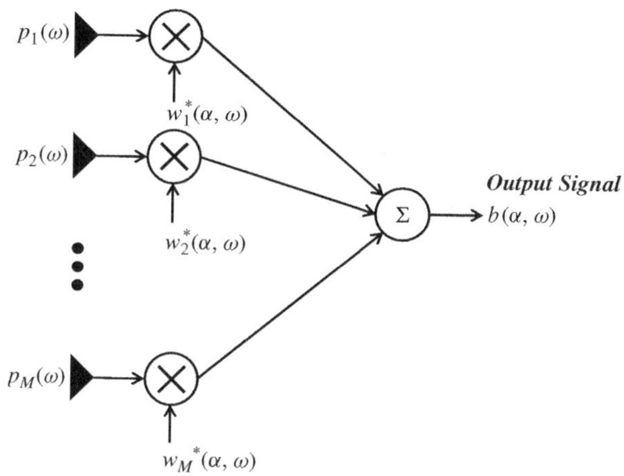

Figure 4.2 General beamforming system

Using the defined scan vector and measured pressure signals, the *beamformer output* can be defined as

$$b(\theta, \omega) = \mathbf{w}(\theta, \omega)^H \mathbf{p}(\omega) \tag{4.4}$$

with superscript H being the Hermitian transpose (complex-conjugate transpose) of a matrix.

The beamformer output of the plane wave model of Equation (4.3) is given by the delayed sum of the microphone array signal, and hence, this type of beamforming is often referred to as a *delay-and-sum technique.*

The expected value of the squared beamformer output is called the *beamforming power*. The beamforming power is a useful measure for visualizing the sound power distribution at the assumed source locations. From Equation (4.4), the beamforming power can be written as

$$\begin{aligned}\beta(\theta, \omega) &= E\left[|b(\theta, \omega)|^2\right] \\ &= \mathbf{w}(\omega)^H \mathbf{R}(\omega)\mathbf{w}(\omega), \qquad \text{where } \mathbf{R}(\omega) = E\left[\mathbf{p}(\omega)\,\mathbf{p}(\omega)^H\right].\end{aligned} \tag{4.5}$$

The expectation operator E is introduced here to express the expected value when the measured signal $p_m(\omega)$ is a random variable.

The beamforming power of Equation (4.5) essentially shows how well the scan vector matches the measured pressure vector of Equation (4.1). As illustrated in Figure 3.3, the scan vector estimates its contribution to the measured pressure $\mathbf{p}$ by changing the parameter, which in this case is the angle θ. The direction of estimation (θ) is often denoted as a *steering angle, bearing angle*, or *look direction.* The vector $\mathbf{w}$ defined for the particular steering angle is a basis function in this case, and its contribution to $\mathbf{p}$ is identified by the inner product of Equation (4.4). Therefore, it is clear that the performance of the beamforming method depends on how well the scan vector matches the measured pressure $\mathbf{p}$.

It is also noteworthy that the structure of $\mathbf{R}$ in Equation (4.5) also determines the performance of the selected beamforming. Each element of $\mathbf{R}$ represents the correlation of two microphone signals in the frequency domain. That is,

$$[\mathbf{R}(\omega)]_{(m,n)} = E\left[p_m(\omega)\, p_n(\omega)^*\right], \tag{4.6}$$

which is equivalent to the correlation of two random process p_m and p_n. Accordingly, the $M \times M$ matrix $\mathbf{R}$ is called a *correlation matrix* of the measured signal. For the special case in which p_m and p_n are zero-mean random variables, Equation (4.6) becomes the covariance, and hence $\mathbf{R}$ is sometimes referred to as a covariance matrix.

In the following sections, we introduce various types of beamformers using different scan vectors and correlation matrices. Representative scan vectors exploiting plane wave and spherical wave models are explained, and their advantages and disadvantages are discussed for different measurement conditions. Various techniques that enhance the beamforming performance by modifying the structure of $\mathbf{R}$, such as the multiple signal classification (MUSIC) [1] method and the estimation of parameters via rotational invariance technique (ESPRIT) [2, 3], are also demonstrated.

4.3 Model-Based Beamforming

4.3.1 Plane and Spherical Wave Beamforming

4.3.1.1 Planar Wavefront Model

The scan vector $\mathbf{w}$ is normally selected based on the model or on an assumption of which type of wavefront will be measured. For example, if the source is far from the array, then the direction of propagation is approximately equal at each sensor, and the propagating field within the array aperture can be regarded as a plane wave. Under such conditions, the plane wave model of Equation (4.3) agrees well with what we actually measure. The complex pressure signal of a plane wave source measured at the mth microphone is given by

$$p_m(\omega) = q(\omega)e^{-i\omega(m-1)(d/c)\sin\theta_0} = q(\omega)h_m(\theta_0, \omega), \tag{4.7}$$

where θ_0 is the incident angle of the plane wave. The amplitude of a plane wave source $q(\omega)$ is often called the *source excitation signal*. The transfer function between the mth microphone signal and the source excitation q is denoted by $h_m(\theta_0, \omega)$. When the measured signal is multiplied by the scan vector of Equation (4.3), the beamformer output in the direction θ can be written as

$$b(\theta, \omega) = \sum_{m=1}^{M} p_m(\omega)w_m(\theta, \omega)^* = \frac{q(\omega)}{M}\sum_{m=1}^{M} e^{-i\omega(m-1)(d/c)(\sin\theta_0-\sin\theta)}. \tag{4.8}$$

Using the trigonometric identity

$$\sum_{m=0}^{M-1} e^{-im\Omega} = \frac{\sin(M\Omega/2)}{\sin(\Omega/2)}e^{-i(M-1)\Omega/2},$$

the beamformer output can be reduced to

$$b(\theta, \omega) = \frac{q(\omega)}{M}\frac{\sin(M(\sin\theta - \sin\theta_0)\varphi/2)}{\sin((\sin\theta - \sin\theta_0)\varphi/2)}e^{i(M-1)(\sin\theta-\sin\theta_0)\varphi/2}, \tag{4.9}$$

where $\varphi = \omega d/c$ is the phase change after the wave has traveled a distance d. If φ is small, that is, d is much smaller than the wavelength, then the beamformer output can be approximated by

$$b(\theta, \omega) \approx q(\omega)\text{sinc}(M(\sin\theta - \sin\theta_0)\varphi/2)e^{i(M-1)(\sin\theta-\sin\theta_0)\varphi/2}. \tag{4.10}$$

The corresponding beamforming power is given by

$$\beta(\theta, \omega) \approx |q(\omega)|^2\text{sinc}^2(M(\sin\theta - \sin\theta_0)\varphi/2). \tag{4.11}$$

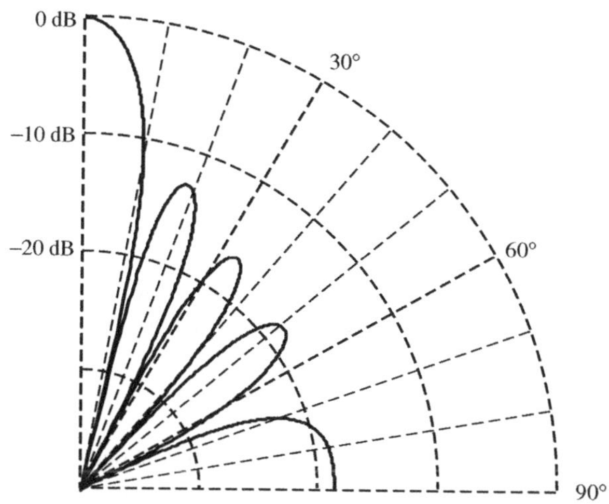

Figure 4.3 Beamforming power distribution

Therefore, the beamforming power is maximized when the steering angle θ of the scan vector coincides with the direction of the plane wave source (Figure 4.3).

4.3.1.2 The Concept of a Virtual Microphone

From Equation (4.10), we can derive the identity

$$b(\theta, \omega) \approx q(\omega) \tag{4.12}$$

when $\theta = \theta_0$. Equation (4.12) implies that the output of the DAS beamformer can be regarded as a source signal. In this regard, the microphone array is considered to be a "virtual microphone" that can reconstruct the source signal as if the microphone is just in front of the source.

4.3.1.3 Normalization of the Scan Vector

As can be seen from Equation (4.11), the scaling factor $1/M$ of a scan vector produces a normalized beamforming power at the peak location ($\theta = \theta_0$). Therefore, the amplitude of the beamformer output is the same as that of the source signal at the peak location. A more generalized expression for the scale factor can be derived by considering a scan vector with arbitrary amplitude.

$$w_m(\theta, \omega) = |w_m(\theta, \omega)| e^{-i\omega(m-1)(d/c)\sin\theta} \tag{4.13}$$

If the steering direction θ is equal to the angle of the true source, then the phase dependency of the beamformer output will be fully compensated. In such a case, the beamformer output is given by

$$b(\theta_0, \omega) = q(\omega) \sum_{m=1}^{M} |w_m(\theta_0, \omega)| = q(\omega) \|\mathbf{w}\|_1, \tag{4.14}$$

where $\| \|_1$ denotes the one-norm of a vector. For a scan vector with $|w_m| = 1$, the one-norm again leads to M. Therefore, the scan vector has to be normalized by $\overline{w} = \|\mathbf{w}\|_1$ in order to have unity gain at the source location. In another definition, the two-norm of a scan vector $\|\mathbf{w}\| = 1/\sqrt{M}$ is used such that the total energy of the beamformer output does not change. The normalization constant can be summarized as follows:

$$\overline{w} = \|\mathbf{w}\|_1 \text{ for preserving the peak amplitude;} \quad \overline{w} = \|\mathbf{w}\| \text{ for preserving the total power.} \tag{4.15}$$

4.3.1.4 Spherical Wavefront Model

If a source is located close to the array, however, the wavefront from the source is perceptively curved when observed over the array. In this case, the spherical wavefront model is well suited to the measurement data. The received sound pressure signal can be written as

$$p_m(\omega) = \frac{q(\omega)e^{i\omega r_m/c}}{r_m}, \tag{4.16}$$

where r_m is the distance between the source and the mth microphone. The weighting function of Equation (4.2) then has to be modified to handle spherical wavefronts. That is,

$$\mathbf{w}(x, y, \omega) = \frac{1}{M}[w_1(x, y, \omega) \quad w_2(x, y, \omega) \quad \cdots \quad w_M(x, y, \omega)]^T, \quad w_m(x, y, \omega) = e^{i\omega r_m/c}, \tag{4.17}$$

where $r_m = \sqrt{(x - x_m)^2 + (y - y_m)^2}$ is the distance between the assumed source location (x, y) and the mth microphone (Figure 4.4). Note that the attenuation factor $1/r_m$ of the spherical wave model is excluded in the scan vector because the beamforming power can be maximized when the phase of the weighting function matches that of the pressure field vector $\mathbf{p}$. If the magnitude of the spherical wavefront model is included, then the beamforming power would be biased by the absolute distance to the modeled source. For example, a sound source far from the microphone array would show a small beamforming power even when the steering position exactly coincides with the true source location.

The beamforming power of the spherical model can be calculated in the same way as the plane wave case. It is interesting to see that even if we handle the same source or sources, the beamforming power can be different depending on the selection of the scan vector. Figure 4.5 demonstrates that what we visualize does depend on what we use: the scan vector or the basis function.

4.3.2 The Array Configuration

What we obtain through the beamforming method is also influenced by what we measure. Therefore, the measurement method, especially the microphone array configuration, is an essential part of the beamformer design. Numerous array configurations have been proposed and used for the beamforming, and the configuration is closely related to the beamforming algorithm to be used.

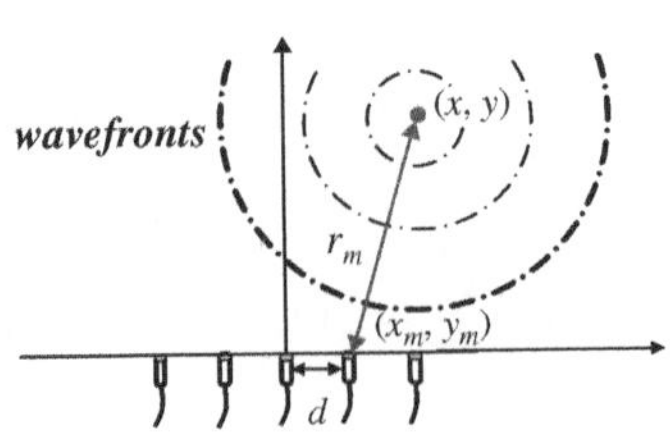

Figure 4.4 Spherical wavefront model

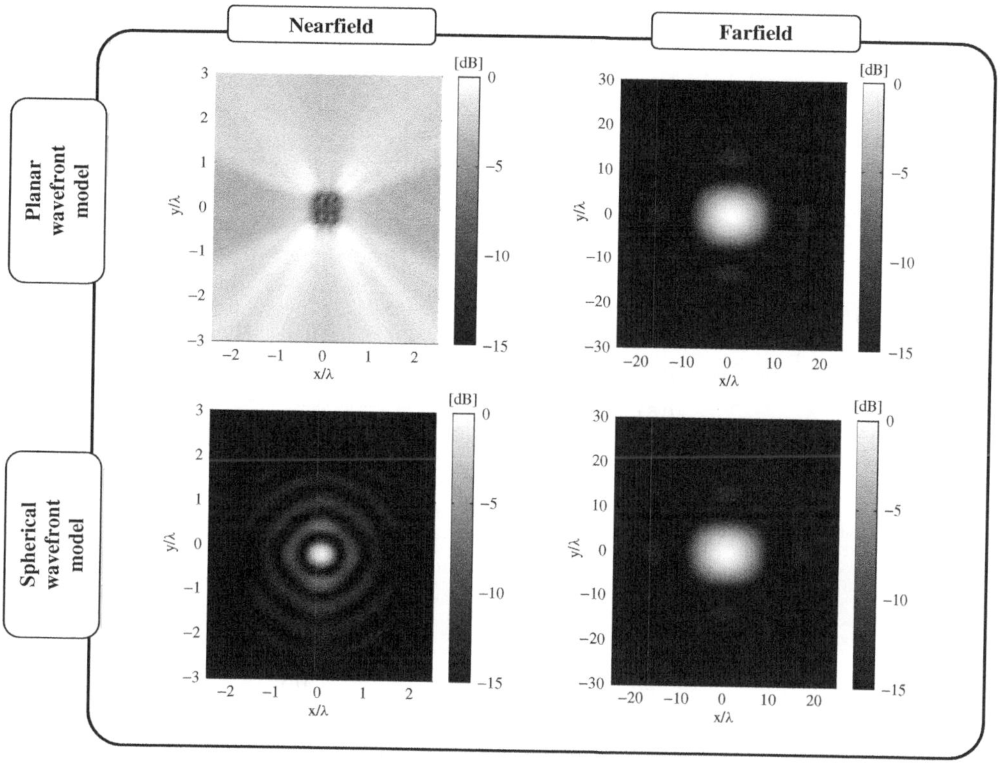

Figure 4.5 Beamforming power distribution obtained from the planar and spherical wavefront models (uniform rectangular array of width: 5λ, height: 6λ, interval: 0.1λ; distance from the source to the array center: nearfield: 0.4λ, farfield: 55λ; distance from the prediction plane to the array: nearfield: 0.35λ, farfield: 54.95λ)

Moreover, the spatial distribution of the sound measured by a microphone array can have various shapes depending on the frequency, aperture size of the array, and sound source location. The performance of the DAS beamformer in relation to these parameters is discussed in Chapter 5. In this chapter, however, we briefly demonstrate how the performance of the DAS beamformer can change in accordance with the microphone array configuration.

Figure 4.6 depicts three exemplary array shapes for the beamforming: a cross-type, grid-type, and curved-spoke-type array. All the arrays consist of 30 microphones and the same aperture size (about 0.8 m).

4.3.2.1 Beamformer Performance

There are numerous measures for evaluating the beamformer performance. Among them, the resolution is an important parameter for estimating the source locations. As can be seen in Equation (4.11), the beamforming power of a finite-sized array has a finite mainlobe width and nonzero sidelobes. In the typical array response in Figure 4.7, the mainlobe location is used as an indicator of the source position, and the width of the mainlobe is related to the resolution of the beamformer. The most frequently used measure is the half-power beamwidth (HPBW), which represents the minimum distance we can distinguish between two closely located sound sources.

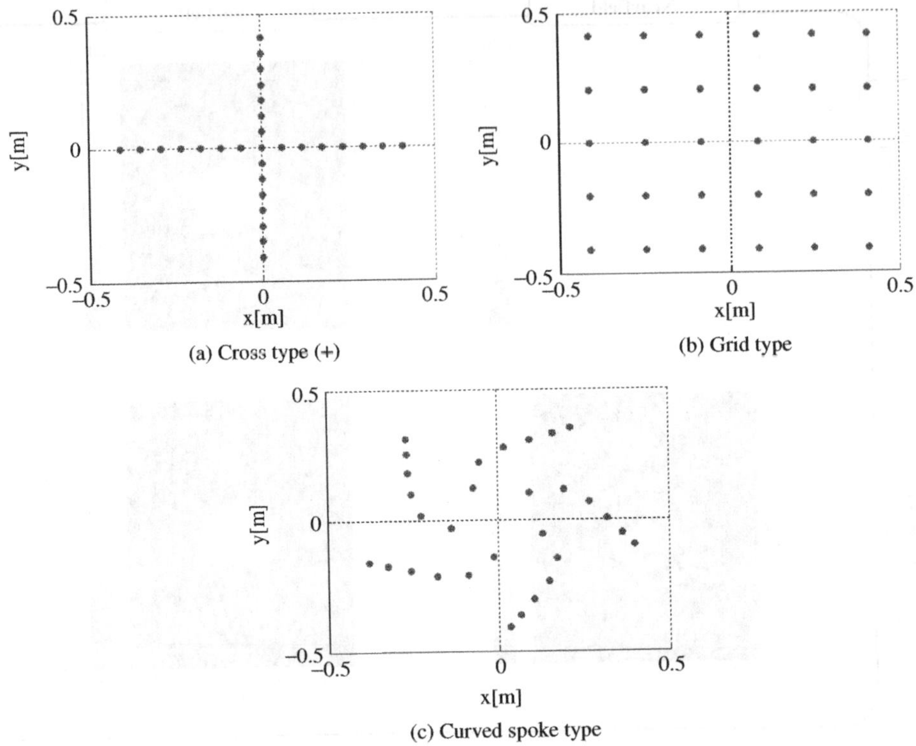

Figure 4.6 (a–c) Three exemplary array shapes

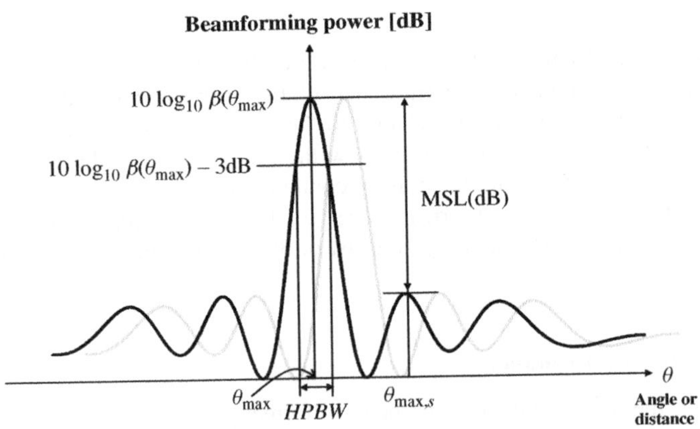

Figure 4.7 The half-power beamwidth (HPBW) and maximum sidelobe level (MSL)

Another popular measure, called the maximum sidelobe level (MSL), is defined as the largest sidelobe magnitude normalized by the peak amplitude of the mainlobe. Therefore, for an ideal beamformer without any sidelobes, the MSL is equal to zero. On the decibel scale, the MSL can be represented as

$$MSL = 10\log_{10}(\beta(\theta_{max})/\beta(\theta_{max,s})), \tag{4.18}$$

where θ_{max} is the angle (or location) of the mainlobe, and $\theta_{max,s}$ denotes the angle of the maximum sidelobe. The MSL represents the ability to detect a secondary sound source of small power in the presence of a strong primary source. If the secondary source power is smaller than the primary source power subtracted by MSL in dB scale, then the secondary source is indistinguishable from the sidelobes of the primary source's beamforming power distribution.

4.3.2.2 Beamforming Power Distribution and Array Configurations

Figures 4.9 and 4.10 show the beamforming power distribution obtained using three types of arrays (Figure 4.6) at 2 and 8 kHz. One monopole source is located at $x_0 = 0\,\text{m}$, $y_0 = 0\,\text{m}$, $z_0 = 3\,\text{m}$ for all simulation conditions, with z_0 being the distance between the measurement and prediction planes (Figure 4.8). The grid array has the smallest beam compared to the curved-spoke- and cross-type arrays. In terms of the MSL, however, the curved-spoke-type array is superior to the others. The MSL of the grid array goes to zero due to spatial aliasing artifacts (Chapter 5, Section 5.3.5) at high frequency (Figure 4.9b). Therefore, to evaluate the array performance, various aspects of the beamforming power have to be considered together.

4.4 Signal-Based Beamforming

The scan vector design we have considered so far is based on the sound field model. However, there is other information we can utilize in designing a scan vector: the measured microphone signals. As can be seen in Equation (4.6), useful information about the spatial variation of the measured sound field is contained within the correlation matrix **R**. This matrix is related to the spatial and statistical correlation of the measured data, since its elements represent the statistical correlation between microphone signals measured at two different locations in space. Accordingly, we can design an improved scan vector by examining the spatial and statistical characteristics of the correlation matrix. It is noteworthy that the correlation between the microphone signals is also influenced by the temporal and spectral characteristics of the source signals. This section describes the performance of sound-source identification according to the temporal or spectral characteristics of the signals by investigating the properties of the correlation matrix in the time and frequency domains.

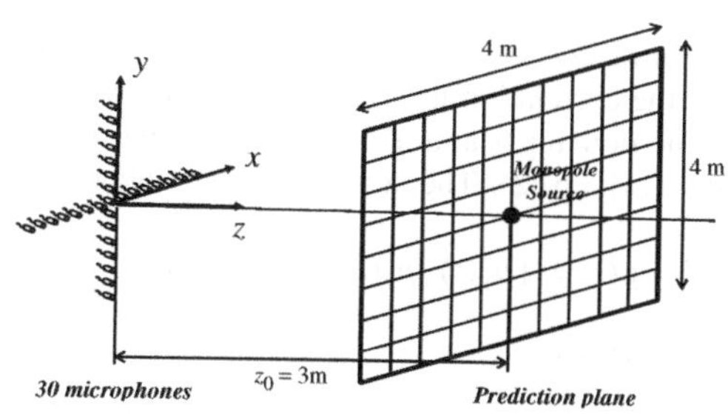

Figure 4.8 Array configuration for the beamforming test (Figure 4.9 and 4.10)

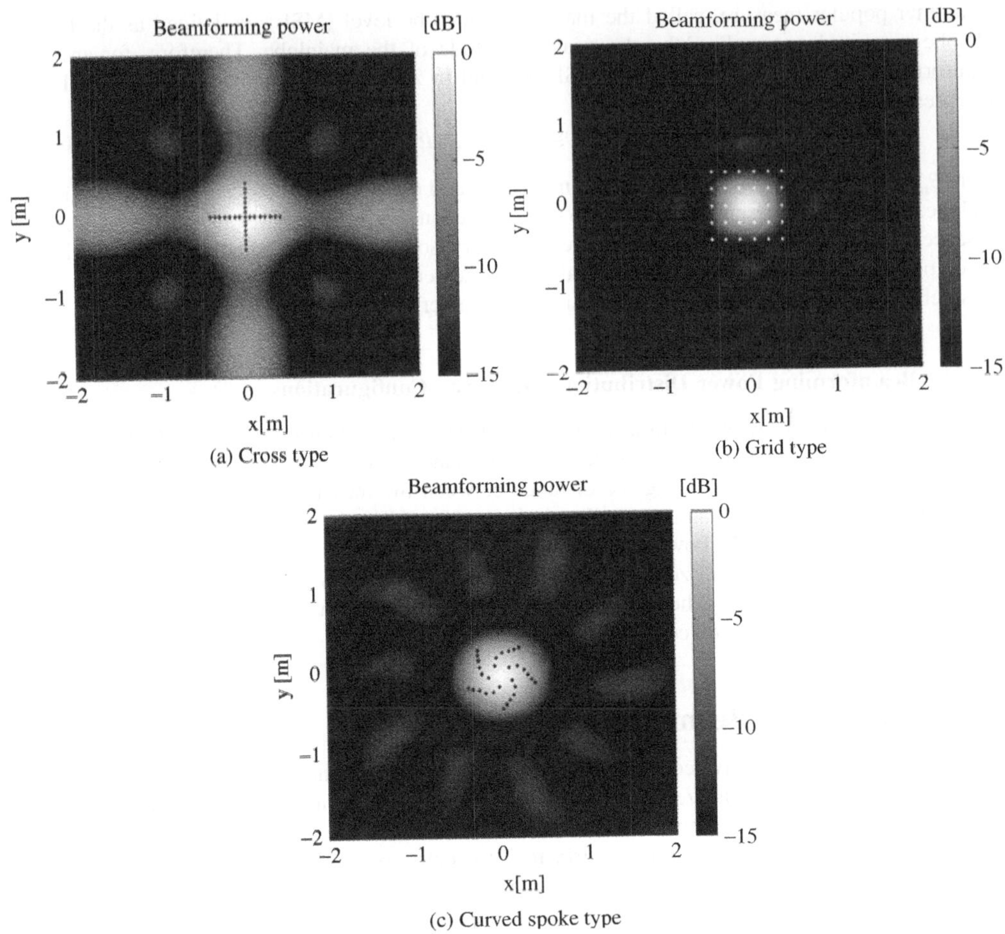

Figure 4.9 Beamforming power distribution in the prediction plane (at 2 kHz)

4.4.1 Construction of Correlation Matrix in Time Domain

4.4.1.1 Generalized Form in Time Domain

Time-domain analysis can be utilized for real-time identification of sound sources. In this section, we examine the properties of the correlation matrix in accordance with the temporal characteristics of the source signals. The time-domain beamformer is the original form of beamforming that has been used since World War I ([4], p. 2). The sound pressure signals received at the microphone array can be expressed as

$$\mathbf{p}(t) = [p_1(t), p_2(t), \cdots, p_M(t)]^T, \tag{4.19}$$

and the scan vector can be written as

$$\mathbf{w}(\theta, t) = [w_1(\theta, t), w_2(\theta, t), \cdots, w_M(\theta, t)]^T.$$

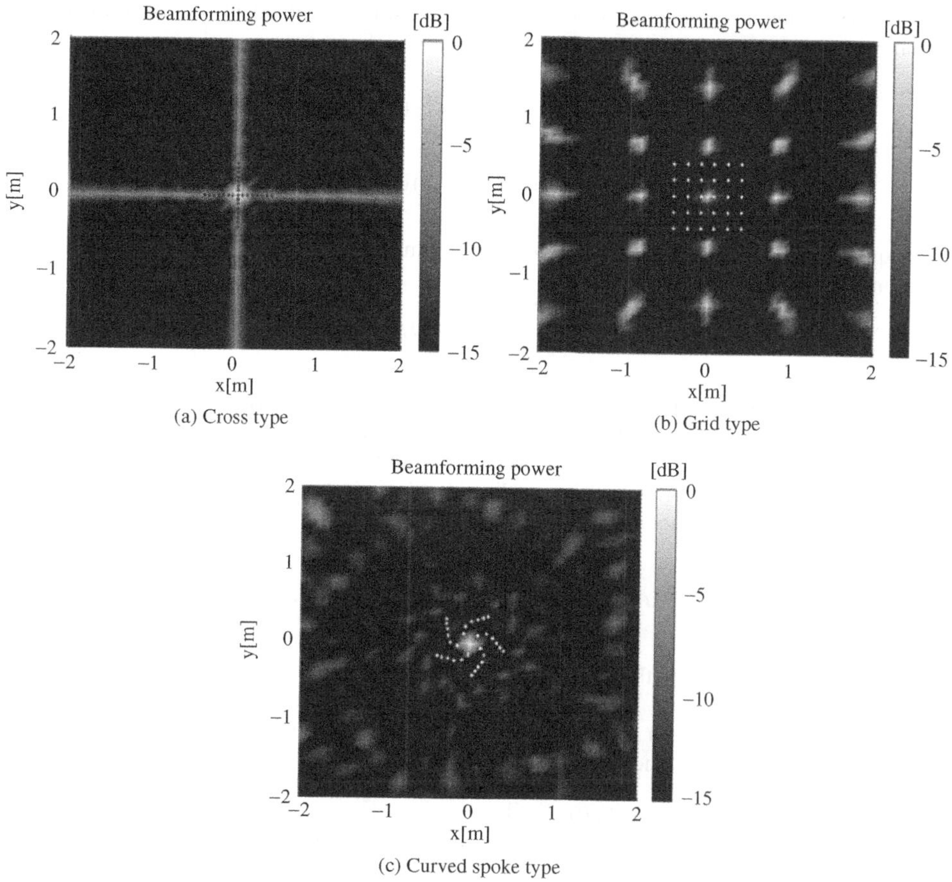

Figure 4.10 Beamforming power distribution in the prediction plane (at 8 kHz)

The time-domain representation of the beamformer output $b(\theta, t)$ in terms of $\mathbf{w}(\theta, t)$ can be derived from the inverse Fourier transform of the frequency-domain expression. The inverse Fourier transform of the frequency-domain beamformer output gives

$$\begin{aligned} b(\theta, t) &= \int_{-\infty}^{\infty} \left[\mathbf{w}(\theta, \omega)^H \mathbf{p}(\omega)\right] e^{-i\omega t} d\omega \\ &= \sum_{m=1}^{M} \int_{-\infty}^{\infty} \left[w_m(\theta, \omega)^* p_m(\omega)\right] e^{-i\omega t} d\omega \\ &= \sum_{m=1}^{M} w_m(\theta, -t) *_t p_m(t), \end{aligned} \tag{4.20}$$

where the expression involving the time-domain convolution, denoted by $*_t$, is equivalent to a temporal correlation of two real-valued signals, which is defined as

$$\begin{aligned} f(-t)*_t g(t) &= \int_{-\infty}^{\infty} f(\tau+t)g(\tau)d\tau \\ &= \int_{-\infty}^{\infty} f(\tau)g(\tau+t)d\tau. \end{aligned} \tag{4.21}$$

Consequently, the general description of the beamformer output is given by the sum of correlations:

$$b(\theta,t) = \sum_{m=1}^{M} \int_{-\infty}^{\infty} [w_m(\theta,\tau)p_m(t+\tau)]d\tau. \tag{4.22}$$

The temporal correlation of Equation (4.20) can be regarded as a pressure signal $p_m(t)$ filtered by a time-reversed scan vector $w_m(\theta,-t)$ in the time domain.

In practice, the filter length of a scan vector is limited, and filtering is thus implemented by discrete-time signal processing. For a discrete time instance ($\tau = n\Delta t$), the integral of Equation (4.22) can be rewritten as

$$b(\theta,t) = \sum_{m=1}^{M}\sum_{n=0}^{N-1} w_m(\theta,n\Delta t)p_m(t+n\Delta t). \tag{4.23}$$

Therefore, one can consider the scan vector $w_m(\theta,n\Delta t)$ to be a finite impulse response (FIR) filter of length N.

To express Equation (4.23) in matrix form, we define two partitioned vectors consisting of $\mathbf{w}(\tau)$ and $\mathbf{p}(t+\tau)$ at each time instance $\tau = n\Delta t$:

$$\widehat{\mathbf{w}}(\theta) = \begin{bmatrix} \mathbf{w}(\theta,0) \\ \mathbf{w}(\theta,\Delta t) \\ \vdots \\ \mathbf{w}(\theta,(N-1)\Delta t) \end{bmatrix}, \quad \widehat{\mathbf{p}}(t) = \begin{bmatrix} \mathbf{p}(t) \\ \mathbf{p}(t+\Delta t) \\ \vdots \\ \mathbf{p}(t+(N-1)\Delta t) \end{bmatrix}. \tag{4.24}$$

Then, the beamformer output can be written as

$$b(\theta,t) = \widehat{\mathbf{w}}(\theta)^T\,\widehat{\mathbf{p}}(t), \tag{4.25}$$

and the corresponding beamforming power can be expressed as

$$\begin{aligned} \beta(\theta,t) &= E\left[|b(\theta,t)|^2\right] \\ &= \widehat{\mathbf{w}}(\theta)^T E\left[\widehat{\mathbf{p}}(t)\,\widehat{\mathbf{p}}(t)^T\right]\widehat{\mathbf{w}}(\theta) \\ &= \widehat{\mathbf{w}}(\theta)^T\,\widehat{\mathbf{R}}(t)\widehat{\mathbf{w}}(\theta). \end{aligned} \tag{4.26}$$

The pressure vector $\widehat{\mathbf{p}}(t)$ has a size of $MN \times 1$, and hence the correlation matrix defined in the time domain is given by

$$\widehat{\mathbf{R}}(t) = E\begin{bmatrix} \mathbf{p}(t)\mathbf{p}(t)^T & \cdots & \mathbf{p}(t)\mathbf{p}(t+(N-1)\Delta t)^T \\ \vdots & \ddots & \vdots \\ \mathbf{p}(t+(N-1)\Delta t)\mathbf{p}(t)^T & \cdots & \mathbf{p}(t+(N-1)\Delta t)\mathbf{p}(t+(N-1)\Delta t)^T \end{bmatrix}. \tag{4.27}$$

The temporal correlation matrix $\widehat{\mathbf{R}}(t) \in \mathbb{R}^{MN \times MN}$ in the time domain is then a combination of many $M \times M$ sub-matrices representing the spatial correlation between two pressure vectors measured at different instances in time ($\mathbf{p}(t)$ and $\mathbf{p}(t + n\Delta t)$). The scan vector $\widehat{\mathbf{w}}(\theta)$ also consists of multiple instances of $\mathbf{w}(\theta, n\Delta t)$.

The overall process for the time-domain beamforming is illustrated in Figure 4.11. The beamforming power of Equation (4.26) is similar to that in the frequency-domain representation, and hence, the time-domain beamforming can be processed in the same way as the frequency-domain beamforming. For any time-variant signal, the correlation matrix becomes a function of time, and hence the scan vector $\widehat{\mathbf{w}}(\theta)$ can be updated to reflect the temporal changes in the correlation matrix.

4.4.1.2 Stationary Random Process

If the measured pressure is a stationary random signal, then the expected value does not change with respect to time t. Accordingly, the time-invariant correlation matrix $\widehat{\mathbf{R}} = \widehat{\mathbf{R}}(t)$ can be derived as

$$\widehat{\mathbf{R}} = \begin{bmatrix} \mathbf{A}_{pp}(0) & \cdots & \mathbf{A}_{pp}((N-1)\Delta t) \\ \vdots & \ddots & \vdots \\ \mathbf{A}_{pp}((N-1)\Delta t)^T & \cdots & \mathbf{A}_{pp}(0) \end{bmatrix}, \tag{4.28}$$

where the auto-correlation matrix $\mathbf{A}_{pp}$ is defined as

$$\mathbf{A}_{pp}(\tau) = E\left[\mathbf{p}(t)\,\mathbf{p}(t+\tau)^T\right]. \tag{4.29}$$

Furthermore, if the measured pressure signal is ergodic, the temporal average can be an accurate estimator of the expected value. That is,

$$\mathbf{A}_{pp}(\tau) = \lim_{T\to\infty} \frac{1}{T}\int_0^T \mathbf{p}(t)\mathbf{p}(t+\tau)^T\,dt. \tag{4.30}$$

where the variable T denotes the averaging period.

4.4.1.3 Delay-and-Sum Beamformer as Wideband Processor

A simple example of the time-domain beamformer can be found in the DAS beamformer (Equation (4.8)) for which the scan vector is composed of simple time delays. For the plane wave model of Equation (4.3), the inverse Fourier transform of the scan vector $w_m(\theta, \omega)$ is given by

$$\begin{aligned} w_m(\theta, t) &= \frac{1}{M}\cdot\frac{1}{2\pi}\int_{-\infty}^{\infty} e^{-i\omega\Delta\tau(m-1)\sin\theta} e^{-i\omega t}\,d\omega \\ &= \frac{1}{M}\delta(t + \Delta\tau(m-1)\sin\theta), \end{aligned} \tag{4.31}$$

which indicates that the measured signal $p_m(t)$ is filtered by a delta function with a certain time delay. The corresponding beamformer output of Equation (4.22), reduced to

$$b(\theta, t) = \frac{1}{M}\sum_{m=1}^{M} p_m(t - \Delta\tau(m-1)\sin\theta), \tag{4.32}$$

shows that the scan vector shifts the array signal p_m with a time delay of $\Delta\tau(m-1)\sin\theta$. Therefore, if the assumed angle θ is the same as the direction of an incoming planar wavefront (Equation (4.7)),

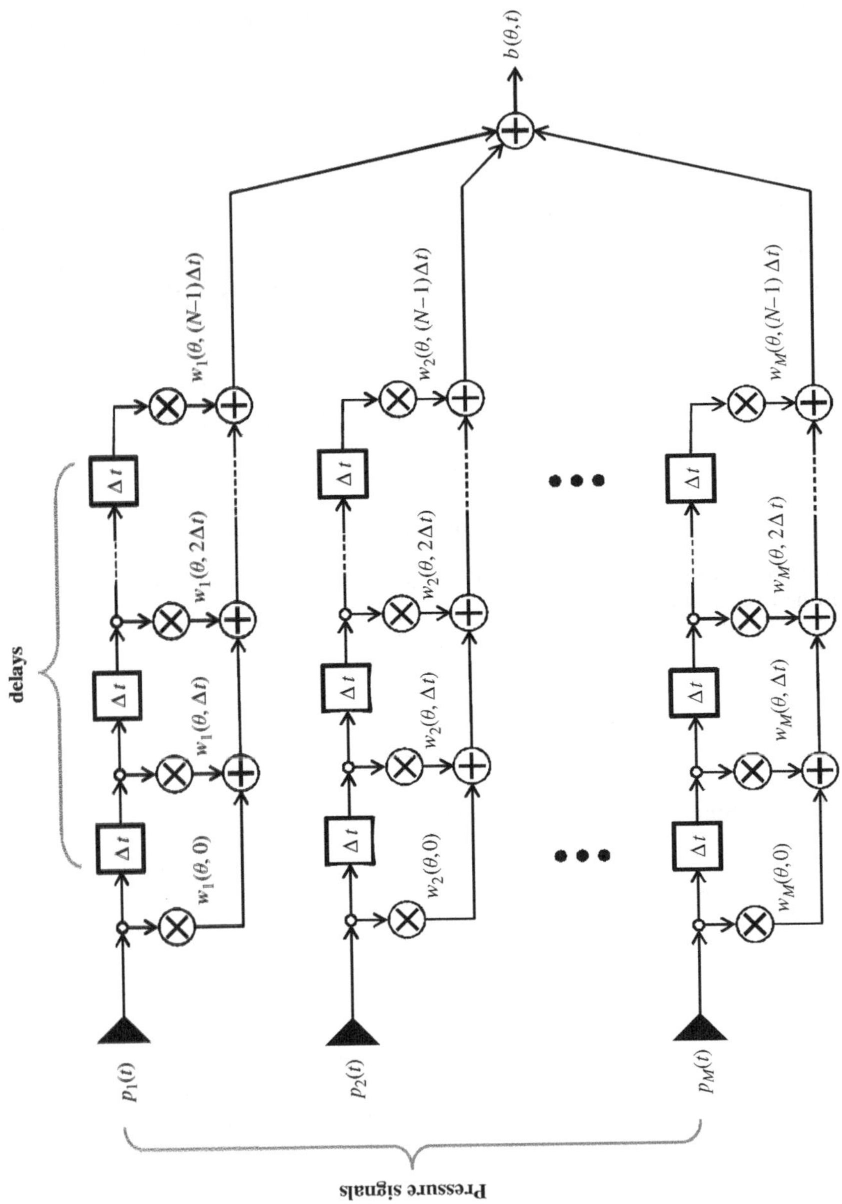

Figure 4.11 General structure of the time-domain beamformer

then all the measured $\mathbf{p}(t)$ signals will be aligned along the time axis. All microphone signals are then in-phase, and hence the beamformer output is maximized. By comparing Equation (4.32) with the general time-domain beamformer of Equation (4.23), we observe that the FIR filter coefficient w_m of the DAS beamformer is given by a pure time delay. In addition, if we use a frequency-independent amplitude weighting a_m, then the scan vector will be

$$\begin{aligned} w_m(\theta, t) &= \frac{1}{M} \cdot \frac{1}{2\pi} \int_{-\infty}^{\infty} a_m e^{-i\omega\Delta\tau(m-1)\sin\theta} e^{-i\omega t} d\omega \\ &= \frac{a_m}{M} \delta(t + \Delta\tau(m-1)\sin\theta), \end{aligned} \tag{4.33}$$

and the corresponding beamformer output can be written as

$$b(\theta, t) = \frac{1}{M} \sum_{m=1}^{M} a_m p_m(t - \Delta\tau(m-1)\sin\theta). \tag{4.34}$$

The time-domain beamformer can also be formulated for the spherical wavefront model. In general, the time-domain representation of beamforming is computationally inefficient, as opposed to the frequency-domain approach that utilizes a fast Fourier transform (FFT), but its formula is straightforward and can be advantageous when only a pure time delay is considered. Another benefit of time-domain processing is addressed in Section 4.7.4, in relation to the temporal characteristics of the source signal.

4.4.2 *Construction of Correlation Matrix in Frequency Domain*

4.4.2.1 Narrowband Description

In principle, the total beamforming power in the time domain is equivalent to that in the frequency domain. This can be verified from Parseval's theorem (Appendix A, Section A.1.1.5):

$$\int_{-\infty}^{\infty} \beta(\theta, t)dt = \frac{1}{2\pi} \int_{-\infty}^{\infty} \beta(\theta, \omega)d\omega. \tag{4.35}$$

If Parseval's theorem is applied to a signal of finite length T, then the equality can be written in terms of a Fourier series, which implies that the temporal mean of the beamforming power can be related to a discrete sum of the beamforming power in the frequency domain. That is,

$$\langle\beta(\theta, t)\rangle_t = \frac{1}{T} \int_{t_0}^{t_0+T} \beta(\theta, t)dt = \sum_{n=-\infty}^{\infty} \beta_n(\theta), \tag{4.36}$$

where $\langle\rangle_t$ indicates the temporal average and $\beta_n(\theta)$ is the square of the Fourier series coefficient given by

$$\beta_n(\theta) = \left| \frac{1}{T} \int_{t_0}^{t_0+T} b(\theta, t) e^{i(2\pi nt/T)} dt \right|^2 = \frac{1}{T} \beta(\theta, \omega = 2\pi n/T).$$

For a signal $p_m(\omega)$ band-limited within $\omega \in [\omega_l, \omega_h]$, the summation of Equation (4.36) can be further reduced to

$$\langle\beta(\theta, t)\rangle_t = \frac{2}{T} \sum_{n=n_l}^{n_h} \beta(\theta, \omega_n = 2\pi n/T), \tag{4.37}$$

with $\omega_l = 2\pi n_l/T$ and $\omega_h = 2\pi n_h/T$. Note that the right-hand side of Equation (4.37) is doubled to express the one-sided spectrum including the negative frequency components. Equation (4.37) implies that the mean of the beamforming power in the time domain is equivalent to the finite sum of the beamforming power at discrete frequencies.

From the summation rule of Equation (4.37), we can see that the correlation matrix in the frequency domain can be calculated as

$$\sum_{n=n_l}^{n_h} \beta(\theta, \omega_n) = \sum_{n=n_l}^{n_h} \left\{ \mathbf{w}(\theta, \omega_n)^H E\left[\mathbf{p}(\omega_n)\,\mathbf{p}(\omega_n)^H\right] \mathbf{w}(\theta, \omega_n) \right\}$$
$$= \left[\sum_{n=n_l}^{n_h} \mathbf{w}(\theta, \omega_n)^H \mathbf{R}(\omega_n) \mathbf{w}(\theta, \omega_n) \right]. \tag{4.38}$$

However, designing the scan vector for every frequency of interest would require significant computational effort. If the frequency bandwidth is narrow enough, then one can design a representative scan vector $\mathbf{w}(\theta, \omega_c)$ for all the frequency components in the finite band. That is,

$$\sum_{n=n_l}^{n_h} \beta(\theta, \omega_n) \approx \mathbf{w}(\theta, \omega_c)^H \left[\sum_{n=n_l}^{n_h} \mathbf{R}(\omega_n) \right] \mathbf{w}(\theta, \omega_c)^H$$
$$\approx \mathbf{w}(\theta, \omega_c)^H \mathbf{R}(\omega_c) \mathbf{w}(\theta, \omega_c)^H, \tag{4.39}$$

where ω_c denotes an arbitrary reference frequency within the band. The correlation matrix $\mathbf{R}(\omega_c)$ of a narrowband signal can be considered as a superposition of matrices defined at multiple frequencies. The overall process to deal with the narrowband signal is hence similar to the single frequency case.

4.4.2.2 Broadband Description

Unlike the narrowband case, the broadband signal cannot be approximated by Equation (4.39). The general equation (Equation (4.38)) states that we need to calculate the beamforming result at each frequency of interest. Moreover, if one attempts to use a high-resolution beamformer that uses a different definition of the beamforming power (such as MUSIC in Section 4.6.2), the summation rule of Equation (4.38) may no longer apply.

To implement a broadband beamformer that can deal with correlation matrices of different frequencies, one can use a frequency correction method (Section 4.7.2). The frequency correction technique attempts to convert the correlation matrices measured at different frequencies to a single reference frequency; however, to that end, one needs to understand how the measured pressure vectors at different frequencies $\mathbf{p}(\omega_n)$ are related to each other. In this regard, prior to studying the correlation matrix for a broadband signal, we first investigate the behavior of the correlation matrix at a single frequency. The general source and noise model for a single frequency is presented, and then the corresponding frequency correction methods are explained.

4.4.3 Correlation Matrix of Multiple Sound Sources

To analyze the characteristics of the correlation matrix $\mathbf{R}$, we consider a general signal model consisting of multiple sound sources and noise. Suppose that K sound sources $(q_1(\omega), q_2(\omega), \cdots, q_K(\omega))$ generate a sound field. The pressure signal measured at the mth microphone can be written as

$$p'_m(\omega) = \sum_{\kappa=1}^{K} p_{m\kappa}(\omega) + n_m(\omega)$$
$$= \sum_{\kappa=1}^{K} h_{m\kappa}(\omega) q_\kappa(\omega) + n_m(\omega), \tag{4.40}$$

where $n_m(\omega)$ is the measurement noise at the microphone and $h_{m\kappa}$ is the transfer function from the κth source to the mth microphone position. If we rewrite Equation (4.40) in matrix form, the measured pressure vector $\mathbf{p}'$ can be described as a sum of the true pressure vector $\mathbf{p}$, which is the pressure signal without noise, and a noise vector $\mathbf{n}$;

$$\mathbf{p}'(\omega) = \mathbf{p}(\omega) + \mathbf{n}(\omega)$$
$$= \mathbf{H}(\omega)\mathbf{q}(\omega) + \mathbf{n}(\omega), \tag{4.41}$$

where the matrix $\mathbf{H} \in \mathbb{C}^{M\times K}$ expresses the transfer function from multiple sources to multiple microphones.

The correlation matrix of the measured pressure signal can then be represented as

$$\mathbf{R}'(\omega) = E\left[\mathbf{p}'(\omega)\,\mathbf{p}'(\omega)^H\right]$$
$$= E\left[(\mathbf{Hq} + \mathbf{n})\,(\mathbf{q}^H\mathbf{H}^H + \mathbf{n}^H)\right]. \tag{4.42}$$

Note that the transfer function $\mathbf{H}$ depends on the propagation model, which is deterministic. The frequency dependency ω is omitted for brevity. If we expand Equation (4.42) term-by-term, the measured correlation matrix $\mathbf{R}'$ can be decomposed into

$$\mathbf{R}' = \mathbf{HR}_{qq}\mathbf{H}^H + \mathbf{R}_{nq}\mathbf{H}^H + \mathbf{HR}_{nq}^H + \mathbf{R}_{nn}, \tag{4.43}$$

where the new correlation matrices are defined by

$$\mathbf{R}_{qq} = E\left[\mathbf{qq}^H\right] : \text{intersource correlation matrix}$$
$$\mathbf{R} = E\left[\mathbf{pp}^H\right] = \mathbf{HR}_{qq}\mathbf{H}^H : \text{intersignal correlation matrix}$$
$$\mathbf{R}_{nq} = E\left[\mathbf{nq}^H\right] : \text{noise–source correlation matrix}$$
$$\mathbf{R}_{nn} = E\left[\mathbf{nn}^H\right] : \text{internoise correlation matrix}. \tag{4.44}$$

Equation (4.43) shows the relationship between the measured correlation $\mathbf{R}'$ and the source–noise correlations. The first correlation matrix $\mathbf{R}_{qq}$ indicates the relationship between sound source signals. Likewise, the internoise correlation matrix $\mathbf{R}_{nn}$ consists of correlations between noise signals included in different microphones. The correlations between the source and noise signals are denoted by $\mathbf{R}_{nq}$.

4.4.3.1 Statistical Matrices

To deal with random signals, it is sometimes beneficial to use statistical vector notation. Consider an ensemble of sample pressure vectors $\{\tilde{\mathbf{p}}'^{(1)}, \cdots, \tilde{\mathbf{p}}'^{(\infty)}\}$ of a stochastic process $\mathbf{p}'$. A statistical matrix that has N sample vectors obtained from this ensemble as its column entries can be defined as

$$\tilde{\mathbf{P}}' = \left[\tilde{\mathbf{p}}'^{(1)}, \cdots, \tilde{\mathbf{p}}'^{(N)}\right], \tag{4.45}$$

where the tilde indicates that the vector or matrix contains more than one statistical sample (observations). In the same manner, the statistical matrices for the noise-free pressure field $\mathbf{p}$ and source signal $\mathbf{q}$ can be defined as

$$\tilde{\mathbf{P}} = [\tilde{\mathbf{p}}^{(1)}, \cdots, \tilde{\mathbf{p}}^{(N)}], \quad \tilde{\mathbf{Q}} = [\tilde{\mathbf{q}}^{(1)}, \cdots, \tilde{\mathbf{q}}^{(N)}]. \tag{4.46}$$

Figure 4.12 illustrates the concept of the statistical matrix. The benefit of using statistical matrices is that the ensemble mean or expectation of Equation (4.42) can be expressed as a matrix multiplication:

$$\mathbf{R}' = \lim_{N\to\infty} \left[\frac{1}{N} \tilde{\mathbf{P}}'\tilde{\mathbf{P}}'^{H} \right]. \tag{4.47}$$

Therefore, this notation is useful in dealing with the statistical properties of the measured signal. In practice, we can attempt to predict the correlation of Equation (4.47) using a finite number of observations (N) and the estimation of $\mathbf{R}'$ can then be written as

$$\hat{\mathbf{R}}' = \frac{1}{N} \tilde{\mathbf{P}}'\tilde{\mathbf{P}}'^{H}, \tag{4.48}$$

where $\hat{\mathbf{R}}'$ indicates the estimation of $\mathbf{R}'$. Likewise, the estimation of intersource correlations ($\hat{\mathbf{R}}_{qq}$) is given by $\tilde{\mathbf{Q}}\tilde{\mathbf{Q}}^H/N$.

4.4.3.2 Incoherent Signals

The correlation matrix can have various forms depending on the coherence between the source and noise signals. Each element of the correlation matrix $\mathbf{R}$ represents how the signals in two different spatial positions are correlated in a statistical manner. For example, suppose that two pressure fields measured by multiple microphones are incoherent; then signals in each row of $\tilde{\mathbf{P}}$ are orthogonal, and the correlation matrix $\mathbf{R}$ becomes a diagonal matrix in which the elements represent the expected signal powers or variances. If all the microphone signals are incoherent with each other, then the correlation matrix is a full-rank matrix, with a rank equal to the number of microphones. The same statement can be made for the intersource correlation matrix $\mathbf{R}_{qq}$; the correlation matrix is full rank if the source signals are incoherent with each other.

4.4.3.3 Incoherent Measurement Noise

If the measurement noise $\tilde{\mathbf{N}}$ is incoherent with the source signal $\tilde{\mathbf{Q}}$, then the cross-correlation matrix (noise–source correlation) becomes zero ($\mathbf{R}_{nq} = 0$). Accordingly, the measured correlation matrix $\mathbf{R}'$ of Equation (4.43) can be reduced to

$$\begin{aligned} \mathbf{R}' &= \mathbf{R} + \mathbf{R}_{nn} \\ &= \mathbf{H}\mathbf{R}_{qq}\mathbf{H}^H + \mathbf{R}_{nn}. \end{aligned} \tag{4.49}$$

The measurement noise thus inevitably affects the beamforming power. Since the internoise correlation $\mathbf{R}_{nn}$ is added to the true correlation matrices, the beamforming power of Equation (4.5) is altered as

$$\begin{aligned} \beta(\theta) &= \beta_{signal}(\theta) + \beta_{noise}(\theta) \\ &= \mathbf{w}(\theta)^H \mathbf{R}\mathbf{w}(\theta) + \mathbf{w}(\theta)\mathbf{R}_{nn}\mathbf{w}(\theta). \end{aligned} \tag{4.50}$$

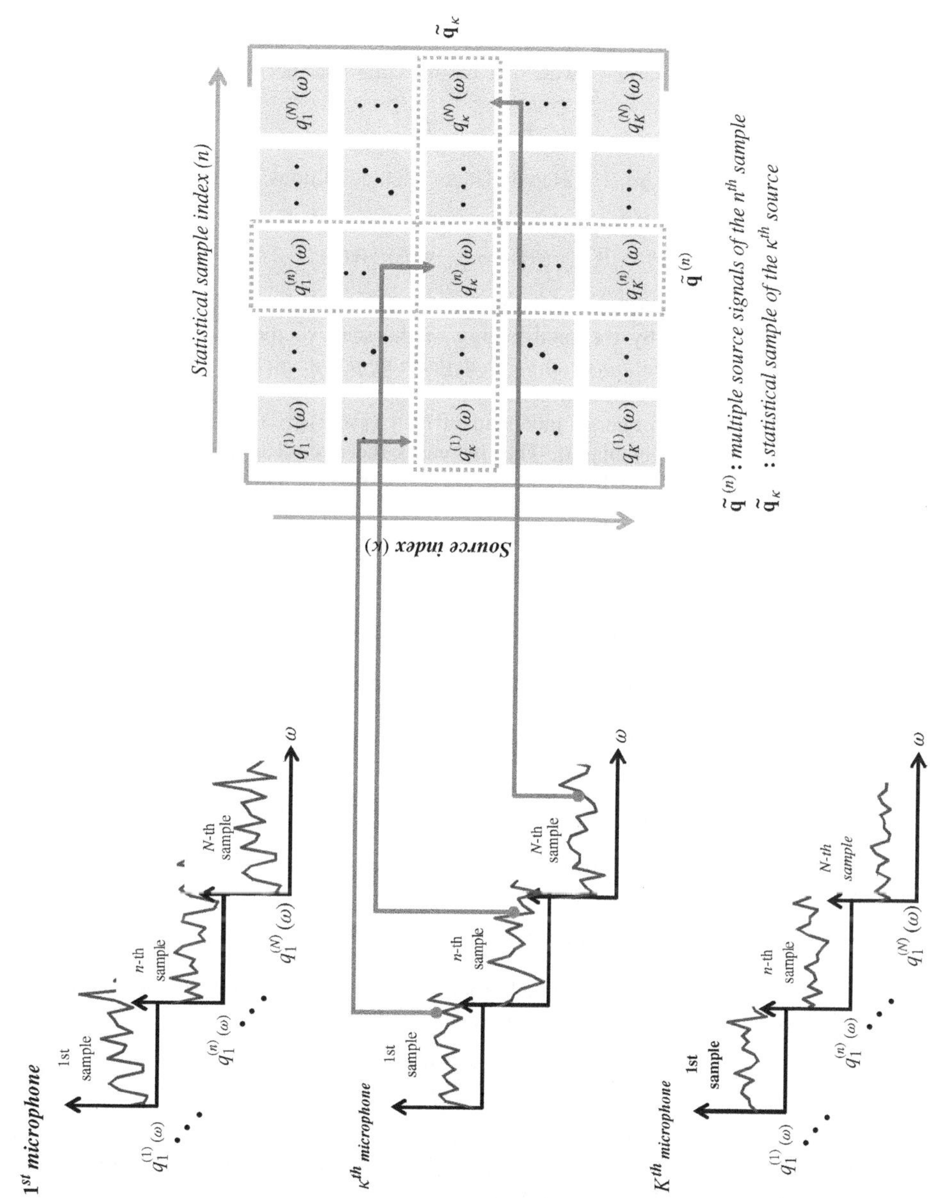

Figure 4.12 Concept of the statistical matrix

The second term $\beta_{noise}(\theta)$ denotes the perturbed beamforming power due to the measurement noise. The matrix product is of quadratic form and always positive; therefore, the perturbation is always additive to the original beamforming power $\beta_{signal}(\theta)$.

If we further assume that the measurement noise is incoherent, then the internoise correlation matrix $\mathbf{R}_{nn}$ will be a diagonal matrix. Accordingly, the effect of noise only appears in the diagonal elements of $\mathbf{R}'$. In such a case, the noise is said to be spatially white (uncorrelated from sensor to sensor). For noise signals of the same power, that is, the same variances across all microphones, Equation (4.49) is given by

$$\mathbf{R}' = \mathbf{R} + \sigma_{noise}^2 \mathbf{I}, \tag{4.51}$$

where σ_{noise}^2 denotes the estimated noise variance. Under such conditions, the perturbed beamforming power can be written as

$$\beta(\theta) = \beta_{signal}(\theta) + \sigma_{noise}^2 \mathbf{w}(\theta)^H \mathbf{w}(\theta). \tag{4.52}$$

In Equation (4.52), it is interesting to see that the perturbation in the beamforming power equals the noise power σ_{noise}^2 amplified by the total energy of the scan vector $\|\mathbf{w}(\theta)\|^2 = \mathbf{w}(\theta)^H \mathbf{w}(\theta)$. Therefore, *the effect of measurement noise on the beamforming power increases in accordance with the energy of the scan vector in a certain direction* θ.

For a scan vector with uniform energy independent of θ ($\|\mathbf{w}(\theta)\|^2 = \text{const}$), the beamforming power of Equation (4.52) is simply biased. This increases the overall level of the beamforming power, and nulls and sidelobes of small magnitude are especially vulnerable to the bias error.

4.4.3.4 Peak-to-Sidelobe Ratio in the Presence of Noise

To examine the effect of noise on the beamforming power in detail, consider a simple planar wavefront model for which the measured pressure field is produced by a single plane wave source:

$$\mathbf{p} = [h_1(\theta_0), \cdots, h_M(\theta_0)]^T q = \mathbf{h}(\theta_0) q. \tag{4.53}$$

Here, the transfer function $h_m(\theta_0)$ follows the definition of Equation (4.7). In this case, the correlation matrix of Equation (4.51) can be written as

$$\mathbf{R}' = |q|^2 \mathbf{h}(\theta_0)\mathbf{h}(\theta_0)^H + \sigma_{noise}^2 \mathbf{I}. \tag{4.54}$$

If we use a scan vector identical to the transfer function, $\mathbf{w}(\theta) = \mathbf{h}(\theta)/M$, then the beamforming power has the following form:

$$\beta(\theta) = |q|^2 |\mathbf{h}(\theta)^H \mathbf{h}(\theta_0)|^2 / M^2 + \sigma_{noise}^2 / M. \tag{4.55}$$

For ease of derivation, we denote the squared product of two transfer functions, ($\mathbf{h}(\theta)$ and $\mathbf{h}(\theta_0)$, normalized by M^2 as a *transfer function similarity*:

$$\cos^2\gamma(\theta) = \frac{|\mathbf{h}(\theta)^H \mathbf{h}(\theta_0)|^2}{M^2}. \tag{4.56}$$

For a planar wavefront of unit amplitude, the variable γ represents the angle between the two transfer function vectors. In terms of the defined similarity measure, beamforming power of the DAS beamformer (Equation (4.55)) can be rewritten as

$$\beta(\theta) = |q|^2 \cos^2\gamma(\theta) + \sigma_{noise}^2 / M. \tag{4.57}$$

When the direction of the scan vector and the source location are the same ($\theta = \theta_0$), the transfer function similarity becomes $\cos^2\gamma(\theta_0) = 1$. In this case, the beamforming power is maximized and given by $|q|^2 + \sigma_{noise}^2/M$. In contrast, when $\mathbf{h}(\theta)^H\mathbf{h}(\theta_0) \approx 0$, the minimum level of the beamformer (σ_{noise}^2/M) is obtained. The minimum level of the beamformer depends on the variance of incoherent noise added to the microphone signal. Therefore, if we define the ratio of the maximum beamforming power to the minimal sidelobe level as the peak-to-sidelobe ratio (PSR), then the PSR can be written as

$$PSR = M\frac{|q|^2}{\sigma_{noise}^2} + 1. \tag{4.58}$$

The PSR can be used to represent the bias level of the beamforming power in the presence of noise. For the transfer function $\mathbf{h}(\theta)$ modeled as a planar wavefront of unit amplitude, $|q|^2/\sigma_{noise}^2$ represents the SNR of a microphone signal, and hence, the PSR can be expressed in terms of SNR:

$$PSR = M \cdot SNR + 1. \tag{4.59}$$

Consequently, the PSR depends on the number of microphone (M) and the SNR of the microphone signal.

Figure 4.13 depicts the beamforming power of a DAS beamformer with respect to the different SNRs. For this simulation, nine microphones arranged with a half wavelength interval on a line is used, and the planar wave propagation model is assumed. It can be seen that increasing the noise level decreases the PSR. In this example, however, each beamformer output is normalized by its peak value, and hence the beamforming power near the peak location is not affected by the increase in the noise variance.

4.4.3.5 Diagonal Loading and Removal

Since the noise correlation matrix of Equation (4.51) is diagonal, one can utilize the "diagonal loading" property to devise a robust beamforming technique. A popular method is to modify the diagonal elements of the correlation matrix $\mathbf{R}'$. For example, a simple diagonal unloading can reduce the bias error due to noise. First, let us describe the diagonal and off-diagonal elements of $\mathbf{R}'$ separately:

$$\mathbf{R}' = \mathbf{R}'_{diag} + \mathbf{R}'_{offdiag}.$$

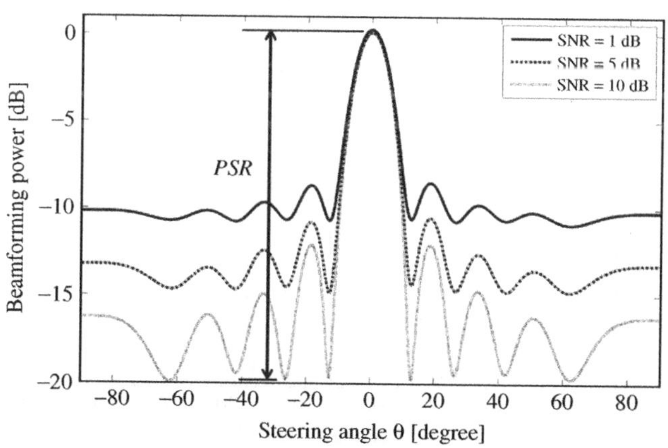

Figure 4.13 Beamforming power of the delay-and-sum beamformer for different SNRs (uniform line array with nine microphones of a half wavelength distance; plane wave model)

One interesting aspect can be observed when the off-diagonal and diagonal matrices of $\mathbf{R}'$ are combined with the scan vector $\mathbf{w}$:

$$\begin{aligned}\beta(\theta) &= \mathbf{w}(\theta)^H \mathbf{R}' \mathbf{w}(\theta) \\ &= \mathbf{w}(\theta)^H (\mathbf{R}'_{offdiag}) \mathbf{w}(\theta) + \mathbf{w}(\theta)^H (\mathbf{R}'_{diag}) \mathbf{w}(\theta).\end{aligned}$$

The beamforming power β is also divided into contributions from the diagonal and off-diagonal components of $\mathbf{R}'$. As we saw in Equation (4.51), the effect of noise only contributes to the diagonal elements of $\mathbf{R}'$. More importantly, if the scan vector amplitude is uniform across all microphones ($|w_m| = const$), then the beamforming power for the diagonal and off-diagonal components of $\mathbf{R}'$ can be written separately as

$$\beta(\theta) = \mathbf{w}(\theta)^H (\mathbf{R}'_{offdiag}) \mathbf{w}(\theta) + \mathrm{Tr}(\mathbf{R}'_{diag}) \|\mathbf{w}(\theta)\|^2 / M, \tag{4.60}$$

where Tr denotes the sum of the diagonal elements of a matrix. In this case, the two-norm of a scan vector $\|\mathbf{w}(\theta)\|^2$ is independent of θ, and hence, the contribution of the diagonal elements to the beamforming power is just the addition of a constant. Accordingly, the diagonal elements just bias the beamforming pattern, which is not helpful in determining the source location. Therefore, removing the diagonal elements of the correlation matrix $\mathbf{R}'$ can reduce the effects of noise. Even without the measurement noise, the removal can reduce the bias in the beamforming power.

However, the peak amplitude of the beamforming power in the source direction ($\theta = \theta_0$) will also be lowered by the removal of the diagonal elements, and we need to rescale the overall beamforming power. For the plane wave model, the peak beamforming power without the diagonal removal (Equation (4.14)) is given by $|q|^2 \|\mathbf{w}\|_1^2$. Since the diagonal removal reduces the beamforming power by $|q|^2 \|\mathbf{w}\|^2$, the scan vector has to be scaled by $\overline{w} = \sqrt{\|\mathbf{w}\|_1^2 - \|\mathbf{w}\|^2}$ to normalize the beamforming power.

Both the planar and spherical wavefront models (Equations (4.7) and (4.17)) satisfy the assumption of a uniform scan vector amplitude ($|w_m| = const$). Figure 4.14 depicts the beamforming power with and without the removal of the diagonal elements. It can be seen that the diagonal removal

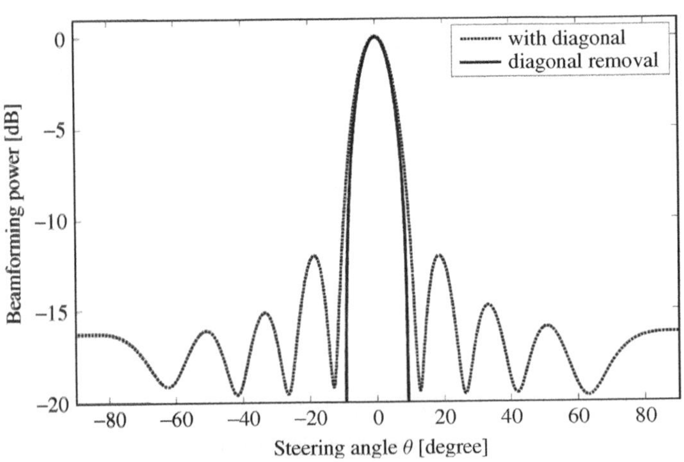

Figure 4.14 Delay-and-sum beamforming with and without the diagonal elements (uniform line array with nine microphones of a half wavelength distance; SNR = 10 dB)

effectively reduces the bias in the beamforming power. In practice, the sidelobes can be negative after the diagonal removal, which will provide incorrect information when the absolute value of the beamforming power is plotted. In this regard, negative values of the beamforming power should be truncated to zero.

4.4.3.6 Noise-Free Measurement and Rank of the Correlation Matrix

There are advanced techniques that can enhance the beamforming performance by modifying the structure of the correlation matrix. As is discussed in the following sections, popular algorithms such as MUSIC or ESPRIT exploit the inverse of the correlation matrix.

The rank of the correlation matrix is an important measure, when the inverse of the correlation matrix is taken into account. All the correlation matrices in Equation (4.44) are $\mathbf{R}_{xx} \in \mathbb{C}^{M\times M}$, but the rank of $\mathbf{R}$ can be less than M. If this is the case, then we can have a rank deficiency.

If the measured signals are free from noise, then the correlation matrix only depends on the correlations between the source signals. That is,

$$\mathbf{R}' = \mathbf{R} = \mathbf{H}\mathbf{R}_{qq}\mathbf{H}^H. \tag{4.61}$$

Then, the beamforming power is the ideal power defined in Equation (4.5), and the performance of the sound source location is then solely determined by the scan vector $\mathbf{w}$. Without measurement noise, the rank of the correlation matrix $\mathbf{R}'$ depends on ranks of the transfer function $\mathbf{H}$ and the intersource correlation matrix $\mathbf{R}_{qq}$. If the source signals are perfectly incoherent with each other, then the off-diagonal elements of $\mathbf{R}_{qq}$ can be reduced to zero. In such a case, Equation (4.61) can be simplified as

$$\mathbf{R} = \mathbf{H}\begin{bmatrix} \sigma_1^2 & 0 & 0 \\ 0 & \ddots & 0 \\ 0 & 0 & \sigma_K^2 \end{bmatrix}\mathbf{H}^H, \tag{4.62}$$

where σ_κ^2 $(\kappa = 1, \cdots, K)$ represents the estimated variance or signal power of the κth source. For example, in the case of K planar sound sources positioned at $[\theta_1, \theta_2, \cdots, \theta_K]$, the pressure field measured by the mth microphone can be written as

$$p_m(\omega) = \sum_{\kappa=1}^{K} e^{-i(m-1)\varphi\sin\theta_\kappa} q_\kappa(\omega), \quad \varphi = \omega d/c \tag{4.63}$$

where q_κ is the complex amplitude of the κth plane wave. The transfer matrix for incoherent plane wave sources is given by

$$[\mathbf{H}]_{(m,\kappa)} = e^{-i(m-1)\varphi\sin\theta_\kappa}.$$

If the source signals are incoherent, then the intersource correlation matrix will be reduced to a diagonal matrix

$$\mathbf{R} = \mathbf{H}\begin{bmatrix} E\left[|q_1|^2\right] & 0 & 0 \\ 0 & \ddots & 0 \\ 0 & 0 & E[|q_K|^2] \end{bmatrix}\mathbf{H}^H, \tag{4.64}$$

which is equivalent to Equation (4.62).

Since the number of sources is usually smaller than the number of microphones ($M > K$), the rank of the transfer matrix $\mathbf{H}$ is limited by the number of sources. However, when the transfer functions of different sources are highly correlated in space, for example, the propagation angles (θ_κ) of a set of plane waves are similar, the rank can be further reduced. Therefore, the spatial separation of sources and the linear independence of the transfer function affect the rank of the correlation matrix.

The practical importance of the incoherent sources assumption (Equation (4.62)) is that the correlation between the measured signal can be expressed as the superposition of correlation matrices of individual sources. If we consider a column vector $\mathbf{h}^{(\kappa)}$ consisting of the transfer function of each source,

$$\mathbf{h}^{(\kappa)} = \left[h_{1\kappa}, \cdots, h_{M\kappa}\right]^T \Rightarrow \mathbf{H} = \left[\mathbf{h}^{(1)}, \cdots, \mathbf{h}^{(K)}\right], \tag{4.65}$$

then Equation (4.62) can be written as

$$\mathbf{R} = \sum_{\kappa=1}^{K} \sigma_\kappa^2 \mathbf{h}^{(\kappa)} \mathbf{h}^{(\kappa)H} = \sum_{\kappa=1}^{K} \mathbf{R}_\kappa. \tag{4.66}$$

where $\mathbf{R}_\kappa$ represents the correlation matrix of the κth sound source

Since the correlation matrix of one source does not interfere with that of another source, the analysis of the beamforming power can be simplified. The beamforming power also is given by a sum of the individual beamforming powers. That is,

$$\begin{aligned} \beta(\theta) &= \sum_{\kappa=1}^{K} \mathbf{w}(\theta) \mathbf{R}_\kappa \mathbf{w}(\theta)^H \\ &= \sum_{\kappa=1}^{K} \beta_\kappa(\theta). \end{aligned} \tag{4.67}$$

where β_κ is the beamforming power of the κth sound source. Therefore, each beamforming power of an individual source can be decomposed, provided that we have sufficient knowledge of the variance of each source. This idea allows us to apply the partial field decomposition technique to the beamforming power (Section 4.7.3).

4.5 Correlation-Based Scan Vector Design

The resolution of the model-based beamforming technique, such as planar or spherical beamforming, is limited by the aperture size with respect to the wavelength. To overcome this resolution limit, super-directive beamforming techniques have been proposed. This breakthrough has been accomplished by analyzing the spatial information measured by the array. In the methods described in this section, the scan vector is designed using information of the measured signals or correlation matrix. The design of the scan vector often involves an optimization technique, which finds a scan vector that maximizes or minimizes a cost function defined in terms of the correlation matrix. In defining the optimization problem, several constraints should be considered together in order to reflect the practical limitations and secure robustness. In what follows, various beamformer designs are introduced, and their principles for dealing with the resolution and robustness are explained.

4.5.1 Minimum Variance Beamformer

When multiple sources impinge on the microphone array, the beamforming power β becomes a mixture of contributions from other sources as well as the contribution from the sound source in the look direction. The minimum variance (MV) beamformer is designed to minimize the undesirable influence from sources in directions other than the look direction θ. To begin, consider a correlation matrix $\mathbf{R}$ constructed from the sound fields of multiple sources in different directions.

The beamforming power from an unknown scan vector $\mathbf{w}$ is given by

$$\beta(\theta) = \mathbf{w}(\theta)^H \mathbf{R} \mathbf{w}(\theta). \tag{4.68}$$

Suppose that the scan vector is normalized to produce a designated beamformer output b_0 in response to the wave arriving from the look direction θ:

$$b(\theta) = \mathbf{w}(\theta)^H \mathbf{h}(\theta) = b_0. \tag{4.69}$$

Here, the vector $\mathbf{h}(\theta)$ denotes the transfer function of a sound source in the look direction θ, which is the same as the scan vector $\mathbf{w}(\theta)$ for the DAS beamformer. However, in this case, the optimal scan vector $\mathbf{w}$ has not yet been determined and is different from $\mathbf{h}$.

In fact, the measured correlation $\mathbf{R}$ and the beamforming power of Equation (4.68) consist of contributions from multiple sources in different directions, so it is necessary to reject the contributions from unwanted sound sources. If such a scan vector exists, then the beamforming power will only contain the contribution from the source in the θ direction. To this end, the MV beamformer introduces an important assumption that the total beamforming power β in the θ direction will be minimized if the contributions of other sources are removed by the successive choice of $\mathbf{w}(\theta)$. The MV problem can thus be defined as

$$\text{Minimize } \beta(\theta) = \mathbf{w}(\theta)^H \mathbf{R} \mathbf{w}(\theta), \text{ subject to } \mathbf{w}(\theta)^H \mathbf{h}(\theta) = b_0. \tag{4.70}$$

In this way, the variance of the beamformer output signal

$$E\left[b(\theta)^* b(\theta)\right] = \beta(\theta) \tag{4.71}$$

is minimized with the linear constraint of Equation (4.69). For this reason, the beamformer is often referred to as the linearly constrained minimum variance (LCMV) beamformer, or simply the Capon beamformer after the name of its inventor. With $b_0 = 1$, the beamformer output in response to the wavefront coming from the look direction will not be distorted; this technique is known as the minimum variance distortionless response (MVDR) beamformer.

The optimal solution $\mathbf{w}$ of Equation (4.70) can be found using the Lagrange multiplier method (Appendix A, Section A.4.3). The equivalent optimization problem can be defined as

$$\text{Minimize } J(\mathbf{w}, \mu) = \mathbf{w}^H \mathbf{R} \mathbf{w} + \mu(\mathbf{w}^H \mathbf{h} - b_0), \tag{4.72}$$

where μ is the Lagrange multiplier. By finding the zero-derivative point of J with respect to $\mathbf{w}$ and μ, we have

$$\mathbf{R}\mathbf{w} + \mu \mathbf{h} = 0 \Rightarrow \mathbf{w} = -\mu \mathbf{R}^{-1} \mathbf{h}, \tag{4.73}$$

$$\mathbf{w}^H \mathbf{h} = b_0 \Rightarrow -\mu^* (\mathbf{h}^H \mathbf{R}^{-1} \mathbf{h}) = b_0. \tag{4.74}$$

Arranging Equation (4.74) to solve for $\mathbf{w}$, the optimal scan vector can be written as

$$\mathbf{w}(\theta) = \frac{\mathbf{R}^{-1}\mathbf{h}(\theta)}{\mathbf{h}(\theta)^H \mathbf{R}^{-1} \mathbf{h}(\theta)} b_0^*, \tag{4.75}$$

and the beamforming power of Equation (4.70) is given by

$$\beta_{MV}(\theta) = \mathbf{w}(\theta)^H \mathbf{R} \mathbf{w}(\theta) = \frac{|b_0|^2}{\mathbf{h}(\theta)^H \mathbf{R}^{-1} \mathbf{h}(\theta)}. \tag{4.76}$$

As can be seen in Equation (4.73), the scan vector is basically equivalent to the least-squares solution ($\mathbf{R}^{-1}\mathbf{h}$; Appendix A, Equation A36), except for the scale factor $-\mu$. The correlation matrix $\mathbf{R}$ is usually rank deficient, since the number of uncorrelated sources (K) is less than the number of microphones. However, in many practical situations, the presence of uncorrelated noise ensures the invertibility of $\mathbf{R}$.

4.5.1.1 Performance of MV Beamformer

To evaluate the performance of the MV beamformer in the presence of measurement noise, we consider the incoherent noise of Equation (4.51) and assume a deterministic plane wave source located in the θ_0 direction (Equation (4.53)). Then, the inverse of the measured correlation matrix $\mathbf{R}'$ can be written as

$$\mathbf{R}'^{-1} = \left(|q|^2\mathbf{h}(\theta_0)\mathbf{h}(\theta_0)^H + \sigma_{noise}^2\mathbf{I}\right)^{-1}$$
$$= \frac{1}{\sigma_{noise}^2}\left(\mathbf{I} - \frac{|q|^2}{M|q|^2+\sigma_{noise}^2}\mathbf{h}(\theta_0)\mathbf{h}(\theta_0)^H\right), \quad (4.77)$$

from the matrix inverse relation of Equation (A37). Assuming that the gain in the look direction is 1 ($b_0 = 1$), the beamforming power of the MV beamformer is given by

$$\beta_{MV}(\theta) = \frac{\sigma_{noise}^2}{M}\cdot\frac{M|q|^2+\sigma_{noise}^2}{M|q|^2\sin^2\gamma(\theta)+\sigma_{noise}^2}. \quad (4.78)$$

Note that the transfer function similarity $\cos^2\gamma(\theta)$ of Equation (4.56) is again involved here. The maximum beamforming power of $|q|^2 + \sigma_{noise}^2/M$ can obtained when $\theta = \theta_0$, and the minimum is given by σ_{noise}^2/M for $\mathbf{h}(\theta)^H\mathbf{h}(\theta_0) \approx 0$. From these results, the PSR can be calculated as

$$PSR = M\frac{|q|^2}{\sigma_{noise}^2} + 1, \quad (4.79)$$

which is the same as the PSR of the DAS beamformer (Equation (4.58)). Therefore, the MV beamformer is not beneficial in removing the bias of the DAS beamformer. Instead, the sidelobe perturbation of the MV beamformer is comparably small. If we compare the beamforming power of MV (Equation (4.78)) and DAS (Equation (4.57)), it can be observed that the decay of beamforming power near the mainlobe region (small $\gamma(\theta)$) is more rapid for the MV beamformer. This characteristic is represented in Figure 4.15, which is produced with the same array configuration and

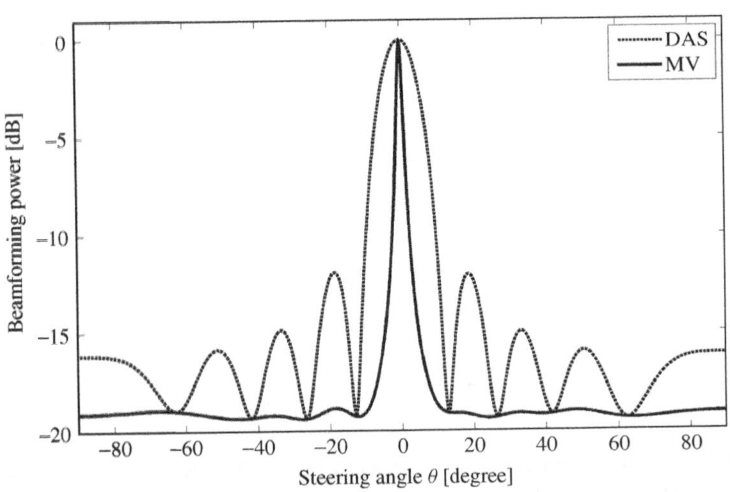

Figure 4.15 Comparison of the delay-and-sum (DAS) and minimum variance (MV) beamformer (the simulation conditions are the same as those in Figure 4.14)

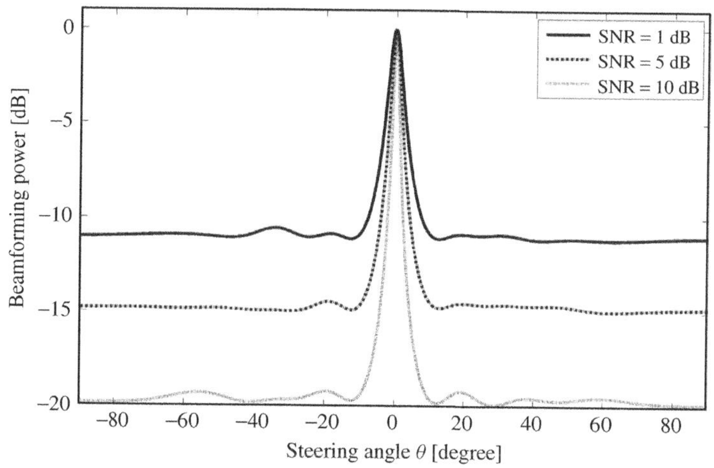

Figure 4.16 Beamforming powers of the MV beamformer with different SNRs

SNR as those in Figure 4.14. The comparison to the DAS beamformer shows that the beamforming power in various directions θ has fewer sidelobes and the width of the mainlobe is much narrower.

This high performance compared with the DAS beamformer is inherited from the data dependency of the scan vector $\mathbf{w}$. Since the scan vector of the MV beamformer incorporates the inverse of the correlation matrix (Equation (4.77)), it adaptively changes according to the structure of the correlation matrix. This is the major difference with the DAS beamformer for which the scan vector does not change in response to the signal characteristics. In Figure 4.16, the beamforming power of the MV beamformer is depicted for different SNR values. It can be seen that the mainlobe width of the MV beamformer adaptively changes in response to the SNR, unlike the DAS beamformer.

Although the MV beamformer has a better source identification performance, the beamforming power may be more sensitive to the variance in the noise. For example, as the noise variance goes to zero, the beamforming power of Equation (4.78) and the inverse of the correlation matrix (Equation (4.77)) become singular. In such a case, any coherent noise in the microphone can easily distort the beamforming power. To obtain a more robust result, it is sometimes desirable to add spatially uncorrelated noise ($\sigma^2\mathbf{I}$) to the correlation matrix $\mathbf{R}$:

$$\mathbf{w}(\theta) = \frac{(\mathbf{R} + \sigma^2\mathbf{I})^{-1}\mathbf{h}(\theta)}{\mathbf{h}(\theta)^H(\mathbf{R} + \sigma^2\mathbf{I})^{-1}\mathbf{h}(\theta)} b_0^*. \tag{4.80}$$

Due to the applied diagonal loading, the inverse of the matrix $\mathbf{R}$ is less influenced by unexpected noise.

4.5.1.2 Generalized LCMV

The single constraint of Equation (4.69) can be generalized to multiple constraints that express more specific control over the beam pattern. For example, if the direction of the undesirable source (θ_u) is known, then one can introduce an additional constraint to produce a zero beamformer output in that direction. The additional constraint, together with the original constraint for the look direction, can be written as

$$\mathbf{w}(\theta)^H \left[\mathbf{h}(\theta)\ \mathbf{h}(\theta_u)\right] = \left[b_0\ 0\right] \Rightarrow \mathbf{w}(\theta)^H\mathbf{C} = \mathbf{b}^T. \tag{4.81}$$

The solution of this generalized LCMV beamformer can be found using the same procedure as that for the single constraint (Equations (4.73) and (4.74)):

$$\mathbf{w}(\theta) = \mathbf{R}^{-1}\mathbf{C}\left[\mathbf{C}^{H}\mathbf{R}^{-1}\mathbf{C}\right]^{-1}\mathbf{b}^{*}.$$

4.5.1.3 Limitations

Despite all these enhancements, the MV beamformer requires one basic assumption to be satisfied: the MV of the beamformer output should be equivalent to the minimal contribution from the sources in other directions. This assumption is valid in the case of incoherent sound sources, for which the beamforming powers of individual sources are always additive (Equation (4.67)), meaning that the MV beamformer can reject undesirable sources.

However, the result is completely different if the sources are partially or completely coherent with each other. In such a case, the beamforming power can be minimal without rejecting the sound fields of the other sources. A certain weighted sum of sound fields from multiple sources, including the one from the desired direction, can minimize the variance without decreasing the contribution from other sources. If we use such a weight as a scan vector, then the beamformer output is minimized not by rejecting the interfering sources but by combining the interfering signals from other directions.

Another important aspect of the MV beamformer is that the transfer function $\mathbf{h}(\theta)$ used for the constraint should be accurate. The correlation matrix $\mathbf{R}$ is what we have measured, but the transfer function $\mathbf{h}(\theta)$ is still based on the model we have assumed. Therefore, an inaccurate model or constraint can lead to a poor result.

The bias of the beamforming power due to the noise is another limitation of the MV beamformer. As explained through Equation (4.79), the PSR of the MV beamformer is not different to that of the DAS beamformer. Nevertheless, the MV beamformer is one of the few super-resolution beamformers that follows the definition of the beamforming power of Equation (4.5). Since it has an explicit form of the scan vector, the beamformer output $b(\theta, \omega)$ can be calculated in the same way as the DAS beamformer. This compatibility with the conventional beamformer is of great advantage when we attempt to reconstruct the source signal $q(\omega)$ from the beamformer output (the virtual microphone of Equation (4.12)).

4.5.2 *Linear Prediction*

Another super-resolution beamformer, based on linear prediction (LP), utilizes the relationship between the spatial Fourier transform of the scan vector $\mathbf{w}$ and the beamformer output. For a scan vector using the planar wavefront model, the output of the DAS beamformer (Equation (4.8)) can be written as

$$b(\theta) = \sum_{m=1}^{M} w_m^* p_m = \frac{1}{M}\sum_{m=1}^{M} p_m e^{i(m-1)\Omega}, \quad \text{where } \Omega = \omega(d/c)\sin\theta. \tag{4.82}$$

It is noteworthy that the beamformer output b is the same as the discrete Fourier transform of the pressure signal p_m, given by

$$P\left(e^{i\Omega}\right) = \frac{1}{M}\sum_{m=1}^{M} p_m e^{i(m-1)\Omega}. \tag{4.83}$$

Therefore, the beamforming power $\beta(\theta)$ of Equation (4.5) can be considered as the *spatial spectrum* of the pressure signal. The objective of beamforming, finding the peak locations of the beamforming power, can then be regarded as finding the peak locations of the spatial spectrum.

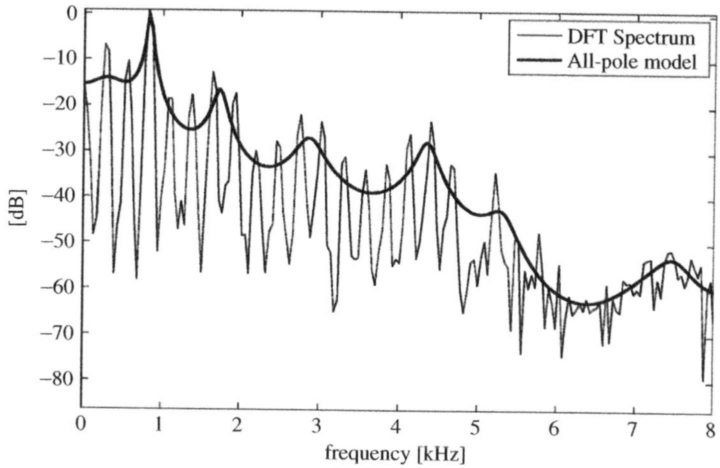

Figure 4.17 Example of spectral estimation by an all-pole model (comparison of the discrete Fourier transform (DFT) and estimated LP spectrum)

The idea behind the method described in this section is to utilize spectral estimation techniques that are superior in finding spectral peaks. The spectral peaks can be presented more distinctly by replacing the discrete Fourier transform, that is, DAS beamforming, with the spectral estimation method (Figure 4.17).

For the identification of the spectral peaks, an all-pole model of a sequence is beneficial because the poles essentially describe the location and magnitude of the spectral peaks. The all-pole model, also called the autoregressive (AR) model, assumes that the frequency response of a sequence can be represented by a combination of poles. The LP method is commonly used to estimate the pole locations.

4.5.2.1 All-Pole Model

To investigate the background of LP, consider the z-transform of a sequence x_n,

$$X(z) = \sum_{n=-\infty}^{\infty} x_n z^{-n}. \tag{4.84}$$

If the sequence x_n follows the all-pole model, then $X(z)$ can be written generally as

$$X(z) = \prod_{j=1}^{N} \frac{X_0}{(1 - a_j z^{-1})}, \tag{4.85}$$

where a_j denotes a pole, and X_0 denotes the general gain factor of a sequence. The number of poles N is often referred to as the prediction order. By setting $z = e^{i\Omega}$, the z-transform of Equation (4.84) can be converted to the discrete Fourier transform, and we can obtain the frequency response of $X(e^{i\Omega})$;

$$X(e^{i\Omega}) = \sum_{n=-\infty}^{\infty} x_n e^{-in\Omega} = \prod_{j=1}^{N} \frac{X_0}{(1 - a_j e^{-i\Omega})}. \tag{4.86}$$

Here, Ω represents the normalized frequency in radians. According to Equation (4.86), the spectrum $|X(e^{i\Omega})|^2$ will show a strong peak whenever Ω is close to the phase of each pole ($\angle a_j$). This explains the behavior of the spectral peak near the resonance frequency. Therefore, by finding the poles of the sequence of x_n, the spectral peaks can be efficiently represented. In practice, however, a sequence x_n may or may not follow the all-pole model of a particular order N, and hence we have to find poles that can approximate a given sequence to the assumed model.

4.5.2.2 Approximation of a Sequence as All-Pole Model

To approximate an arbitrary sequence to an all-pole model, it is necessary to understand the recursive nature of the all-pole model. By expanding the denominator of Equation (4.85), the all-pole model can be rewritten as

$$X(z) = \frac{X_0}{1 - \sum_{m=1}^{N} s_m z^{-m}} \quad \text{or} \quad X\left(e^{i\Omega}\right) = \frac{X_0}{1 - \sum_{m=1}^{N} s_m e^{-im\Omega}}. \tag{4.87}$$

Equation (4.87) can be rearranged to give

$$X(z) = X_0 + X(z) \sum_{m=1}^{N} s_m z^{-m}, \tag{4.88}$$

of which the inverse z-transform gives

$$x_n = X_0 + \sum_{m=1}^{N} s_m x_{n-m}. \tag{4.89}$$

Accordingly, if a sequence x_n perfectly follows an all-pole model of order N, then its current value can be estimated as a linear combination of its previous samples x_{n-m}.

LP utilizes this recursive relation to estimate the coefficients s_m. From Equation (4.89), the nth sample of an arbitrary sequence x_n can be estimated from its past samples. The estimated sample $\hat{x}_n$ can be written as

$$\hat{x}_n = X_0 + \sum_{m=1}^{N} s_m x_{n-m}. \tag{4.90}$$

If the sequence x_n is different from the all-pole model of specified order, then there must be a *prediction error* between the true and estimated values:

$$\begin{aligned} \varepsilon_n &= x_n - \hat{x}_n \\ &= x_n - X_0 - \sum_{m=1}^{N} s_m x_{n-m}. \end{aligned} \tag{4.91}$$

The coefficients s_m are determined such that the prediction error ε_n can be minimized. However, we note that the overall gain X_0 of Equation (4.86) is not important in finding the peaks of the spectrum. Thus, the prediction of x_n and the corresponding prediction error are often defined as

$$\hat{x}_n = \sum_{m=1}^{N} s_m^* x_{n-m}, \tag{4.92}$$

$$\varepsilon_n = x_n - \hat{x}_n = x_n - \sum_{m=1}^{N} s_m x_{n-m}. \tag{4.93}$$

The overall gain X_0 is now embedded in the error ε_n. For the ideal condition in which the measured pressure signal perfectly follows the all-pole model, the overall gain can be estimated from the prediction error as

$$X_0 = \varepsilon_n. \tag{4.94}$$

Consequently, the LP problem can be summarized as

$$\underset{s_m}{\text{Minimize}}\, |\varepsilon_n|^2 = \left| x_n - \sum_{m=1}^{N} s_m x_{n-m} \right|^2. \tag{4.95}$$

4.5.2.3 Beamforming as a LP Problem

Before going further, let us examine how the beamforming problem can be described as an LP problem. By setting the prediction order to $N = M - 1$ and substituting

$$x_{n-m} = p_{M-m}, \quad s_m = -w_{M-m}^* \tag{4.96}$$

into Equation (4.93), the prediction error can be rewritten as

$$\begin{aligned} \varepsilon_M &= p_M - \hat{p}_M \\ &= p_M + \sum_{m=1}^{M-1} w_{M-m}^* p_{M-m} \\ &= \sum_{m=1}^{M} w_m^* p_m, \quad \text{where } w_M = 1. \end{aligned} \tag{4.97}$$

Therefore, the beamforming problem is converted to an LP problem that predicts the data of the Mth microphone using the other microphone signals. By solving the LP problem, that is, by minimizing ε_M, we can design a scan vector w_m that accurately predicts the peak locations. (Note that the Mth element of a scan vector is constrained to $w_M = 1$.) The corresponding spectrum of the all-pole model can be derived from Equation (4.87) as

$$P_{AP}\left(e^{i\Omega}\right) = \frac{P_0}{1 + \sum_{m=1}^{M-1} w_{M-m}^* e^{-im\Omega}}. \tag{4.98}$$

where P_0 denotes the overall gain of the all-pole model P_{AP}.

The optimal coefficient w_m^* minimizing the prediction error ε_M can be derived in a least-squares sense. First, Equation (4.97) can be described in vector form as

$$\varepsilon_M = \mathbf{w}^H \mathbf{p}, \quad \text{with } w_M = 1.$$

Hence, the error minimization problem can be defined as

$$\text{Minimize } E\left[|\varepsilon_M|^2\right] = E\left[\left|\mathbf{w}^H \mathbf{p}\right|^2\right] = \mathbf{w}^H \mathbf{R} \mathbf{w}, \text{ subject to } \mathbf{w}^H \boldsymbol{\delta}_M = 1, \tag{4.99}$$

where $\boldsymbol{\delta}_M$ is a column vector whose elements are zero except for the Mth element ($[\boldsymbol{\delta}_M]_m = \delta_{mM}$).

Since the constrained optimization problem of Equation (4.99) is equivalent to Equation (4.70), the optimal coefficient **w** can be derived as

$$\mathbf{w} = \frac{\mathbf{R}^{-1}\boldsymbol{\delta}_M}{\boldsymbol{\delta}_M^T\mathbf{R}^{-1}\boldsymbol{\delta}_M}. \tag{4.100}$$

From Equation (4.99), we can see how the LP technique is different from the MV beamformer. The cost function is the same, but the constraint is different. The constraint imposed here is not related to the beamformer output but for approximating the given microphone signals as the all-pole model. Accordingly, we observe that the scan vector of Equation (4.100) is not a function of the look direction and hence not related to the directional estimation. The directional estimation is accomplished only after we inspect the spatial spectrum of the all-pole model (Equation (4.98)). The spectrum of an all-pole model can be described as

$$\begin{aligned}|P(e^{i\Omega})|^2 &= \frac{P_0{}^2}{\left|w_M^* + w_{M-1}^* e^{-i\Omega} + \cdots + w_1^* e^{-i(M-1)\Omega}\right|^2} \\ &= \frac{P_0{}^2}{\left|w_M^* e^{i(M-1)\Omega} + w_{M-1}^* e^{i(M-2)\Omega} + \cdots + w_1^*\right|^2}.\end{aligned} \tag{4.101}$$

Inserting $\Omega = \omega(d/c)(\sin\theta)$ into Equation (4.101) yields the LP beamforming power with respect to θ:

$$\beta_{LP}(\theta) = \left|P\left(e^{i\omega(d/c)(\sin\theta)}\right)\right|^2 = \frac{|\mathbf{w}^H\mathbf{p}|^2}{|\mathbf{w}^H\mathbf{h}(\theta)|^2}, \tag{4.102}$$

where the vector **h** is given by

$$\mathbf{h}(\theta) = \left[1, e^{i\Omega}, \cdots, e^{i(M-1)\Omega}\right]^T, (\Omega = \omega(d/c)(\sin\theta)) \tag{4.103}$$

and the overall gain P_0 is obtained from Equation (4.94).

From the optimal scan vector Equation (4.100), the beamforming power of Equation (4.102) can be written as

$$\beta_{LP}(\theta) = \frac{\boldsymbol{\delta}_M^T\mathbf{R}^{-1}\boldsymbol{\delta}_M}{|\boldsymbol{\delta}_M^T\mathbf{R}^{-1}\mathbf{h}(\theta)|^2}. \tag{4.104}$$

The all-pole model discussed so far only considers the prediction of the Mth microphone from the other microphone signals. Assuming that the magnitude of a target spectrum is invariant to the circular shift of the microphone signal, it is also possible to construct the LP coefficient **w** that predicts the m_0th microphone signal from the other microphone signals. In such a case, the beamforming power of Equation (4.104) is given by

$$\beta_{LP}(\theta) = \frac{\boldsymbol{\delta}_{m0}^T\mathbf{R}^{-1}\boldsymbol{\delta}_{m0}}{\left|\boldsymbol{\delta}_{m0}^T\mathbf{R}^{-1}\mathbf{h}(\theta)\right|^2}. \tag{4.105}$$

Although the magnitude of the target spectrum is shift invariant, that of the AR model is not. Therefore, in practice, the choice of m_0 can lead to significant differences in the predicted pole locations. A popular choice of m_0 is either the final element (the Mth microphone) or an element near the phase center of the microphone array [4].

4.5.2.4 Performance of LP Method

It has been shown by Burg [5] that the reciprocal of the MV beamformer is equivalent to the sum of the reciprocals of the LP beamformer up to the $M-1$ th order ($M-1$ poles). That is,

$$\frac{1}{\beta_{MV}(\theta)} = \sum_{m=1}^{M-1} \frac{1}{\beta_{LP,m}(\theta)}, \tag{4.106}$$

where $\beta_{LP,m}$ denotes the beamforming power of the mth order LP estimator. Since the low-order estimations of the LP beamformer are averaged to constitute the MV beamformer, the resolution of the MV beamformer is lower than that of the LP beamformer.

In Figure 4.18, an example of the LP method is shown with the result obtained from the MV technique. The LP is advantageous near the peak locations, and the effect of noise is decreased across all steering angles.

If we consider the planar wavefront model with uncorrelated measurement noise (Equation (4.54)), the beamforming power of Equation (4.105) can be written as

$$\beta_{LP}(\theta) = \sigma_{noise}^2 \frac{\left(1 - \dfrac{|q|^2}{M|q|^2 + \sigma_{noise}^2}\right)}{\left|1 - \dfrac{M|q|^2}{M|q|^2 + \sigma_{noise}^2}\cos^2\gamma(\theta)\right|^2}. \tag{4.107}$$

The maximum and minimum beamforming powers can be obtained when $\cos^2\gamma(\theta) = 1$ and $\cos^2\gamma(\theta) \approx 0$, respectively. The PSR of the LP can be derived as

$$PSR = \left(\frac{M|q|^2}{\sigma_{noise}^2} + 1\right)^2, \tag{4.108}$$

which is the squared value of the MV beamformer (Equation (4.79)). Accordingly, the minimum sidelobe level of the LP method shown in Figure 4.18 is half of that of the MV beamformer in dB scale.

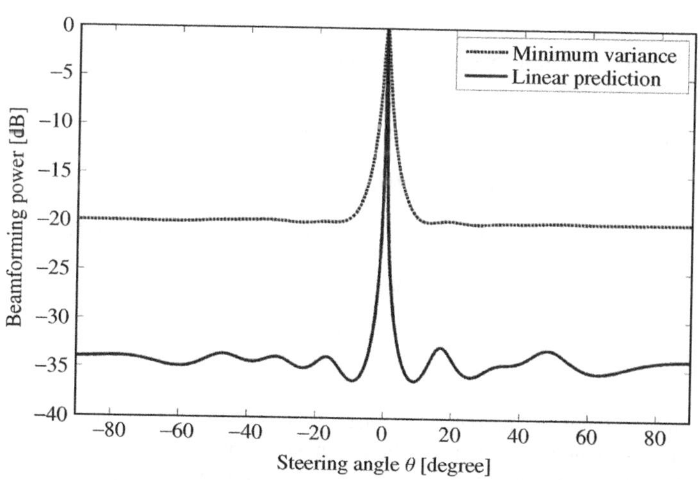

Figure 4.18 Comparison of the LP and MV beamformers (the simulation conditions are the same as those in Figure 4.14)

4.6 Subspace-Based Approaches

4.6.1 *Basic Principles*

The correlation matrix constructed from a real measurement contains information about the source signals and noise. If we can separate the source and noise contributions, then the beamforming performance would be further enhanced. Subspace methods attempt to separate the contributions by utilizing some prior knowledge or by introducing practically reasonable assumptions. In what follows, it is assumed that the noise and source signals measured at the microphones are incoherent. If we reintroduce the statistical vector notation of Equation (4.46), the incoherent condition can be expressed as

$$\tilde{\mathbf{P}}\tilde{\mathbf{N}}^H = 0. \tag{4.109}$$

Then, the two vector spaces spanned by the rows of $\tilde{\mathbf{P}}$ and $\tilde{\mathbf{N}}$ are orthogonal to each other. This "statistical" orthogonality allows us to separate the signal and noise contributions.

A *signal subspace* can be defined in terms of the columns of $\tilde{\mathbf{P}}$. Since each column of $\tilde{\mathbf{P}}$ represents the spatial sound distribution, the basis functions spanning the signal subspace are *spatial vectors*, which represent the possible shapes of the pressure fields at microphone positions. Likewise, the *noise subspace* refers to the vector space spanned by the columns of $\tilde{\mathbf{N}}$.

It should be emphasized, however, that in general the signal subspace is not orthogonal to the noise subspace, that is, $\tilde{\mathbf{P}}^H\tilde{\mathbf{N}} \neq 0$. This is because the dimension of the column space is limited by the number of microphones M. Suppose that the total dimensions of the incoherent sources (K) and noise (N) exceeds the number of microphones ($K + N > M$). Then, the signal and noise subspaces should share the same dimension $\mathbb{C}^{M\times 1}$, and it is thus impossible to discern the signal and noise subspaces. Accordingly, the *row spaces* of $\tilde{\mathbf{P}}$ and $\tilde{\mathbf{N}}$ are orthogonal if they are incoherent, but their column spaces are not. In other words, the spatial distributions of the signal and noise are not orthogonal.

Therefore, some prior knowledge is required to separate the signal and noise subspaces. In what follows, several methods to separate the signal and noise subspaces from the measured correlation matrix $\mathbf{R}$ are presented. It is shown that the signal and noise subspaces can be decomposed under some limited conditions.

4.6.1.1 Incoherent Noise Model with Equal Variances

To begin, consider incoherent measurement noise signals of equal variance σ^2_{noise}. From Equation (4.51), the noise correlation is given by the identity matrix,

$$\mathbf{R}'(\omega) = \mathbf{R}(\omega) + \sigma^2_{noise}(\omega)\mathbf{I}.$$

We then consider an eigenvalue analysis of the signal correlation matrix $\mathbf{R}$. That is,

$$\mathbf{R}\mathbf{u}_j = D_j\mathbf{u}_j \Rightarrow \quad \begin{aligned} \mathbf{R}' &= \mathbf{U}\mathbf{D}\mathbf{U}^H + \sigma^2_{noise}\mathbf{I} \\ &= \sum_{j=1}^{M} D_j\mathbf{u}_j\mathbf{u}_j^H + \sigma^2_{noise}\mathbf{I}, \end{aligned} \tag{4.110}$$

where $\mathbf{D}$ represents a diagonal matrix consisting of eigenvalues (D_j), and $\mathbf{U}$ is the eigenvector matrix, the columns of which are given by the eigenvectors $\mathbf{u}_j$.

For a finite number of incoherent sources ($K < M$), the eigenvalue D_j is nonzero only for $j = 1, \cdots, K$. Since the eigenvectors $\mathbf{u}_j$ for $j = 1, \cdots, K$ span the column space of $\mathbf{R} = \tilde{\mathbf{P}}\tilde{\mathbf{P}}^H$, they become the basis functions of the signal subspace. In practice, an eigenvalue analysis of **R** is impossible, since we cannot separate the signal correlation matrix **R** from the measured **R**′. However, it is noteworthy that the eigenvector $\mathbf{u}_j$ of the signal correlation matrix **R** satisfies the following characteristic equation:

$$\mathbf{R}'\mathbf{u}_j = \left[D_j + \sigma_{noise}^2\right]\mathbf{u}_j. \tag{4.111}$$

Equation (4.111) shows that the eigenvector $\mathbf{u}_j$ of the signal subspace is the eigenvector of the measured correlation matrix **R**′ as well. As a result, we can find the eigenvector of the signal subspace from the measured correlation matrix without separating out the internoise correlation matrix. If we know the number of incoherent sources (K), then the eigenvectors corresponding to the K largest eigenvalues of **R**′ become the orthonormal basis vectors spanning the signal subspace.

It is interesting to see that the same characteristic equation can be set for the noise subspace:

$$\mathbf{R}_{nn}\mathbf{u}_j = \sigma_{noise}^2\mathbf{u}_j. \tag{4.112}$$

Accordingly, the eigenvectors $\mathbf{u}_j$ for $j = 1, \cdots, K$ are the basis functions for both the signal and noise subspaces. This is a special property that only can be obtained for incoherent measurement noise. If the internoise correlation matrix is not diagonal, Equation (4.112) may not hold.

In addition, the noise power $\sigma_{noise}^2(\omega)$ can also be extracted; the $M - K$ smallest eigenvalues of **R**′ are all equal to $\sigma_{noise}^2(\omega)$ (from $D_j = 0$ for $j > K$).

4.6.1.2 Separation of Signal and Noise Subspaces

From the eigenvalue analysis, we can separate the vector spaces spanned by the signals and noise. First, consider the eigenvalue analysis of the measured correlation matrix **R**′. That is,

$$\mathbf{R}' = \mathbf{U}'\mathbf{D}'\mathbf{U}'^H, \tag{4.113}$$

where **U**′ and **D**′ are the eigenvector and eigenvalue matrices of **R**′, respectively. Since the columns of **U**′ are orthogonal, the correlation matrix can be separated into two parts:

$$\mathbf{R}' = \mathbf{U}_{sn}\mathbf{D}_{sn}\mathbf{U}_{sn}^H + \mathbf{U}_n\mathbf{D}_n\mathbf{U}_n^H. \tag{4.114}$$

Here, $\mathbf{U}_{sn}$ and $\mathbf{D}_{sn}$ are the eigenvectors and eigenvalues corresponding to the K largest eigenvalues of **R**′, which are given by

$$\mathbf{U}_{sn} = \left[\mathbf{u}_1, \cdots, \mathbf{u}_K\right], \left[\mathbf{D}_{sn}\right]_{(\kappa,\kappa)} = \left[\mathbf{D}'\right]_{(\kappa,\kappa)} (\kappa = 1, \cdots, K), \tag{4.115}$$

and $\mathbf{U}_n$ and $\mathbf{D}_n$ are those defined from the remaining eigenvalues:

$$\mathbf{U}_n = \left[\mathbf{u}_{K+1}, \cdots, \mathbf{u}_M\right], \left[\mathbf{D}_n\right]_{(j,j)} = \left[\mathbf{D}'\right]_{(K+j,K+j)} (j = 1, \cdots, M - K). \tag{4.116}$$

As shown in Equation (4.111), the columns of $\mathbf{U}_{sn}$ are orthonormal basis vectors for both the noise and signal subspaces (*signal-plus-noise subspace*). Since the eigenvector $\mathbf{U}_{sn}$ of $\mathbf{R} = \tilde{\mathbf{P}}\tilde{\mathbf{P}}^H$ is a left singular vector of the matrix $\tilde{\mathbf{P}}$ (Equation (A65)), it can be said that $\mathbf{U}_{sn}$ and $\tilde{\mathbf{P}}$ span the same subspace. By contrast, the eigenvector and eigenvalue $\mathbf{U}_n$ and $\mathbf{D}_n$ only span the noise subspace.

In the diagonal elements of $\mathbf{D}_{sn}$, the eigenvalues of the signal and noise subspaces are mixed. However, the eigenvalue of the noise subspace can be calculated from the diagonal of $\mathbf{D}_n$, and thus we can separate the eigenvalues and construct the true correlation matrix **R**. Denoting the eigenvalues of the noise subspace ($\mathbf{D}_n$) as σ_{noise}^2, the free correlation matrix is given by

$$\mathbf{R} = \mathbf{U}_{sn}(\mathbf{D}_{sn} - \sigma_{noise}^2\mathbf{I})\mathbf{U}_{sn}^H, \tag{4.117}$$

and the noise correlation matrix can be written as

$$\mathbf{R}_n = \sigma_{noise}^2\mathbf{U}\mathbf{I}\mathbf{U}^H.$$

Consequently, we can perfectly decompose the correlation matrices **R** and $\mathbf{R}_n$ provided the measurement noise is incoherent. Applying the scan vector to the true correlation matrix **R** of Equation (4.117) will give a noise-free beamforming power:

$$\beta(\theta) = \mathbf{w}(\theta)^H\mathbf{R}\mathbf{w}(\theta). \tag{4.118}$$

Figure 4.19 shows the DAS beamforming results before and after applying the subspace separation. The bias of the sidelobes due to incoherent noise with a SNR of 0 dB can be removed by filtering out the noise subspace using Equation (4.117). However, the noise subspace separation is only effective for removing the noise and is not related to the resolution of the beamforming power.

4.6.1.3 Generalized Form of Subspace Separation

In practice, the assumption of incoherent noise signals cannot be applied, partly because of the coherent noise or inaccuracy in estimating the correlation matrix. Nevertheless, the signal subspace can be decomposed if the noise correlation is known or can be measured.

Suppose that the normalized noise covariance is known and given by $\mathbf{R}_{nn}$, without its variance σ_{noise}^2 being known. The measured correlation matrix $\mathbf{R}'$ from the microphones can be written as

$$\mathbf{R}'(\omega) = \mathbf{R}(\omega) + \sigma_{noise}^2\mathbf{R}_{nn}(\omega). \tag{4.119}$$

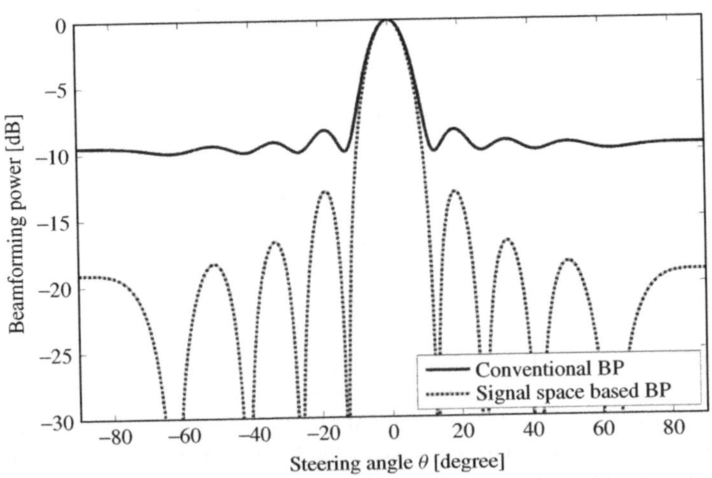

Figure 4.19 Delay-and-sum beamforming with and without noise subspace separation (uniform line array with nine microphones of a half wavelength distance; plane wave model; SNR = 0 dB)

Then consider a generalized eigenvalue problem such as

$$\mathbf{R}\mathbf{u}_j = D_j \mathbf{R}_{nn} \mathbf{u}_j. \tag{4.120}$$

Because the signal correlation matrix has rank K, there are eigenvectors that correspond to the zero eigenvalue $D_j = 0$ for $j > K$.

It can be seen that the eigenvector satisfying Equation (4.120) is also the generalized eigenvector of Equation (4.119), which fulfills

$$\mathbf{R}'\mathbf{u}_j = (D_j + \sigma_{noise}^2)\mathbf{R}_{nn} \mathbf{u}_j. \tag{4.121}$$

Therefore, the K generalized eigenvectors of Equation (4.121) calculated from the known noise covariance matrix $\mathbf{R}_{nn}$ will give the orthogonal basis functions for describing the signal subspace. The eigenvectors for $j > K$ make $D_j = 0$ and give σ_{noise}^2 as the eigenvalues. Accordingly, with prior knowledge of the covariance matrix, we can separate the signal space and estimate the variance of the noise signals.

The covariance matrix of a signal is, in practice, not easy to measure a priori. If we apply the incoherent noise model to the correlated noise signals, then the estimated eigenvalues and eigenvectors of the signal subspace will be distorted. The number of incoherent sound sources is also hard to predict when two source signals are partially coherent or their spatial distributions are similar.

4.6.2 MUSIC Beamformer

So far, we have discussed methods to separate the signal and noise subspaces, and it was shown that noise-free beamforming can be realized from prior knowledge of the noise covariance based on the fact that the eigenvectors corresponding to the K largest eigenvalues of the measured correlation matrix $\mathbf{R}'$ ($\mathbf{U}_{sn}$ of Equation (4.114)) are the basis vectors constituting the signal subspace. Those eigenvectors are orthogonal to the remaining eigenvectors ($\mathbf{U}_n$) spanning the noise subspace.

The MUSIC method utilizes the orthogonal relationship between these two eigenvector groups. However, it focuses more on the second group of eigenvectors ($\mathbf{U}_n$), which tells us whether a given vector does or does not belong to the signal subspace. Since $\mathbf{U}_n$ is orthogonal to the basis vectors of the signal subspace, the beamforming power calculated from the correlation matrix consisting of these eigenvectors,

$$\beta_n(\theta) = \mathbf{w}(\theta)^H \mathbf{U}_n \mathbf{U}_n^H \mathbf{w}(\theta), \tag{4.122}$$

will be minimal for a scan vector $\mathbf{w}(\theta)$ belonging to the signal subspace. This is the essential assumption of the MUSIC method for realizing super-resolution. Using Equation (4.122), the eigenvector correlation matrix can be defined as

$$\mathbf{R}_{EV}^{-1} = \mathbf{U}_n \mathbf{U}_n^H. \tag{4.123}$$

Because β_n is minimal at the source location, we take the inverse of the beamforming power with $\mathbf{R}_{EV}^{-1}$ to indicate the source location. That is,

$$\beta_{EV}(\theta) = \frac{1}{\mathbf{w}(\theta)^H \mathbf{R}_{EV}^{-1} \mathbf{w}(\theta)}. \tag{4.124}$$

Accordingly, $\beta_{EV}(\theta)$ has a maximum value when the scan vector $\mathbf{w}$ belongs to the signal subspace. The beamforming power of Equation (4.124) is often denoted as a *MUSIC spectrum.*

The beamforming power of the MUSIC, MV, and LP methods are compared in Figure 4.20. The ripples in the sidelobes due to the all-pole modeling disappear in the MUSIC spectrum. Like the LP technique, MUSIC utilizes the modified definition of the beamforming power, and an estimation of the beamformer output is not possible. Therefore, it can be used for finding the locations of sound sources, but the source signals at the source locations cannot be reconstructed.

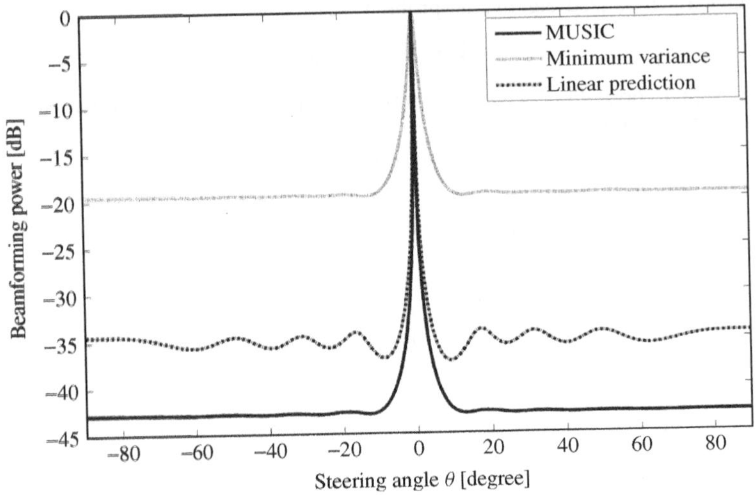

Figure 4.20 Comparison of the MUSIC, MV, and LP beamformers (the simulation conditions are the same as those in Figure 4.14)

4.6.2.1 Error due to the Sound Field Model

Unlike the MV or LP methods, the scan vector **w** of the MUSIC method is based on the model and not optimally determined. The MUSIC method requires that the scan vector **w** is orthogonal to the noise subspace when the steering angle θ coincides with the true source location θ_0. But this orthogonality condition cannot be achieved if the sound field model used for the scan vector **w** is different from the true sound propagation. For example, consider the case in which the scan vector designed from the plane wave model is used to predict spherical wave propagation. Since the signal subspace of the spherical wavefront has an amplitude factor of $1/r$ and a phase factor of $(\omega r/c)$, the scan vector of the planar wavefront may not belong to the signal subspace and can overlap with the noise subspace. For this reason, the peaks in the MUSIC spectrum can be biased [6].

To describe the errors associated with the phase difference between the spherical and plane wave models, simulations were conducted in the second quadrant, in which the angle between the source and mid position of array is between -90° and 0° (Figure 4.21). As illustrated in Figure 4.21, the time delay (τ_{peff}) of the measured sound pressure between the central position of the aperture and that of the mth microphone when the incoming wave is assumed to be planar can be expressed as

$$\tau_{peff} = -\frac{1}{2c}(m-1)d\sin\theta_0. \tag{4.125}$$

If the incoming wavefront is spherical, then the time delay (τ_{eff}) associated with the measured sound pressure can be readily obtained as

$$\tau_{eff} = \frac{(r_m - r_c)}{c} = \frac{1}{c}\left[\left(r_c^2 + \left(\frac{(m-1)d}{2}\right)^2 - r_c(m-1)d\sin\theta_0\right)^{1/2} - r_c\right], \tag{4.126}$$

where r_m is the distance to the mth microphone and r_c is the distance from a point source to the center of the array.

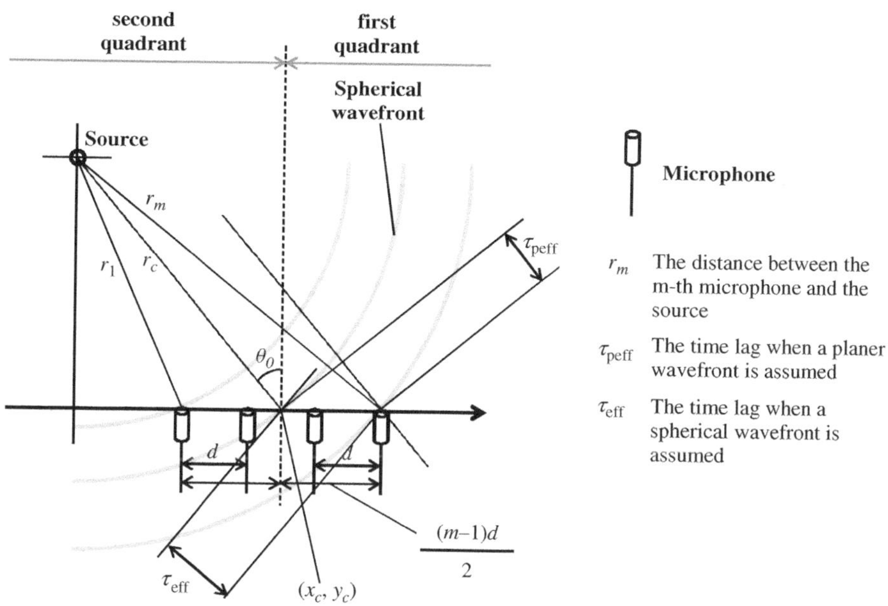

Figure 4.21 Typical wave propagation mode and a line array; propagation of plane and spherical waves

As the distance between the source and receiver goes to infinity, the time delay of the spherical wave model converges to the delay of the planar wavefront, that is,

$$\tau_{eff} \approx \frac{1}{c}\left[r_c\left(1 - \frac{(m-1)\,d}{r_c}\sin\theta_0\right)^{1/2} - r_c\right] \approx -\frac{1}{2c}(m-1)d\sin\theta_0 = \tau_{peff}. \qquad (4.127)$$

From this observation, one can see that the estimation error based on the plane wave model is a function of the geometrical parameters of the array: the total number of microphones M, the steering (bearing) angle θ between the sound source and aperture location, and the distance r_c between the sound source and microphone array.

4.6.2.2 Effects of the Source

The effects of a spherical wavefront on the estimation errors are shown through simulations of a 340-Hz point sound source. The distance r_c was set to 2.0 m. The percentage error associated with the estimation of the source angle is shown in Figure 4.22 (DAS) and Figure 4.23 (MUSIC). For these simulations, the measured sound pressure signals at each microphone are assumed to be planar. An increase in the number of microphones (or decrease in the distance between adjacent microphones), when the aperture size is kept constant at 2.0λ, is related to the error reduction due to an increase in the signal summation. Compared with the DAS beamformer, the errors in the MUSIC beamformer are much larger but tend to be zero when the bearing angle approaches 90° because in this special case the wavefronts received at each microphone can be regarded as planar wavefronts except for the amplitude factor of a spherical wave. We also note that there are rapid fluctuations in the errors associated with the MUSIC method. This can be readily understood

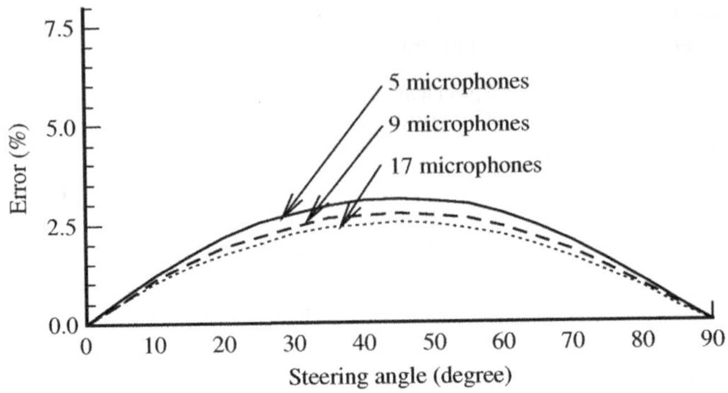

Figure 4.22 Prediction errors of the delay-and-sum beamforming method as a function of the steering angle for different numbers of microphones (aperture size: 2.0 λ; source frequency: 340 Hz)

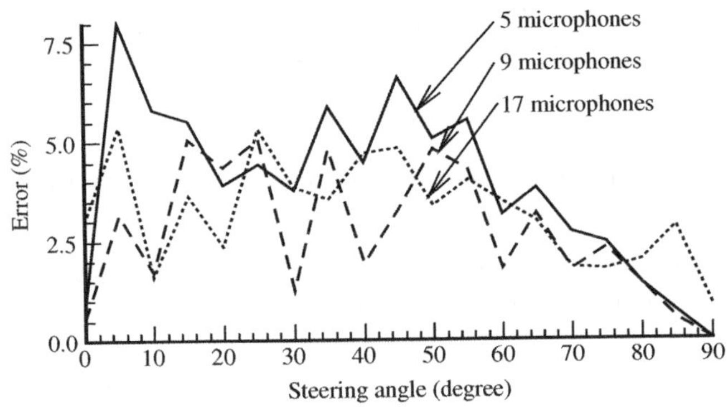

Figure 4.23 The prediction errors of the MUSIC algorithm as a function of the steering angle for different numbers of microphones (aperture size: 2.0 λ; source frequency: 340 Hz)

by realizing that the inner product of the noise subspace and scan vector can be regarded as a $(M-1)$th-order polynomial with respect to the bearing angle, and may produce many zeros when M is increased. Figures 4.22 and 4.23 also indicate that an increase in the number of microphones does not greatly improve the performance of the source angle estimation.

4.6.2.3 Effects of the Distance r_c

Another parameter associated with errors is, of course, the distance between the source and the microphone array. The aperture size used in the simulation is kept constant at 2.0λ for consistent comparisons. The effect of the distance is investigated for cases where r_c is equal to 2.0λ, 4.0λ, and 6.0λ. The results using the DAS beamforming and the MUSIC methods illustrate that an increase in r_c decreases the overall error level (Figures 4.24 and 4.25, respectively). These results are expected since the wavefront becomes more planar as r_c/λ increases.

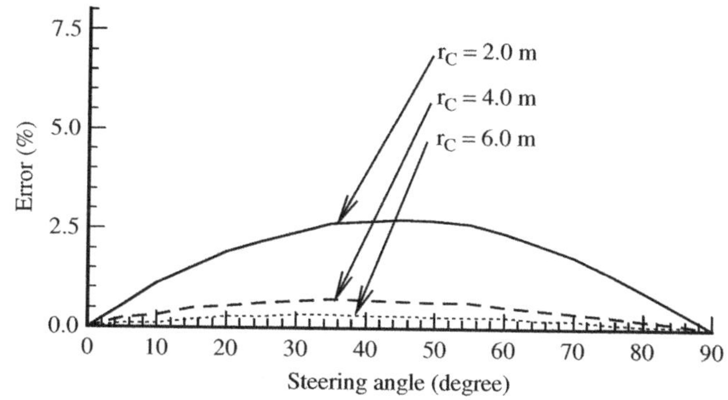

Figure 4.24 Errors due to an increase in the radius r_c as a function of the steering angle using the delay-and-sum beamforming (source frequency: 340 Hz)

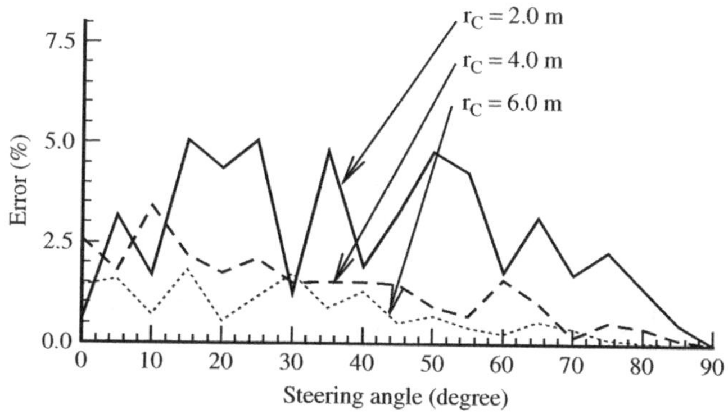

Figure 4.25 Errors due to an increase in the radius r_c as a function of the steering angle using the MUSIC power spectrum (source frequency: 340 Hz)

4.6.2.4 Effects of the Total Aperture Size

The errors associated with the assumption of a planar wavefront are increased by an increase in the aperture size, since an increase in the aperture length means an increase in the sphericity, that is, the total curvature of the spherical wavefront observed by the aperture. Figures 4.26 and 4.27 show the results based on the DAS beamforming and MUSIC methods when the number of microphones is increased, but the distance between the microphones is fixed to a quarter wavelength. The larger aperture size produces a larger error in the estimated source angle due to the large deviation from the plane wave model.

The estimation errors associated with the planar wavefront assumption mainly decrease as the bearing angle approaches 90° and as the distance between the source and microphone array is increased because the spherical wavefronts approach those of plane waves in both cases. These observations confirm our expectation that the errors associated with the spherical wavefronts are functions of the array parameters and the location of the spherical sound source, and therefore they

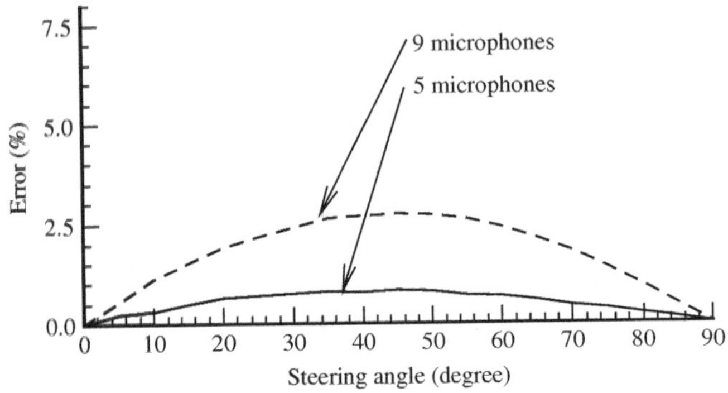

Figure 4.26 Prediction errors using the delay-and-sum beamforming due to total aperture size ($d = \lambda/4$; source frequency: 340 Hz)

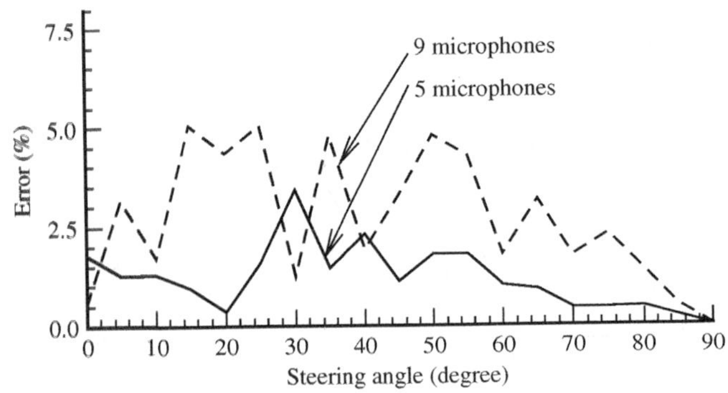

Figure 4.27 Prediction errors using the MUSIC algorithm due to total aperture size ($d = \lambda/4$; source frequency: 340 Hz)

cannot be reduced by the methods based on a plane wave model. Thus, one needs to use a modified scan vector to resolve spherical wavefronts.

4.6.2.5 Spherical MUSIC

The spherical MUSIC method [6] utilizes the scan vector designed for the spherical propagation model. The scan vector for the spherical wave model can be defined as

$$\mathbf{w}(x, y) = \left[\frac{1}{r_1}e^{i\omega r_1/c}, \frac{1}{r_2}e^{i\omega r_2/c}, \ldots, \frac{1}{r_M}e^{i\omega r_M/c}\right]^T,$$

where $r_m = \sqrt{(x - x_m)^2 + (y - y_m)^2}$ is the distance between the assumed source position (x, y) and the microphone position (x_m, y_m). When the assumed source position coincides with the true position, the orthogonality conditions between the noise subspaces and scan vector $\mathbf{w}$ can be satisfied, and therefore the location of the spherical sound source can be found. Note that the scan vector has a factor of $1/r_m$, which is different from the weighting vector for the spherical

DAS beamformer of Equation (4.17). This is motivated by the fact that the eigenvectors $\mathbf{U}_n = [\mathbf{u}_{K+1}, \mathbf{u}_{K+2}, \cdots, \mathbf{u}_M]$ of the noise subspace for the spherical wavefront also have a distance dependency of $1/r_m$. Figure 4.28 illustrates the array and sound source configuration used in the simulation. The performance of the spherical DAS beamforming and the spherical MUSIC model are shown in Figures 4.29 and 4.30, respectively. As can be seen, the spherical MUSIC model shows

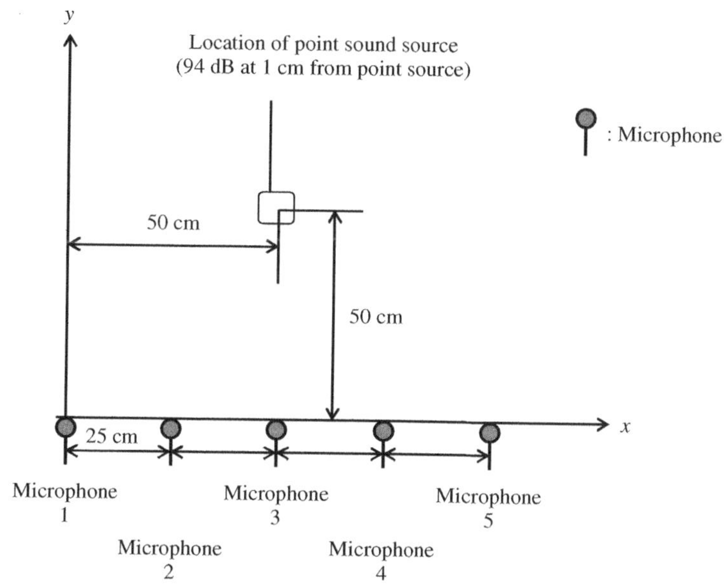

Figure 4.28 Location of the source and positions of the microphones (source frequency: 340 Hz; distance between microphones: 25 cm ($= \lambda/4$); source position: (0.5 m, 0.5 m))

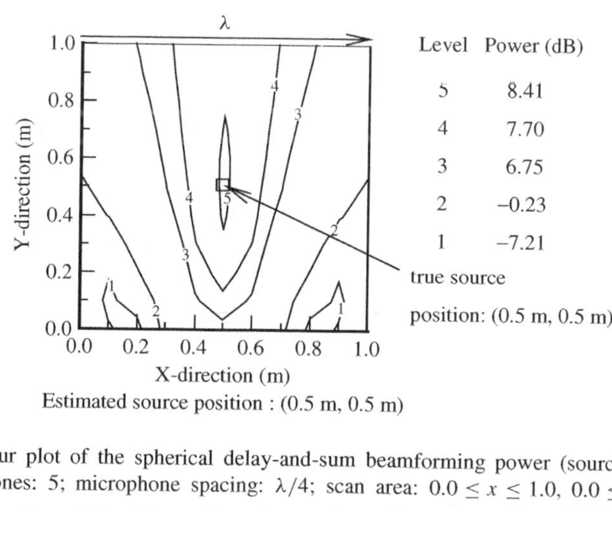

Figure 4.29 Contour plot of the spherical delay-and-sum beamforming power (source frequency: 340 Hz; number of microphones: 5; microphone spacing: $\lambda/4$; scan area: $0.0 \leq x \leq 1.0$, $0.0 \leq y \leq 1.0$, $\Delta x = 0.1$, $\Delta y = 0.1$)

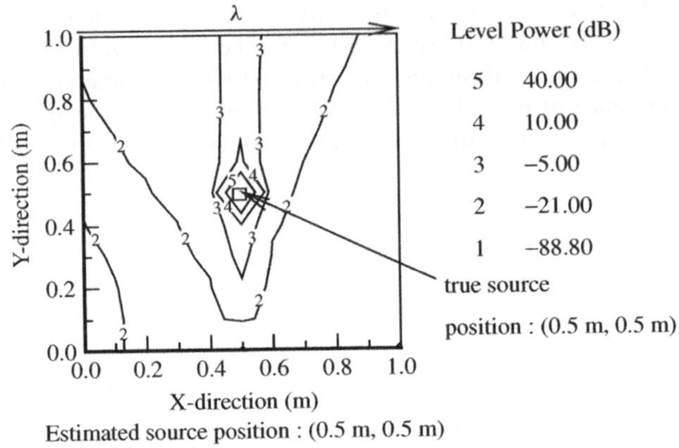

Figure 4.30 Contour plot of the spherical MUSIC beamforming power (source frequency: 340 Hz; number of microphones: 5; microphone spacing: $\lambda/4$; scan area: $0.0 \leq x \leq 1.0$, $0.0 \leq y \leq 1.0$, $\Delta x = 0.1$, $\Delta y = 0.1$)

an improved resolution in both the x and y directions compared with those obtained by the DAS beamformer.

4.6.3 ESPRIT

The ESPRIT is one popular modification of the subspace-based approaches. The basic idea of this technique is to partition the microphone array into two subarrays and exploit the relationship between the two subarrays for an accurate estimation of the source location.

To do this, we first identify the microphone array as a combination of two identical subarrays displaced by a known amount Δd. For example, a line array along the x axis with an intersensor interval d can be regarded as two subarrays separated by $\Delta d = d$ (Figure 4.31a). The measured

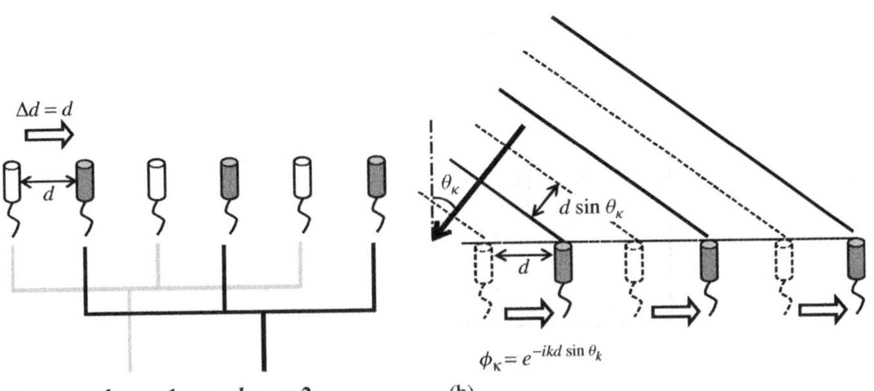

Figure 4.31 Concept of subarrays (a) construction of two subarrays from the single line array (b) transfer function change due to the position shift

pressure signal of Equation (4.41) can then be rewritten as

$$\mathbf{p}'(\omega) = \mathbf{H}(\omega)\mathbf{q}(\omega) + \mathbf{n}(\omega)$$
$$= \begin{bmatrix} \mathbf{H}_1(\omega) \\ \mathbf{H}_2(\omega) \end{bmatrix} \mathbf{q}(\omega) + \begin{bmatrix} \mathbf{n}_1(\omega) \\ \mathbf{n}_2(\omega) \end{bmatrix}. \tag{4.128}$$

The second transfer function $\mathbf{H}_2 \in \mathbb{C}^{M/2\times K}$ is measured at the locations displaced by Δd from the first subarray of which the transfer function is $\mathbf{H}_1 \in \mathbb{C}^{M/2\times K}$. If the measurement is taken under free-field conditions, the displacement in the measurement positions can be regarded as a shift of the source location in the opposite direction (shift invariance). Let us denote the transfer function change due to an equivalent shifted source location as

$$[\mathbf{H}_2]_{(m,\kappa)} = [\mathbf{H}_1]_{(m,\kappa)}\phi_\kappa. \tag{4.129}$$

Note that the change in the transfer function only depends on the source index κ. For instance, the plane wave model with the true source in the direction θ_κ will give $\phi_\kappa = e^{-ik\Delta d \sin\theta_\kappa}$ (Figure 4.31b). It is remarkable that the parameter ϕ_κ contains the essential information necessary for detecting the source location θ_κ. The objective of the ESPRIT is to estimate the parameter ϕ_κ without a priori information and using only the transfer functions of the two subarrays. In matrix form, the transfer functions $\mathbf{H}_1$ and $\mathbf{H}_2$ are related by

$$\mathbf{H}_2(\omega) = \mathbf{H}_1(\omega)\mathbf{\Phi}(\omega), \mathbf{H}(\omega) = \begin{bmatrix} \mathbf{H}_1(\omega) \\ \mathbf{H}_1(\omega)\mathbf{\Phi}(\omega) \end{bmatrix}, \tag{4.130}$$

where the parameter matrix $[\mathbf{\Phi}]_{\kappa,\kappa} = \phi_\kappa$ is diagonal and represents the change in the transfer function due to the subarray displacement.

The diagonal elements of the parameter matrix $\mathbf{\Phi}$ can be found from an eigenvalue analysis. Consider the eigenvalue analysis of $\mathbf{R}'$ in Equation (4.114). Since the eigenvector $\mathbf{U}_{sn} \in \mathbb{C}^{M\times K}$ and the transfer function $\mathbf{H} \in \mathbb{C}^{M\times K}$ span the same subspace, there exists an invertible transform $\mathbf{T}$ that gives

$$\mathbf{U}_{sn} = \mathbf{HT} \quad \text{or} \quad \begin{bmatrix} \mathbf{U}_{sn1}(\omega) \\ \mathbf{U}_{sn2}(\omega) \end{bmatrix} = \begin{bmatrix} \mathbf{H}_1(\omega) \\ \mathbf{H}_1(\omega)\mathbf{\Phi}(\omega) \end{bmatrix}\mathbf{T}, \tag{4.131}$$

where $\mathbf{U}_{sn1} \subset \mathbb{C}^{M/2\times K}$ and $\mathbf{U}_{sn2} \in \mathbb{C}^{M/2\times K}$ denote partitioned matrices expressing the eigenvectors at the two subarray positions. From Equation (4.131), we can construct the following relations:

$$\mathbf{U}_{sn1}(\omega) = \mathbf{H}_1(\omega)\mathbf{T}$$
$$\mathbf{U}_{sn2}(\omega) = \mathbf{H}_1(\omega)\mathbf{\Phi}(\omega)\mathbf{T}. \tag{4.132}$$

Eliminating $\mathbf{H}_1$ from Equation (4.132) gives

$$\mathbf{U}_{sn2} = \mathbf{U}_{sn1}\mathbf{\Psi}, \quad \text{with}\,\mathbf{\Psi} = [\mathbf{T}^{-1}\mathbf{\Phi}\mathbf{T}]. \tag{4.133}$$

Equation (4.133) implies that, for any square matrix $\mathbf{\Phi}$ and invertible matrix $\mathbf{T}$, the two matrices $\mathbf{\Phi}$ and $\mathbf{T}^{-1}\mathbf{\Phi}\mathbf{T}$ share the same eigenvalues. Therefore, by finding $\mathbf{\Psi}$ and its eigenvalues, we can estimate the directional parameters $\mathbf{\Phi}$.

4.6.3.1 TLS-ESPRIT

There are various ways to construct $\boldsymbol{\Psi}$ in practice; under ideal conditions, $\boldsymbol{\Psi}$ can be directly derived from the left pseudo inverse of $\mathbf{U}_{sn1}$:

$$\boldsymbol{\Psi} = (\mathbf{U}_{sn1})^{+}\mathbf{U}_{sn2} = (\mathbf{U}_{sn1}^{H}\mathbf{U}_{sn1})^{-1}\mathbf{U}_{sn1}^{H}\mathbf{U}_{sn2}. \tag{4.134}$$

The matrix represents a solution to reconstruct $\mathbf{U}_{sn2}$ with a minimal error of $\varepsilon = \|\mathbf{U}_{sn1}\boldsymbol{\Psi} - \mathbf{U}_{sn2}\|_F$ ($\|\|_F$: Frobenius norm). However, in many practical situations, we have errors in both eigenvector estimates $\mathbf{U}_{sn1}$ and $\mathbf{U}_{sn2}$. In such a case, the total least-squares (TLS; Appendix A, Section A.4.6) approach is more suitable for solving the problem. Using the eigenvector errors $\Delta\mathbf{U}_{sn1}$ and $\Delta\mathbf{U}_{sn1}$, the TLS problem can be defined as

$$\text{Minimize}\left\|\left[\Delta\mathbf{U}_{sn1} \quad \Delta\mathbf{U}_{sn2}\right]\right\|_F, \text{subject to}\,(\mathbf{U}_{sn1} + \Delta\mathbf{U}_{sn1})\boldsymbol{\Psi} = \mathbf{U}_{sn2} + \Delta\mathbf{U}_{sn2}. \tag{4.135}$$

The solution $\boldsymbol{\Psi}$ minimizing the TLS error Equation (4.135) is given by Equation (A78) in Appendix A, Section A.4.6.

$$\boldsymbol{\Psi} = -\mathbf{Y}_1\mathbf{Y}_2^{-1},$$

where $\mathbf{Y} = \left[\mathbf{Y}_1^T \;\; \mathbf{Y}_2^T\right]^T$ is the matrix consisting of the K smallest eigenvectors of $\left[\mathbf{U}_{sn1} \;\; \mathbf{U}_{sn2}\right]^H \left[\mathbf{U}_{sn1} \;\; \mathbf{U}_{sn2}\right]$. Then, the eigenvalues of $\boldsymbol{\Psi}$ give an estimate of the directional parameter $\boldsymbol{\Phi}$.

4.7 Wideband Processing Technique

As explained in Section 4.4.2, the most basic approach for wideband beamforming is to sum all the beamforming powers processed for the discrete frequencies of interest. However, when one attempts to use an optimization technique such as the MV beamformer, the wideband processing requires the matrix inversion to be calculated at each frequency. This may be impractical considering the computational effort required for matrix inversion. In what follows we discuss the unification of correlation matrices at discrete frequencies into a single matrix. This is possible by introducing a transformation of the measurement space to a beam space that is frequency invariant.

4.7.1 Frequency-Domain Approach: Mapping to the Beam Space

Consider the total beamforming power of the broadband signals given in Equation (4.38). The summation of the beamforming power for discrete frequencies ω_n is given by

$$\sum_{n=n_l}^{n_h} \beta(\theta, \omega_n) = \left[\sum_{n=n_l}^{n_h} \mathbf{w}(\omega_n)^H \mathbf{H}(\omega_n)\mathbf{R}_{qq}(\omega_n)\mathbf{H}(\omega_n)^H \mathbf{w}(\omega_n)\right]. \tag{4.136}$$

Under the condition that the propagation matrix $\mathbf{H}$ is of rank K, there exists a mapping $\mathbf{X} \in L \times M$ that converts the matrix $\mathbf{H}$ of a certain frequency ω_n to a unified, frequency-independent domain ζ such that

$$\mathbf{X}(\zeta\,|\omega_n)\mathbf{H}(\omega_n) = \mathbf{H}(\zeta). \tag{4.137}$$

Using the mapping $\mathbf{X}$, we can convert Equation (4.136) to a single problem:

$$\mathbf{R}_\zeta = \sum_{n=n_l}^{n_h} (\mathbf{X}(\zeta\,|\omega_n)\mathbf{H}(\omega_n))\mathbf{R}_{qq}(\omega_n)(\mathbf{H}(\omega_n)^H \mathbf{X}(\zeta\,|\omega_n)^H)$$

$$\beta(\theta) = \mathbf{w}_\zeta^H \mathbf{R}_\zeta \mathbf{w}_\zeta, \tag{4.138}$$

where $\mathbf{w}_\zeta$ is a unified scan vector defined in the frequency-independent domain. Therefore, one can apply the narrowband beamforming techniques to the wideband problem. The mapping $\mathbf{X}$ transforms the microphone signals to another dimension, and hence the beamforming is processed in a domain different from the *element space* of the microphones.

Among the many possible mapping functions, one popular mapping is the focusing operator consisting of scan vectors. For example, the mapping matrix $\mathbf{X}$ can be defined in terms of the scan vector of the planar or spherical wave model:

$$\mathbf{X}(\theta|\omega_n) = \mathbf{W}(\omega_n)^H = \begin{bmatrix} \mathbf{w}(\theta_1, \omega_n)^H \\ \vdots \\ \mathbf{w}(\theta_L, \omega_n)^H \end{bmatrix}. \quad (4.139)$$

The mapping defined in this way transforms the microphone signals to multiple beamformer output signals $b(\theta_l, \omega)$ which are functions of the angle θ_l $(l = 1, \cdots, L)$. Figure 4.32 illustrates this concept. Summing the beamformer outputs of different frequencies gives the unified correlation matrix of Equation (4.138), which represents the broadband correlation between beamformer outputs at different angles. We can then consider each beamformer output as a signal measured by a *virtual microphone* that picks up sound from a single direction θ_l. The optimal weights $\mathbf{w}_\theta$ can be designed to further increase the beamforming performance of the virtual microphone array. The signal space consisting of the virtual microphone signals is often referred to as the *beam space*. This space usually has fewer dimensions (L) than the element space (M), and hence can be effective in reducing the complexity of the wideband processing.

Another commonly used mapping is the *frequency-focusing* matrix $\mathbf{X}(\omega_c|\omega_n)$ that transforms a microphone signal of frequency ω_n to that of ω_c. If we can design such a transform, the correlation matrices of different frequencies can be treated as a single frequency datum, which allows us to

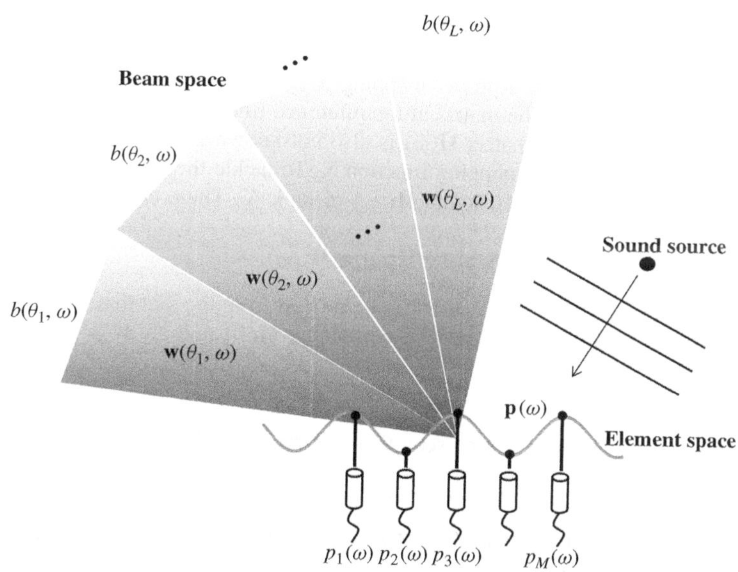

Figure 4.32 Concept of the beam and element spaces

directly apply the narrowband processing technique to a wideband problem. The coherent subspace method (CSM) explained in the following section is one well-known frequency-focusing technique.

4.7.2 Coherent Subspace Method (CSM)

In the CSM [7], a transformation matrix is defined such that the correlation matrices of different frequencies ω_n are converted to the matrices for a single reference frequency ω_c. For example, if a planar wavefront is impinging on the linear microphone array, then the transfer function given by

$$h_m(\omega) = e^{-i\omega(m-1)(d/c)\sin\theta_0}$$

can be converted to another frequency ω_c:

$$\begin{aligned} h_m(\omega_c) &= e^{-i\omega_c(m-1)(d/c)\sin\theta_0} \\ &= e^{-i(\omega_c-\omega)(m-1)(d/c)\sin\theta_0} \cdot e^{-i\omega(m-1)(d/c)\sin\theta_0} \\ &= X_m(\omega_c|\omega)h_m(\omega). \end{aligned} \tag{4.140}$$

Here, X_m is a mapping function that maps the transfer function at frequency ω to the reference frequency ω_c. For the wideband signal consisting of multiple frequency components, the mapping of Equation (4.140) can be written as

$$\mathbf{R}_{\omega_c} = \sum_n \mathbf{X}(\omega_c|\omega_n)\mathbf{h}(\omega_n)E\left[q(\omega_n)^* q(\omega_n)\right]\mathbf{h}(\omega_n)^H \mathbf{X}(\omega_c|\omega_n)^H, \tag{4.141}$$

where $\mathbf{X}$ is a diagonal matrix with $[\mathbf{X}]_{(m,m)} = e^{-i(\omega_c-\omega_n)(m-1)(d/c)\sin\theta_0}$. This mapping to the single frequency ω_c is often called "focusing" of a correlation matrix. The focusing procedure enables the unified scan vector to be designed only for the frequency ω_c. However, to design a proper mapping function X_m, the direction of a plane wave θ_0 has to be known. Moreover, the mapping $\mathbf{X}$ cannot be constructed without prior knowledge of the number and propagating directions of the sound sources, especially when multiple plane waves or multiple sound sources are involved.

For a more detailed analysis of the multiple sources case, let us consider the general equation of Equation (4.49). Our aim is to find a proper mapping $\mathbf{X}$ that transforms the matrix $\mathbf{H}(\omega) \in \mathbb{C}^{M\times K}$ of rank K (number of sources) to the matrix at the reference frequency $\mathbf{H}(\omega_c)$. Since the directions of the sources are not known, the matrix $\mathbf{H}(\omega)$ is also not known a priori. For a given correlation matrix $\mathbf{R}$, we need to find a proper mapping function $\mathbf{X}$. To tackle the given problem, first suppose that we are aware of two matrices $\mathbf{H}(\omega)$ and $\mathbf{H}(\omega_c)$ of rank K. Then, non-unique transformations $\mathbf{X}(\omega_c|\omega)$ exist such that

$$\mathbf{X}(\omega_c|\omega)\mathbf{H}(\omega) = \mathbf{H}(\omega_c). \tag{4.142}$$

As addressed by Wang and Kaveh [7], this can be proved from the fact that an arbitrary full-rank matrix $\mathbf{H}_f(\omega) \in \mathbb{C}^{M\times M}$ can be constructed by combining $\mathbf{H}(\omega)$ and an $M \times (M-K)$ matrix $\mathbf{Z}$. That is,

$$\mathbf{H}_f(\omega) = [\mathbf{H}(\omega)\,\mathbf{Z}(\omega)]. \tag{4.143}$$

Then, from Equation (4.142), it can be seen that the transform $\mathbf{X}(\omega_c|\omega)$ satisfies the relation

$$\mathbf{X}(\omega_c|\omega)\mathbf{H}_f(\omega) = \mathbf{H}_f(\omega_c), \tag{4.144}$$

with $\mathbf{H}_f(\omega_c) = [\mathbf{H}(\omega_c)\,\mathbf{Z}(\omega_c)]$ and $\mathbf{X}\mathbf{Z}(\omega) = \mathbf{Z}(\omega_c)$. Since $\mathbf{H}_f$ is of full rank, the transform $\mathbf{X}$ can be any kind of matrix given by

$$\mathbf{X}(\omega_c|\omega) = \mathbf{H}_f(\omega_c)\mathbf{H}_f(\omega)^{-1}. \tag{4.145}$$

The question we now have is how to construct the rank-K transfer matrix $\mathbf{H}(\omega)$. In the CSM, the concept of a beam space (Equation 4.139) is utilized to obtain a rough estimate of $\mathbf{H}(\omega)$. A rough directional estimation for multiple incidence angles (θ_l, $l = 1, \cdots, M$) is performed using the scan matrix of Equation (4.139), and K columns of $\mathbf{W}(\omega)$ that correspond to the preliminary directional estimations are collected to constitute $\mathbf{H}(\omega)$. Next, $M - K$ auxiliary angles that do not overlap with the preliminary estimation are chosen to build a matrix $\mathbf{Z}(\omega)$. Once these transfer function estimations are completed, the transform matrix can be built from Equations (4.144) and (4.145). The transform matrix is then used to "focus" the correlation matrix $\mathbf{R}$ of frequency ω to the reference frequency ω_c. By focusing all the correlation matrices of multiple frequencies (ω_n), we can obtain a unified correlation matrix of a single frequency ω_c:

$$\mathbf{R}_{\omega_c} = \sum_n \mathbf{X}(\omega_c|\omega_n)\mathbf{R}(\omega_n)\mathbf{X}(\omega_c|\omega_n)^H. \tag{4.146}$$

We can then apply the eigenvalue analysis to $\mathbf{R}_{\omega_c}$ so that the subspace-based techniques (e.g., MUSIC) can be used.

With CSM, the MUSIC beamformer will produce a more accurate directional estimation for wideband sources. In Figure 4.33, the difference between two MUSIC beamformers with and without CSM is shown. Here, "without CSM" denotes the MUSIC beamformer calculated from the correlation matrix based on the narrowband assumption of Equation (4.39). The comparison shows that the CSM can provide better resolution for the broadband noise source. In practice, however, the choice of $\mathbf{Z}(\omega)$ plays an important role for improving the beamformer performance. Depending on $\mathbf{Z}(\omega)$, the transform can yield a undesired mapping of plane waves, or the matrix $\mathbf{X}$ can be singular due to the ill-conditioning of $\mathbf{H}_f(\omega)^{-1}$. Detailed design examples of the transform $\mathbf{X}$ can be found from other literatures [7, 8].

4.7.3 *Partial Field Decomposition in Beam Space*

The correlation- and subspace-based approaches explained so far rely on the spatial distribution of the microphone array signal. As can be seen in Equation (4.45), the array signal has both spatial

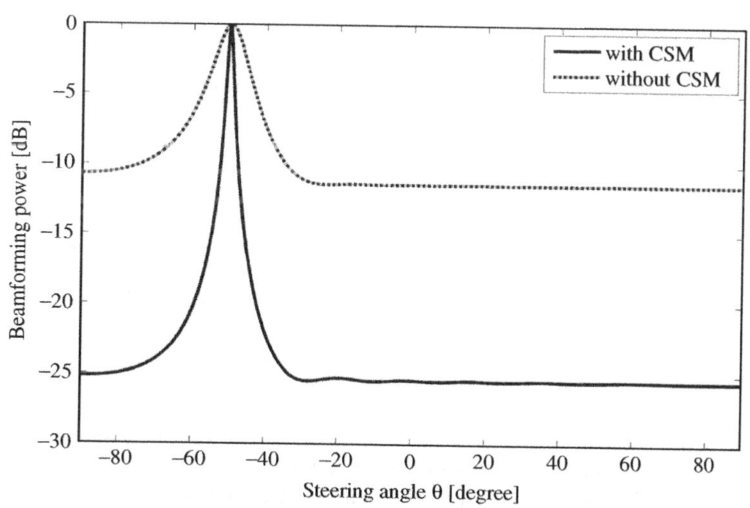

Figure 4.33 MUSIC beamforming power of a bandlimited white noise source, with and without CSM ($\theta_0 = -50°$, f = 500 Hz to 1 kHz). The simulation conditions are the same as those in Figure 4.14

and statistical distributions, and hence, the statistical property of the measured signal can also be utilized. For instance, the beamformer output of each incoherent sound source can be extracted using the statistical independence. Similar to the partial field decomposition technique used in acoustic holography, the contribution of multiple independent sources can be identified in the beamforming without prior information or an eigenstructure analysis. For ease of interpretation, the technique is described in terms of the statistical vector notation of Equation (4.45).

4.7.3.1 Beamforming in Multiple Incoherent Sources

Consider incoherent sound sources positioned at different locations in space. The statistical matrix $\tilde{\mathbf{P}}$ of Equation (4.45) is transformed into the beam space by multiplying by the scan vector matrix of Equation (4.139):

$$\tilde{\mathbf{B}} = \mathbf{W}^H\tilde{\mathbf{P}} = \mathbf{W}^H\mathbf{H}\tilde{\mathbf{Q}}. \tag{4.147}$$

The beamformer output $\tilde{\mathbf{B}} \in \mathbb{C}^{L\times N}$ defined in the beam space consists of an ensemble of N beamformer outputs for L scanning directions. From Equations (4.48) and (4.5), Equation (4.147) can be rewritten to express the estimated beamforming power $\hat{\boldsymbol{\beta}}$

$$\hat{\boldsymbol{\beta}} = \mathrm{diag}\left[\tilde{\mathbf{B}}\tilde{\mathbf{B}}^H\right], \tag{4.148}$$

which indicates that the beamforming power is equivalent to the autospectrum of the beamformer output.

Using Equations (4.147) and (4.148), the beamforming power can be represented in terms of the spectrum of each source signal;

$$\hat{\boldsymbol{\beta}} = \mathrm{diag}\left[\mathbf{W}^H\mathbf{H}\mathbf{R}_{qq}\mathbf{H}^H\mathbf{W}\right]. \tag{4.149}$$

Because the sound sources are incoherent, the intersource correlation matrix $\mathbf{R}_{qq}$ is diagonal. From Equation (4.66), the beamforming power of Equation (4.149) can be expressed as

$$\hat{\boldsymbol{\beta}} = \mathrm{diag}\left[\mathbf{W}^H\left(\sum_{\kappa=1}^{K}\sigma_\kappa^2\mathbf{h}^{(\kappa)}\mathbf{h}^{(\kappa)H}\right)\mathbf{W}\right] = \sum_{\kappa=1}^{K}\sigma_\kappa^2\mathrm{diag}\left[\mathbf{W}^H\mathbf{h}^{(\kappa)}\mathbf{h}^{(\kappa)H}\mathbf{W}\right]. \tag{4.150}$$

Equation (4.150) shows that the beamforming power consists of each source contribution, and this conversely states that an individual beamforming power can be estimated separately.

4.7.3.2 Decomposition of Each Source Contribution

The separation of the beamforming power can be accomplished by decomposing the beamformer output of each sound source. Suppose that we can estimate the source signal $\tilde{\mathbf{Q}}$. Then, the beamformer output of multiple sound sources, given by Equation (4.147), can be separated by considering the projection in beam space. For example, the beamformer output for the estimation of the κth source signal ($\tilde{\mathbf{q}}_\kappa$) can be extracted as

$$\tilde{\mathbf{B}}_\kappa = \mathbf{W}^H\mathbf{H}\left(\boldsymbol{\delta}_\kappa\tilde{\mathbf{q}}_\kappa\right), \quad \left(\tilde{\mathbf{B}}_\kappa \in \mathbb{C}^{L\times N}\right) \tag{4.151}$$

where the statistical vector $\tilde{\mathbf{q}}_\kappa = [q_\kappa^{(1)}, \cdots, q_\kappa^{(N)}]$ consists of the statistical samples (an ensemble) of the κth source signal, and $\boldsymbol{\delta}_\kappa$ is a column vector defined by $[\boldsymbol{\delta}_\kappa]_{(j)} = \delta_{\kappa j}$. From the incoherent source assumption, the Kronecker delta vector $\boldsymbol{\delta}_\kappa$ can be rewritten as

$$\boldsymbol{\delta}_\kappa \approx \frac{\tilde{\mathbf{Q}}\tilde{\mathbf{q}}_\kappa^H}{\|\tilde{\mathbf{q}}_\kappa\|^2}. \tag{4.152}$$

Substituting Equation (4.152) into Equation (4.151) leads to

$$\tilde{\mathbf{B}}_\kappa \approx \mathbf{W}^H\mathbf{H}\left(\frac{\tilde{\mathbf{Q}}\tilde{\mathbf{q}}_\kappa^H}{\|\tilde{\mathbf{q}}_\kappa\|^2}\tilde{\mathbf{q}}_\kappa\right) = \frac{\tilde{\mathbf{B}}\tilde{\mathbf{q}}_\kappa^H\tilde{\mathbf{q}}_\kappa}{\tilde{\mathbf{q}}_\kappa\tilde{\mathbf{q}}_\kappa^H}. \tag{4.153}$$

The last expression of Equation (4.153) can be regarded as the projection of the beamformer output $\tilde{\mathbf{B}}$ onto the vector space spanned by the vector $\tilde{\mathbf{q}}_\kappa$;

$$\tilde{\mathbf{B}}_\kappa \approx \frac{\tilde{\mathbf{B}}\tilde{\mathbf{q}}_\kappa^H}{\|\tilde{\mathbf{q}}_\kappa\|}\frac{\tilde{\mathbf{q}}_\kappa}{\|\tilde{\mathbf{q}}_\kappa\|} = \operatorname{proj}_{\tilde{\mathbf{q}}_\kappa}(\tilde{\mathbf{B}}). \tag{4.154}$$

Consequently, if we have accurate information about the κth source signal ($\tilde{\mathbf{q}}_\kappa$), then its beamformer output $\tilde{\mathbf{B}}_\kappa$ can be extracted from the output $\tilde{\mathbf{B}}$ of multiple sources. The corresponding beamforming power $\hat{\boldsymbol{\beta}}_\kappa$ can also be calculated by inserting Equation (4.154) into (4.148).

The key problem in decomposing the beamformer output is estimating an accurate source signal $\tilde{\mathbf{q}}_\kappa$ from the array measurement. To that end, we first analyze the beam space to find a dominant source location. The conventional beamforming technique is applied to get a preliminary beamforming power distribution, and then the beamformer output in the direction of the maximum beamforming power is taken as the source signal of the first sound source. This is based on the assumption that the beamformer output of a source will be maximized near the source location and will be identical to the source signal multiplied by some linear transform T_κ:

$$\tilde{\mathbf{q}}_1' = [\tilde{\mathbf{B}}]_{(l_1,:)} \approx T_1\tilde{\mathbf{q}}_1, \tag{4.155}$$

where l_1 denotes the index of the steering angle that gives the maximum beamforming power. To see how this assumption is validated, let us consider the plane wave propagation model of Equation (4.10) for the transfer function $\mathbf{H}$ and scan vector $\mathbf{W}$. For multiple sources, Equation (4.10) can be written as

$$\begin{aligned} b(\theta) &= \sum_{m=1}^{M}\sum_{\kappa=1}^{K} e^{i\omega(m-1)(d/c)(\sin\theta-\sin\theta_\kappa)}q_\kappa \\ &\approx \sum_{\kappa=1}^{K} q_\kappa \operatorname{sinc}(M(\sin\theta-\sin\theta_\kappa)\varphi)e^{i(M-1)(\sin\theta-\sin\theta_\kappa)\varphi}. \end{aligned} \tag{4.156}$$

If the sources are well separated in space, the sinc function of Equation (4.156) will be a maximum at the location of the source with the largest amplitude, and the contribution of the other sources will be negligible;

$$\begin{aligned} b(\theta_1) &= q_1 + \sum_{\kappa\neq 1} q_\kappa \operatorname{sinc}(M(\sin\theta_1-\sin\theta_\kappa)\varphi/2)e^{i(M-1)(\sin\theta_1-\sin\theta_\kappa)\varphi/2} \\ &\approx q_1. \end{aligned} \tag{4.157}$$

Therefore, the beamformer output at the location of maximum beamforming power can be regarded as the source signal q_1. The multiplication by any constant T_1 in Equation (4.155) does not affect the projection of Equation (4.154), and hence the beamformer output $\tilde{\mathbf{B}}_1$ of the first source can be estimated using the approximated source signal $\tilde{\mathbf{q}}_1'$.

Since we have assumed that the sources are incoherent, the beamformer output of the other sources will be orthogonal to the estimated beamformer output. The orthogonal vector space spanned by the other sources is given by subtracting the beamformer output of the first source:

$$\tilde{\mathbf{B}}_{\perp 1} \approx \tilde{\mathbf{B}} - \frac{\tilde{\mathbf{B}}\tilde{\mathbf{q}}_1'^H}{\|\tilde{\mathbf{q}}_1'\|}\frac{\tilde{\mathbf{q}}_1'}{\|\tilde{\mathbf{q}}_1'\|}. \tag{4.158}$$

The beamformer output of the second source can be obtained by repeating the procedure of Equation (4.155); the peak location of the beamformer output $\tilde{\mathbf{B}}_{\perp 1}$ is selected as the signal reference of the second source ($\tilde{\mathbf{q}}_2'$), and the projection of $\tilde{\mathbf{B}}_{\perp 1}$ onto $\tilde{\mathbf{q}}_2'$ gives the relevant beamformer output:

$$\tilde{\mathbf{B}}_2 \approx \mathrm{proj}_{\tilde{\mathbf{q}}_2'}(\tilde{\mathbf{B}}_{\perp 1}). \tag{4.159}$$

The contribution from the remaining sources can be determined by repeating the application of Equations (4.158) and (4.159). The partial field decomposition of the beamforming can be summarized as follows:

- Obtain the total beamformer output $\tilde{\mathbf{B}}$ from the beamforming as shown in Equation (4.147).
- Obtain the reference signal $\tilde{\mathbf{q}}_1' = [\tilde{\mathbf{B}}]_{(l_1,:)}$ at the location of maximum beamforming power ($l_1 = \arg\max_l\{\hat{\boldsymbol{\beta}}[l]\}$).
- Separate out the beamformer output of the first source by using Equation (4.154) and determine the residual beamformer output from Equation (4.158).
- Estimate the beamforming power of the first source.
- Repeat Steps 2–4 using the residual beamformer output.

4.7.3.3 Example of Partial Field Decomposition

In Figure 4.35, we show an example of the partial field decomposition for two incoherent monopole sources. A planar microphone array with a spacing of 1/4λ and 25 measurement points are considered as illustrated in Figure 4.34. The distance between the array and source plane is one wavelength. Two monopole sources of the same power are located at (−0.5 λ, 0) and (λ, 0).

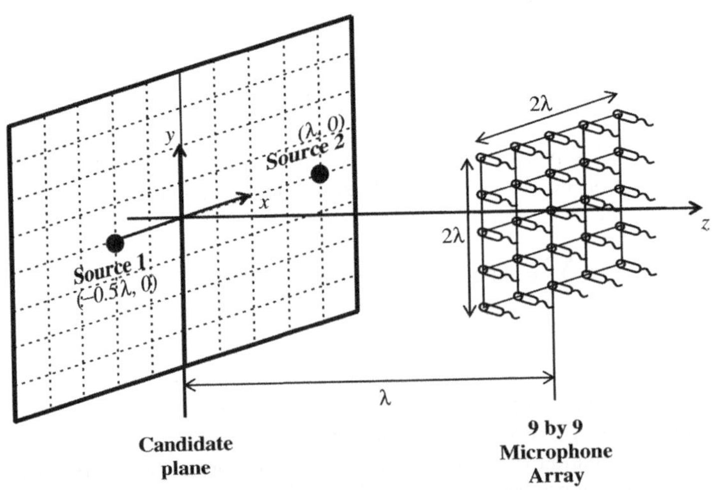

Figure 4.34 Simulation conditions for the partial field decomposition

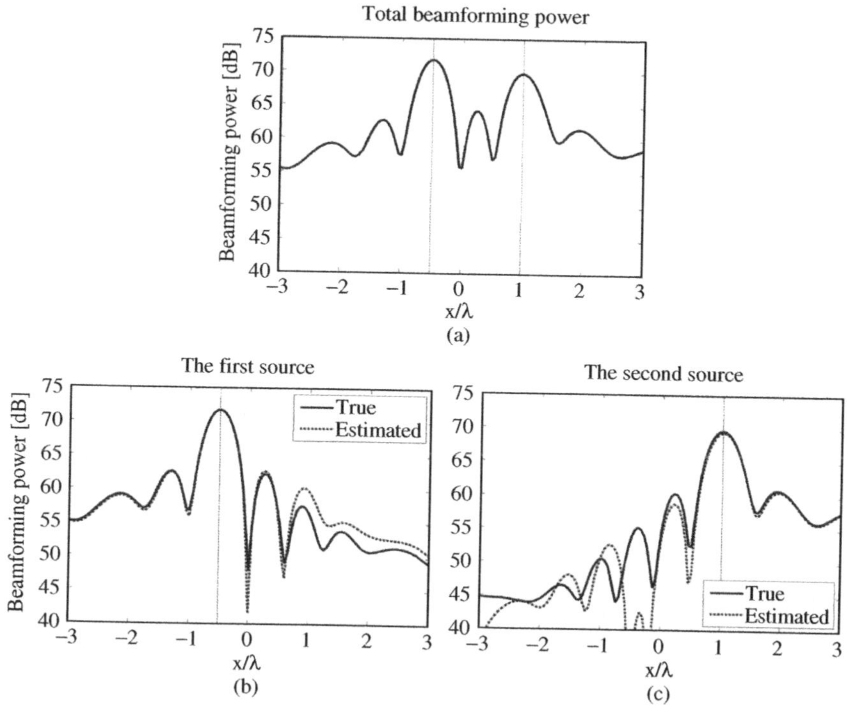

Figure 4.35 Partial field decomposition of two sound sources separated by 1.5 λ (a) total beamforming power, (b) contribution of the first source, and (c) contribution of the second source

Figure 4.35a shows the total beamforming power of two monopoles on the $y = 0$ line. The two source contributions are shown separately in Figure 4.35b,c. In this example, we can see that the contribution of the first sound source is dominant at the peak position of the total beamforming power. Taking the beamformer output at this peak position as a reference, the contribution of the first source is extracted from the total beamformer output. In Figure 4.35b,c, the true beamforming power is calculated from a single monopole source, and the estimated result is obtained by applying the partial field decomposition. The result shows that the difference between the true and estimated beamforming power is very small near the sound source location.

4.7.3.4 Effect of Source Distance

The estimation error will be greater when the beamforming powers of the first and second sources overlap at the location of maximum beamforming power. In general, the contribution of the second source can be measured at the peak location because of the limited resolution of the DAS beamformer. In Figure 4.36, the separation performance of the two incoherent sound sources is examined for a closer source distance of 0.4λ. The separation is inaccurate, and from this we can recognize that the partial field decomposition is not useful for enhancing the beamforming resolution.

4.7.3.5 Advantage of Partial Field Decomposition

The real advantage of the partial field decomposition lies in its ability to find low-level sources hidden in the total beamforming power. In the example in Figure 4.37, the power ratio of the first

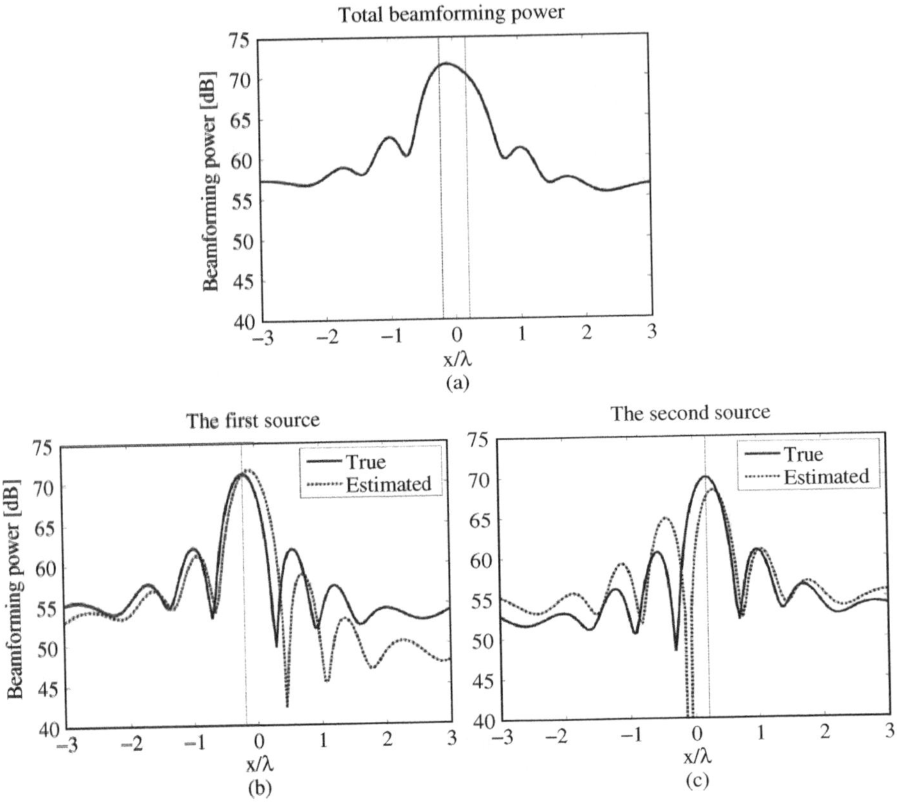

Figure 4.36 Partial field decomposition of two sound sources separated by 0.4 λ (a) total beamforming power, (b) contribution of the first source, and (c) contribution of the second source

source, Figure 4.37b, and the second source, Figure 4.37c, is 12 dB, and the beamforming power of the second source is buried under that of the first source. It can be observed that the partial field decomposition can separate the beamforming power of the second source even when it is a source of small power completely hidden in the beamforming power of the first source.

The partial field decomposition utilizes and produces beam space data (the beamformer output), and thus, it can be combined with the MV beamformer. By using a super-resolution algorithm, a better estimate of the source signals is possible, which yields a clearer separation of each source contribution.

4.7.4 Time-Domain Technique

The wideband processing methods discussed so far extend the narrowband processing techniques using the conversion to beam space. However, beamforming can also be implemented in the time domain. In principle, the frequency- and time-domain beamformers produce similar beamforming powers: Parseval's theorem (Appendix A, Section A.1.1.5) states that the overall power summation in the frequency domain is equivalent to the sum in the time domain. This is generally true when the source signal is stationary and all time signals are converted to the frequency domain. However, the

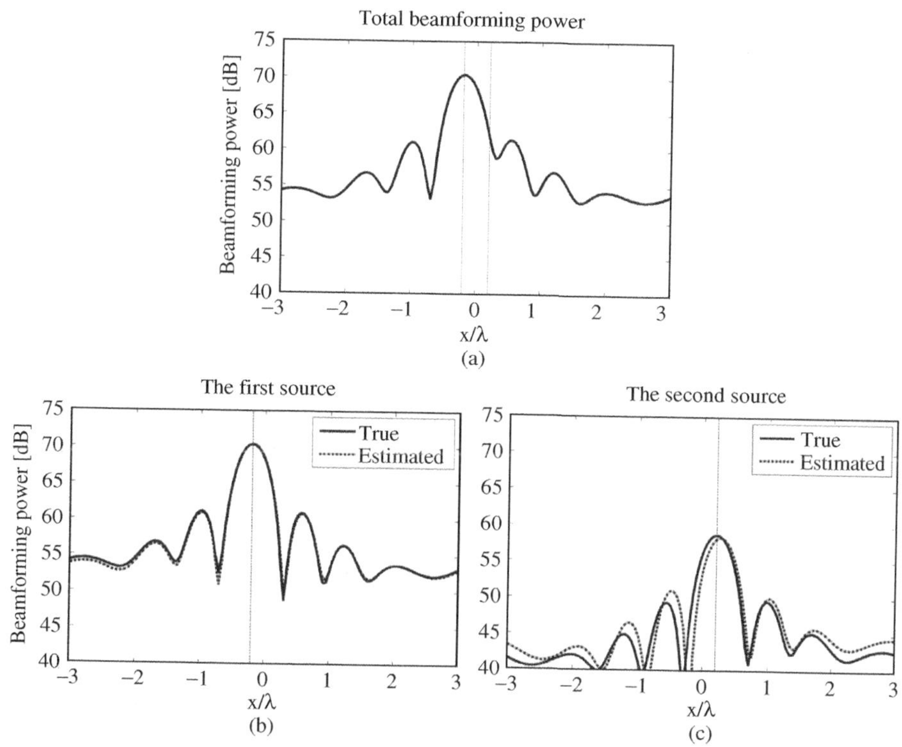

Figure 4.37 Partial field decomposition of two sound sources separated by 0.4 λ (a) total beamforming power, (b) contribution of the first source, and (c) contribution of the second source

time-domain processing can be selectively applied to the temporal region with a high SNR, which can be advantageous in reducing the effect of noise.

4.7.4.1 Equivalence of the Time- and Frequency-Domain Beamformers

For example, suppose that a planar wave source radiates a sound signal, and stationary noise is mixed with the measurement. Let the power of the source signal be given by

$$E\left[q^2(t)\right] = Q^2. \tag{4.160}$$

It is assumed that the input of each channel is corrupted by zero-mean additive white Gaussian noises $n_m(t)$, which are incoherent with each other and have equal variance (σ^2):

$$E\left[n_m(t)\, n_{m'}(t)\right] = \sigma^2 \delta_{mm'}. \tag{4.161}$$

Then the SNR of the measured microphone signals (SNR_{input}) can be written as

$$SNR_{\text{input}} = \frac{Q^2}{\sigma^2}. \tag{4.162}$$

When the beamformer delays are properly adjusted to the sound source location, the maximum beamformer output of the planar wavefront model is given by

$$\beta_{\max} = Q^2. \tag{4.163}$$

From Equation (4.32), the beamformer output and power due to the noise signal can be calculated as

$$\begin{aligned}
b_{noise}(\theta, t) &= \frac{1}{M}\sum_{m=1}^{M} n_m \left(t - \frac{d}{c}(m-1)\sin\theta\right). \\
\beta_{noise} &= E[b_{noise}^2(\theta, t)] \\
&= E\left[\frac{1}{M^2}\sum_{m=1}^{M} n_m^2(t)\right] \\
&= \sigma^2/M. \qquad (4.164)
\end{aligned}$$

Accordingly, the SNR of the beamforming power in the peak direction (SNR_{output}) can be derived as

$$SNR_{\text{output}} = \frac{\beta_{\max}}{\beta_{noise}} = \frac{MQ^2}{\sigma^2}. \qquad (4.165)$$

Then, the array gain (ARG), defined as the ratio of the input and output SNRs, becomes

$$ARG \triangleq \frac{SNR_{\text{output}}}{SNR_{\text{input}}} = M. \qquad (4.166)$$

As a result, the ARG of a stationary noise signal is proportional to the number of microphones (M). The result remains the same in both the frequency and time domains.

On the other hand, if the source signal is transient, then the SNR varies with respect to time. As a simple example, consider an impulsive source signal of peak amplitude A and duration Δt. For simplicity, we assume a rectangular pulse whose area under the time signal $q(t)$ is given by

$$\int_{-\Delta t/2}^{\Delta t/2} q(t)dt = A\Delta t. \qquad (4.167)$$

The transient signal power is a function of time and can be written as

$$q(t)^2 = \begin{cases} A^2 & \text{if } |t| < \Delta t/2 \\ 0 & \text{otherwise.} \end{cases} \qquad (4.168)$$

With the same incoherent noise of variance σ^2 at the microphones, the input SNR of an impulsive signal is given by

$$SNR_{input}(t) = \frac{q(t)^2}{\sigma^2}. \qquad (4.169)$$

The output of the DAS beamformer can also be calculated as

$$b_{\max}(t) = q(t), \quad \beta_{\max}(t) = q^2(t), \qquad (4.170)$$

$$\beta_{noise} = \sigma^2/M. \qquad (4.171)$$

The output SNR is equal to

$$SNR_{output}(t) = Mq(t)^2/\sigma^2, \qquad (4.172)$$

which gives an ARG of $ARG = M$. However, when the observation time of the beamformer output is not within $|t| < \Delta t/2$, the output SNR is zero, and hence the ARG goes to zero. One can use the root-mean-square (RMS) value of the signal power, instead of $q(t)^2$, to obtain a time-invariant ARG. If we measure $q(t)$ for a finite duration τ, then the RMS signal power is equal to

$$q_{rms}^2 = \frac{1}{\tau}\int_{-\tau/2}^{\tau/2} q(t)^2 dt = \frac{1}{\tau}\int_{-\Delta t/2}^{\Delta t/2} q(t)^2 dt = \frac{A^2 \Delta t}{\tau} \tag{4.173}$$

which is invariant with respect to the observation time. The ARGs calculated from the RMS power is $ARG = M$, which is equivalent to the case of a stationary random signal (Equation (4.166)). Therefore, the time-domain processing is not beneficial in view of the ARG. However, the time-varying output SNR of Equation (4.172) can be much higher than the SNR of the stationary source signal Equation (4.165) if we select the proper time interval for $q(t)^2$. It is evident that the output SNR would increase at the peak location of $q(t)$, and hence, we can obtain a higher output SNR if $q(t)$ is an impulsive signal. The next section illustrates how we can utilize this idea for an impulsive source signal.

4.7.4.2 Evaluation of Time-Domain Beamformer by Peak Value

For an impulsive signal, the performance of the time-domain beamformer can be enhanced by introducing a different definition of the beamforming power. First, consider the output of a time-domain beamformer, which is given by

$$b_{signal}(\theta, t) = \frac{1}{M}\sum_{m=1}^{M} q\left(t - \frac{d}{c}(m-1)(\sin\theta - \sin\theta_0)\right) \quad \text{for a source signal,} \tag{4.174}$$

$$b_{noise}(\theta, t) = \frac{1}{M}\sum_{m=1}^{M} n_m\left(t - \frac{d}{c}(m-1)\sin\theta\right) \quad \text{for measurement noise.} \tag{4.175}$$

The beamformer output b_{noise} is the mean of the incoherent random noise and hence will converge to zero. In view of this, the beamformer output can be more advantageous than the beamforming power in reducing the effect of measurement noise.

For ease of derivation, suppose a continuous distribution of microphones along the x axis. Then, the beamformer output can be expressed as

$$b_{signal}(\theta, t) = \frac{1}{L}\int_0^L q\left(t - \frac{x}{c}(\sin\theta - \sin\theta_0)\right)dx, \tag{4.176}$$

where L is the aperture length of the microphone array. It is noteworthy that the integral of Equation (4.176) can be regarded as the time average of the source signal $q(t)$:

$$\begin{aligned} b_{signal}(\theta, t) &= \frac{c}{L}\int_0^{L/c} q(t - \tau \cdot a(\theta))d\tau, \quad a(\theta) = \sin\theta - \sin\theta_0 \\ &= \frac{1}{T(\theta)}\int_0^{T(\theta)} q(t - \tau')d\tau', \quad T(\theta) = La(\theta)/c. \end{aligned} \tag{4.177}$$

We also note that the beamformer output is the result of the *moving average* (MA) of the source signal $s(t)$ with a temporal window of size $T(\theta) = La(\theta)/c$ (Figure 4.38). The size of the temporal window is proportional to the array length L and increases as the difference between the steering angle θ and the source location increases. The smallest window size is applied when the steering angle θ of the beamformer is the same as the source location θ_0 ($a(\theta) = 0$), which leads to the beamformer output $b_{signal}(\theta_0, t) = q(t)$.

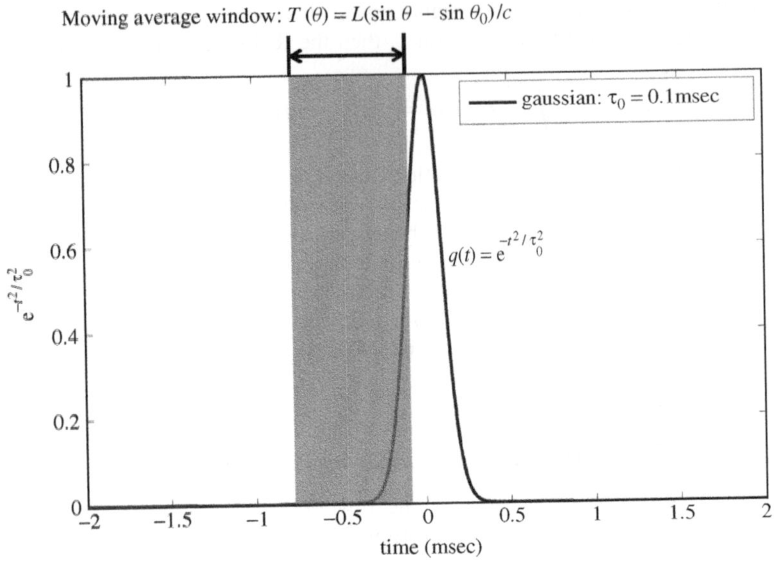

Figure 4.38 Time-domain beamforming represented as an MA process (source signal: Gaussian function)

The peak value of the MA is a useful measure of the location of an impulsive source:

$$b_{peak}(\theta) = \underset{t}{\text{Max}}\left[\left|b_{signal}(\theta, t)\right|\right]. \tag{4.178}$$

The peak value of the MA decreases with an increase in the window length $T(\theta)$. For example, if the source signal $q(t)$ is an ideal Dirac delta function ($q(t) = \delta(t)$), then the peak value goes to infinity at the source location ($\theta = \theta_0$) or is otherwise $1/T(\theta)$. Therefore, we can use the peak value $b_{peak}(\theta)$, instead of the beamforming power, to estimate the source location.

As another example, consider a Gaussian function $q(t) = e^{-t^2/2\tau_0^2}$ (Figure 4.38). The peak value is equal to 1 at the source location and equal to the error function at all other locations. The error function is defined as

$$\begin{aligned}\text{erf}(x) &= \frac{2}{\sqrt{\pi}}\int_0^x e^{-t^2}dt \\ &= \frac{2}{\sqrt{\pi}}\left(x - \frac{x^3}{3} + \frac{x^5}{10} - \frac{x^7}{42} + \cdots\right),\end{aligned} \tag{4.179}$$

and hence the MA (Equation (4.177)) of the Gaussian $q(t) = e^{-t^2/2\tau_0^2}$ is given by

$$b_{signal}(\theta, t) = \sqrt{\frac{\pi}{2}}\frac{\tau_0}{T(\theta)}\left[\text{erf}\left(\frac{1}{\sqrt{2}}\frac{t}{\tau_0}\right) - \text{erf}\left(\frac{1}{\sqrt{2}}\frac{t - T(\theta)}{\tau_0}\right)\right]. \tag{4.180}$$

Therefore, the beamformer output is related to the difference in the error function (Figure 4.39). If the window length T is much smaller than the variance τ_0 of the Gaussian function, then the beamformer output of Equation (4.180) behaves like the derivative of the error function (which is

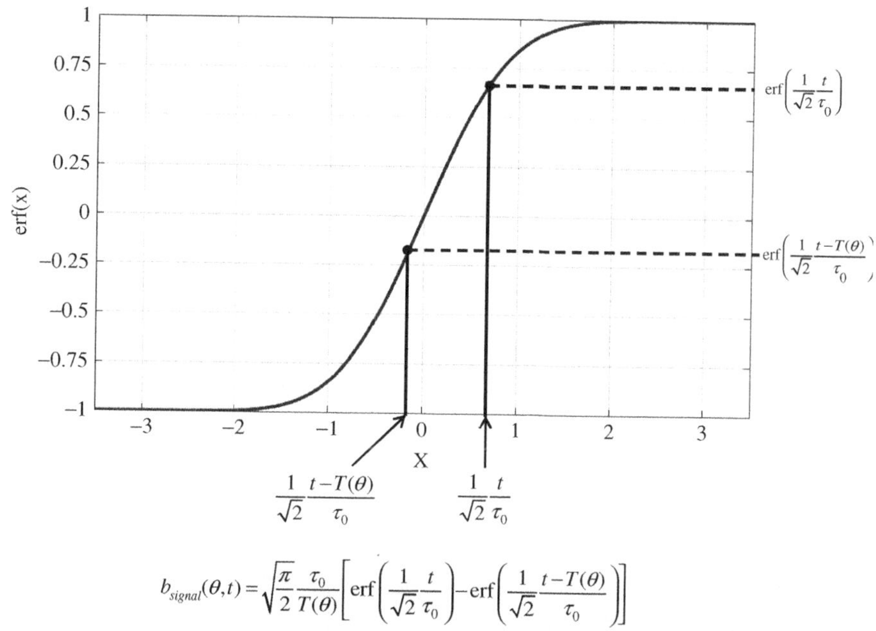

Figure 4.39 The error function erf(x)

Gaussian). Accordingly, the peak value of the beamformer output resembles a Gaussian function near the source location. On the other hand, if the window length T is large, then the difference in the error function saturates at 2 (Figure 4.39). Consequently, the peak value converges to

$$b_{peak}(\theta) \approx \begin{cases} 1 & for \quad \theta = \theta_0 \\ \sqrt{2\pi}\dfrac{\tau_0}{T(\theta)} & for \quad \theta \neq \theta_0 \end{cases}, \tag{4.181}$$

as the steering angle moves further away from the source location. This value can be likened to the sidelobe level of the beamforming power. It can be observed that a sharp signal $q(t)$ with a small τ_0 or a large window length $T(\theta)$ gives a lower sidelobe level. The sidelobe level decreases with the impulsiveness of a time signal and is not affected by measurement noise, and thus the time-domain beamforming evaluated by the peak value of the beamformer output is useful for analyzing impulsive sound sources. Figure 4.40 demonstrates the performance of the peak estimation of time-domain beamformer as compared to the RMS estimation.

4.7.4.3 Example: Impulsive Source Localization in Anechoic Chamber

The next example shows an experiment applying the time-domain beamforming to an impulsive sound source. The experimental setup consists of signal generating and measurement devices as shown in Figure 4.41a,b. Gaussian white noise emanated from four loudspeakers, and an air-gun was shot to produce the impulsive sound (Figure 4.41a). A curved-spoke array composed of 30-channel microphones was used for the measurement. The experiment was conducted in an anechoic chamber with a cut-off frequency of 100 Hz. Figure 4.42a depicts the recorded signal measured

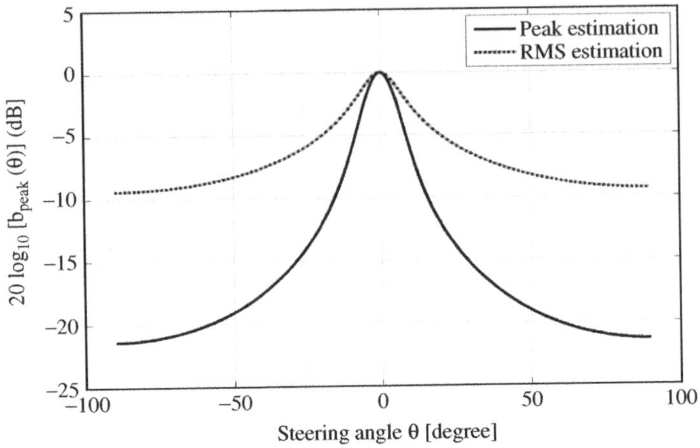

Figure 4.40 Comparison of the peak and RMS estimators of the time-domain beamforming ($L = 1\,\text{m}$, $\tau_0 = 0.1\,\text{ms}$)

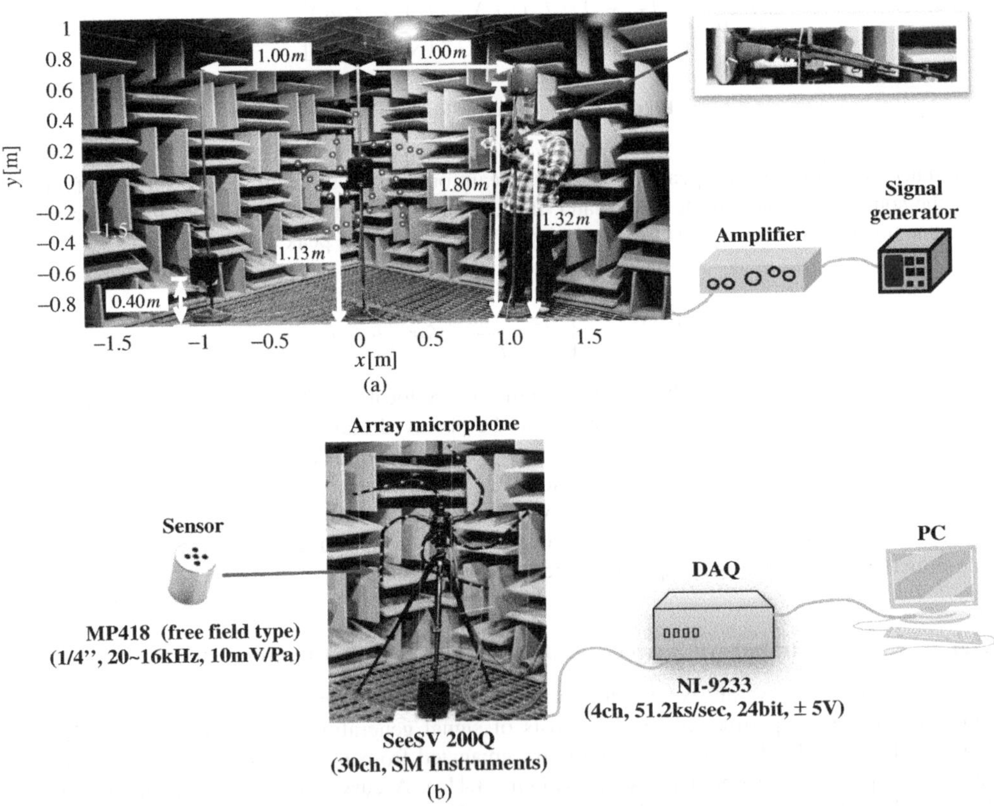

Figure 4.41 Experimental setup for the time-domain beamforming; four loudspeakers producing Gaussian white noise are placed together with an impulsive sound source (gunshot) (a) configuration of sound sources in the anechoic chamber and (b) measurement devices

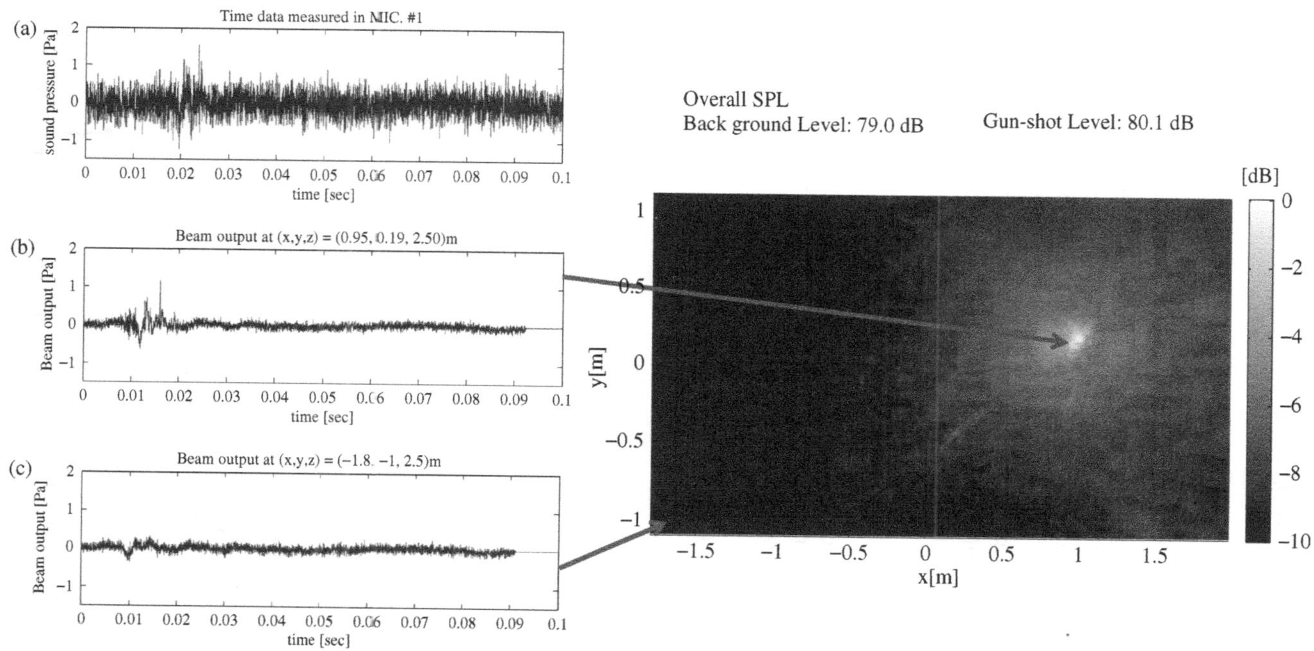

Figure 4.42 The time-domain beamforming for an impulsive sound source (a) microphone signal with noise, (b) beamformer output signal at the source location, and (c) beamformer output signal at a location other than the source

by the microphone array. Figure 4.42b,c show the beam outputs at the positions of maximum and minimum beamforming power. The maximum beamformer output position corresponds to the location of the impulsive source. Figure 4.42a,c also show that the SNR of the beamformer output increased at the source location as compared with that of the microphone signal or the beamformer output for other directions.

4.7.5 Moving-Source Localization

For a moving source whose position is continuously changing with respect to time, the output of the time-domain beamformer will only show a blurred image of the sound source along the source trajectory. To find an accurate position of a moving source, the beamformer has to be modified such that moving-source effects, such as the Doppler effect, can be compensated.

To begin, consider a source signal q defined in space and time. For a moving source, the source signal is represented as a multiplication of the time signal $q_0(\tau_s)$ and the Dirac delta function expressing the change in the source position with respect to time:

$$q(\mathbf{r}_s, \tau_s) = q_0(\tau_s)\delta(\mathbf{r}_s - \mathbf{r}_0(\tau_s)). \tag{4.182}$$

The source signal then propagates in space and time. The wavefront emitted at time τ_s and position $\mathbf{r}_s$ arrives at the observer position $\mathbf{r}$ after a time delay of $t = \tau_s + |\mathbf{r} - \mathbf{r}_s|/c$ (Figure 4.43), and hence the transfer function of a moving source is given by

$$h(\mathbf{r}, t|\mathbf{r}_s, \tau_s) = \frac{\delta(t - \tau_s - |\mathbf{r} - \mathbf{r}_s|/c)}{4\pi|\mathbf{r} - \mathbf{r}_s|} \tag{4.183}$$

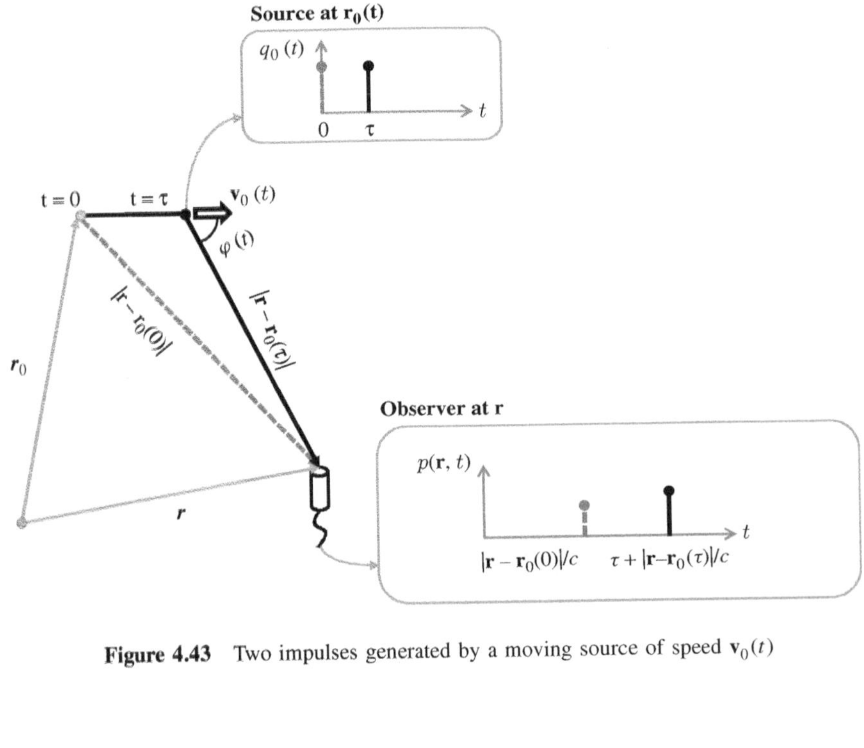

Figure 4.43 Two impulses generated by a moving source of speed $\mathbf{v}_0(t)$

for a monopole sound source. The sound field produced by a moving monopole can be obtained by integrating the product of Equations (4.182) and (4.183) over space and time. That is,

$$p(\mathbf{r}, t) = \int_{-\infty}^{\infty} \int_{V} \frac{q_0(\tau_s)\delta(\mathbf{r}_s - \mathbf{r}_0(\tau_s))\delta(t - \tau_s - |\mathbf{r} - \mathbf{r}_s|/c)}{4\pi |\mathbf{r} - \mathbf{r}_s|} dV(\mathbf{r}_s) d\tau_s$$
$$= \int_{-\infty}^{\infty} \frac{q_0(\tau_s)\delta(t - \tau_s - |\mathbf{r} - \mathbf{r}_0(\tau_s)|/c)}{4\pi |\mathbf{r} - \mathbf{r}_0(\tau_s)|} d\tau_s. \quad (4.184)$$

The integral with respect to τ_s can be simplified by exploiting the following Dirac delta identity [9]:

$$\delta[g(\tau)] = \sum_j \frac{\delta(\tau)}{|dg/d\tau|_{\tau=\tau_j}}, \quad \text{where } g(\tau_j) = 0, \quad (4.185)$$

which is a generalization of $\delta(a\tau) = \delta(\tau)/a$ (Appendix A, Section A.2).

From Equation (4.185), the integral of Equation (4.184) can be replaced by a finite sum:

$$p(\mathbf{r}, t) = \sum_j \frac{q_0(\tau_j)}{4\pi |\mathbf{r} - \mathbf{r}_0(\tau_j)|} \frac{1}{|1 - M_0(\tau_j)|}, \quad (4.186)$$

where the τ_j are the roots of

$$\tau_j = t - |\mathbf{r} - \mathbf{r}_0(\tau_j)|/c. \quad (4.187)$$

The relative velocity of the source in the direction of the observer M_0 (in terms of the Mach number) is given by

$$M_0(\tau_j) = -\frac{1}{c} \frac{d\left|\mathbf{r} - \mathbf{r}_0\left(\tau_s\right)\right|}{d\tau_s}\Bigg|_{\tau_s=\tau_j}$$
$$= \frac{\mathbf{v}_0(\tau_j)}{c} \cdot \frac{\mathbf{r} - \mathbf{r}_0(\tau_j)}{|\mathbf{r} - \mathbf{r}_0(\tau_j)|}$$
$$= \frac{|\mathbf{v}_0(\tau_j)|}{c} \cos\varphi(\tau_j), \quad \textit{where} \quad \mathbf{v}_0(\tau_j) = \frac{d\mathbf{r}_0}{d\tau_s}\Bigg|_{\tau_s=\tau_j}. \quad (4.188)$$

If we only consider a single τ_j and neglect the $1/|1 - M_0(\tau_j)|$ term, then Equation (4.186) just explains how the monopole sound field emitted by the source at time τ_j propagates to the listener position at time t. Here, the measurement time t is often referred to as the *reception time*, and τ_j is called the *emission time*. The amplitude of the wavefront decays according to the distance between the position where the wavefront is excited ($\mathbf{r}_s(\tau)$) and the observation point $\mathbf{r}$ (Figure 4.43).

The extra magnitude scaling of $1/|1 - M_0(\tau_j)|$ of Equation (4.186) is due to the time-varying delta function $\delta(t - \tau_s - |\mathbf{r} - \mathbf{r}_0(\tau_s)|/c)$ of Equation (4.184). The physical meaning of this scale factor can be understood by comparing the Dirac delta functions at two different observation points (observer I and II of Figure 4.44). Note that the delta function can be derived from reducing the duration of a rectangular pulse to zero while its area is held constant. Since the delta functions at the positions of observer I and II have different arguments with respect to τ_s, the amplitudes of the Dirac delta function (or the area of a rectangular pulse of infinitesimal time duration) are different. If we consider the density of discrete pulses of Figure 4.44 as the total area of the delta function, we can infer that the pulse density at observer II will be higher than at observer I. Therefore, the area under the same infinitesimal time duration will be larger at observer II, and the corresponding amplitude of the delta function will be higher. This kind of change in the amplitude that occurs due to a moving (convective) monopole is called *convective Doppler amplification*.

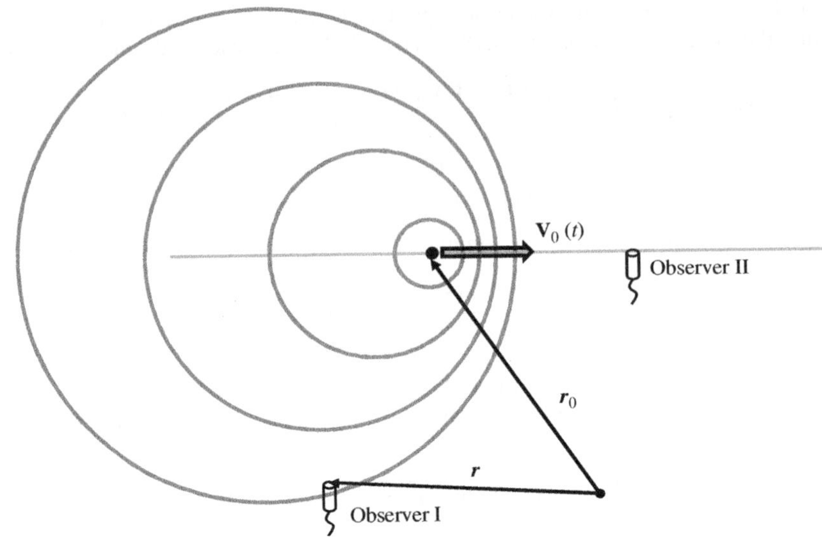

Figure 4.44 Propagation of discrete pulses produced by a moving source

4.7.5.1 Conversion between Reception and Emission Times

The conversion between the reception and emission times can be accomplished through Equation (4.187). For example, if a source is moving in the positive x direction at a constant speed v_0, then the source and receiver positions can be described as

$$\mathbf{r}_0 = [v_0\tau_j, 0, 0], \mathbf{r} = [x, y, z],$$

and the roots of Equation (4.187) can be found by solving the following equation with respect to τ_j:

$$\tau_j = t - \sqrt{(x - v_0\tau_j)^2 + y^2 + z^2}/c. \tag{4.189}$$

The solutions of Equation (4.189) are given by

$$\tau_{1,2} = \frac{ct - M_a x \pm \sqrt{(x - cM_a t)^2 + (1 - M_a^2)(y^2 + z^2)}}{c(1 - M_a^2)}, \quad \text{with } M_a = v_0/c. \tag{4.190}$$

For a source moving at a speed less than the speed of sound ($M_a < 1$), the positive root cannot be a physical solution since the emission time is delayed more than the reception time. As a consequence, the emission time can be related to the reception time as

$$\tau = \frac{ct - M_a x - \sqrt{(x - cM_a t)^2 + (1 - M_a^2)(y^2 + z^2)}}{c(1 - M_a^2)}, \tag{4.191}$$

and the measured pressure field at $\mathbf{r}$ is simply given by

$$p(\mathbf{r}, t) = \frac{q_0(\tau)}{4\pi\,|\mathbf{r} - \mathbf{r}_0(\tau)|}\frac{1}{|1 - M_0(\tau)|}. \tag{4.192}$$

Therefore, the moving source produces two types of distortions as compared to a stationary source. The first is the change in the arrival time of a source signal $q_0(\tau)$, and the second is Doppler amplification. For linear motion of the source, M_0 is given by

$$M_0(\tau) = -\frac{1}{c}\frac{d}{d\tau}(\sqrt{(x - v_0\tau)^2 + y^2 + z^2})$$
$$= M_a \frac{(x - v_0\tau)}{\sqrt{(x - v_0\tau)^2 + y^2 + z^2}}. \quad (4.193)$$

Consequently, a source approaching the microphone ($x > v_0\tau$) produces an increased measured amplitude, and a source moving away from the microphone decreases the measured amplitude.

4.7.5.2 De-Dopplerization

The measured pressure signal of Equation (4.192) is distorted by the movement of the source. The compensation of the distortion, often called de-Dopplerization, is accomplished by transforming the signal measured by an microphone array fixed in space to one that moves with the same speed as the source. The concept of de-Dopplerization is illustrated in Figure 4.45. In this section, we investigate how the pressure signal of a microphone moving with the source can be reconstructed as if the measurement was taken by a microphone fixed in space.

De-Dopplerization can be implemented as follows: first, using the relation of Equation (4.192), the pressure data measured by a fixed microphone array is converted to the source signal $q_0(\tau)$. Next, from the estimated source signal, the pressure signal at a microphone array moving with the source is reconstructed.

For the estimation of the source signal, we first need to relate the temporal axis defined for the reception time with that defined for the emission time. If the temporal axis for the emission time is defined with a regular sampling interval of Δt, it can be written as

$$\tau[n] = n\Delta t, n = 0, \cdots, N - 1, \quad (4.194)$$

and we can calculate the location of a sound source $\mathbf{r}_0(\tau)$ and $M_0(\tau)$ over the defined temporal axis. The source signal $q(\tau[n])$, which is unknown, can be derived from Equations (4.186) and

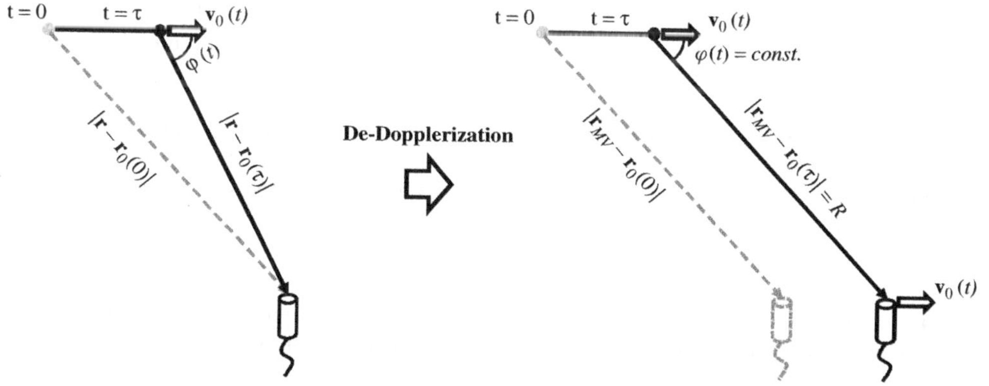

Figure 4.45 Conceptual drawing of de-Dopplerization

(4.187). The reception time can be expressed in terms of the emission time as

$$t = \tau + |\mathbf{r} - \mathbf{r}_0(\tau)|/c, \tag{4.195}$$

and the pressure signal takes the form

$$p(\mathbf{r}, \tau[n] + |\mathbf{r} - \mathbf{r}_0(\tau[n])|/c). \tag{4.196}$$

However, the pressure signal $p(\mathbf{r}, t)$ measured by the fixed microphone array might have uniform temporal sampling with respect to the reception time t and not to the emission time τ. Therefore, to calculate Equation (4.196), the measured data has to be resampled. A resampling technique with a fractional delay filter [10] can be employed for this procedure. Once the pressure signal at the reception time is obtained, then the source signal $q(\tau)$ can be derived from Equation (4.192). That is,

$$q_0(\tau) = 4\pi |\mathbf{r} - \mathbf{r}_0(\tau)| \, |1 - M_0(\tau)| \, p(\mathbf{r}, \tau + |\mathbf{r} - \mathbf{r}_0(\tau)|/c). \tag{4.197}$$

We now have an estimation of the source signal but no information about the location of the source. To estimate the source location, we reconstruct the pressure signal at the location of a microphone moving with the sound source. By denoting the position of the moving microphone as $\mathbf{r} = \mathbf{r}_{MV}(\tau)$, the pressure signal measured by the moving microphone can be written as

$$p(\mathbf{r}_{MV}(\tau), t) = \frac{q_0(\tau)}{4\pi |\mathbf{r}_{MV}(\tau) - \mathbf{r}_0(\tau)|}. \tag{4.198}$$

Note that the Doppler amplification $1/|1 - M_0(\tau)|$ disappears because the relative motion between the microphone and moving source is zero (Figure 4.45). Without this relative motion, the distance between the microphone and source $|\mathbf{r}_{MV}(\tau) - \mathbf{r}_0(\tau)| = R$ is time invariant, and the relation between the reception and emission times is just given by a pure time delay, that is,

$$t = \tau + R/c. \tag{4.199}$$

To summarize, the de-Dopplerized microphone signal can be written as

$$p(\mathbf{r}_{MV}(\tau), \tau + R/c) = |1 - M_0(\tau)| \frac{|\mathbf{r} - \mathbf{r}_0(\tau)|}{R} p(\mathbf{r}, \tau + |\mathbf{r} - \mathbf{r}_0(\tau)|/c). \tag{4.200}$$

4.7.5.3 Beamforming with De-Dopplerization

The de-Dopplerization procedure described in Equation (4.200) assumes a single source and single microphone. Extending the concept to multiple sources and microphones, beamforming of a moving object can be accomplished.

Suppose a number of grid points representing the candidate source locations are attached to a frame with a center $\mathbf{r}_0(\tau)$ that moves with a speed $\mathbf{v}_0$ (Figure 4.46). Denoting each position of the candidate source as $\mathbf{r}_\kappa(\tau) = \mathbf{r}_0(\tau) + \Delta\mathbf{r}_\kappa$ ($\kappa = 1, \cdots, K$), the source signal of the κth source can be estimated from the mth microphone signal $p(\mathbf{r}_m, t)$. From (4.197), the source signal estimated by the mth microphone ($q_{m\kappa}$) is given by

$$q_{m\kappa}(\tau) = 4\pi |\mathbf{r}_m - \mathbf{r}_\kappa(\tau)| \, |1 - M_{m\kappa}(\tau)| \, p(\mathbf{r}_m, \tau + |\mathbf{r}_m - \mathbf{r}_\kappa(\tau)|/c), \tag{4.201}$$

where $\mathbf{r}_m$ is the position of the mth microphone. The variable $M_{m\kappa}$ represents the relative speed of the κth sound source with respect to the mth microphone and is given by

$$M_{m\kappa}(\tau) = \frac{|\mathbf{v}_0(\tau)|}{c} \cos\varphi_{m\kappa}(\tau),$$

with $\varphi_{m\kappa}$ being the angle between the direction of velocity vector $\mathbf{v}_0$ and $\mathbf{r}_m - \mathbf{r}_\kappa(\tau)$.

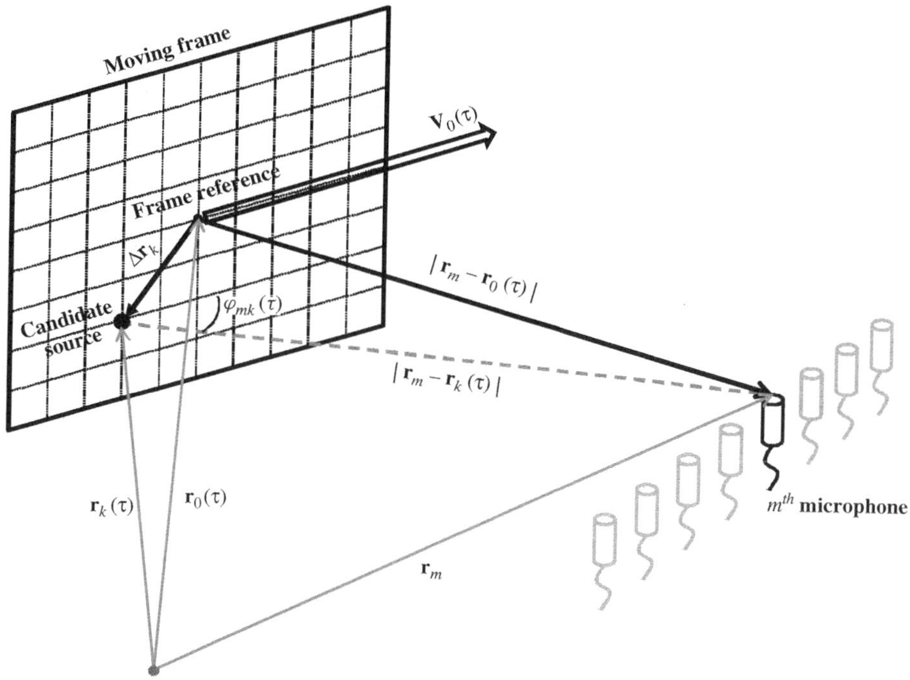

Figure 4.46 Beamforming problem with sources on a moving frame

Since we have K source signals each of which is estimated from each microphone, the de-Dopplerized source signal calculated from multiple microphones are averaged to yield the beamformer output:

$$b(\Delta\mathbf{r}_\kappa, \tau) = \frac{1}{M}\sum_{m=1}^{M} q_{m\kappa}(\tau). \tag{4.202}$$

This procedure can be likened to the DAS beamformer of Equation (4.32). The major difference with the stationary case is that the extra time delay due to the source movement and the Doppler amplification are compensated.

4.7.5.4 Example: Moving-Source Localization

Figure 4.47 shows an experiment setup for localizing the impulsive sound source of a moving vehicle. The vehicle moves from right to left with a constant velocity ($v_0 = 60\,\text{km/h}$), and the measurement is triggered by two photo sensors located 5 m from both sides of the array. To generate the impulsive sound, four spikes were installed on the front tire. The measurement was taken with a curved-spoke-type microphone array (30 channels).

In Figure 4.48, the peak value of the beamformer output (Equation (4.178)) presented on a decibel scale is overlapped with an image of the vehicle. The beamformer output shows that the estimated source position coincides with the contact point of the spikes and the ground. The beamformer output signal reconstructed at the source location is similar to the periodic impulse signal. The result demonstrates that the location of a moving source can be accurately found, provided the velocity profile of the moving object is known.

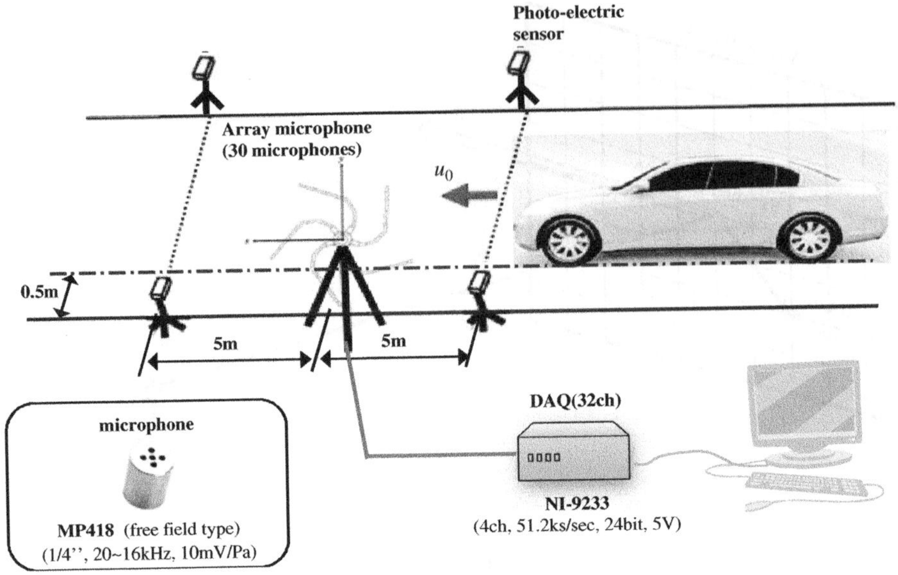

Figure 4.47 Experimental setup for moving-source localization

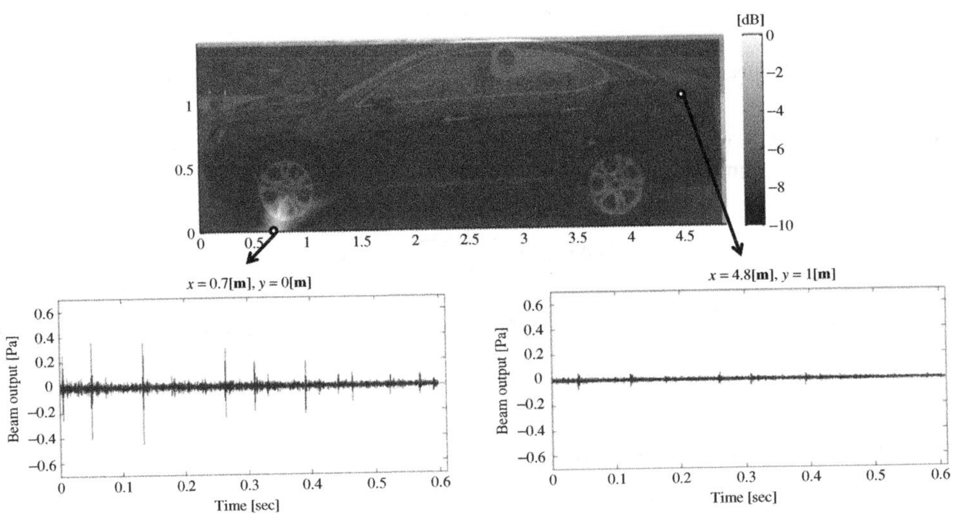

Figure 4.48 Time-domain beamforming with de-Dopplerization

4.8 Post-Processing Techniques

4.8.1 Deconvolution and Beamforming

In this section, we deal with post-processing techniques based on the image processing technique of deconvolution to enhance the obtained beamforming image. To begin, we assume a single frequency component of the measured sound field, and hence, the frequency dependency ω will be omitted.

Suppose that we have monopole sound sources distributed on a source plane $Sq(\mathbf{r}_s)$, then the final goal of beamforming is to find the exact location of the sound sources by predicting the distribution of q. When a single monopole source at $\mathbf{r}_s$ is excited, then a sound field is generated and propagates to the location of the microphone array. The pressure field due to a monopole source can be expressed in terms of the free-field Green's function from $\mathbf{r}_s$ to the measurement position $\mathbf{r}_m$ ($g_f(\mathbf{r}_m|\mathbf{r}_s)$). If all the sound fields from all monopoles are superposed, then the sound field can be described as

$$p(\mathbf{r}_m) = \int_S g_f(\mathbf{r}_m|\mathbf{r}_s)q(\mathbf{r}_s)dS(\mathbf{r}_s), \tag{4.203}$$

where the measurement position $\mathbf{r}_m$ is in the finite measurement plane S_m ($\mathbf{r}_m \in S_m$).

To predict the excitation function q from the measurement taken at S_m, the measured pressure field is multiplied by a scan vector to produce the beamformer output. Our aim is to predict the source distribution in the source plane S, and hence the scan vector for predicting the source excitation in S can be denoted as $w(\mathbf{r}_m|\mathbf{r})^*$;

$$b(\mathbf{r}) = \int_{S_m} w(\mathbf{r}|\mathbf{r}_m)p(\mathbf{r}_m)dS_m, \tag{4.204}$$

where the prediction point is again in the source plane $\mathbf{r} \in S$.

The whole beamforming process from the source excitation $q(\mathbf{r}_s)$ to the beamformer output $b(\mathbf{r})$ can be summarized as

$$b(\mathbf{r}) = \int_{S_m}\int_S [w(\mathbf{r}_m|\mathbf{r})^* g_f(\mathbf{r}_m|\mathbf{r}_s)q(\mathbf{r}_s)]dS(\mathbf{r}_s)dS_m(\mathbf{r}_m). \tag{4.205}$$

Therefore, the excitation function $q(\mathbf{r}_s)$ experiences two transforms when being converted to the beamformer output $b(\mathbf{r})$.

The transform of Equation (4.205) can be further simplified if the shift invariant properties of the scan vector and free-field Green's function are exploited. The shift invariance, denoted as

$$w(\mathbf{r}_m|\mathbf{r})^* = w(\mathbf{r}_m - \mathbf{r})^*, \quad g_f(\mathbf{r}_m|\mathbf{r}_s) = g_f(\mathbf{r}_m - \mathbf{r}_s), \tag{4.206}$$

implies the functions w and g_f are only a function of the relative distance between two positions, rather than the absolute position $\mathbf{r}_s$ or $\mathbf{r}_m$. It can be easily shown that the free-field Green's function

$$g_f(\mathbf{r}_m|\mathbf{r}_s) - \frac{e^{ik|\mathbf{r}_m - \mathbf{r}_s|}}{4\pi|\mathbf{r}_m - \mathbf{r}_s|}$$

satisfies shift invariance, and the scan vector derived from the planar or spherical wavefront model also fulfills this property. For example, the plane wave model with a wavenumber vector $\mathbf{k}$ can be written as a function of $\mathbf{r}_m - \mathbf{r}$:

$$w(\mathbf{r}_m|\mathbf{r})^* = e^{i\mathbf{k}\cdot(\mathbf{r}_m - \mathbf{r})^*}. \tag{4.207}$$

If the two transforms satisfy shift invariance (Equation (4.206)), the whole beamforming process of Equation (4.205) can be described as two *spatial convolutions*:

$$\begin{aligned} b(\mathbf{r}) &= \int_{S_m}\int_S [w(\mathbf{r}_m|\mathbf{r})^* g_f(\mathbf{r}_m - \mathbf{r}_s)q(\mathbf{r}_s)]dS(\mathbf{r}_s)dS_m(\mathbf{r}_m) \\ &= w(\mathbf{r})^* *_s (g_f(\mathbf{r}) *_s q(\mathbf{r})) \\ &= (w(\mathbf{r})^* *_s g_f(\mathbf{r})) *_s q(\mathbf{r}), \end{aligned} \tag{4.208}$$

where $*_s$ denotes the *spatial convolution operator* defined by

$$f(\mathbf{r})*_s g(\mathbf{r}) = \int_S f(\mathbf{r}-\mathbf{r}')g(\mathbf{r}')dS(\mathbf{r}'). \tag{4.209}$$

It is noteworthy that the whole process to convert the excitation function to the beamformer output is given by the spatial convolution. Therefore, the beamforming process of Equation (4.208) can be regarded as a system with an impulse response function $psf(\mathbf{r})$ given by

$$psf(\mathbf{r}) = w(\mathbf{r})^* *_s g_f(\mathbf{r}). \tag{4.210}$$

When the excitation q is fed into the system, it is convolved with the impulse response $psf(\mathbf{r})$ to produce an output $b(\mathbf{r})$:

$$b(\mathbf{r}) = psf(\mathbf{r}) *_s q(\mathbf{r}). \tag{4.211}$$

Therefore, the unknown input to the system (q) can be readily found from the given beamformer output b by solving an inverse problem. Since the output of the system is equal to the spatial convolution of the impulse response and input signals, the inverse problem can be thought as a deconvolution of the impulse response from the output signals. That is,

$$q(\mathbf{r}) = psf(\mathbf{r}) *_s^{-1} b(\mathbf{r}), \tag{4.212}$$

where $*_s^{-1}$ denotes the deconvolution operator.

In solving the inverse problem, the *spatial impulse response* of a beamforming system $psf(\mathbf{r})$ plays an important role. The spatial impulse response represents the output when a point excitation

$$q(\mathbf{r}_s) = \delta(\mathbf{r}_s - \mathbf{r}_0) \tag{4.213}$$

is given as an input. In optics, the spatial impulse response $psf(\mathbf{r})$ expresses how much a point "input" is blurred by the imaging system, and hence, is commonly referred to as the point spread function (PSF).

The deconvolution in the spatial domain is equivalent to a spectral division in the wavenumber domain. That is,

$$Q(k_x, k_y) = \frac{B(k_x, k_y)}{PSF(k_x, k_y)}, \tag{4.214}$$

where Q, B, and PSF are the spatial Fourier transforms of q, b, and psf, respectively. This inverse problem, however, is often ill-posed when the array aperture size is small, because most wavefronts arriving from different directions look similar on the small aperture. To examine the ill-posedness in detail, consider the spatial Fourier transform of the PSF. From Equation (4.210), $PSF(k_x, k_y)$ can be expanded to

$$PSF(k_x, k_y) = W(k_x, k_y)G_f(k_x, k_y). \tag{4.215}$$

where W and G_f denotes the spatial Fourier transform of the scan vector w and the free-field Green's function g_f, respectively. For the beamforming, the measurement is usually taken in the farfield region away from the source, and hence the Green's function described in the wavenumber domain is bandlimited to within $\sqrt{k_x^2 + k_y^2} \le k$. The inverse of the band-limited wavenumber spectrum tends to diverge due to the small stop-band spectrum.

It is also important to note that the array aperture is truncated in space due to the finite aperture size of the microphone array. Accordingly, the spatial Fourier transform of the Green's function measured by the microphone array is, in fact, equivalent to a convolution with a sinc function in the wavenumber domain. That is,

$$g_{ft}(x, y|x_s, y_s) = g_f(x, y|x_s, y_s)\text{rect}(x/L)\text{rect}(y/L)$$
$$G_{ft}(k_x, k_y|x_s, y_s) = \frac{L^2}{4\pi^2} G_f(k_x, k_y|x_s, y_s) *_{k_x} \text{sinc}(k_x L/2) *_{k_y} \text{sinc}(k_y L/2), \quad (4.216)$$

where $\text{rect}(x/L)$ is a rectangular function of length L (Equation (A6)), and $*_k$ is the convolution operator defined in the wavenumber domain. Equation (4.216) implies that the wavenumber spectrum is spread out by the sinc function. Because of this spreading in the wavenumber domain, only a small fraction of the original wavenumber spectrum remains within the bandlimited spectrum. This phenomenon is the so-called diffraction limit (Chapter 5, Section 5.3.3). The small fraction of the remaining spectrum contains little information, and the deconvolution of Equation (4.214) performs poorly in finding the original $Q(k_x, k_y)$ without recovering the truncated data in $\sqrt{k_x^2 + k_y^2} > k$.

4.8.2 Nonnegativity Constraint

The ill-conditioning of the inverse problem only can be resolved by adding extra constraints. The deconvolution methods explained in the following sections use different approaches based on nonnegativity constraints. The nonnegativity assumption is common in image signal processing for which the measured data is always nonnegative. However, for the complex amplitudes of Equation (4.211) such a nonnegativity constraint is far from realistic.

To use the nonnegativity constraint with the beamforming problem, we will consider the beamforming power, rather than the beamformer output, and the power of the source signal. To this end, let us return to the beamforming power expressed in terms of the correlation matrix. From Equations (4.5) and (4.44), the beamforming power can be written as

$$\beta(x, y) = \mathbf{w}(x, y)^H \mathbf{G}\mathbf{R}_{qq}\mathbf{G}^H \mathbf{w}(x, y). \quad (4.217)$$

Note that the transfer matrix $\mathbf{H}$ of Equation (4.44) is replaced by $\mathbf{G}$ of which elements are given by the free-field Green's function, because we deal with the free-field radiation from point sources.

If the source signals at multiple positions are assumed to be incoherent with each other, then the correlation matrix $\mathbf{R}_{qq}$ will be a diagonal matrix. In such a case, Equation (4.217) can be rewritten as

$$\begin{aligned}\beta(x, y) &= (\mathbf{w}(x, y)^H \mathbf{G})\mathbf{R}_{qq}(\mathbf{w}(x, y)^H \mathbf{G})^H \\ &= \left[(\mathbf{w}(x, y)^H \mathbf{G}) \circ (\mathbf{w}(x, y)^H \mathbf{G})^*\right] \text{diag}(\mathbf{R}_{qq}),\end{aligned} \quad (4.218)$$

where $\circ$ denotes the element-wise product of two vectors of the same size (Hadamard product; $[\mathbf{A} \circ \mathbf{B}]_{nm} = [\mathbf{A}]_{nm}[\mathbf{B}]_{nm}$). To deal with the beamforming power at multiple prediction points $(x_\kappa, y_\kappa) \in S$, let us define the beamforming power vector $\boldsymbol{\beta}$ and scan matrix $\mathbf{W}$ as follows:

$$\begin{aligned}\boldsymbol{\beta} &= [\beta(x_1, y_1), \cdots, \beta(x_K, y_K)]^T \\ \mathbf{W} &= [\mathbf{w}(x_1, y_1), \cdots, \mathbf{w}(x_K, y_K)].\end{aligned} \quad (4.219)$$

The scan matrix $\mathbf{W} \in \mathbb{C}^{M \times K}$ consists of column vectors $\mathbf{w}(x_\kappa, y_k)$ for multiple prediction points.

From Equation (4.218), the beamforming power at multiple points can be described as

$$\boldsymbol{\beta} = [(\mathbf{W}^H \mathbf{G}) \circ (\mathbf{W}^H \mathbf{G})^*] \mathbf{s}, \quad (4.220)$$

where $\mathbf{s} = \text{diag}(\mathbf{R}_{qq})$ is a column vector containing each source power as its elements. It is noteworthy that the beamforming power $\boldsymbol{\beta}$ and source powers $\mathbf{s}$ are all nonnegative. In addition, the transfer function between these two quantities, defined by

$$\mathbf{T} = \left[(\mathbf{W}^H\mathbf{G}) \circ (\mathbf{W}^H\mathbf{G})^*\right], \tag{4.221}$$

is also nonnegative. Consequently, we can obtain a linear equation with the nonnegative variables

$$\boldsymbol{\beta} = \mathbf{Ts}, \tag{4.222}$$

and the matrix $\mathbf{T}$ becomes the PSF of the given problem.

Since a prediction point (x_κ, y_κ) is on the same grid as the sources (Figure 4.49), the scan vector $\mathbf{W}$ will be the same as $\mathbf{G}$ as long as the assumed wavefront model is perfect;

$$\begin{aligned}\boldsymbol{\beta} &= [(\mathbf{G}^H\mathbf{G}) \circ (\mathbf{G}^H\mathbf{G})^*]\mathbf{s} \\ &= \mathbf{Ts}.\end{aligned} \tag{4.223}$$

The beamforming problem is now converted to the inverse problem consisting of nonnegative input, output, and PSFs (Figure 4.50). The nonnegativity makes the inverse problem (Equation (4.223)) well conditioned (e.g., [11]). However, the conventional least-squares approach cannot reflect the nonnegativity constraint. Popular methods such as nonnegative least-squares (NNLS) approach and the deconvolution approach for the mapping of acoustic sources (DAMAS) employ iterative search algorithms to solve the inverse problem with a nonnegativity constraint.

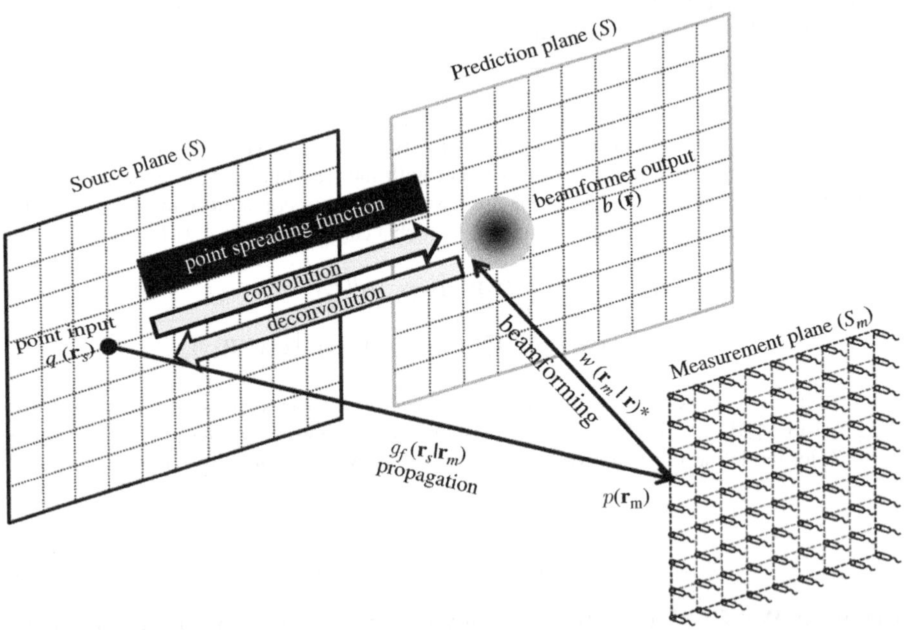

Figure 4.49 Concept of convolution and deconvolution for the beamforming problem (prediction plane = source plane)

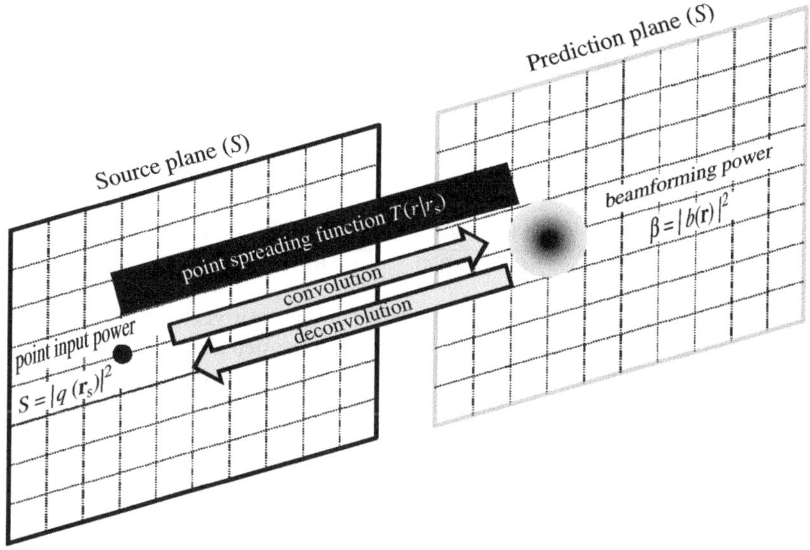

Figure 4.50 Concept of deconvolution with nonnegative variables and a point spread function (prediction plane = source plane)

4.8.3 Nonnegative Least-Squares Algorithm

The NNLS approach attempts direct minimization of the error between the beamforming power map and the source strength **s**, which is defined by

$$\text{Minimize}\, \|\boldsymbol{\varepsilon}\| = \|\boldsymbol{\beta} - \mathbf{Ts}\|, \text{ subject to}\, \mathbf{As} \geq \mathbf{c}. \tag{4.224}$$

In general, **A** is a constraint matrix for the generalized linear equality constraints $\mathbf{As} - \mathbf{c} \geq 0$, but for the NNLS, $\mathbf{A} = \mathbf{I}$ and $\mathbf{c} = \mathbf{0}$ such that $\mathbf{s} \geq 0$. It was shown by Lawson and Hanson [12] that the NNLS problem can be solved by iteration and that the iteration always converges and terminates.

A variety of methods have been developed to solve the NNLS problems, and what is introduced here is a very simple technique called the first-order gradient-projection algorithm [16]. This technique aims to find the minimum of Equation (4.224) in the direction for which the error gradient is negative. If we denote the source strength at the nth iteration as $\mathbf{s}^{(n)}$, the gradient of squared error with respect to $\mathbf{s}^{(n)}$ is given by

$$\Delta\boldsymbol{\varepsilon}^{(n)} = \frac{\partial(\boldsymbol{\varepsilon}^{(n)T}\boldsymbol{\varepsilon}^{(n)})}{\partial\mathbf{s}^{(n)}} = 2\mathbf{T}^T(\mathbf{Ts}^{(n)} - \boldsymbol{\beta}). \tag{4.225}$$

The gradient determines the search direction for a given input $\mathbf{s}^{(n)}$. If the error is increasing, that is, the gradient is positive with respect to the κth element of $\mathbf{s}^{(n)}$ ($s_\kappa^{(n)}$), and if the solution $s_\kappa^{(n)}$ is on the boundary of the feasible region ($s_\kappa^{(n)} = 0$) defined by the constraint, the search in that direction is cancelled. That is,

$$\Delta s_\kappa^{(n)} = \begin{cases} \min\left\{0,\, \Delta\varepsilon_\kappa^{(n)}\right\} & \text{if} \quad s_\kappa^{(n)} = 0 \\ \Delta\varepsilon_\kappa^{(n)} & \text{otherwise.} \end{cases}$$

Next, we need to determine the step size, that is, the amount by which the solution will be perturbed to the new direction. There are various methods for determining the step size; however, one simple way is to examine the amount of output perturbation. The perturbation of the output $\boldsymbol{\beta}$ in the direction of $\Delta\mathbf{s}^{(n)}$ is given by

$$\Delta\boldsymbol{\beta}^{(n)} = \mathbf{T}\Delta\mathbf{s}^{(n)}. \tag{4.226}$$

Then, the step size (μ) is determined by the size and direction of the current error:

$$\mu^{(n)} = \frac{\Delta\boldsymbol{\beta}^{(n)} \cdot \boldsymbol{\varepsilon}^{(n)}}{\Delta\boldsymbol{\beta}^{(n)} \cdot \Delta\boldsymbol{\beta}^{(n)}} = \frac{\Delta\boldsymbol{\beta}^{(n)} \cdot (\mathbf{T}\mathbf{s}^{(n)} - \boldsymbol{\beta})}{\Delta\boldsymbol{\beta}^{(n)} \cdot \Delta\boldsymbol{\beta}^{(n)}}. \tag{4.227}$$

The step size is equivalent to the projection of the current error vector $\boldsymbol{\varepsilon}^{(n)}$ onto the perturbation occurring due to the new search direction $\Delta\mathbf{s}^{(n)}$. Therefore, the step size will be greater if the error of the current step is large or if the error direction coincides with the direction of the perturbation $\Delta\boldsymbol{\beta}^{(n)}$.

The new candidate for the source strength is then given by

$$\hat{\mathbf{s}}^{(n+1)} = \mathbf{s}^{(n)} - \mu^{(n)}\Delta\mathbf{s}^{(n)}. \tag{4.228}$$

However, the new solution at each iteration should satisfy the nonnegativity constraint. Therefore, the source strength is set to zero if it is negative. That is,

$$s_\kappa^{(n+1)} = \max(\hat{s}_\kappa^{(n+1)}, 0). \tag{4.229}$$

The source strength $\mathbf{s}^{(n+1)}$ obtained from Equation (4.229) becomes the new solution for the $(n+1)$th iteration. Figure 4.51 shows the beamforming power deconvolved by NNLS technique. In this simple case with two monopole sources, the resolution of beamforming is dramatically enhanced only within two iteration steps.

4.8.4 DAMAS

The DAMAS [13–15] is an alternative method that also utilizes the iterative procedure to solve the deconvolution problem. The main difference with the NNLS formula is that a single iteration consists of element-wise updates with nonnegativity constraints applied. Suppose that a given search algorithm sequentially updates the source strength in increasing order of j ($j = 1, \cdots, K$). When the source strengths up to the $(j-1)$th element are updated in the $(n+1)$th iteration, the error in the jth source strength can be written as (from Equation (4.224))

$$\varepsilon_j^{(n)} = \underbrace{\sum_{\kappa=1}^{j-1} T_{j\kappa}s_\kappa^{(n+1)}}_{\text{updated}} + \underbrace{T_{jj}\hat{s}_j^{(n+1)}}_{\text{to be determined}} + \underbrace{\sum_{\kappa=j+1}^{K} T_{j\kappa}s_\kappa^{(n)}}_{\text{not updated}} - \beta_j \tag{4.230}$$

where the jth source strength $\hat{s}_j^{(n+1)}$ is now to be calculated. The new source strength $\hat{s}_j^{(n+1)}$ is determined such that the error $\varepsilon_j^{(n)}$ can be minimized. Forcing the error to be zero ($\varepsilon_j^{(n)} = 0$) gives

$$\hat{s}_j^{(n+1)} = \frac{1}{T_{jj}}\left[\beta_j - \sum_{\kappa=1}^{j-1} T_{j\kappa}s_\kappa^{(n+1)} - \sum_{\kappa=j+1}^{K} T_{j\kappa}s_\kappa^{(n)}\right] \tag{4.231}$$

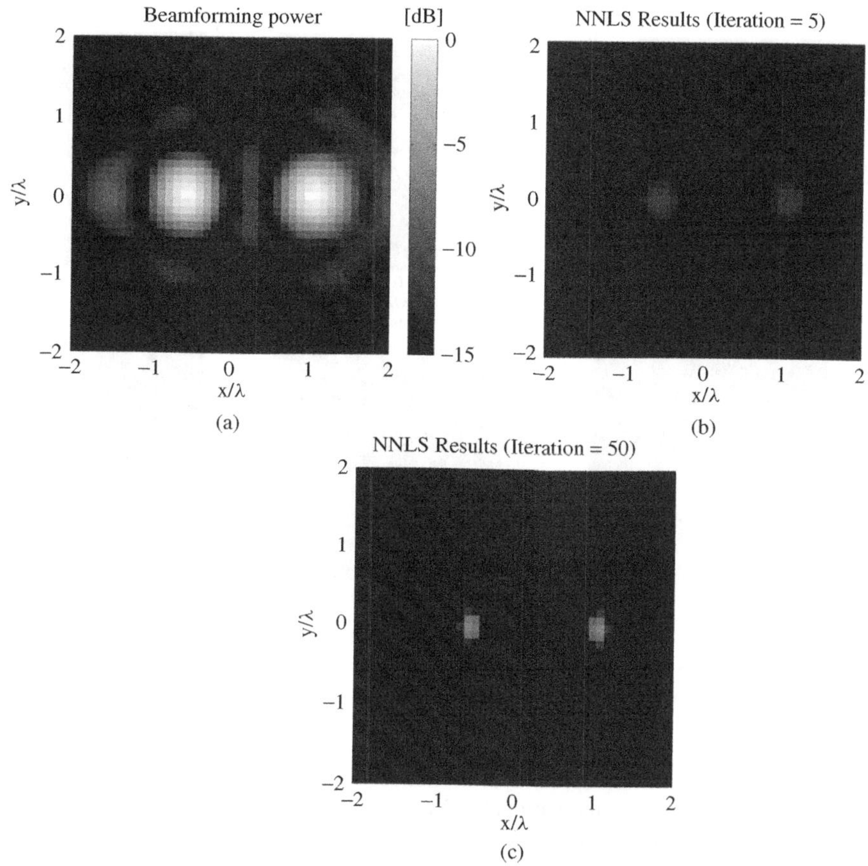

Figure 4.51 NNLS example: (a) original beamforming power and the deconvolved result after the (b) first iteration and (c) second iteration (the simulation conditions are the same as those in Figure 4.34)

or

$$\hat{s}_j^{(n+1)} = s_j^{(n)} - \frac{1}{T_{jj}}\left[\beta_j - \sum_{\kappa=1}^{j-1} T_{j\kappa}s_\kappa^{(n+1)} - \sum_{\kappa=j}^{K} T_{j\kappa}s_\kappa^{(n)}\right].$$

In addition, the nonnegativity constraint has to be satisfied for the source strength $\mathbf{s}$, and hence the updated source strength is given by

$$s_j^{(n+1)} = \max\left(\hat{s}_j^{(n+1)}, 0\right). \tag{4.232}$$

For the first iteration ($n = 1$), the initial value of $s_\kappa^{(1)}$ can be set as zero. After the first iteration, the calculation direction is reversed, that is, the coefficient $s_\kappa^{(2)}$ is evaluated from $\kappa = K$ to $\kappa = 1$. The next iteration starts again at $\kappa = 1$. It has been shown that this alternating sweep approach can provide images with fewer oscillations than a one-way update [15]. Figure 4.52 demonstrates the performance of the DAMAS in comparison to the DAS beamforming. The beamforming power at each iteration step shows that the deconvolution of the PSF can be quickly accomplished with a few iteration steps.

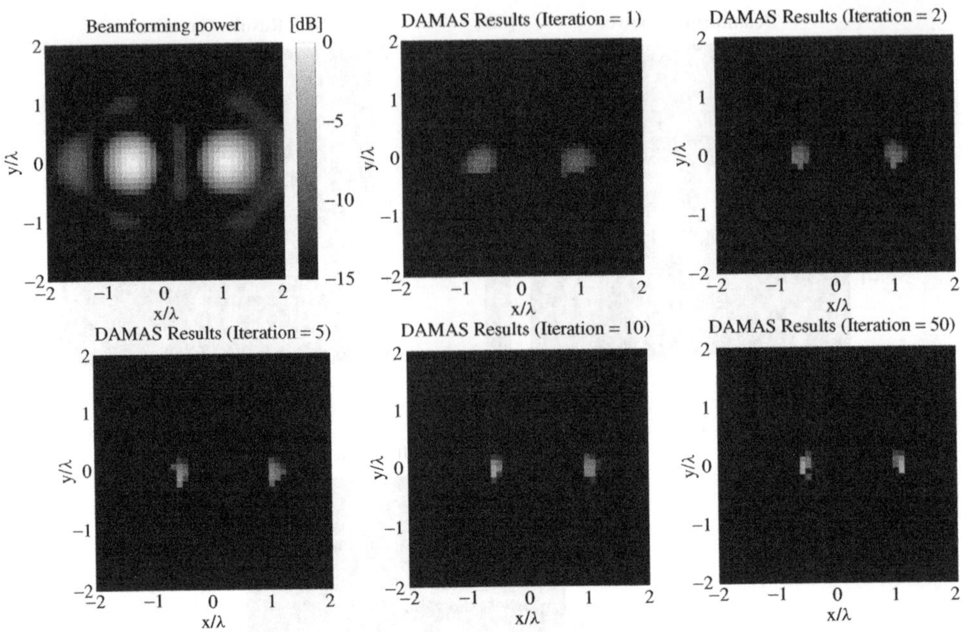

Figure 4.52 (a–f) The DAMAS applied to two point sources separated by 1.5 λ (the simulation conditions are the same as those in Figure 4.34)

Both the DAMAS and NNLS approaches do not require the shift-invariance assumption of the transfer function **T**, but their computational effort is relatively high compared with conventional beamforming. There are faster and lighter algorithms (e.g., FFT-NNLS [16] or DAMAS2 [17]) that exploit Fourier-based iterations, but these techniques assume shift invariance of the transfer function.

References

1. Schmidt, R.O. (1986) Multiple emitter location and signal parameter estimation. *IEEE Transactions on Antennas Propagation*, **AP-34**, 276–280.
2. Paulraj, A., Roy, R. and Kailath, T. (1986) A subspace rotation approach to signal parameter estimation. *Proceedings of the IEEE*, **74**(7), 1044–1046.
3. Roy, R. and Kailath, T. (1989) ESPRIT – estimation of signal parameters via rotation invariance techniques. *IEEE Transactions on Acoustics, Speech and Signal Processing*, **37**(7), 984–995.
4. Johnson, D.H. and Dudgeon, D.E. (1993) *Array Signal Processing: Concepts and Techniques*, Prentice-Hall, Englewood Cliffs, NJ.
5. Burg, J.P. (1972) The relationship between maximum entropy and maximum likelihood spectra. *Geophysics*, **37**(2), 375–376.
6. Choi, J.W. and Kim, Y.H. (1995) Spherical beam-forming and MUSIC methods for the estimation of location and strength of spherical sound sources, *Mechanical Systems and Signal Processing*, **9**(5), 569–588.
7. Wang, H. and Kaveh, M. (1985) Coherent signal-subspace processing for the detection and estimation of angles of arrival of multiple wide-band sources. *IEEE Transactions on Acoustics, Speech and Signal Processing*, **33**(4), 823–831.

8. Choi, J.W. and Kim, Y.H. (1995) Estimation of locations and strengths of broadband planar and spherical noise sources using coherent signal subspace. *Journal of the Acoustical Society of America*, **98**(4), 2082–2093.
9. Dowling, A.P. and Williams, J.E.F. (1983) *Sound and Sources of Sound*, John Wiley & Sons, Inc., New York.
10. Farrow, C.W. (1988) A continuously variable digital delay element. *Proceedings of the IEEE International Symposium on Circuits and Systems*, **3**, 2641–2645.
11. Bardsley, J., Merikoski, J. and Vio, R. (2008) The stabilizing properties of nonnegativity constraints in least-squares image reconstruction. *International Journal of the Pure Applied Mathematics*, **43**(1), 95–109.
12. Lawson, C.L. and Hanson, R.J. (1974) *Solving Least Squares Problems*, Prentice-Hall, Englewood Cliffs, NJ.
13. Brooks, T.F. and Humphreys, W.M. (2006) A deconvolution approach for the mapping of acoustic sources (DAMAS) determined from phased microphone arrays. *Journal of Sound and Vibration*, **294** (4), 856–879.
14. Brooks, T.F. and Humphreys, W.M. (2005) Three-dimensional application of DAMAS methodology for aeroacoustic noise source definition. Proceedings of the 11th AIAA/CEAS Aeroacoustics Conference, 2005–2960.
15. Brooks, T.F. and Humphreys, W.M. (2006) Extension of DAMAS phased array processing for spatial coherence determination (DAMAS-C). Proceedings of the 12th AIAA/CEAS Aeroacoustics Conference, 2006–2654.
16. Ehrenfried, K. and Koop, L. (2007) Comparison of iterative deconvolution algorithms for the mapping of acoustic sources. *American Institute of Aeronautics and Astronautics Journal*, **45**(7), 1584–1595.
17. Dougherty, R.P. (2005) Extension of DAMAS and benefits and limitations of deconvolution in beamforming. Proceedings of the 11th AIAA/CEAS Aeroacoustics Conference, 2005–2961.

Part III

Sound Manipulation

Throughout Part II, various sound visualization techniques that utilize a microphone array have been explained. Interestingly, replacing a microphone array with a loudspeaker array makes it possible to generate a sound field in a desired region, where the visualization or prediction of the sound field is made. Figure III.1 demonstrates that the prediction of the sound field is possible in both directions: backward as well as forward. On the same basis, we can argue that a loudspeaker array can generate a desired sound field in the direction of interest (e.g., [1]). Hereafter, generating a desired sound field in a selected region is referred to as "sound manipulation."

From among the many problems considered in research fields related to sound manipulation, two representative problems are discussed in this book, namely, the sound field reproduction and focusing problems. In the sound field reproduction problem, a loudspeaker array, such as that shown in Figure III.2a, is controlled to replicate a desired sound field from a virtual sound source in space and time (Figure III.2b). From the Kirchhoff-Helmholtz (KH) integral, it is evident that we can generate a desired sound field using monopole and dipole sources distributed over a surface. For example, in Figure III.1, sound sources completely covering a planar surface can mimic the wavefront radiated from a sunflower behind them. The sound field reproduction problem is closely connected to the acoustic holography problem discussed in Chapter 3, in that both rely on the KH integral to find the unknown source strength distribution over a surface using the measured data or the given desired field.

On the other hand, the sound focusing problem deals with the method to concentrate the sound field only within a selected region in space. It can thus be considered as the counterpart of the beamforming technique that also pursues the focusing of beamforming power only in the direction or area of interest. For example, let us again discuss the simple beamforming problem shown in Figure III.3a. If we consider a sound source driven by unit source strength and located at $\mathbf{r}_s$, then the pressure measured at the nth microphone location ($\mathbf{r}_m^{(n)}$) is given by the transfer function between the source and the microphone ($h(\mathbf{r}_m^{(n)}|\mathbf{r}_s)$). The beamforming output $b(\mathbf{r}_s)$ at the source location is then predicted by calculating the inner product of the scan vector $w(\mathbf{r}_m^{(n)})$ and the transfer function. In the sound focusing problem, the source strength or excitation signals $q(\mathbf{r}_s^{(\kappa)})$ are emitted from multiple sound sources. The radiated pressure field can also be expressed as an inner product of the excitation and transfer function (Figure III.3b). Since the reciprocity relation holds for the transfer function of the linear sound wave propagation, the problem of designing a scan vector w is similar to the problem of determining the optimal source excitation signal q.

Sound Visualization and Manipulation, First Edition. Yang-Hann Kim and Jung-Woo Choi.

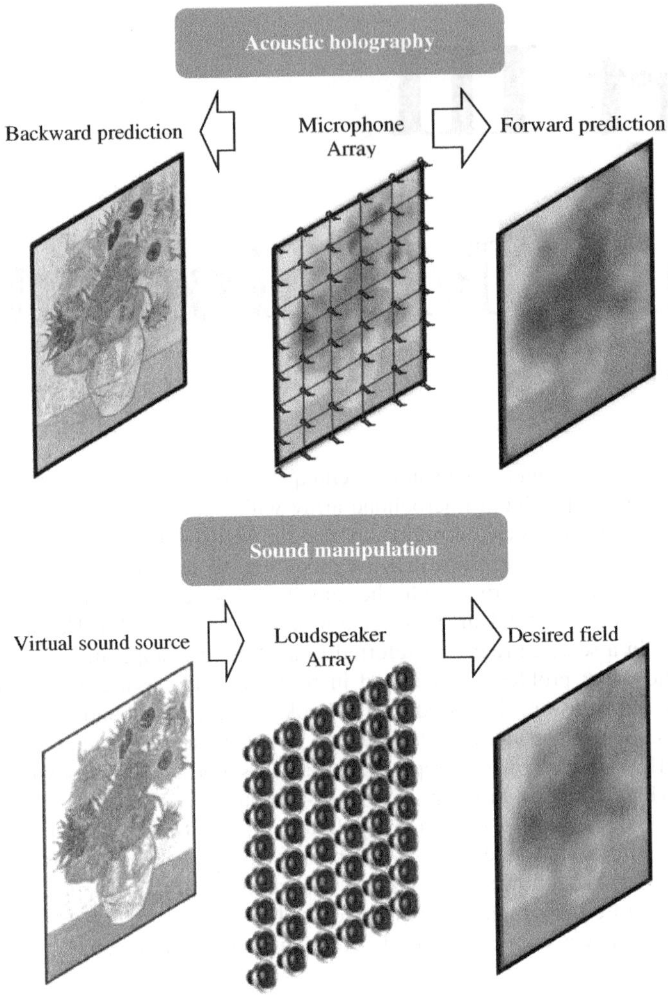

Figure III.1 Conceptual drawings illustrating the similarity between acoustic holography and sound field reproduction. (Modified image from the original version of [2].)

In solving the sound field reproduction and focusing problems, a sound field is generated with an insufficient number of transducers, that is, with a finite number of loudspeakers; therefore, the problem of sound manipulation is also an inverse problem, similar to the sound visualization problem. The desired sound field can fit either the parts of the elephants or those of greyhounds, and hence, the illustration in Figure III.3 also holds for sound manipulation. Therefore, the selection of the basis function determines the quality of sound manipulation. As noted in the sound visualization problem, "the generation of the sound field completely depends on how well the basis function mimics reality." The selection of the basis function of the sound manipulation depends on the characteristics of the desired sound field and the acoustic relations between the sources and the region.

Despite these similarities, there are prominent differences between sound visualization and manipulation problems. For the sound visualization problem, we can use various propagation operators

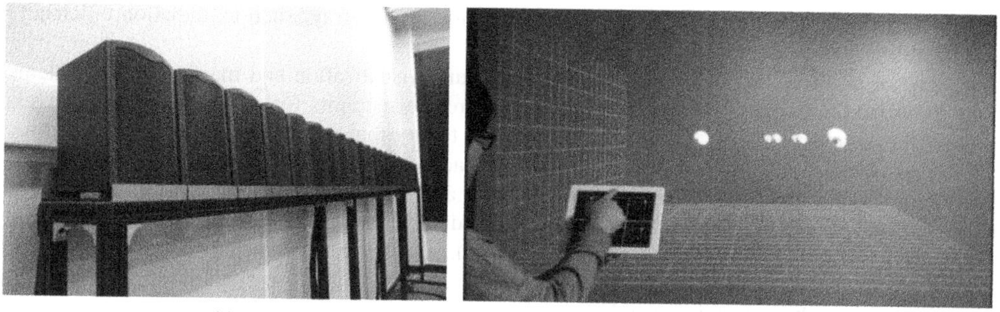

(a) (b)

Figure III.2 (a) Line loudspeaker array for sound field reproduction and (b) rendering of virtual sound sources. (Sound manipulation system installed at the Center for Noise and Vibration Control, KAIST, 2012.)

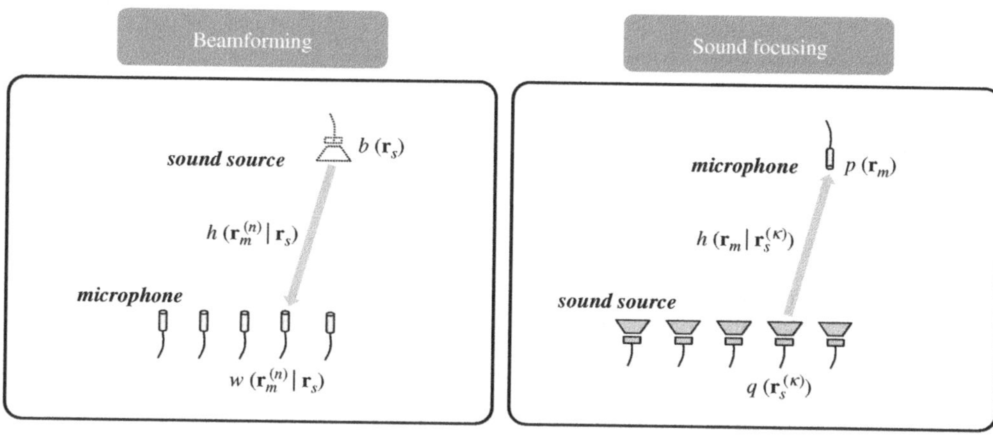

$$b(\mathbf{r}_s) = \frac{1}{N}\sum_{n=1}^{N} h(\mathbf{r}_m^{(n)}|\mathbf{r}_s) w(\mathbf{r}_m^{(n)})^*$$

(a)

$$p(\mathbf{r}_m) = \sum_{\kappa=1}^{K} h(\mathbf{r}_m|\mathbf{r}_s^{(\kappa)})\, q(\mathbf{r}_s^{(\kappa)})$$

(b)

Figure III.3 Representation of (a) beamforming and (b) sound focusing as a dual problem

or transfer functions, and we can even modify the operator to achieve a better prediction result. For example, we can employ various Green's functions as a propagator of acoustic holography. In case of beamforming, we can modify the structure of a correlation matrix to enhance the resolution of a beamforming image (e.g., diagonal removal technique), or we can introduce another definition of beamforming power (e.g., linear prediction, MUSIC (multiple signal classification)). For the sound manipulation problem, however, the transfer function is determined by the radiation characteristics and positions of loudspeakers. Accordingly, the transfer function cannot be changed once the reproduction room and type of loudspeaker have been selected. Hence, the sound manipulation problem has a strong physical constraint for realizing the desired sound field. Nevertheless, it does not mean that the manipulation problem is always more difficult to solve. In the sound visualization problem, the measured microphone signals are often corrupted by interfering sources and background noises, whereas signals fed to loudspeakers are less influenced by noise. Apart from noise, there are complicated issues involved with loudspeakers constituting an array, such as radiation efficiency, manufacturing variances, and distortions.

In view of all these, it can be summarized that sound visualization and manipulation problems share the same theoretical background but have different constraints for realizing their objectives. In the following chapters, we focus on delivering the theory that can be complementarily used for both sound visualization and manipulation. Chapter 5 addresses the sound focusing problems. From obtaining a desired directivity by using a loudspeaker array its dependency on array characteristics and associated wavelengths are extensively discussed in terms of the concepts of addition (+), subtraction (−), multiplication (×), and division (÷), which are fundamental operators used for shaping the sound energy distribution in space. These operations are also essential to the design of a scan vector for a beamforming problem.

Chapter 6 describes how an inverse problem is defined and why the problem is often ill-posed. The knowledge on the inverse problem might be helpful to those who study acoustic holography. The possible means to reproduce sound fields of interest are then introduced, with techniques to overcome issues related to ill-posedness. The method needs to be changed depending on what one desires to reproduce. The well-known wave field synthesis (WFS) [3, 4] and mode-matching approaches [5, 6] are introduced. The methods used to solve the exterior and interior problems of the KH integral equation are also described.

References

1. Berkhout, A.J. (1988) A holographic approach to acoustic control. *Journal of the Audio Engineering Society*, **36**(12), 977–995.
2. Kim, Y.H. (2004) Can we hear the shape of a noise source? The 18th International Congress on Acoustics (ICA 2004), Kyoto International Conference Hall, Japan, April 4–9.
3. Berkhout, A.J. de, Vries, D., and Vogel, P. (1993) Acoustic control by wave field synthesis. *Journal of the Acoustical Society of America*, **93**(5), 2764–2778.
4. Start, E.W. (1997) Direct sound enhancement by wave field synthesis. PhD thesis. Delft University of Technology.
5. Gerzon, M. (1973) Periphony: with-height sound reproduction, *Journal of the Audio Engineering Society*, **21**(1), 2–10.
6. Poletti, M. (2005) Three-dimensional surround sound systems based on spherical harmonics, *Journal of the Audio Engineering Society*, **53**(11), 1004–1025.

5

Sound Focusing

5.1 Introduction

The aim of sound focusing techniques is to focus sound energy within a selected region using transducer arrays. The sound energy, that is, the acoustic potential energy, is a basic acoustic variable that determines the magnitude of sound. Examples of controlling acoustic potential energy can be found from ancient times. The Pantheon in Rome (27 BC) and other examples of renaissance architecture utilized a self-focusing structure to deliver sound from one place to another (Figure 5.1a). In modern concert halls, line arrays are widely used to evenly distribute sound energy over a long distance (Figure 5.1b). In this case, the purpose of sound focusing is to decrease the energy decay rate along the propagation distance. Another function of a loudspeaker array is to increase the clarity of sound by increasing the magnitude of the direct wave and decreasing the reflections. In this case, the sound beam generated by an array is focused in order to decrease the reflections from unwanted directions. Besides these examples, there are numerous applications of loudspeaker arrays.

The sound focusing by loudspeaker arrays can be regarded as a counterpart to the beamforming technology, which is addressed in Chapter 4. In this chapter, though, a similar problem is explained from completely different perspectives. The basic theory is explained in terms of the wavenumber domain, and the major artifacts related to the array arrangement, for example, the diffraction limit due to the finite size of the array, are addressed. Popular techniques such as delay-and-sum (DAS), frequency-invariant, and minimum variance beamformers are explained, and this explanation is extended to include super-directivity techniques that can be used to overcome diffraction limits.

The principle of sound focusing can be described in terms of the four basic operators (+, −, ×, and ÷) of array signal processing.

First, the summing operator (+) describes the classical DAS technique, which generates a beam pattern using the constructive interference of the sound field generated by array transducers. The basic theory of the DAS technique can be considered against the same background as the problem of microphones. In this chapter, however, we attempt to explain the problem in wavenumber domain. This approach makes it possible to explain the problem of sound focusing in terms of a Fourier transform.

The differential operator (−) is a technique that can be used to generate a narrow beam pattern without restriction by the array size limit. This principle is explained using an array that consists of two elements; the general behavior of differential sources, beginning from a dipole source, is addressed.

Sound Visualization and Manipulation, First Edition. Yang-Hann Kim and Jung-Woo Choi.

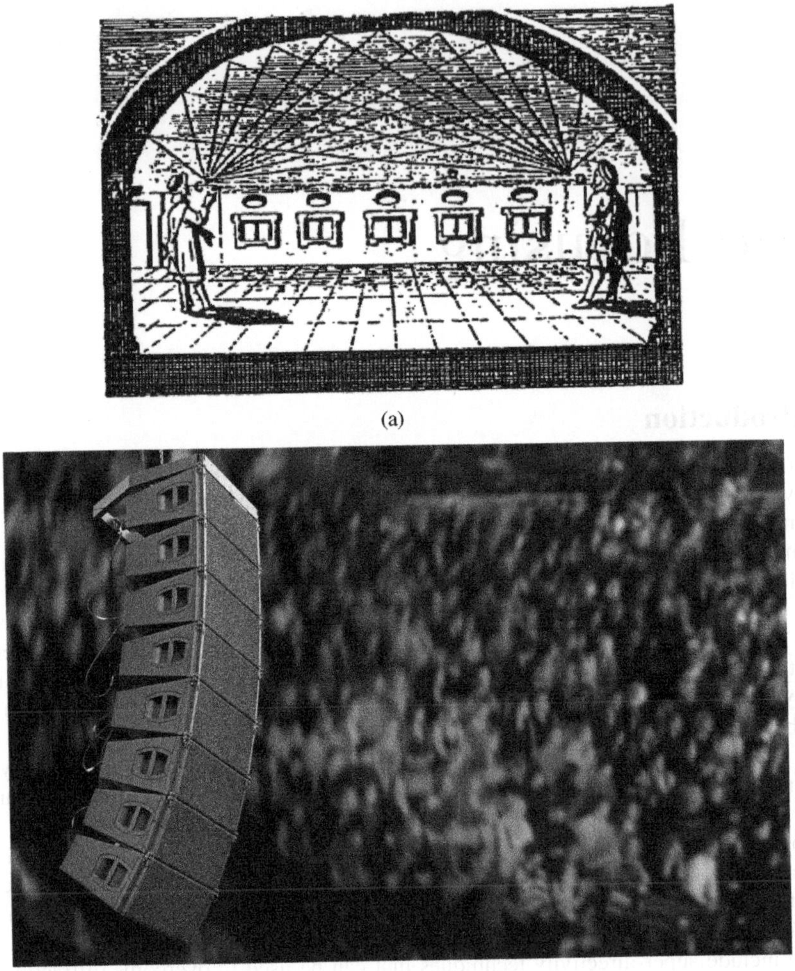

(a)

(b)

Figure 5.1 (a) Image of Renaissance acoustics from Musurgia Universalis (Athanasius Kircher, 1650). (b) Line array loudspeakers for million-person audience (© Fotosearch.kr, 2013)

The product theorem (×) is a convenient tool for combining multiple beam patterns. From the Fourier transform relation of the farfield beam pattern, it is shown how complex beam patterns or multidimensional beams can be designed from simple 1D beam patterns.

The ratio operator (÷) involves optimization techniques for generating sound beams with different constraints or criteria. This enables us to find an optimal beam pattern for loudspeaker arrays that have complex geometries. We also introduce the concept of regional sound control, which can be used to focus sound over a zone with an arbitrary shape. It is demonstrated how optimization with ratios can be applied to regional control, and with enhanced flexibility.

The explanations of all these methods begin with the simple case of a 1D array. A line array or a simple two- or three-loudspeaker array is the most fundamental element that constitutes the complex arrays. For even the simplest array, however, the propagation of sound follows the governing equation. To explain the sound propagation from an array, the early part of this chapter is devoted to the basic principles that relate the input signal to the arrays and the resultant beam patterns. The principles of sound focusing are explained in terms of the simplified form of the Rayleigh integral for farfield listeners.

5.2 Descriptions of the Problem of Sound Focusing

5.2.1 *Free-Field Radiation from Loudspeaker Arrays*

For loudspeaker arrays, the problem of radiation can be characterized by defining the relation between the input and the output. Since the problem of sound focusing is related to control of the sound pressure level (SPL), it is logical to consider the sound pressure or acoustic potential energy that is directly connected to the SPL.

Moreover, if we neglect the diffraction from the loudspeaker enclosure, the velocity distribution on the vibrating surfaces of loudspeaker arrays can be considered to be a combination of many monopole sources radiating in a free-field. In this regard, the distribution of monopole excitation in space can be regarded as inputs to the system, and the challenge is to then design the monopole excitation function that gives the desired sound pressure field.

The sound pressure field produced by multiple monopole sources can be described by the integral equation that relates the monopole excitation to the sound pressure field. In this regard, we begin our discussion by introducing the Rayleigh's first integral. Most of descriptions used in this chapter are derived in frequency domain, and hence, the frequency dependency ω will be omitted unless noted otherwise.

5.2.1.1 Array Aperture and Excitation Function

The term *aperture* of an array refers to the spatial region that is activated in order to transmit or receive sound waves. In loudspeaker arrays, an aperture is the area that consists of the loudspeaker diaphragms from which electric signals are converted to acoustic pressure or velocity.

If we assume that the loudspeakers constituting an array can be regarded as monopole sources, the aperture can be defined as the region in which the spatial distribution of monopole excitation function is nonzero. Then the pressure field produced in free-field can be described by Rayleigh's first integral as

$$p(\mathbf{r}) = \int_S g_f(\mathbf{r} - \mathbf{r}_s) q(\mathbf{r}_s) dS = \int_S \frac{e^{ikR}}{4\pi R} q(\mathbf{r}_s) dS, \tag{5.1}$$

where g_f denotes the free-field Green's function and q is the monopole excitation. The distance $R = |\mathbf{r} - \mathbf{r}_s|$ is defined as the distance between the source and listener.

In general, an *aperture function* represents the spatial distribution of a signal that excites an aperture. Hence, the monopole excitation q is denoted as the aperture *excitation function or excitation pattern*. For a line array defined on a single axis (Figure 5.2a), the excitation function can be defined in terms of a single variable ($x_s \in L$). For the 2D array configuration (Figure 5.2b), the excitation function is defined over a surface or area (S).

5.2.2 *Descriptions of a Sound Field Depending on the Distance from the Array*

The sound radiation characteristics of an array are quite different, depending upon the location of the listener and the aperture size. Basically, three regions exist, in which an array generates three distinct radiation patterns (Figure 5.3).

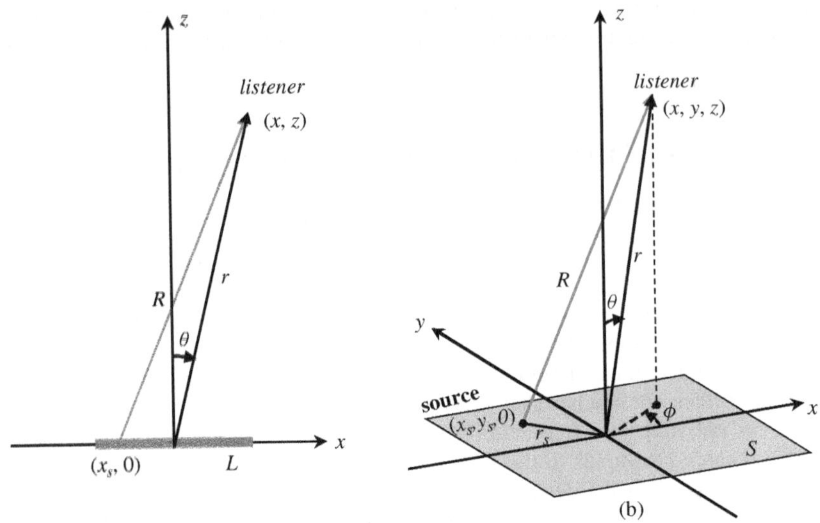

Figure 5.2 Definition of the coordinate system: (a) 1D array and (b) 2D array

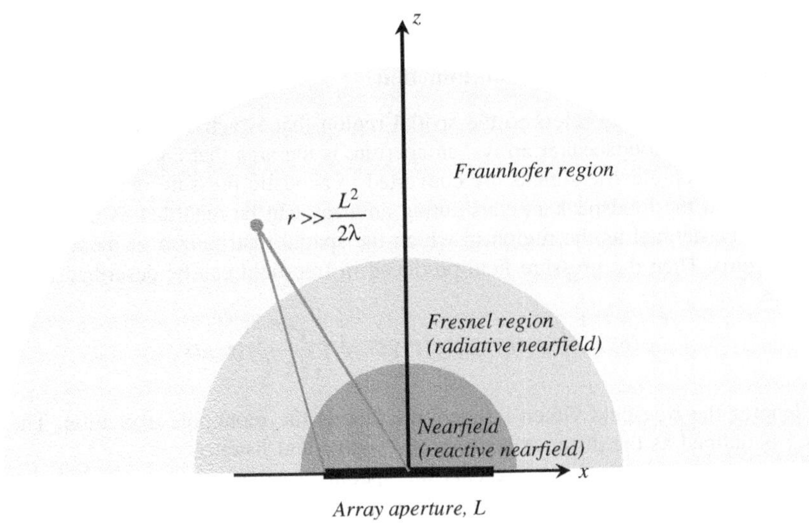

Figure 5.3 Schematic of field regions for an array with aperture size L

The first region is the farfield area, which is termed the Fraunhofer region. This region indicates the acoustic farfield, in which the distance from an array is much greater than the array size. As will be shown later, the sound field in the Fraunhofer region can be simply interpreted in terms of a Fourier transform. The next region, called the Fresnel region, is located more closely to the loudspeaker array. In this region, the analysis of a sound field is no simple matter, but the sound field can be described by modifying the expression for the Fraunhofer region. The third region, which is extremely near to the array, is where the evanescent wavenumber components are present. The analysis and synthesis of a sound beam pattern is only possible by incorporating optimization techniques using knowledge of the transfer functions.

In these three regions, the Rayleigh integral can be approximated differently. All three descriptions are different forms of the Rayleigh integral equation, which are approximated by the geometric relations between the listener and the array.

5.2.3 Fresnel Approximation

5.2.3.1 Derivation of the Fresnel Approximation

A representative formula is the case in which the observer is at a small angle from the radiator. The Fresnel approximation can be derived in Cartesian coordinates. The distance R is now expressed in Cartesian coordinates as

$$\begin{aligned} R &= \sqrt{(x-x_s)^2+(y-y_s)^2+z^2} \\ &= z\sqrt{1+\frac{(x-x_s)^2+(y-y_s)^2}{z^2}} \quad \text{or} \quad r\sqrt{1+\frac{x_s^2+y_s^2-2xx_s-2yy_s}{r^2}}, \end{aligned} \tag{5.2}$$

where r is the distance from the origin to the observer $\left(r=\sqrt{x^2+y^2+z^2}\right)$.

This can be expanded in a Taylor series as

$$\begin{aligned} R &= z\sqrt{1+\frac{a^2}{z^2}} \\ &= z\left[1+\frac{1}{2}\left(\frac{a^2}{z^2}\right)-\frac{1}{8}\left(\frac{a^2}{z^2}\right)^2+\cdots\right], \end{aligned} \tag{5.3}$$

where $a=\sqrt{(x-x_s)^2+(y-y_s)^2}$. If a is much smaller than the propagation distance z, then R can be approximated using only the first two terms of Equation (5.3).

$$R \approx z+\frac{(x-x_s)^2+(y-y_s)^2}{2z}, \tag{5.4}$$

or in terms of the listener distance r, this approximation can be written as

$$R \approx r+\frac{x_s^2+y_s^2-2xx_s-2yy_s}{2r}. \tag{5.5}$$

This small a/z condition can hold true for a small aperture size or for a small observation angle from the aperture. Such a condition can be made possible when the higher order term of Equation (5.3) is negligible in kR. That is,

$$\frac{k}{8}\left(\frac{a^4}{z^3}\right) \ll 2\pi. \tag{5.6}$$

From the Fresnel approximation, if we neglect the small variations in magnitude ($z \approx r$), the monopole radiation can then be reduced to

$$g_f(x-x_s, y-y_s, z) \approx \frac{e^{ikz}}{4\pi z}e^{ik[(x-x_s)^2+(y-y_s)^2]/2z}. \tag{5.7}$$

Using this approximation, we can then rewrite the Rayleigh integral as

$$p(x, y, z) = \frac{e^{ikz}}{4\pi z}\iint q(x_s, y_s, 0)e^{ik[(x-x_s)^2+(y-y_s)^2]/2z}dx_s dy_s. \tag{5.8}$$

As an alternative that involves distance r instead of z, this integral can be rewritten as

$$p(x, y, z) = \frac{e^{ikr}}{4\pi r}\iint q(x_s, y_s, 0)e^{ik[x_s^2+y_s^2-2xx_s-2yy_s]/2r}dx_s dy_s \tag{5.9}$$

from the description of Equation (5.5).

One interesting point regarding Equation (5.9) is that the integral is a Fourier transform of q multiplied by a Gaussian phase window $\gamma(x_s, y_s, r) = e^{ik(x_s^2+y_s^2)/2r}$. That is,

$$p(x, y, z) = \frac{e^{ikr}}{4\pi r}\iint [q(x_s, y_s, 0)\gamma(x_s, y_s, r)]e^{-ik[\frac{x}{r}x_s+\frac{y}{r}y_s]}dx_s dy_s. \tag{5.10}$$

In spherical coordinates with $x = r\cos\theta\sin\phi$, $y = r\sin\theta\sin\phi$, this can be rewritten as

$$p(r, \theta, \phi) = \frac{e^{ikr}}{4\pi r}\iint [q(x_s, y_s, 0)\gamma(x_s, y_s, r)]e^{-i(k_x x_s+k_y y_s)}dx_s dy_s, \tag{5.11}$$

with $k_x = k\cos\theta\sin\phi$, $k_y = k\sin\theta\sin\phi$. Therefore, the pressure response is related to the wavenumber spectrum of $q\gamma$.

$$\begin{aligned} p(r, \theta, \phi) &= \frac{e^{ikr}}{4\pi r}\mathcal{F}_s[q(x_s, y_s, 0)\gamma(x_s, y_s, r)] \\ &= \frac{e^{ikr}}{4\pi r}Q(k_x, k_y, 0) *_{k_x} *_{k_y} \Gamma(k_x, k_y, r), \end{aligned} \tag{5.12}$$

where Γ is a spatial Fourier transform of γ ($\Gamma = \mathcal{F}_s[\gamma]$), and $*_{k_x} *_{k_y}$ denotes the two dimensional convolution in the wavenumber domain. Since multiplication in the space domain is equivalent to convolution in the wavenumber domain, the Fourier transform of the velocity distribution (i.e., wavenumber spectrum $Q(k_x, k_y, 0)$ is convolved with the Fourier transform of the Gaussian phase window (which is also Gaussian; see Table A.1 in Appendix A). Therefore, the wavenumber spectrum Q is blurred because of the convolution with Γ. This blurred spectrum can then be directly connected to the pressure field from the relation $k_x = k\cos\theta\sin\phi$, $k_y = k\sin\theta\sin\phi$.

Another variation of Equation (5.12) with respect to z can be derived from Equation (5.8) as

$$\begin{aligned} p(x, y, z) &= \frac{e^{ikz}}{4\pi z}\gamma(x, y, z)\mathcal{F}_s[q(x_s, y_s, 0)\gamma(x_s, y_s, z)] \\ &= \frac{e^{ikz}}{4\pi z}\gamma(x, y, z)[Q(k_x, k_y, 0) *_{k_x} *_{k_y} \Gamma(k_x, k_y, z)] \end{aligned} \tag{5.13}$$

for $k_x = kx/z, k_y = ky/z$. This formula is beneficial for predicting the sound field in Cartesian coordinates.

5.2.3.2 Fresnel Approximation in Wavenumber Domain

Another way to obtain the Fresnel approximation formula is to derive it in the wavenumber domain. Let us assume that the wavenumber spectrum of the excitation pattern $Q(k_x, k_y, 0)$ is band-limited to the low wavenumber region $\left(\sqrt{k_x^2 + k_y^2} \ll k\right)$. This is equivalent to having plane waves propagating within a small angle.

$$k_z = \sqrt{k^2 - k_x^2 - k_y^2} \approx k - \frac{k_x^2 + k_y^2}{2k}. \tag{5.14}$$

The benefit of this approximation is that the wavenumber spectrum of the free-field Green's function $G_f(k_x, k_y, z)$ can be separated into two independent functions of k_x and k_y.

$$G_f(k_x, k_y, z) \approx \left(\frac{ie^{ikz}}{2k} e^{-i\frac{k_x^2}{2k}z} e^{-i\frac{k_y^2}{2k}z} \right). \tag{5.15}$$

The monopole field, which is given by the inverse Fourier transform of G_f, can then be expressed in the space domain as

$$\begin{aligned} g_f(x - x_s, y - y_s, z) &\approx \frac{1}{4\pi^2} \iint G_f(k_x, k_y, z) e^{i[k_x(x-x_s)+k_y(y-y_s)]} dk_x dk_y \\ &\approx \frac{ie^{ikz}}{2k} \left(\frac{1}{2\pi} \int_{-\infty}^{\infty} e^{-i\frac{k_x^2}{2k}z} e^{ik_x(x-x_s)} dk_x \right) \left(\frac{1}{2\pi} \int_{-\infty}^{\infty} e^{-i\frac{k_y^2}{2k}z} e^{ik_y(y-y_s)} dk_y \right) \\ &\approx \frac{e^{ikz}}{4\pi z} e^{ik[(x-x_s)^2+(y-y_s)^2]/2z}. \end{aligned} \tag{5.16}$$

It is noteworthy that we can obtain the same Fresnel representation of the monopole radiation (Equation (5.7)) using the small propagation angle approximation in the wavenumber domain. Explaining the wavenumber spectrum as a Gaussian function is helpful when dealing with a Gaussian beam problem.

5.2.4 *Farfield Description of the Rayleigh Integral (Fraunhofer Approximation)*

5.2.4.1 Derivation of the Fraunhofer Approximation

The Fraunhofer region begins at a farther distance from the loudspeaker array.

$$\begin{aligned} R &\approx r\sqrt{1 + \frac{x_s^2 + y_s^2 - 2xx_s - 2yy_s}{r^2}} \\ &\approx r + \frac{x_s^2 + y_s^2}{2r} - \frac{xx_s + yy_s}{r}. \end{aligned} \tag{5.17}$$

If the quadratic term of this first-order approximation can be negligible, that is, the aperture size $L = 2\sqrt{x_s^2 + y_s^2}$ is much smaller than the listener distance r, then the distance between the listener and a source (Equation (5.5)) can be further simplified as

$$R \approx r - \frac{xx_s + yy_s}{r}. \tag{5.18}$$

The monopole radiation observed at the listener distance r can then be written as

$$g_f(x - x_s, y - y_s, z) \approx \frac{e^{ikr}}{4\pi r} e^{-ik(xx_s+yy_s)/r}. \tag{5.19}$$

This farfield approximation for a small aperture, called the Fraunhofer approximation, directly connects the pressure field at z and the wavenumber spectrum of the excitation ($Q(k_x, k_y)$). From Equations (5.19) and (5.1), we obtain

$$p(x, y, z) \approx \frac{e^{ikr}}{4\pi r} \iint q(x_s, y_s, 0) e^{-i[(k\frac{x}{r})x_s + (k\frac{y}{r})y_s]} dx_s dy_s$$
$$\approx \frac{e^{ikr}}{4\pi r} Q\left(k\frac{x}{r}, k\frac{y}{r}\right). \tag{5.20}$$

Therefore, the wavenumber spectrum at $(k_x, k_y) = (kx/r, ky/r)$ is directly mapped to the pressure at (x, y, z). The wavenumber spectrum is stretched by a factor of x/r and y/r, which is inversely proportional to the distance. Therefore, as the distance r increases, only low wavenumber components can be observed.

In spherical coordinates with $(x, y, z) = (r\cos\phi\sin\theta, r\sin\phi\sin\theta, r\cos\theta)$, Equation (5.20) can be rewritten as

$$p(r, \theta, \phi) \approx \frac{e^{ikr}}{4\pi r} \int_{L_y} \int_{L_x} q(x_s, y_s, 0) e^{-ik(\cos\phi\sin\theta x_s + \sin\phi\sin\theta y_s)} dx_s dy_s$$
$$\approx \frac{e^{ikr}}{4\pi r} Q(k\cos\phi\sin\theta, k\sin\phi\sin\theta). \tag{5.21}$$

Therefore, the Rayleigh integral can be transformed to a Fourier integral. The Rayleigh integral in the Fraunhofer region expresses that the farfield pressure at angle (θ, ϕ) is equivalent to the wavenumber spectrum of the velocity distribution, with coordinate mapping $(k_x, k_y) = (k\sin\theta\cos\phi, k\sin\theta\sin\phi)$. Since the pressure at r can be computed by a Fourier transform of the surface velocity distribution, this approximation greatly reduces the computational load required to predict the farfield pressure from a given excitation q.

For purposes of brevity, we define the pressure field normalized by monopole radiation as a *normalized radiation pattern*, or simply, a *radiation pattern*.

$$b(\theta, \phi) = p(r, \theta, \phi) \Big/ \left(\frac{e^{ikr}}{4\pi r}\right). \tag{5.22}$$

From Equation (5.21), the normalized radiation pattern can be related to the wavenumber spectrum by

$$b(\theta, \phi) = Q(k\cos\phi\sin\theta, k\sin\phi\sin\theta). \tag{5.23}$$

5.2.4.2 Rayleigh Distance

The farfield assumption used to obtain the Fraunhofer approximation formula can be represented by a single measure. The magnitude of the quadratic term in the first-order approximation of kR (Equation (5.17)) is given by

$$k\frac{x_s^2 + y_s^2}{2r} = \pi\frac{L^2}{4r\lambda}. \tag{5.24}$$

The condition in which this quadratic term can be neglected can be written as

$$\pi\frac{L^2}{4r\lambda} \ll \frac{\pi}{2}. \tag{5.25}$$

Accordingly, we can define a distance in order to apply the Fraunhofer approximation:

$$r_R \equiv \frac{L^2}{2\lambda} \ll r. \tag{5.26}$$

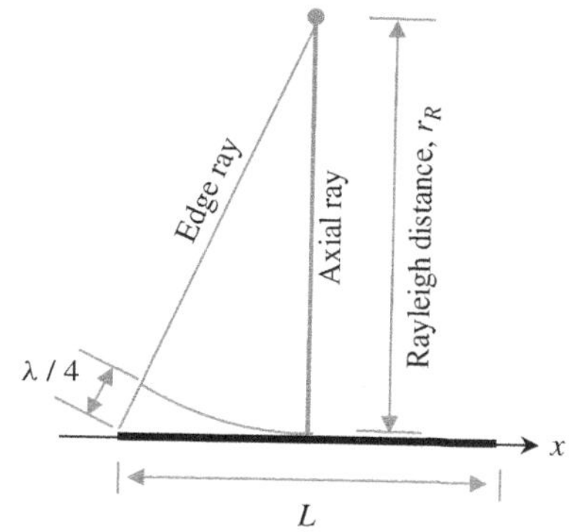

Figure 5.4 Definition of the Rayleigh distance

The distance r_R is called the *Rayleigh distance*. An alternative definition of the Rayleigh distance is given by the axial distance from a radiating aperture to a point at which the path difference between the axial ray and the edge ray is $\lambda/4$ (Figure 5.4).

For an array of aperture size L, the condition $r \gg r_R$ defines the farfield, and $r \sim r_R$ defines the nearfield. Another useful measure is a Fresnel number, which is defined as

$$N_F \equiv \frac{L^2}{r\lambda}. \tag{5.27}$$

When $N_F \geq 1$, the wave is considered to be in the nearfield, and the Fresnel approximation can be applied in order to calculate the radiation.

5.2.5 *Descriptors of Directivity*

5.2.5.1 Acoustical Axis (Beam Axis)

In the farfield, the directivity of the sound source does not change with respect to the distance (r) from the source. The amplitude of the sound pressure can be expressed as

$$p(r, \theta, \phi) = p_{ax}(r)DF(\theta, \phi), \tag{5.28}$$

where θ and ϕ are the angular directions in spherical coordinates. $p_{ax}(r)$ is called the on-axis pressure, and its direction is called the acoustical axis (Figure 5.5). In general, this indicates the

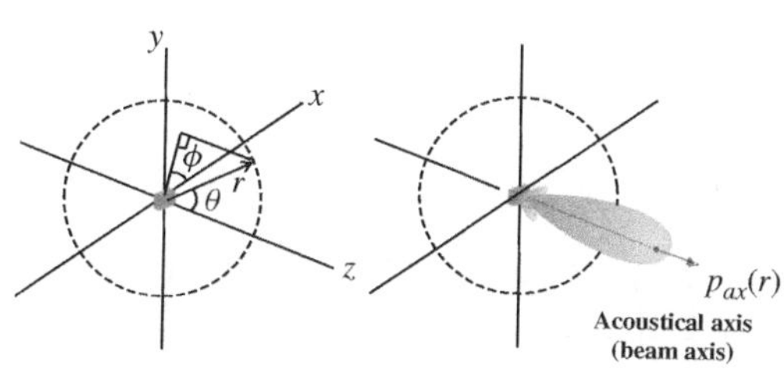

Figure 5.5 Acoustical axis

direction in which the sound source emits the maximum power, while the *on-axis pressure* is the maximum sound pressure radiated by the sound source to the distance r. The angular dependency $DF(\theta, \phi)$ is called the directional factor.

5.2.5.2 Directivity, Array Gain (D)

Directivity or the directivity factor of a sound source is given by the ratio between the intensity of a given sound source and that of an omnidirectional sound source that emits the same acoustic power as the source of interest. That is,

$$D(\theta, \phi) = \frac{|p|^2}{\langle |p|^2 \rangle_{avg}} = \frac{I}{I_{avg}}, \tag{5.29}$$

where $\langle \rangle_{avg}$ is the spatial averaging operator over a spherical surface given by

$$\langle |p(r, \theta, \phi)|^2 \rangle_{avg} = \frac{1}{4\pi} \int_0^{2\pi} \int_0^{\pi} |p(r, \theta, \phi)|^2 \sin\theta d\theta d\phi. \tag{5.30}$$

The average intensity I_{avg} can be regarded as the intensity radiated by an *omnidirectional sound source* that has the same acoustic power as the sound source of interest. Therefore, the directivity on the acoustical axis has the minimum value of 1 in the case of spherical sources. Frequently, the direction (θ, ϕ) is omitted and the directivity on the acoustical axis is used as a representative value. (For example, D = 100 in the case of Figure 5.6.) The directivity is also known as the *array gain*.

5.2.5.3 Directional Factor (DF)

The directional factor (Figure 5.7) is normalized by $p_{ax}(r)$, the on-axis pressure, and always has the maximum value of 1, while the maximum value of the directivity factor is determined by the type of sound source. The directional factor $DF(\theta, \phi)$ is given by

$$DF(\theta, \phi) = \frac{|p(r, \theta, \phi)|}{|p_{ax}(r)|}. \tag{5.31}$$

5.2.5.4 Directivity Index (DI)

The directivity index represents the directivity factor (D) on a log scale (Figure 5.8):

$$DI = 10\log_{10} D. \tag{5.32}$$

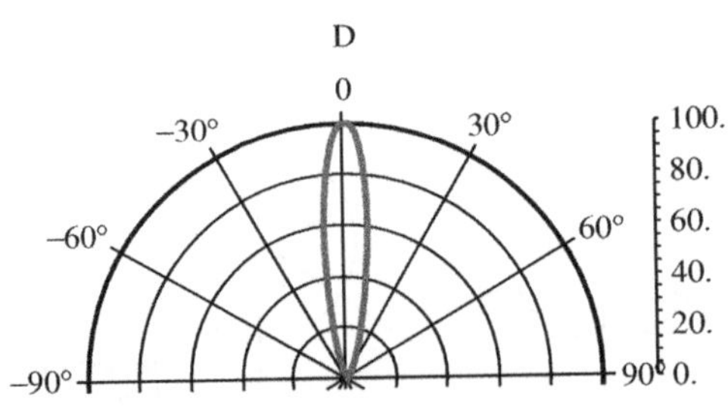

Figure 5.6 Directivity D of the sound field radiated by an infinitely baffled circular piston with $ka = 10$

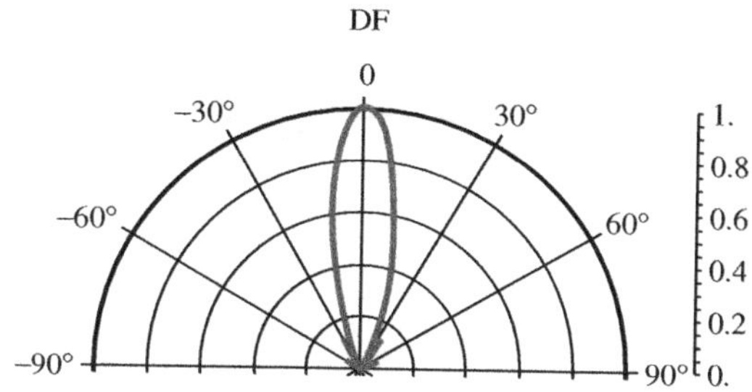

Figure 5.7 Directional factor, *DF* of the sound field radiated by an infinitely baffled circular piston with $ka = 10$

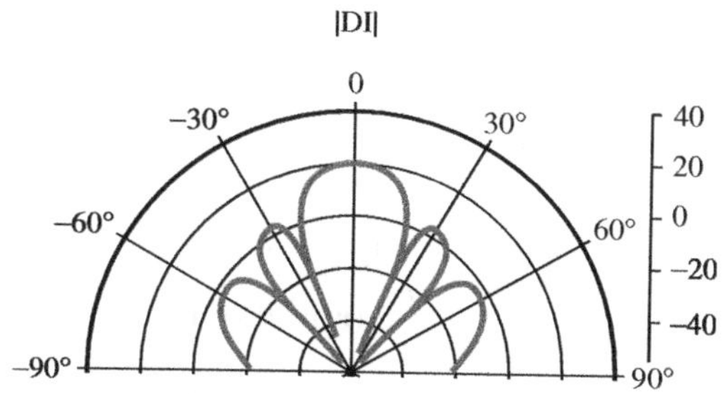

Figure 5.8 Directivity index, *DI* of the sound field radiated by an infinitely baffled circular piston with $ka = 10$

Assuming that the power radiated by an omnidirectional sound source is the same as the total power radiated by the sound source with respect to distance *r*, we can estimate the directivity index by the spatial average of the power on the measurement surface. That is,

$$DI = 10\log_{10}\left(\frac{I}{I_{avg}}\right), \tag{5.33}$$

where I_{avg} denotes the spatially averaged intensity on the surface, $S(= 4\pi r^2)$.

5.2.5.5 Beam Pattern (*BP*)

The beam pattern represents the directional factor $DF(\theta, \phi)$ on a log scale (Figure 5.9), and is defined as the ratio between the relative intensity and the intensity on the acoustical axis. That is,

$$BP(\theta, \phi) = 10\log_{10}\left(\frac{I(r, \theta, \phi)}{I_{ax}(r)}\right) = 20\log_{10} DF(\theta, \phi). \tag{5.34}$$

5.2.5.6 White Noise Gain (WNG)

In the practical realization of loudspeaker arrays, implicit errors or noises can be involved. For the microphone array problem, measurement noises are included in the measured microphone signals. The sensitivity to the noise or error is often evaluated by the white noise gain (WNG), which is defined as the array gain (directivity) for the incoherent white noise. Since the array gain

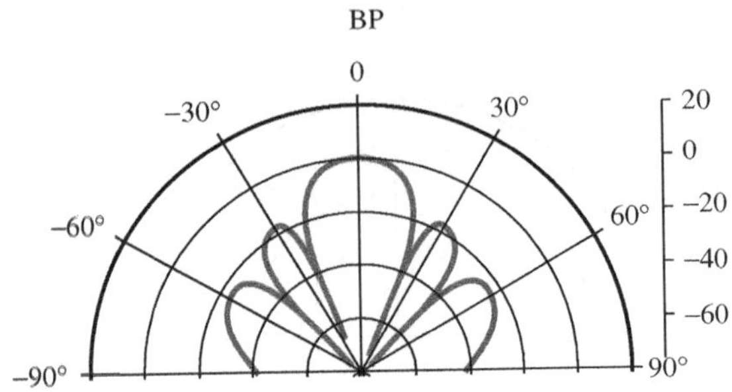

Figure 5.9 Beam pattern, *BP* of the sound field radiated by an infinitely baffled circular piston with $ka = 10$

represents the signal-to-noise ratio (SNR) enhancement obtained by using an array, low WNG indicates that a beamformer can be easily disturbed by the noise on the input, that is, microphone signals. In the loudspeaker array problem, however, such input noise is not significant. Instead, the WNG is related to the energy efficiency of an array system. The WNG for the loudspeaker array problem can be defined as

$$WNG = \frac{|p(\theta,\phi)|^2}{\int_S |q(\mathbf{r}_s)|^2 dS(\mathbf{r}_s)} \tag{5.35}$$

where the numerator denotes the total input power from the loudspeakers distributed on the surface S. Equation (5.35) represents how much acoustic energy can be generated with respect to the total power of the excitation signal. Therefore, the loudspeaker array with higher WNG can produce more sound pressure with the same input. The WNG can also be used as a robustness measure, since any pressure perturbation due to the error in the transfer function or the input signal might be smaller than the original pressure signal $p(\theta,\phi)$ in the high WNG system.

5.3 Summing Operator (+)

5.3.1 Delay-and-Sum Technique

The DAS technique refers to a focusing technique that generates a sound beam mainly by using the constructive interference at the focal point. In DAS technique, the excitation pattern is weighted and delayed to generate multiple wavefronts from the loudspeakers that meet at the steered direction almost in-phase. In this regard, the DAS technique mainly aims to generate a sound beam by summing all the radiations from loudspeakers on the aperture.

The Rayleigh integral itself shows that the summation of sound fields from multiple monopoles can produce a radiation pattern in space. For the sake of simplicity, consider a line array (Figure 5.2a) located on the x axis, and a listener positioned at the $\phi = 0$ plane. Rayleigh's first integral can then be written as

$$p(r,\theta) = \int_{-\infty}^{\infty} \frac{e^{ikR}}{4\pi R} q(x_s) dx_s. \tag{5.36}$$

By applying the Fraunhofer approximation, we can rewrite Equation (5.36) as

$$p(r,\theta) \approx \frac{e^{ikr}}{4\pi r} \int_{-\infty}^{\infty} q(x_s) e^{-ik\sin\theta x_s} dx_s. \tag{5.37}$$

The radiation pattern $p(r, \theta)$ generated by the function $q(x_s)$, which is slowly varying in space, is obtained by summing all the excitation q delayed by $\tau = x_s \sin\theta/c$. As was mentioned earlier in Section 5.2.4, the radiation pattern can be also derived from its wavenumber spectrum $Q(k_x)$ that is defined as a spatial Fourier transform of the excitation function q.

$$Q(k_x) = \mathcal{F}_s\left[q\left(x_s\right)\right] = \int_{-\infty}^{\infty} q(x_s)e^{-ik_x x_s}dx_s. \tag{5.38}$$

From Equation (5.23), we can relate the wavenumber spectrum $Q(k_x)$ and the normalized radiation pattern by

$$b(\theta) = Q(k\sin\theta) = \int_{-\infty}^{\infty} q(x_s)e^{-ikx_s\sin\theta}dx_s. \tag{5.39}$$

This mapping is essentially related to the Fourier transform. Accordingly, we can utilize various properties of the Fourier transform to design a desired radiation pattern $b(\theta)$. The whole mapping procedure is summarized in Figure 5.10.

5.3.2 *Beam Shaping and Steering*

There can be various forms of excitation $q(x_s)$; however, a basic form to consider first is the excitation with a position-dependent time delay $\tau(x_s)$ and an amplitude weighting $w(x_s)$.

$$q(x_s) = w(x_s)e^{i\omega\tau(x_s)} = w(x_s)e^{ikc\tau(x_s)}. \tag{5.40}$$

If a linearly progressive delay

$$\tau(x_s) = \alpha x_s/c, \qquad (c: \text{speed of sound}) \tag{5.41}$$

is used, then the Fourier transform of the excitation $q(x_s)$ can be expressed as

$$Q(k_x) = W(k_x) *_{k_x} \delta(k_x - k\alpha) = W(k_x - k\alpha), \tag{5.42}$$

where $W(k_x)$ is the Fourier transform of the weighting function $W(k_x)$. From Equation (5.39), we can express the normalized radiation pattern $b(\theta)$ by

$$b(\theta) = W(k(\sin\theta - \alpha)). \tag{5.43}$$

When Equation (5.43) is compared to Equation (5.39), it can be observed that the shape of a radiation pattern is determined by the spectrum of the weighting function $W(k_x)$ (Figure 5.11a). On

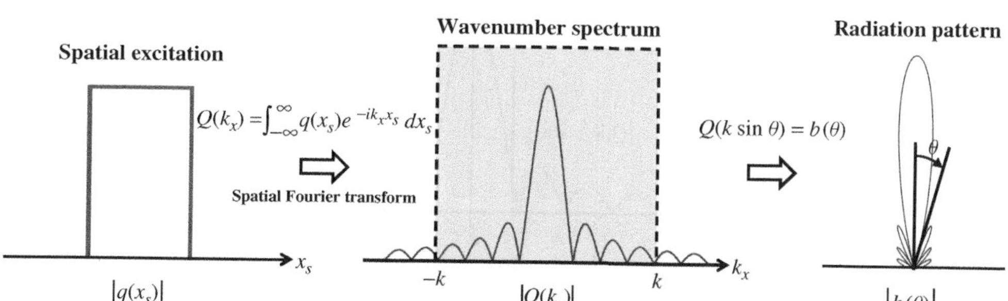

Figure 5.10 Relations between the spatial excitation, wavenumber spectrum, and radiation pattern

the other hand, the progressive time delay $\tau(x_s)$ simply shifts the wavenumber spectrum from $W(k_x)$ to $W(k_x - k\alpha)$ (Figure 5.11b). The direction of the mainlobe occurs when $k_x = k \sin\theta = k\alpha$, and hence the progressive time delay steers the sound beam to

$$\theta' = \sin^{-1}(\alpha). \quad (5.44)$$

It is also noteworthy that the steering angle θ' is independent of frequency. Therefore, one can control the direction of a sound beam across all frequencies simply by applying a progressive time delay to the excitation function.

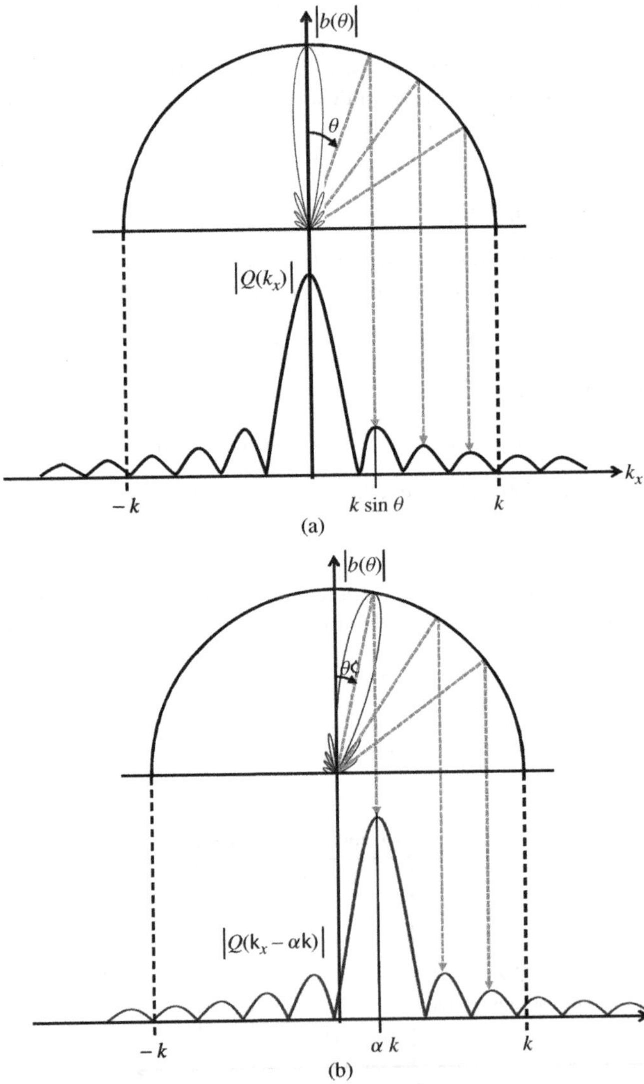

Figure 5.11 Mapping relation of the wavenumber spectrum and the radiation pattern of (a) centered beam and (b) steered beam by progressive time delay $\tau(x_s) = \alpha x_s / c$

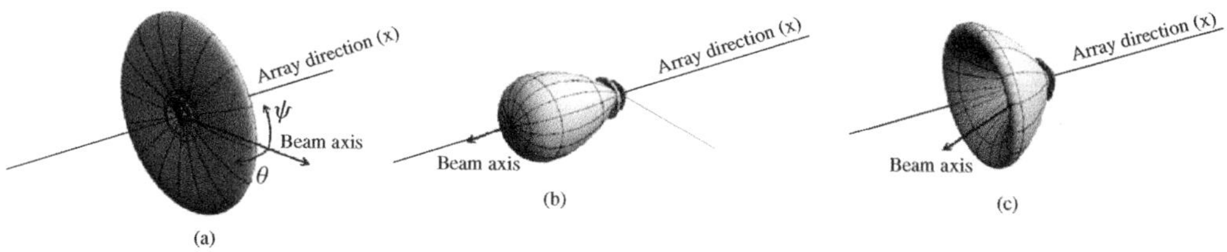

Figure 5.12 Sound beams generated by 1D array: (a) broadside beam ($\theta_{ax} = 0$), (b) endfire beam ($\theta_{ax} = \pm 90°$), and (c) steered beam ($\theta_{ax} = 45°$)

5.3.2.1 Broadside and Endfire Radiation Patterns

When the steered direction of a beam axis is normal to the distribution of an array, the sound beam is called a *broadside* beam. In the case that the beam axis is parallel to the direction of array arrangement, an *endfire* beam is formed. With a 1D array, however, the corresponding radiation pattern is invariant with respect to the elevation angle (ψ). Figure 5.12 illustrates the broadside, endfire, and steered radiation patterns generated by a 1D array.

From the Equations (5.41) and (5.43) for steering the radiation pattern, we can generate an endfire radiation pattern with beam axis $\theta_{ax} = \pm 90°$ by applying the progressive time delay

$$\tau_{endfire}(x_s) = \pm x_s / c, \tag{5.45}$$

which is equivalent to the travel time of a sound wave propagating along the direction of the array. Each element of an array has a different time delay according to its location on the array.

5.3.3 Wavenumber Cone and Diffraction Limit

The mapping relation between the observer angle θ and the wavenumber k_x (Equation (5.39)) implies that the radiation pattern depends on the excitation frequency, even with the same excitation function $q(x)$ and wavenumber spectrum $Q(k_x)$. In detail, the argument $k \sin\theta$ in Equation (5.39) can vary only within $[-k, k]$; hence, the observer can only see a part of the wavenumber spectrum, which is truncated at the wavenumber $\pm k$. The wavenumber spectrum beyond $k_x > k = \omega/c$ is the nearfield component (evanescent spectrum), and hence, it does not propagate to the farfield. The cut-off wavenumber k decreases as the frequency decreases, and only a small portion of the wavenumber spectrum is visible at a low frequency.

Figure 5.13 shows the propagating part of the wavenumber spectrum in $k - k_x$ space. It can be seen that the propagating region of the wavenumber spectrum is proportional to the frequency $\omega = ck$, forming the triangular shape. When visualized for the general 2D spectrum (k_x, k_y), this region has the shape of a cone. The propagating region, called the *wavenumber cone*, expresses the propagating wavenumber components with respect to the frequency (Figure 5.14).

5.3.3.1 Finite Aperture and Diffraction Limit

The relation between the farfield radiation pattern and the wavenumber spectrum implies that the radiation pattern is a function of the frequency and aperture size. For an array with a finite aperture size L ($|x_s| \le L/2$), the integration interval of Equation (5.38) is limited to

$$Q(k_x) = \int_{-L/2}^{L/2} q(x_s) e^{-ik_x x_s} dx_s. \tag{5.46}$$

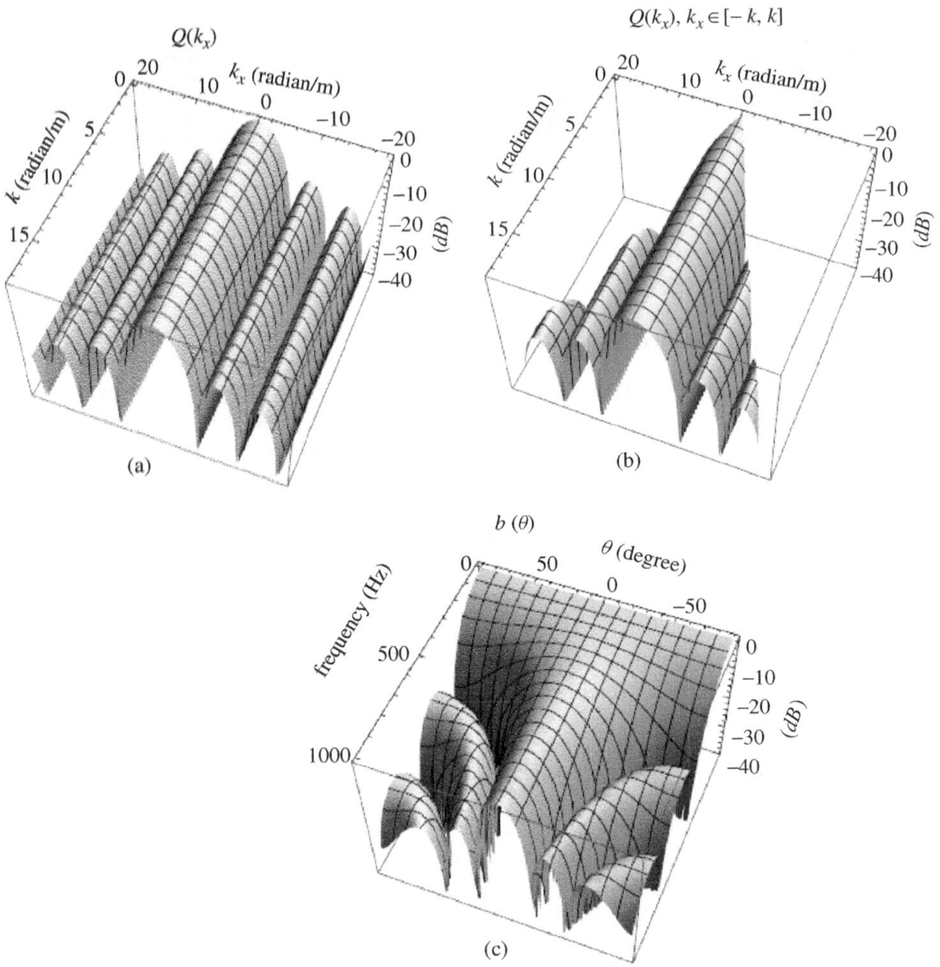

Figure 5.13 (a) Wavenumber spectrum $Q(k_x)$ of a finite uniform excitation $q(x_s) = 1, x_s \in [-0.5, 0.5]$, (b) band-limited wavenumber spectrum within $|k_x| \leq k$, and (c) corresponding radiation pattern $b(\theta)$ across frequencies

Let us consider a uniform excitation ($q(x) = 1/L$ for $|x| \leq L/2$, $q(x) = 0$ elsewhere). The finite integral gives

$$Q(k_x) = \frac{1}{L}\int_{-L/2}^{L/2} e^{-ik_x x_s} dx_s$$
$$= \text{sinc}(k_x L/2), \tag{5.47}$$

and the radiation pattern of Equation (5.39) can be derived by

$$b(\theta) = \text{sinc}\left(\frac{kL}{2}\sin\theta\right). \tag{5.48}$$

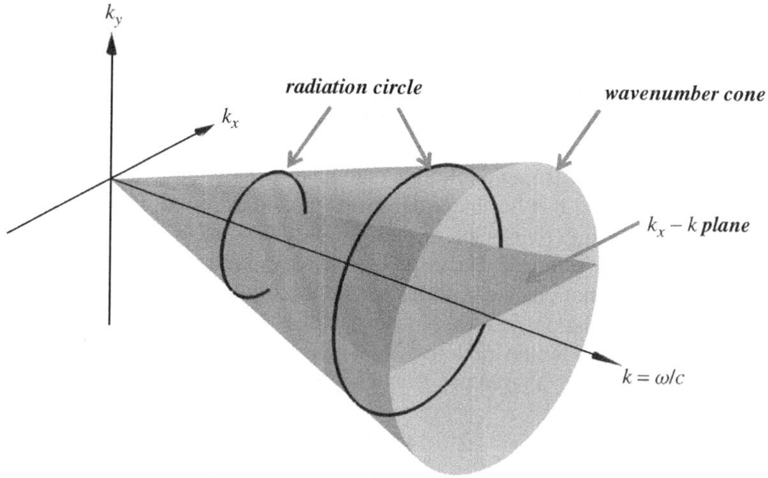

Figure 5.14 Wavenumber cone and radiation circle

The radiation pattern is generally represented by the sinc function. It has a maximum value in a direction perpendicular to the array distribution ($\theta = 0$) and gradually decreases, showing oscillatory behavior with increasing θ. Since the argument of the sinc function is scaled by kL, the aperture size with respect to the wavelength is associated with the size of the sound beam.

There are various measures for expressing the size of the mainlobe, but the half-power beamwidth (HPBW) and the null-to-null beamwidth are widely used. In the case of uniform excitation, the half-power point of a sinc function takes place at

$$\frac{kL}{2}\sin\left(\frac{\theta_{HPBW}}{2}\right) = 0.443\pi \quad \text{or} \quad \theta_{HPBW} = 2\sin^{-1}\left(0.443\frac{\lambda}{L}\right) \text{ (rad)}. \tag{5.49}$$

For large kL, the HPBW may be small ($\theta_{HPBW} \ll 1$). Then, from $\sin\theta \cong \theta$, Equation (5.49) can be reduced to

$$\theta_{HPBW} \approx 0.886\frac{\lambda}{L} \text{ (rad)}. \tag{5.50}$$

The null-to-null width refers to the angular width spanned by the first nulls of the radiation pattern. The first null of the sinc function occurs at

$$\theta_{null} \cong \frac{\lambda}{L} \text{(rad)} \tag{5.51}$$

for small angles, and the null-to-null width is $2\theta_{null}$.

The beamwidth is proportional to the wavelength and is inversely proportional to the aperture size. Therefore, in order to generate a sharp radiation pattern, the aperture size should be large compared to the wavelength. The theoretical HPBW that can be obtained by uniform excitation and an aperture size of one wavelength is about 50°.

The limit of the beamwidth (or beam resolution) due to a finite aperture size is commonly called the *diffraction limit*. This term stems from the optical resolution limit induced by the finite size of a diffraction slit. It should be noted that the diffraction limit is a physical constraint due to the propagation of sound waves. The diffraction limit acts as a physical wavenumber filter whose cut-off is given by the radiation circle ($k = \omega/c$).

5.3.4 Frequency Invariant Radiation Pattern

The physical importance of the diffraction limit is that the directivity pattern of a constant excitation varies with the frequency, even if it has the same wavenumber spectrum. In some applications, this frequency dependency of a directivity pattern needs to be minimized, in order to have a constant radiation pattern across wide frequency bands. The frequency invariant radiation pattern refers to a radiation pattern that does not change with respect to the frequency.

A frequency invariant radiation pattern can be designed using the Fourier transform relation between the excitation and radiation patterns. To begin, let us assume that we have a desired radiation pattern b_0 designed for a reference frequency ω_0

$$b_0(\theta) = b(\theta, \omega_0) \tag{5.52}$$

and that we want to have the same radiation pattern b_0 across all frequencies ($b(\theta, \omega) = b_0(\theta)$).

A reference wavenumber spectrum $Q_0(k_x)$ can then be calculated, which gives the desired radiation pattern b_0 at the frequency $\omega_0 = ck_0$. From Equation (5.39), we have

$$Q_0(k_x) = Q_0(k_0 \sin\theta) = b_0(\theta). \tag{5.53}$$

Then, by stretching the reference wavenumber spectrum in the k_x direction by a factor of k_0/k, we can derive the wavenumber spectrum for a different frequency ω. That is,

$$Q(k_x, \omega) = Q_0\left(\frac{k_0}{k} \cdot k_x\right) = Q_0(k_0 \sin\theta) = b_0(\theta). \tag{5.54}$$

It should be noted that a wavenumber spectrum that is defined in this way gives a radiation pattern that only depends on θ, irrespective of the frequency change. The corresponding excitation pattern can be derived by taking the inverse Fourier transform of Equation (5.54). That is,

$$q(x_s, \omega) = \mathcal{F}_s^{-1}\left[Q_0\left(\frac{k_0}{k}k_x\right)\right]. \tag{5.55}$$

Since the inverse Fourier transform of a scaled function $F(ak_x)$ is $f(x/a)/a$, the excitation pattern can be simply written as

$$q(x_s, \omega) = \frac{k}{k_0} q_0\left(\frac{k}{k_0}x_s\right), \text{ or } q(x_s, \omega) = \frac{\omega}{\omega_0} q_0\left(\frac{\omega}{\omega_0}x_s\right) \tag{5.56}$$

where q_0 is the reference excitation pattern given by the inverse Fourier transform of Q_0. It can be seen that the frequency invariant excitation can be derived by stretching and scaling the function q_0 with a factor of ω/ω_0. Because the scale factor is proportional to the frequency, we need a relatively large aperture size in the low frequency region in order to maintain the frequency invariance of a radiation pattern.

Figure 5.15 illustrates an example of a frequency invariant radiation pattern. In order to design a frequency invariant beam, a rectangular excitation is configured for $f = 1$ kHz. The excitation of other frequencies is calculated according to Equation (5.56). The wavenumber spectrums before and after truncation by the wavenumber cone are presented. The resultant radiation pattern $b(\theta)$

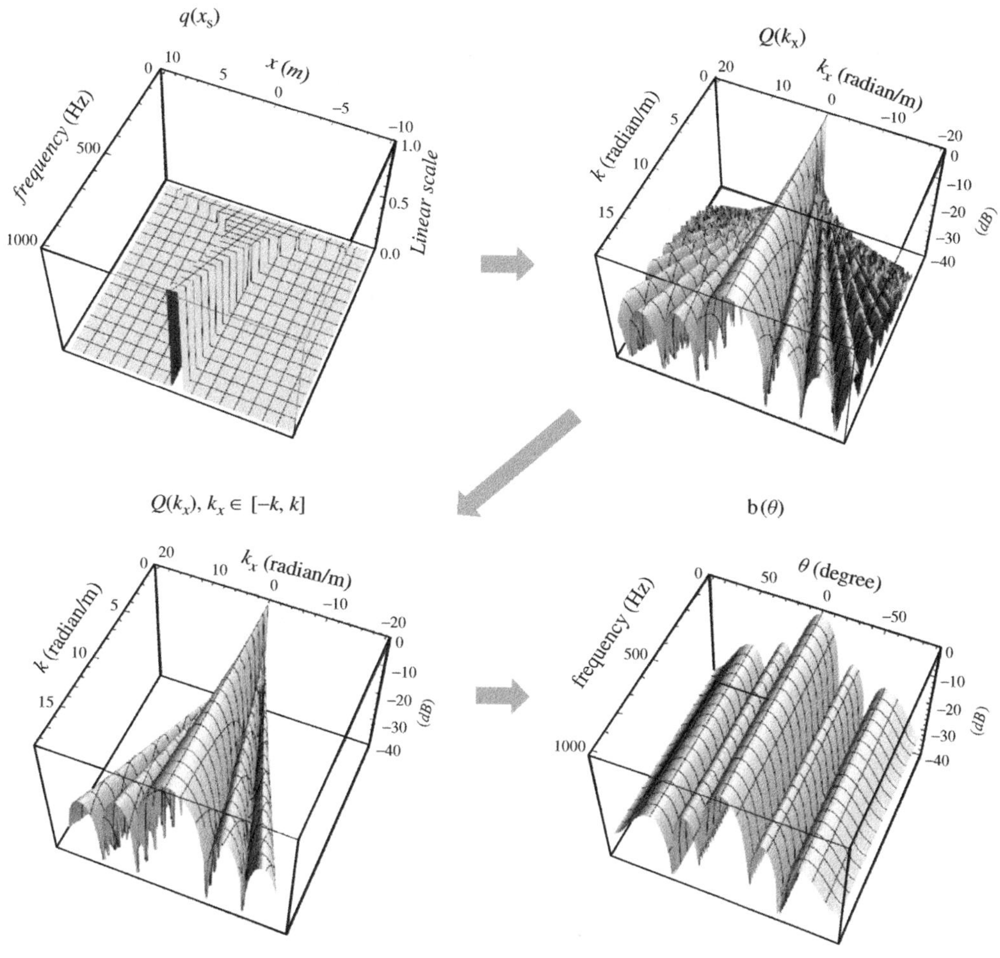

Figure 5.15 Examples of design of a frequency invariant radiation pattern

shows the radiation pattern that does not change with the frequency. However, in obtaining this radiation pattern, the aperture size of $q(x_s)$ increases exponentially in the low frequency region.

5.3.5 Discrete Array and Grating Lobes

Thus far, we have dealt with continuous arrays in which the excitation function is also continuous in space. In practice, a loudspeaker array is implemented with a finite number of loudspeakers that are discretely distributed in space. This discrete arrangement in space, that is, spatial sampling, generates an artifact in the produced radiation pattern.

For a discrete source distribution that is equally spaced with distance Δx, the excitation function also becomes discrete in space. This discrete excitation can be likened to a continuous excitation multiplied by a periodic delta function.

$$q_D(x_s) = q(x_s) \sum_{n=-\infty}^{\infty} \delta(x_s - n\Delta x). \tag{5.57}$$

The Fourier transform of the excitation signal is given by (Appendix A, Equation A10)

$$Q_D(k_x) = Q(k_x) *_{k_x} \frac{1}{\Delta x} \sum_{m=-\infty}^{\infty} \delta\left(k_x - \frac{2\pi}{\Delta x} m\right). \tag{5.58}$$

A convolution with periodic impulses results in a periodic wavenumber spectrum $Q_D(k_x)$. This periodic spectrum has multiple peaks that are analogous to the aliasing phenomenon of discrete-time signal processing, and hence it can be regarded as a spatial aliasing effect. If the peaks exist within $k_x \in [-k, k]$ (Figure 5.16), we can observe multiple lobes that are no smaller than the mainlobe. These extra lobes, which are referred to as *grating lobes*, are usually much greater than the sidelobes, and they constitute significant artifacts that cannot be neglected. Using Equation (5.39), we can express the radiation pattern with grating lobes as

$$b_D(\theta) = \frac{1}{\Delta x} \sum_{m=-\infty}^{\infty} Q\left(k \sin\theta - \frac{2\pi}{\Delta x} m\right). \tag{5.59}$$

If we consider a steered beam to an angle θ_0, then a grating lobe is visible if

$$k \sin\theta = k \sin\theta_0 + \frac{2\pi}{\Delta x} m, \quad (m = \pm 1, \pm 2, \cdots) \tag{5.60}$$

where the location of each grating lobe is given by

$$\theta_{gr}^{(m)} = \sin^{-1}\left(\sin\theta_0 + \frac{\lambda}{\Delta x} m\right) = \sin^{-1}\left(\sin\theta_0 + \frac{\Delta k_x}{k} m\right). \tag{5.61}$$

Here, Δk_x denotes the sampling wavenumber, which is defined as $\Delta k_x = 2\pi/\Delta x$. In terms of the interval Δx, the condition for preventing a grating lobe is given by

$$\frac{\Delta x}{\lambda} \le \frac{1}{1 + |\sin\theta_0|}.$$

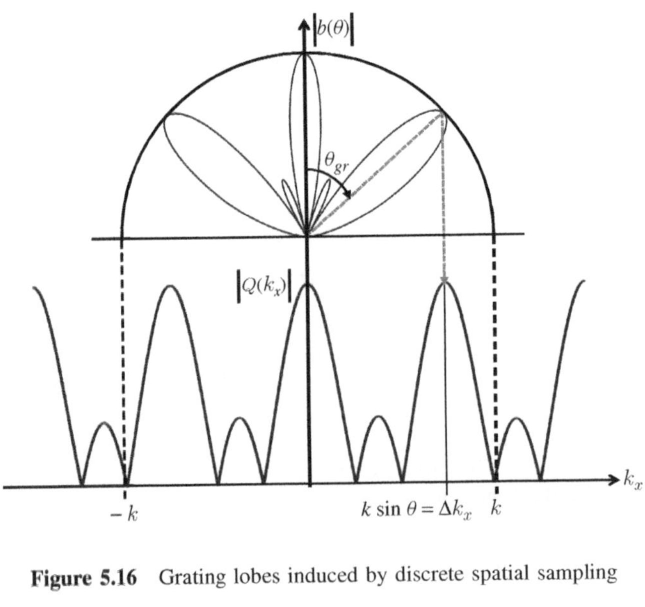

Figure 5.16 Grating lobes induced by discrete spatial sampling

Since the wavenumber distance between these multiple peaks is Δk_x, dense spatial sampling can reduce the artifact to some extent. For the broadside beam ($\theta_0 = 0$) in Figure 5.12, the first grating lobe can be detected at $\pi/2$ when $\Delta x = \lambda$ or $\Delta k_x = k$. For the endfire beam ($\theta_0 = \pi/2$) in Figure 5.12, the first grating lobe becomes visible at $-\pi/2$ when $\Delta x = \lambda/2$. Therefore, if a sound beam is allowed to be steered, the source interval should satisfy $\Delta x \leq \lambda/2$ in order to prevent grating lobes.

The condition that produces a grating lobe is thus related to the spatial sampling interval Δx and the frequency $\omega = ck$. A grating lobe is more visible in the high frequency region or with a larger sampling interval.

5.3.5.1 Discrete Array Design by Digital Filter Design Method

Consider a finite line array consisting of point sources arranged with the same interval (Δx). For discrete source positions $x_s^{(\kappa)} = (\kappa - (K+1)/2)\Delta x$ ($\kappa = 1, \cdots, K$), the wavenumber spectrum of the excitation function $q(x_s^{(\kappa)})$ can be rewritten as

$$Q(k_x) = \left(e^{ik_x \Delta x (K-1)/2}\right) \sum_{\kappa=1}^{K} q(x_s^{(\kappa)}) e^{-ik_x \Delta x (\kappa - 1)}. \tag{5.62}$$

For this finite and discrete source arrangement, the relation of Equation (5.62) can be expressed in terms of z-transform. Inserting $z = e^{ik_x \Delta x}$ into Equation (5.62) gives

$$Q(z) = z^{(K-1)/2} \sum_{n=0}^{K-1} q[n] z^{-n} \tag{5.63}$$

Note that the definition of z is different from that of the coordinate system (x, y, z). The problem can now be approached using common finite impulse response (FIR) filter design methods, for example, by window-based designs such as the Kaiser or Chebyshev window. Figure 5.17

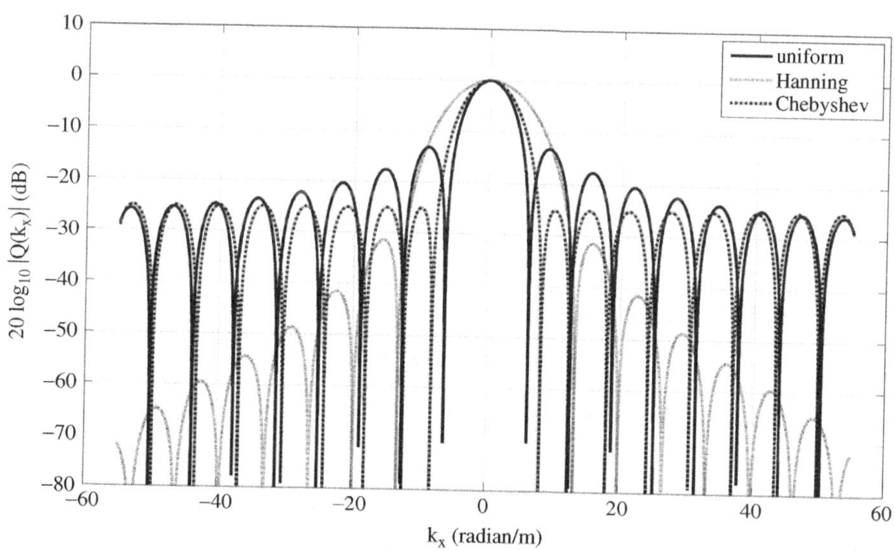

Figure 5.17 Wavenumber spectrums of various discrete window functions

depicts the wavenumber spectrums and corresponding radiation patters of the uniform, Hanning, and Chebyshev window.

5.4 Product Theorem (×)

The Fourier transform relation between the excitation pattern and the wavenumber spectrum states that we can utilize various useful properties of the Fourier transform. Among these, the convolution-multiplication relation between the spatial data and the wavenumber spectrum allows us to design complex radiation patterns using a simple multiplicative operation.

5.4.1 *Convolution and Multiplication of Sound Beams*

One useful characteristic of a Fourier transform is that the convolution in the space domain is represented as a product in the wavenumber domain (Appendix A, Section A.1.1.4). That is,

$$\mathcal{F}_s\left[q_1(x) *_x q_2(x)\right] = Q_1(k_x)Q_2(k_x). \tag{5.64}$$

where $*_x$ represents the spatial convolution with respect to x. From the mapping of the wavenumber spectrum and the radiation pattern (Equation (5.39)), the following also holds true:

$$b_1(\theta)b_2(\theta) = Q_1(k\sin\theta)Q_2(k\sin\theta). \tag{5.65}$$

This relation implies that we can generate a radiation pattern derived from the product of two different radiation patterns by convolving their excitation patterns in the space domain. This particular relation is denoted as the product theorem for the beam synthesis problem.

5.4.1.1 Array and Element Directivities

One popular use of the product theorem is to predict the total directivity when an array is built of directive elements. For example, consider the excitation pattern q_D that gives a radiation pattern with grating lobes (Equation (5.57)). In the previous example, all the elements constituting the array were assumed to be a point source, and for this reason, the series of the Delta function was considered. In this example, we consider a finite line source whose excitation value is nonzero over a finite length dL.

$$\begin{aligned} q_E(x_s) &= \frac{1}{dL},\ for\ x_s \in \left[-\frac{dL}{2}, \frac{dL}{2}\right]. \\ &= 0,\ otherwise \end{aligned} \tag{5.66}$$

From Equation (5.47), the wavenumber spectrum is given by

$$Q_E(k_x) = \mathrm{sinc}(k_x dL/2). \tag{5.67}$$

The radiation pattern of the single element then leads to

$$b_E(\theta) = \mathrm{sinc}(kdL\sin\theta/2). \tag{5.68}$$

If we replace all the point sources with this finite line source, then the total excitation pattern of such an array is given as the convolution of q_E and q_D (Figure 5.18).

$$q_T(x_s) = q_E(x_s) *_{x_s} q_D(x_s). \tag{5.69}$$

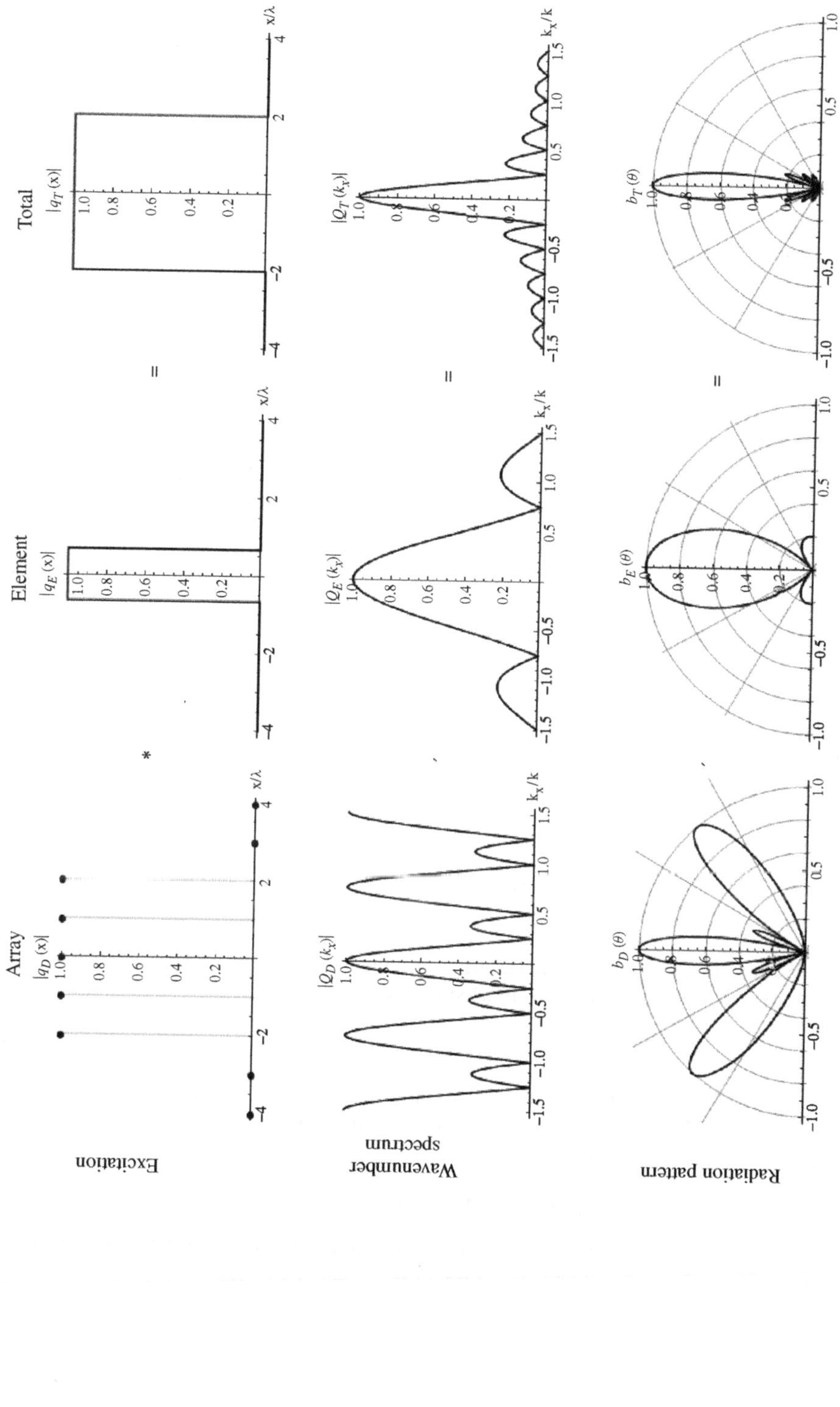

Figure 5.18 Product theorem applied to a discrete array consisting of piston-like elements

The corresponding wavenumber spectrum is given by

$$Q_T(k_x) = \mathcal{F}_s\left[q_E\left(x_s\right) *_{x_s} q_D(x_s)\right] = Q_E(k_x)Q_D(k_x). \tag{5.70}$$

The final radiation pattern $b_T(\theta)$ can be expressed as a product of the element's radiation pattern and a discrete array's radiation pattern.

$$b_T(\theta) = b_E(\theta)b_D(\theta) = \mathrm{sinc}(kdL\sin\theta/2)b_D(\theta). \tag{5.71}$$

Since the spectral envelope of the sinc function of Equation (5.67) monotonically decreases with increasing k_x, this limits the wavenumber components that can propagate. As a result, the element's radiation pattern b_E suppresses the grating lobes of the discrete array ($b_D(\theta)$). In this regard, the finite line source can be considered as a *spatial anti-aliasing filter* that truncates the periodic wavenumber spectrum.

Figure 5.18 shows how finite line sources can reduce grating lobes due to discrete sampling. In practice, a finite line source can be likened to a piston source whose height in the y direction is small but which has a considerable aperture in the x direction. Therefore, it is beneficial to use loudspeakers with a large radiating surface in order to have reduced grating lobes.

5.4.1.2 Multidimensional Array

Using the product theorem, we can design a two-dimensional array by combining two 1D arrays lying in different axes. For example, the 1D source excitation along the x axis can be described in two-dimensional space (x_s, y_s) as

$$q_x(x_s, y_s) = f_x(x_s)\delta(y_s). \tag{5.72}$$

For an array distributed in the y direction, we can write

$$q_y(x_s, y_s) = \delta(x_s)f_y(y_s). \tag{5.73}$$

We can now construct a 2D array from the convolution of two 1D source excitations. The spatial convolution of q_x and q_y gives

$$\begin{aligned} q_{xy}(x_s, y_s) &= q_x(x_s, y_s) *_{x_s} *_{y_s} q_y(x_s, y_s) \\ &= \int_{-\infty}^{\infty}\int_{-\infty}^{\infty} q_x(x_s', y_s')q_y(x_s - x_s', y_s - y_s')dx_s'dy_s' \\ &= f_x(x_s)f_y(y_s). \end{aligned} \tag{5.74}$$

This 2D array can be regarded as a 1D array arranged in the x axis whose elements have the same directivity in y direction. The two-dimensional wavenumber spectrum of Equation (5.74) can be written as

$$Q_{xy}(k_x, k_y) = F_x(k_x)F_y(k_y), \tag{5.75}$$

where $Q_{xy} = \mathcal{F}_s[q_{xy}]$ is the 2D spatial Fourier transform of the 2D excitation function q_{xy}, and $F_x = \mathcal{F}_s[f_x]$, $F_y = \mathcal{F}_s[f_y]$. Consequently, the total radiation pattern is given as the product of two radiation patterns in the x and y directions.

$$b_{xy}(\theta, \psi) = b_x(\theta)b_y(\psi), \tag{5.76}$$

where b_{xy} is the radiation pattern obtained from the 2D wavenumber spectrum Q_{xy} and b_x, b_y represent radiation patterns generated by arrays in the x and y directions, respectively. Note that the beam pattern is described in terms of the azimuthal (θ) and elevational (ψ) angles, which differ from the definition of spherical coordinates (θ, φ), in order to emphasize the rotational invariance of the beam pattern.

Figure 5.19 shows the two-dimensional radiation pattern with respect to the azimuthal (θ) and elevational (ψ) angles. The original radiation pattern $b_x(\theta)$, which is omnidirectional in the ψ direction, is modified such that the directivity in the ψ direction is also enhanced. This is accomplished by replacing the element of the array with a finite line element that can control the radiation pattern in the y direction. The resultant 2D array generates a radiation pattern that is focused in both directions.

However, this does not mean that any 2D radiation pattern can be decomposed in terms of two 1D patterns. In fact, this example represents a very special case of a 2D radiation pattern; a more general description of a 2D radiation pattern is given by the 2D Fourier transform of the two dimensional excitation. In spherical coordinates of Equation (5.21), the radiation pattern is given by

$$\begin{aligned} Q(k_x, k_y) &= \int_{-\infty}^{\infty}\int_{-\infty}^{\infty} \left[q\left(x_s, y_s\right) e^{-i(k_x x_s + k_y y_s)}\right] dx_s dy_s \\ b(\theta, \phi) &\approx Q(k\cos\phi\sin\theta, k\sin\phi\sin\theta). \end{aligned} \tag{5.77}$$

5.4.2 *On-Axis Pressure Response*

On-axis responses of broadside line arrays along the distance z are shown in Figure 5.20. In this example, line arrays with different aperture sizes are driven by the hanning window function. In the farfield (Fraunhofer region), the radial decay rate of an on-axis response is given by $1/r$, as

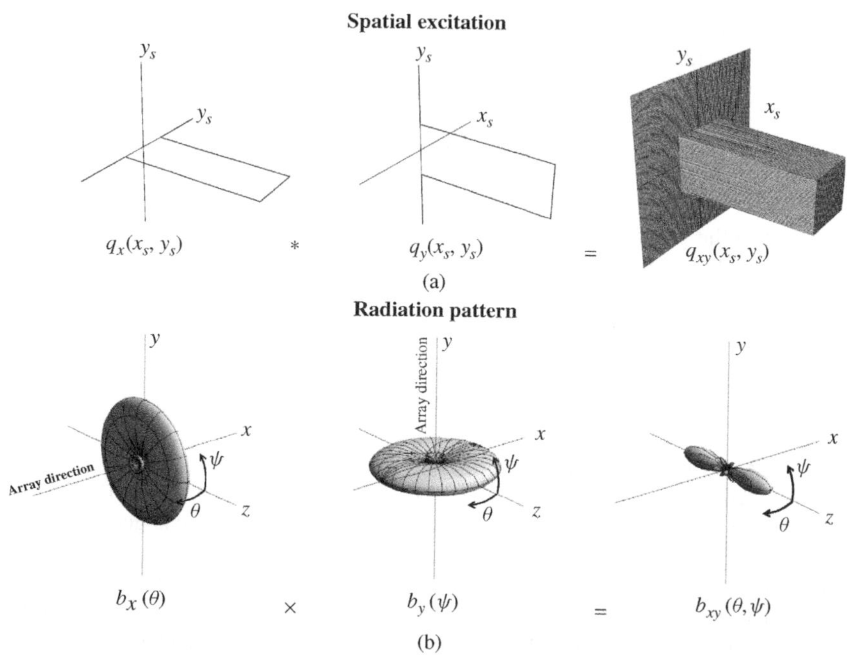

Figure 5.19 Two-dimensional radiation pattern generated by the convolution of two 1D excitation patterns. (a) Spatial excitation function. (b) Radiation pattern

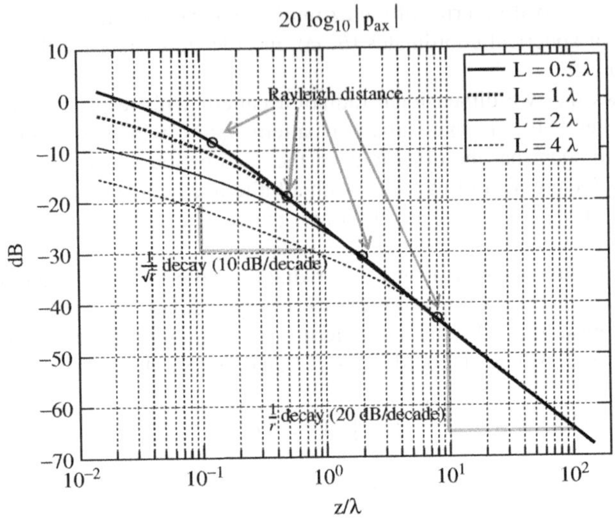

Figure 5.20 On-axis pressure response of line arrays driven by 1D Hanning window of different length *L*

can be seen in Equation (5.21). On the other hand, the nearfield decay is much slower than the farfield decay.

In the nearfield, the listener is close to the array, so most of the contribution of the Rayleigh integral comes from the portion of the array that is close to the listener. Therefore, the integral over a finite array aperture is similar to the integral over an infinite length. The Rayleigh integral over an infinite line array of unit excitation is given by

$$p(0,0,z) = \int_{-\infty}^{\infty} \frac{e^{ik\sqrt{x_s^2+z^2}}}{4\pi\sqrt{x_s^2+z^2}} dx_s = \frac{i}{4} H_0^{(1)}(kz), \tag{5.78}$$

which is likened to the point source propagating in 2D space or the line source in 3D space. This behavior can be explained by using the stationary phase approximation discussed in Appendix B, Section B.2.2. Since the phase change of the integrand is rapid enough for the listener's position close to the array, we can use Equation (B115) to obtain

$$p(0,0,z) \approx \sqrt{\frac{2\pi}{ik}} \left(\frac{e^{ik\sqrt{z}}}{\sqrt{z}} \right) \tag{5.79}$$

for $y = y_s = 0$. Accordingly, the nearfield decay rate can be approximated as $1/\sqrt{z}$. The change in the decay rate from $1/\sqrt{z}$ to $1/z$ can be observed near the Rayleigh distance. Figure 5.20 shows the variation of the Rayleigh distance and the decay ratio on the beam axis according to changes in aperture size.

However, 2D arrays exhibit a decay rate that is different from that of line arrays in the nearfield (Figure 5.21). The Rayleigh integral over an infinite planar surface transforms the problem into one of one-dimensional wave propagation. Consequently, the beam propagates like a plane wave in the nearfield from an array, which has an extremely slow decay rate.

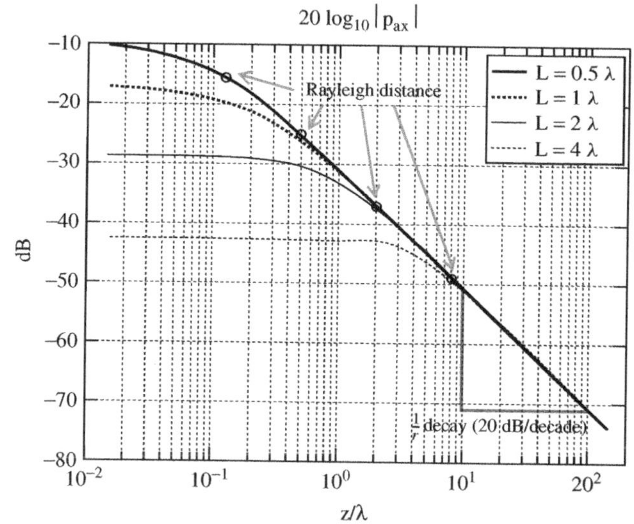

Figure 5.21 On-axis pressure response of planar arrays driven by 2D Hanning window of different length L

5.5 Differential Operator and Super-Directivity (−)

In the DAS technique, the HPBW is inversely proportional to the aperture size (Equation (5.50)); hence, it is nearly impossible to generate directivity with a point source whose aperture size is much smaller than the wavelength. Let us recall, however, the directional point source such as a point dipole and quadruple sources. One can observe that the behavior of these point sources cannot be explained in terms of the DAS excitations. How then can these point sources produce such directional radiation patterns with small aperture size?

In this section, we focus on arrays whose aperture size is much smaller than or comparable to the wavelength. For this low-frequency region in which the diffraction limit becomes problematic, we need special types of excitation functions. These excitation patterns, called *differential patterns*, are the basic form of *super-directivity*, which can overcome the diffraction limit.

5.5.1 *Endfire Differential Patterns*

Endfire differential sources generate a sound beam in a direction parallel to the array distribution. For example, a dipole source that is analogous to two compactly located monopoles generates a figure-eight beam pattern on a line connecting these two monopole sources. By investigating these basic sources, we can understand how differential patterns can be implemented.

In this section, we study simple combinations that consist of two monopoles that are compactly located in space. However, to emphasize the endfire arrangement, the sources are arranged along the z-axis, as shown in Figure 5.22. With this configuration, the listener's position on the z-axis is given by $z = r\cos\theta$, and the Fraunhofer approximation (Equation (5.20)) for the z-axis can be rewritten as

$$p(r,\theta,\phi) \approx \frac{e^{ikr}}{4\pi r} Q(k\cos\theta). \tag{5.80}$$

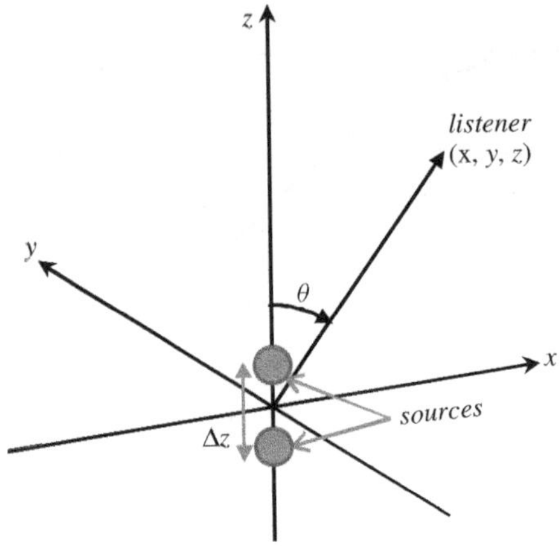

Figure 5.22 Source configurations for endfire differential patterns

5.5.1.1 Dipole

A dipole can be produced from two monopole sources whose signs are opposite to each other. Therefore, the excitation of a dipole source can be expressed in terms of two delta functions with opposite signs, provided that the separation distance Δz of the delta functions is much smaller than the wavelength ($\Delta z \ll \lambda$).

$$q(z_s) = -\delta\left(z_s + \frac{\Delta z}{2}\right) + \delta\left(z_s - \frac{\Delta z}{2}\right). \tag{5.81}$$

Since the aperture size is small, we can still use the Fraunhofer approximation to derive the radiation pattern of a dipole source. First, if we take a Fourier transform of $q(z_s)$, the wavenumber spectrum is given by

$$\begin{aligned} Q(k_z) &= \int_{-\infty}^{\infty} \left[-\delta\left(z_s + \Delta z/2\right) + \delta(z_s - \Delta z/2)\right] e^{-ik_z z_s} dz_s \\ &= -2i \sin(k_z \Delta z/2). \end{aligned} \tag{5.82}$$

Using $k_z = k\cos\theta$, the radiation pattern of a dipole can be derived as

$$b(\theta) = -2i \sin(k\Delta z \cos\theta/2). \tag{5.83}$$

However, the aperture size of a point dipole source is much smaller than the wavelength, so if we use $\sin x \approx x$ for small x, we obtain

$$\begin{aligned} b(\theta) &\approx -i(k\Delta z)\cos\theta \\ &\approx -i\omega\tau_0 \cos\theta, \end{aligned} \tag{5.84}$$

where $\tau_0 = \Delta z/c$. Therefore, the dipole source has the first-order cosine directivity, with the angle θ being evaluated from a direction normal to the beam axis.

In order to understand these dynamics in detail, we need to analyze the wavenumber spectrum of the dipole excitation. Figure 5.23 shows the spatial excitation, wavenumber spectrum, and normalized radiation pattern of a dipole source. In practice, the wavenumber spectrum is neither

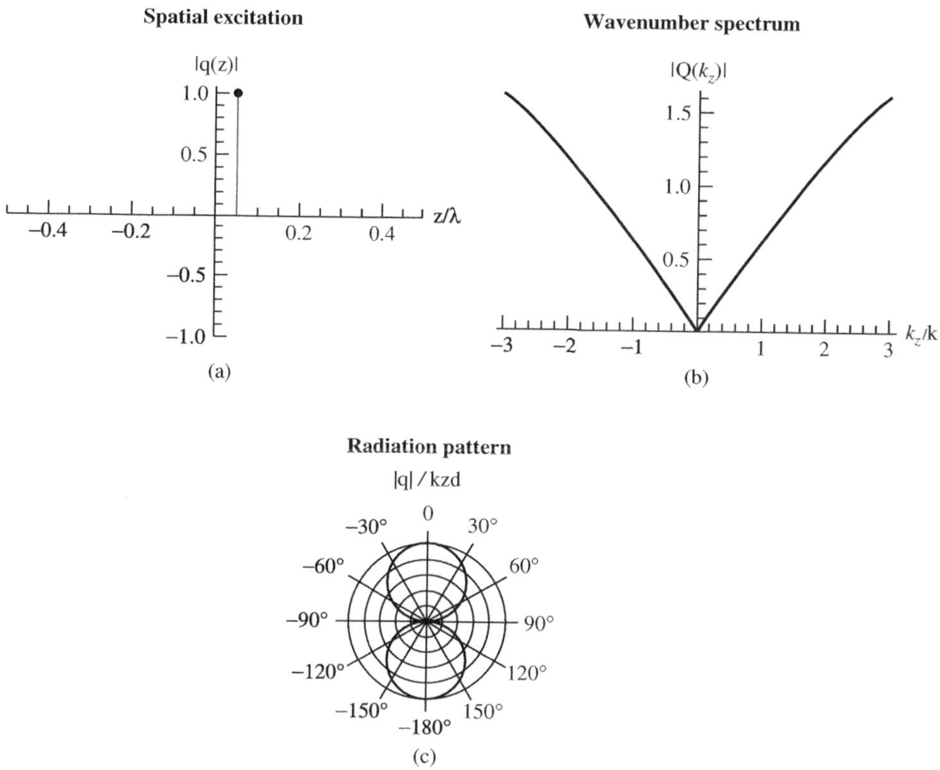

Figure 5.23 Dipole radiation: (a) excitation, (b) wavenumber spectrum, and (c) normalized radiation pattern

focused nor sharp, since $\sin(k_z \Delta z/2)$ is a periodic function. However, it should be noted that the bandwidth of the visible region, which is below the cut-off wavenumber ($\pm k$), is small, so most of the wavenumber spectrum is invisible in the farfield. Therefore, what we can observe as a radiation pattern is a small portion of the spectrum having a narrow trough at the center. Consequently, the radiation pattern shows two focused beams at $k_z = \pm k$ or $\theta = 0°, 180°$.

We can now appreciate one important finding from the dipole case: the dipole generates directivity by *concealing* most of its undesirable spectrum within the evanescent region and by radiating only a small portion of the spectrum that is desirable in the farfield. This explains how compact multipole sources can generate directivity even with their small aperture size.

Unlike the DAS pattern obtained by summing the sound fields of multiple sources, the radiation pattern of a dipole is obtained by subtracting the sound fields of two sources. In general, the radiation pattern produced by the subtraction of sound fields is denoted as a *differential pattern*, and the compact source generating such a pattern is called a *differential source*. Since the case of two sources discussed in this section can generate an effective differential pattern in endfire directions, they are denoted as *endfire differentials*.

5.5.1.2 General First-Order Differential Source

The dipole source emits sound into the on-axis in opposite directions. However, we can also generate asymmetric radiation patterns. Let us recall that the wavenumber spectrum can be translated horizontally by applying a *progressive time delay* to the excitation pattern (Equation (5.40)). The same principle holds true for the wavenumber spectrum of differential sources. Since only two sources are being considered here, the progressive time delay can be described as a delay $\tau = \alpha\tau_0$ that is assigned to one of the excitation signals. Here, the *time delay ratio* α represents how much extra delay is included, as compared to the time required for the sound to travel between the sources ($\tau_0 = \Delta z/c$). The corresponding phase shift at frequency ω is given by $\omega\tau = \alpha k\Delta z$, and we can write the general differential excitation as

$$q(z_s) = -\delta(z_s + \Delta z/2)e^{i\alpha k\Delta z} + \delta(z_s - \Delta z/2). \tag{5.85}$$

The complete wavenumber spectrum can then be obtained as follows:

$$\begin{aligned} Q(k_z) &= \int_{-\infty}^{\infty} \left[-\delta\left(z_s + \Delta z/2\right) e^{i\alpha k\Delta z} + \delta(z_s - \Delta z/2)\right] e^{-ik_z z_s} dz_s \\ &= -e^{ik_z\Delta z/2} e^{i\alpha k\Delta z} + e^{-ik_z\Delta z/2} \\ &= e^{i\alpha k\Delta z/2}[-2i\sin((k_z + \alpha k)\Delta z/2)]. \end{aligned} \tag{5.86}$$

It should be noted that the wavenumber spectrum of the dipole source (Equation (5.82)) is now shifted by $-\alpha k$. Based upon the assumption of a compact source (small $k\Delta z$), we can apply the same approximation ($\sin(x) \approx x$) that is used for the dipole source. If we neglect the very small phase shift ($e^{i(\alpha k\Delta z/2)}$), Equation (5.86) can be reduced to

$$Q(k_z) \approx -i(k_z + \alpha k)\Delta z, \tag{5.87}$$

and the radiation pattern is given by

$$\begin{aligned} b(\theta) &\approx -i(k\Delta z)(\alpha + \cos\theta) \\ &\approx -i\omega\tau_0(\alpha + \cos\theta). \end{aligned} \tag{5.88}$$

The maximum of the radiation pattern given by Equation (5.88) depends on the time delay ratio α, so we can introduce a normalized excitation pattern as

$$q_{norm}(z_s) = \frac{1}{|\alpha| + 1}\left[-\delta\left(z_s + \Delta z/2\right) e^{i\alpha k\Delta z} + \delta(z_s - \Delta z/2)\right]. \tag{5.89}$$

The corresponding normalized radiation pattern is given by

$$b(\theta) \approx -i\omega\tau_0 \frac{(\alpha + \cos\theta)}{|\alpha| + 1}. \tag{5.90}$$

Compared to the DAS pattern, the shifted wavenumber spectrum of a dipole does not change the direction of the beam axis. Therefore, the sound beams from Equation (5.90) are all endfire patterns. The shape of a beam changes for different values of α, and the beam becomes asymmetric in the front (θ) and rear side ($\pi - \theta$). The reason for this is that the hidden wavenumber spectrum of a dipole ($|k_z| > k$) intrudes in the visible region due to the progressive delay $\alpha\tau_0$ (Figure 5.25).

5.5.1.3 Reference Time Delay

As can be seen in Equation (5.85), the differential source is characterized by a negative sign with an additional time delay. Here, the basic time delay with $\alpha = 1$ ($\tau_0 = \Delta z/c$) expresses the time required for a sound wave to traverse the source interval Δz. Throughout this chapter, we denote the time delay τ_0 as a *reference time delay*.

5.5.1.4 Monopole-to-Dipole Amplitude Ratio

The radiation pattern given by Equation (5.88) can be interpreted as a summation of monopole (α) and dipole ($\cos\theta$) radiation patterns. Therefore, the first-order differentials can generally be produced by mixing monopole and dipole sources that have different amplitude ratios. The *time delay ratio* α controls the relative contribution of the monopole and dipole patterns. The third column in Figure 5.25 shows the variation of the monopole-to-dipole ratio for different kinds of differential sources.

5.5.1.5 On-Axis Pressure of the Differential Excitation Pattern

One prominent difference of the endfire differential pattern from the DAS pattern is that it is independent of frequency. As long as the assumption of a compact source ($\Delta z \ll \lambda$) holds true, the $\cos\theta$ directivity can be maintained throughout a wide frequency band. However, there is a clear disadvantage associated with the enhanced directivity. The amplitude of the sound field radiation is reduced by a factor of $k\Delta z$ (Equation (5.88)). The amplitude of a differential source linearly decreases with decreasing ω, and the low frequency sound can be inaudible. The amplitude variation with a rate of $k\Delta z = \omega\tau_0$ is likened to that of a first-order high-pass filter.

Figure 5.24 shows the on-axis response of a dipole source implemented by two dipole sources. We can observe that increasing the source interval can be helpful for increasing the low-frequency response, but an increase in the source interval would easily violate the assumption of a small $k\Delta z$ at high frequencies. This causes the radiation pattern itself to be broken, and the on-axis response is distorted as well. Therefore, the source interval should be carefully selected according to the desired frequency range. The high-pass characteristic can then be equalized by passing the input signal to the first-order low-pass filter that has a frequency response of $(-i\omega)^{-1}$. This equalization, however, leads to a reduction in the overall pressure level across all frequencies.

5.5.1.6 Cardioid

A representative example of the differential source is a *cardioid pattern* that is generated by $\alpha = 1$ (Figure 5.25). In a cardioid pattern, the contributions of the monopole and dipole sources are equal;

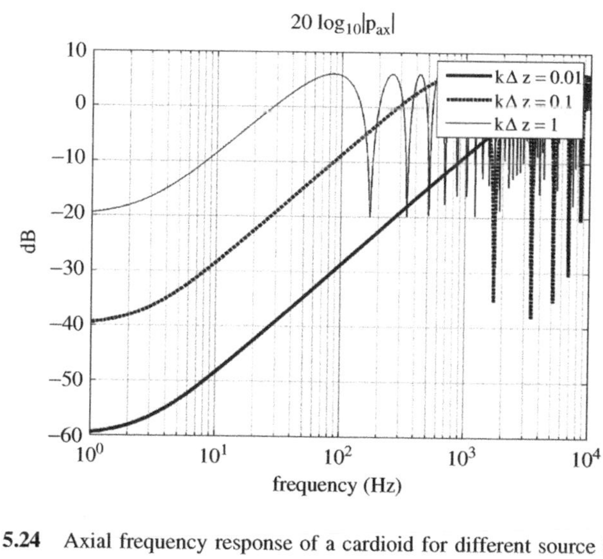

Figure 5.24 Axial frequency response of a cardioid for different source intervals

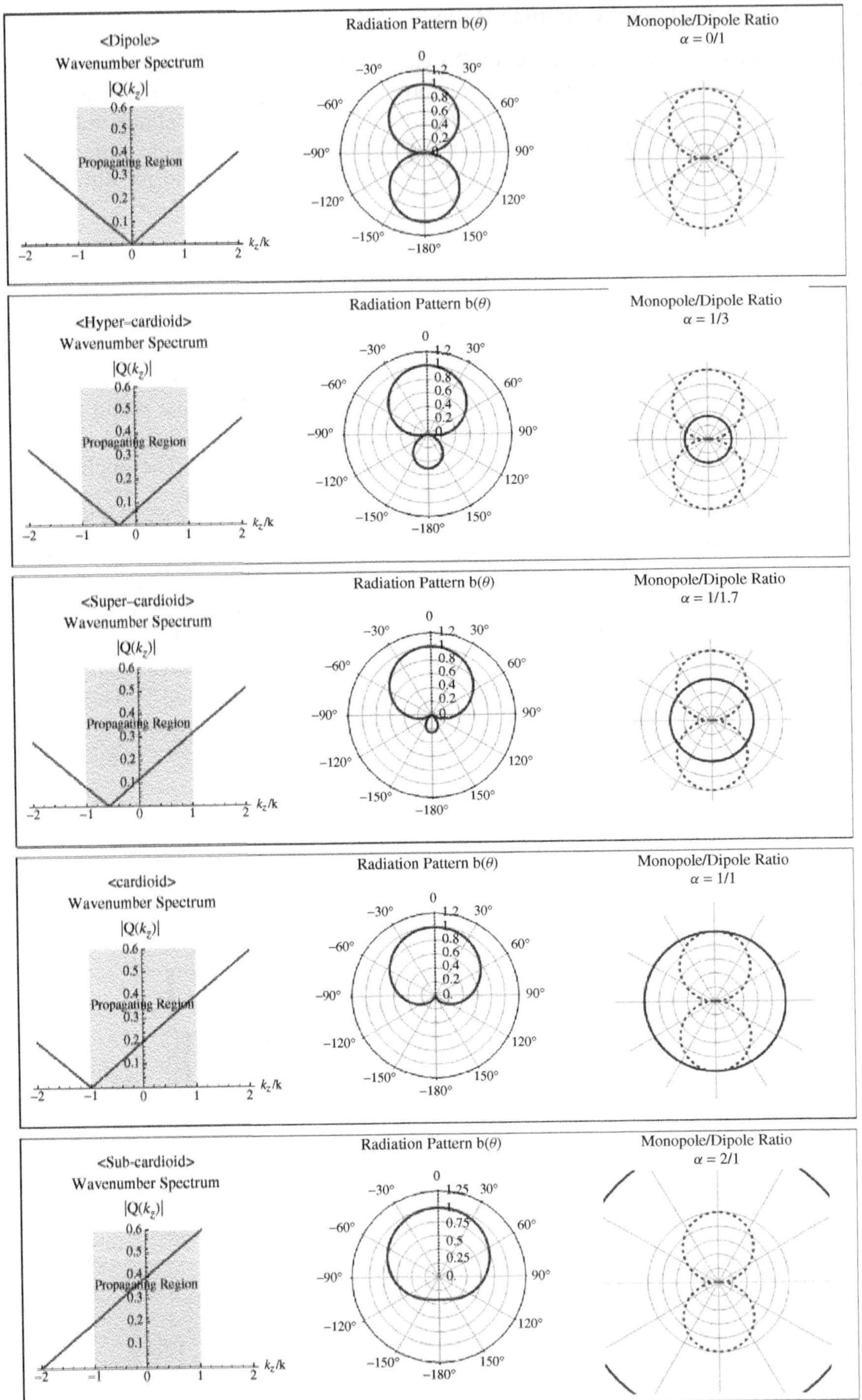

(a) (b) (c)

Figure 5.25 Differential sources: (a) wavenumber spectrum, (b) radiation pattern, and (c) monopole/dipole ratio hence the radiation pattern has a maximum value at $\theta = 0$ and a null value when $\theta = \pi$.

$$b(\theta) \approx -i\omega\tau_0(1 + \cos\theta). \tag{5.91}$$

The time delay corresponding to $\alpha = 1$ is given by the reference delay τ_0. Therefore, delaying one excitation signal by the wave propagation time required to travel Δz and changing the sign of the signal generates a cardioid pattern.

5.5.1.7 Hyper-Cardioid

For α less than 1, it is possible to generate differential patterns with a higher directivity. Among these, a pattern with the maximum directivity factor (D) is called hyper-cardioid. From the radiation pattern of and the definition of directivity (Equation (5.29)), the directivity factor can be written as

$$\begin{aligned} D &= \frac{4\pi(\alpha+1)^2}{2\pi\int_0^{\pi}(\alpha+\cos\theta)^2\sin\theta d\theta} \\ &= \frac{(\alpha+1)^2}{(\alpha^2+1/3)}. \end{aligned} \tag{5.92}$$

By differentiating the directivity factor with respect to α, we can derive the time delay ratio that gives the maximum directivity:

$$\alpha_{hyper} = 1/3. \tag{5.93}$$

Therefore, the maximum directivity factor

$$D = 4 \quad \text{or} \quad DI = 10\log_{10}D = 6\,\text{dB} \tag{5.94}$$

can be obtained by adding a short time delay $\alpha_{hyper}\tau_0$ to one of the two sources. Since the time delay ratio α also represents the monopole-to-dipole amplitude ratio, the dipole amplitude is three times greater than that of the monopole for a hyper-cardioid.

5.5.1.8 Super-Cardioid

Another special pattern of practical importance is the super-cardioid pattern, which has to do with maximizing the energy ratio between the frontal ($\theta \le \pi/2$) and rear ($\pi/2 < \theta \le \pi$) radiations. To maximize the energy ratio, a front-back ratio (FBR) can be defined as

$$\begin{aligned} FBR &= \frac{\int_0^{\pi/2}(\alpha+\cos\theta)^2\sin\theta d\theta}{\int_{\pi/2}^{\pi}(\alpha+\cos\theta)^2\sin\theta d\theta} \\ &= \frac{3\alpha^2+3\alpha+1}{3\alpha^2-3\alpha+1}. \end{aligned} \tag{5.95}$$

Accordingly, the maximum value of the FBR is given for

$$\alpha_{super} = \frac{1}{\sqrt{3}}. \tag{5.96}$$

As can be seen in Figure 5.25, the super-cardioid pattern has a smaller sidelobe in the rear half as compared to the frontal half-plane.

5.5.1.9 Sub-Cardioid

It is noteworthy that the radiation pattern with $\alpha > 1$ does not have a null value. This is the result when the contribution of the monopole is greater than that of the dipole. The broad radiation pattern produced by $\alpha > 1$ is referred to as a sub-cardioid.

5.5.1.10 Null Direction

The null direction ($b(\theta_{null}) = 0$) of the endfire differential pattern is given by

$$\theta_{null} = \pi - \cos^{-1}(\alpha). \tag{5.97}$$

For a time delay ratio with a magnitude of less than one ($|\alpha| < 1$), we can always observe a null in the differential pattern. For example, the null direction of a dipole ($\alpha = 0$) is at $\theta_{null} = \pi/2$. As the time delay ratio α increases, the null direction moves toward π (cardioid) because of the shift in the wavenumber spectrum. Therefore, by changing α, one can design a radiation pattern that has a null in the desired direction. This property is useful for generating a quiet region in a selected direction, while producing sound in the direction of beam axis.

5.5.1.11 Half-Power Beamwidth (HPBW)

The beam axis of the first-order differential pattern occurs at $\theta = 0$, and the HPBW (θ_{HPBW}) of a differential pattern can be derived from

$$\frac{\alpha + \cos(\theta_{HPBW}/2)}{|\alpha| + 1} = \frac{1}{\sqrt{2}} \Rightarrow \theta_{HPBW} = 2\cos^{-1}\left[\frac{1}{\sqrt{2}}\left(1 - \alpha(\sqrt{2} - 1)\right)\right]. \tag{5.98}$$

As shown in Figure 5.26, the HPBW is proportional to the time delay ratio α. From a comparison of various differential patterns with different α (Figure 5.25), it can be deduced that the differential source of a narrow HPBW has a greater contribution of dipoles (lower α).

5.5.2 *Combination of Delay-and-Sum and Endfire Differential Patterns*

In 1972, Olsen [1] proposed a double-layer loudspeaker system that combines the DAS and differential arrays (Figure 5.27). It is basically a line array in which each element consists of two loudspeakers operating in a differential mode. The broadside directivity of the loudspeaker system can be controlled via the DAS pattern, whereas the front-back directivity of the array can also be controlled by changing the delay between the inputs fed to the two loudspeakers that comprise each array element (Figure 5.28).

This principle is one example of the multidimensional array described in Figure 5.19. The main difference, though, is that the double layer is arranged in the z-direction, such that the endfire differential can be implemented. The use of a differential loudspeaker enables a greater reduction of the sound radiated to the rear side if the elements are driven for a cardioid radiation pattern. These advantages are accompanied by a loss of WNG, but the combination with the DAS pattern that usually has a high WNG compensates for this loss to a certain degree.

5.5.3 *Broadside Differential Pattern*

From the previous examples of the endfire differential, we can understand that a differential array generates a narrow radiation pattern by concealing part of the wavenumber spectrum in the

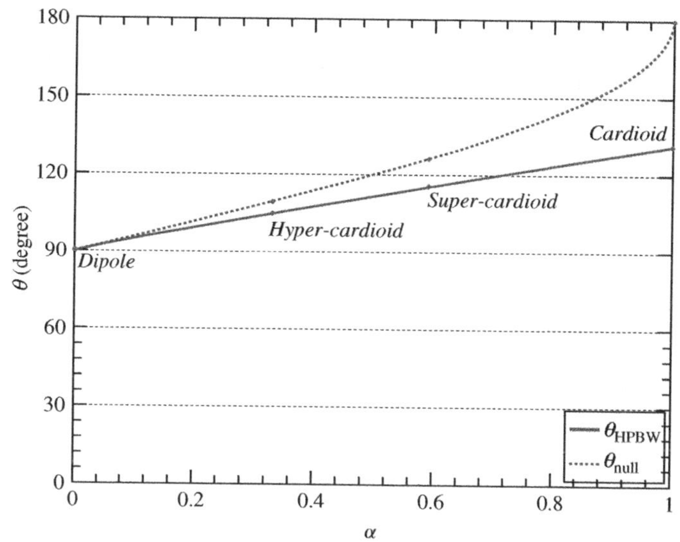

Figure 5.26 Variation of the HPBW and null direction according to changes in the time delay ratio (α)

evanescent region. A broadside beam pattern can also be generated by following the same principle. We next discuss the most basic excitation needed to obtain a broadside differential pattern.

5.5.3.1 General Formula

By definition, a broadside pattern has a symmetric radiation for $\pm\theta$. Accordingly, its excitation function should also be symmetric. However, two symmetric point sources only can produce a uniform excitation. In this regard, we need at least three point sources in order to generate a broadside differential pattern. The symmetric excitation of three elements (Figure 5.29) separated by Δx in the x direction can generally be written as

$$q(x_s) = \delta(x_s + \Delta x) - [2\cos(\beta k \Delta x)]\,\delta(x_s) + \delta(x_s - \Delta x), \tag{5.99}$$

where β is the tuning parameter for the broadside beam pattern. The parameter β determines the contribution of the center element that has the negative sign with respect to the others.

The wavenumber spectrum of this excitation function and its corresponding radiation pattern are given by

$$\begin{aligned} Q(k_x) &= \mathcal{F}_s\left[q(x)\right] \\ &= 2\left[\cos\left(k_x \Delta x\right) - \cos(\beta k \Delta x)\right], \end{aligned} \tag{5.100}$$

$$b(\theta) = 2\left[\cos\left(k \Delta x \sin\theta\right) - \cos(\beta k \Delta x)\right]. \tag{5.101}$$

For $k\Delta x \ll 1$, the cosine term can be expanded in terms of a Taylor series ($\cos x = 1 - x^2/2 + \cdots$), and the radiation pattern of Equation (5.101) can be approximated to obtain

$$b(\theta) \approx (k\Delta x)^2(\beta^2 - \sin^2\theta), \tag{5.102}$$

which has a second-order differential pattern. Figure 5.30 shows the broadside differential pattern for various values of β.

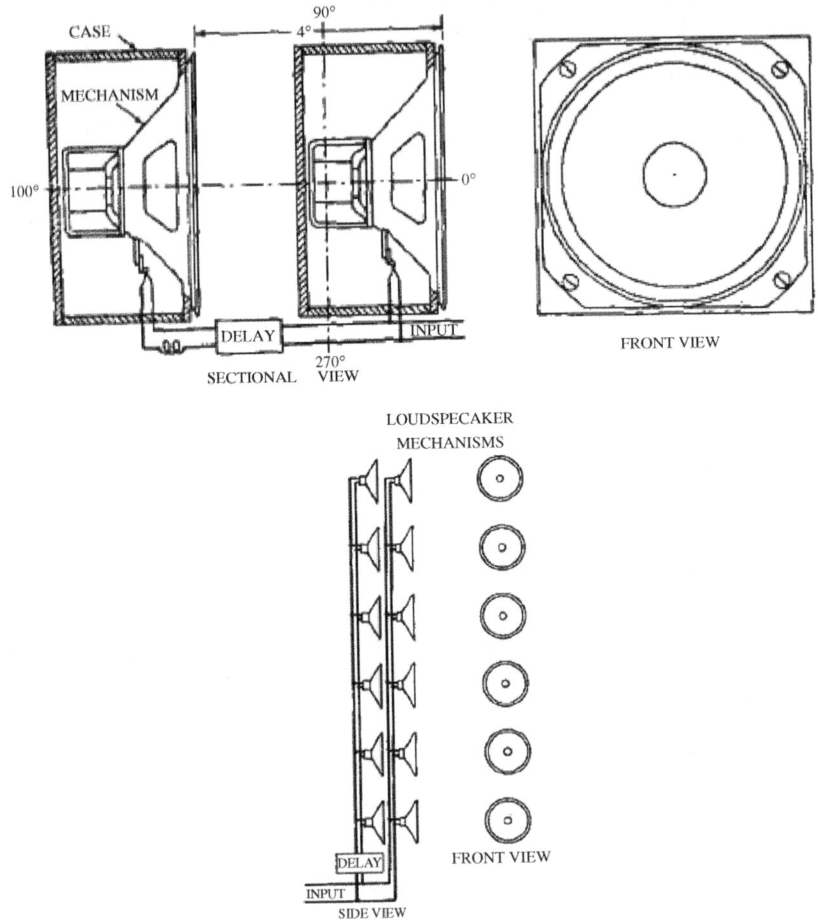

Figure 5.27 Olsen's gradient loudspeakers. Reproduced from H. F. Olson. Gradient Loudspeakers. J. Audio Eng. Soc., vol.21, no.2, pp. 86–93, March 1973

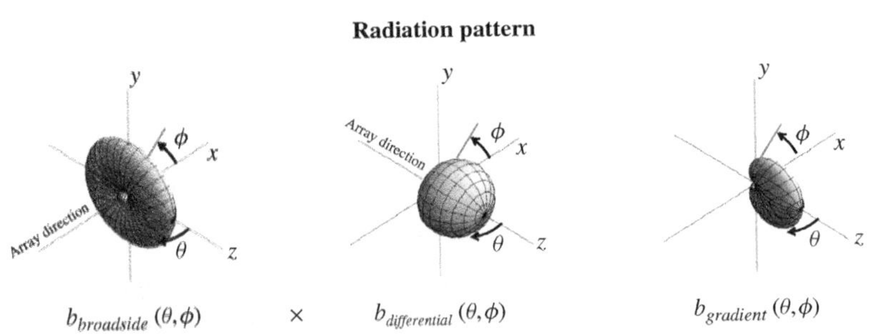

Figure 5.28 Example of a gradient array. A broadside radiation pattern is combined with a cardioid beam to generate a multidimensional radiation pattern

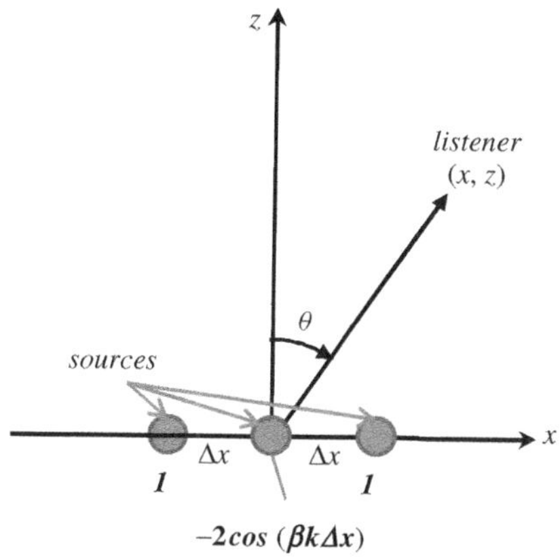

Figure 5.29 Source arrangement for a broadside differential beam

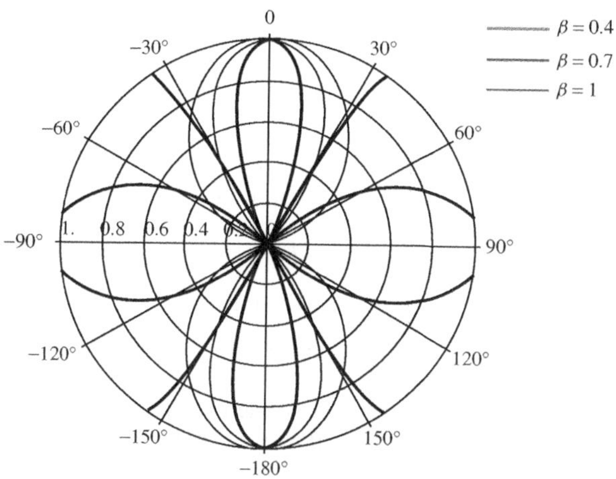

Figure 5.30 Broadside differential patterns of various values of β. (Normalized by the on-axis pressure.)

5.5.3.2 HPBW

The super-directive nature of the broadside differential can be easily observed by calculating the HPBW, which can be derived as

$$\theta_{HPBW} \approx 2\sin^{-1}\left[\beta\left(1-\frac{1}{\sqrt{2}}\right)\right]. \tag{5.103}$$

Figure 5.31 depicts the θ_{HPBW} with respect to β. The radiation pattern is sharper as β approaches zero. The case of $\beta = 0.5\pi/k\Delta x$ turns off the center element of the array, and is thus equivalent to the DAS pattern produced by two sources that are $2\Delta x$ apart. Considering that values of β will be very large (because of a small $k\Delta x$), a small HPBW with small β represents the super-directive nature of compactly located sources.

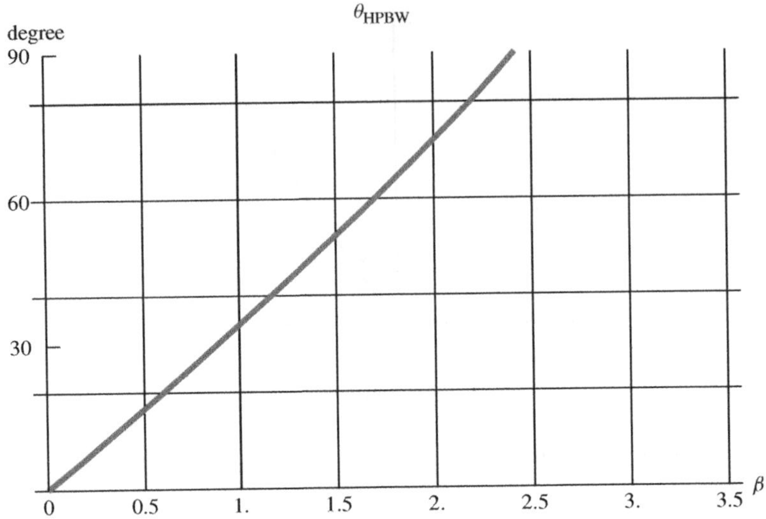

Figure 5.31 HPBW of a broadside differential beam

5.5.3.3 Null Direction

Super-directivity can also be observed through the zero-crossing point in the wavenumber spectrum. The first zero-crossing point in the wavenumber spectrum is given by

$$k_x = \beta k, \tag{5.104}$$

and the corresponding null direction θ_{null} can be written as

$$\theta_{null} = \sin^{-1}\beta. \tag{5.105}$$

Therefore, the tuning parameter β is directly connected to the zero-crossing point. Because the zero crossing point is also proportional to the HPBW, we can directly narrow down the beamwidth by decreasing β.

The special case of $\beta = 1$ corresponds to the null direction $\theta = \pi/2$, in which the cut-off wavenumber and the zero-crossing point coincide. The directivity pattern is given by

$$b(\theta) = (k\Delta x)^2 \cos^2\theta, \tag{5.106}$$

and we can obtain the second-order cosine directivity.

The main difference between the broadside and endfire differentials lies in the frequency dependency of the excitation pattern. In the case of an endfire differential, the phase of the excitation pattern is linearly proportional to the frequency; hence, the beam pattern can be generated by adding a constant time delay. For a broadside differential, however, the amplitude of the center element depends on the frequency. Therefore, a frequency-dependent weight must be applied, and the weight of the center element must be decreased in order to maintain frequency invariance at higher frequencies. For example, if we wish to have a constant radiation pattern

$$b(\theta) \approx (k\Delta x)^2 (\beta_0^2 - \sin^2\theta) \tag{5.107}$$

with a fixed value of β_0, then the relative weight of the center element

$$|q(0)| = 2|\cos(\beta_0 k\Delta x)| \tag{5.108}$$

has to decrease with increasing $\omega = ck$.

5.5.3.4 Radiation Efficiency and WNG

The directivity of this differential pattern originates from the $\sin^2\theta$ term in Equation (5.102). However, the higher directivity does not come without a price. Figure 5.32 shows the wavenumber spectrum $Q(k_x)$ for different values of β. The wavenumber spectrum has high sidelobes that increase with β, and thus in fact, it is not a narrow spectrum at all. However, if the first zero-crossing point of the spectrum is greater than the cut-off wavenumber k, the sidelobes become invisible in the farfield. Therefore, the same principle that governs the endfire differentials serves to resolve the diffraction limit in the broadside case.

A broadside differential beam is sensitive to β, and its radiation efficiency is usually lower than that of the DAS pattern. The low radiation efficiency of a differential pattern is due to a highly amplified evanescent field that does not contribute to the farfield radiation. A pattern's efficiency can be evaluated in terms of the WNG, which represents how much sound energy is radiated in the beam axis direction with respect to the total input power. The on-axis response at $\theta = 0$ or $k_x = 0$ is given by

$$b(\theta = 0) = \beta(k\Delta x)^2. \tag{5.109}$$

The WNG can be calculated from dividing the square of the on-axis response by the squared sum of the input:

$$WNG = \frac{|b(0)|^2}{\displaystyle\int_{-\infty}^{\infty} |q(x_s)|^2 dx_s} = \frac{\beta^2(k\Delta x)^4}{2 + 4\cos^2(\beta k\Delta x)}. \tag{5.110}$$

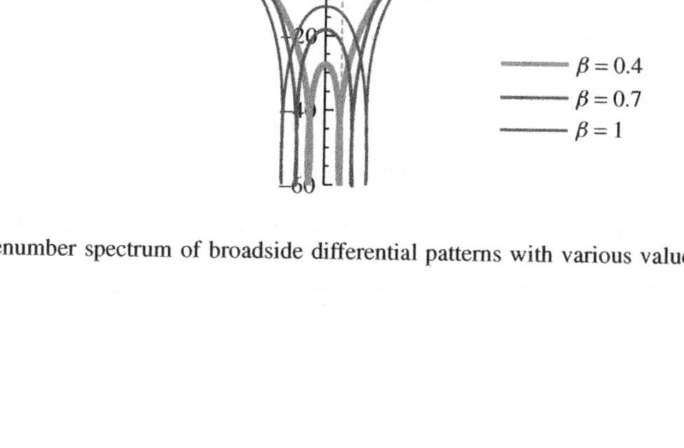

Figure 5.32 Wavenumber spectrum of broadside differential patterns with various values of β ($k\Delta x = 0.5$)

Again, for small $k\Delta x \ll 1$, the ratio in Equation (5.110) can be approximated as

$$WNG \approx \frac{\beta^2 (k\Delta x)^4}{6}, \tag{5.111}$$

which is much smaller than the efficiency ratio of the uniform excitation case ($\mathbf{q} = \begin{bmatrix}1 & 1 & 1\end{bmatrix}^T$)

$$WNG = 3. \tag{5.112}$$

Because the WNG of a broadside differential pattern is even worse than that of the first-order endfire differential pattern, it has to be used with care. It is also proportional to β^2, which means that the broader beam has better efficiency.

5.5.3.5 Maximum Sidelobe Level (MSL)

The broadside differential array achieves super-directivity by concealing its amplified sidelobes within the wavenumber spectrum. Hence, the ratio of the amplitude of the mainlobe to the maximum sidelobe level (MSL) can be a useful measure for evaluating its efficiency.

From Equation (5.100), the maximum value of the wavenumber spectrum is given by

$$\text{Max}\left[\left|Q\left(k_x\right)\right|\right] = 2\left[1 + \cos\left(\beta k \Delta x\right)\right], \tag{5.113}$$

and this gives the expression of the MSL:

$$MSL = \frac{|Q(0)|}{\text{Max}\left[\left|Q\left(k_x\right)\right|\right]} = \frac{|1 - \cos(\beta k \Delta x)|}{|1 + \cos(\beta k \Delta x)|}, \tag{5.114}$$

which is always smaller than 1, indicating that the level of the mainlobe is always lower than that of the sidelobe, as can be seen in Figure 5.32. The MSL is proportional to β, and hence, a beam pattern with a wider HPBW has a lower sidelobe level (higher MSL in dB).

5.5.4 *Combination of the Delay-and-Sum and Broadside Differential Patterns*

The broadside differential pattern is advantageous for generating a sharp beam pattern with a very small aperture size. The case of two or three elements presented in this section is the simplest form of broadside super-directivity, and higher-order differential patterns with a greater number of sources are possible in theory. In many cases, however, the broadside differential pattern has too low WNG to be used alone. In practice, therefore, it is common to use it with a DAS pattern in order to obtain a higher WNG with moderate directivity.

For a line array whose elements are equally spaced, a combination of the DAS and broadside differential patterns can be written as

$$q(x_s) = q_{BD}(x_s) *_{x_s} q_{DAS}(x_s), \tag{5.115}$$

where q_{BD} denotes the broadside differential excitation (Equation (5.99)), and the DAS excitation q_{DAS} is given by

$$q_{DAS}(x_s) = \delta(x_s - \Delta x/2) + \delta(x_s + \Delta x/2). \tag{5.116}$$

The combination of q_{BD} and q_{DAS} can be expanded as

$$q(x_s) = \delta(x_s + 1.5\Delta x) + [1 - 2\cos(\beta k \Delta x)][\delta(x_s + 0.5\Delta x) + \delta(x_s - 0.5\Delta x)] \\ + \delta(x_s - 1.5\Delta x), \quad (5.117)$$

and its corresponding wavenumber spectrum can be approximated as

$$Q(k_x) \approx 2(k\Delta x)^2(\beta^2 - k_x^2/k^2). \quad (5.118)$$

This equation implies that the wavenumber spectrum of the five-element pattern (Equation (5.118)) is identical to that of the three-element differential pattern, except that its magnitude is doubled. The WNG of the combined pattern is given by

$$WNG \approx \frac{\beta^2(k\Delta x)^4}{2}, \quad (5.119)$$

which is increased by a factor of 3 as compared to the case of the broadside differential in Equation (5.111). Therefore, a combination of the differential and summing patterns can be used to design an efficient super-directive beam.

5.6 Optimization with Energy Ratios (÷)

The basic theories of sound focusing are useful for designing a desired beam pattern. In many practical problems, however, an array can have a complex shape, or the design constraint imposed on the array cannot be explicitly expressed in simple mathematics. In such cases, it can be rather difficult to apply mathematical theory directly.

The numerical optimization technique can be utilized to resolve complex problems of sound focusing. The basic principle of the optimization technique is to find the best solution that can maximize or minimize a given *cost function*, within a feasible region that satisfies several *constraints*. An optimization problem with cost functions and constraints can often be expressed as a *ratio* of two different quantities. For instance, the *directivity index* can be regarded as an example of the ratio.

5.6.1 Problem Statement

To define an optimization problem, let us consider a general description of a sound field with an arbitrary arrangement of discrete loudspeakers. Let us also assume that there is a multichannel system driving the multiple loudspeaker or sources, as shown in Figure 5.33. An input (source) signal $s(\omega)$ of frequency ω is fed into the system, and it passes through a multichannel filter that controls the multiple sources. For the sake of brevity, we assume that the input signal $s(\omega)$ is unity; also, the frequency description ω will be omitted throughout this section. The filter coefficients of the multichannel filter can then be regarded as input signals of the multiple sources (Figure 5.34).

Let us next consider K sound sources fixed in space, and denote their locations as $\mathbf{r}_s^{(\kappa)}$ ($\kappa = 1, \ldots, K$). It is noteworthy that there is no restriction or constraint on the location or radiation characteristics of the control sources. The entire space can be either bounded or unbounded by an arbitrary boundary condition.

If we denote a filter coefficient that drives the control source located at $\mathbf{r}_s^{(\kappa)}$ as $q^{(\kappa)}$, the total pressure field can be written as a superposition of the sound fields generated by multiple sources.

$$p(\mathbf{r}) = \sum_{\kappa=1}^{K} h(\mathbf{r}|\mathbf{r}_s^{(\kappa)})q^{(\kappa)}, \quad (5.120)$$

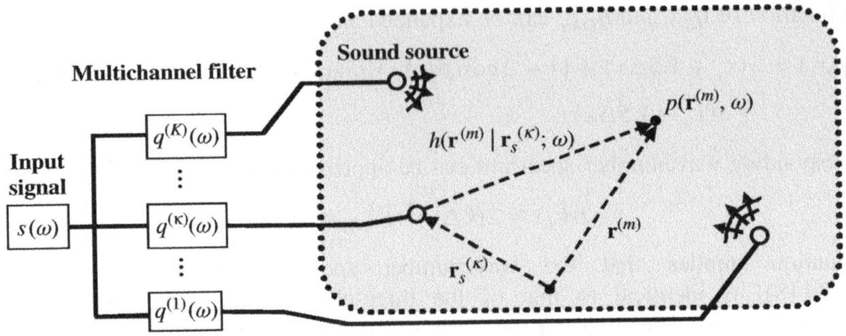

Figure 5.33 Schematic of multichannel control problem ($\mathbf{r}_s^{(\kappa)}$: location of the κth source, $\mathbf{r}^{(m)}$: field position, $h(\mathbf{r}^{(m)}|\mathbf{r}_s^{(\kappa)}; \omega)$: transfer function between the mth field position and κ^{th} source, $p(\mathbf{r}^{(m)}, \omega)$: pressure at the field position)

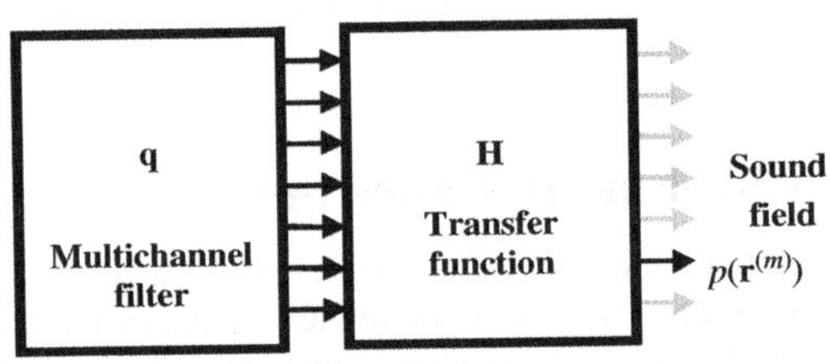

Figure 5.34 General structure of multichannel system (pure-tone case with unit input s)

where $h(\mathbf{r}|\mathbf{r}_s^{(\kappa)})$ represents the complex magnitude of a transfer function between the κth source and the listener at position $\mathbf{r}$. The transfer function can be replaced by the free-field Green's function $g_f(\mathbf{r}|\mathbf{r}_s^{(\kappa)})$ of Equation (5.1), for loudspeakers of omnidirectional radiation patterns emitting sound in the free-field condition.

Considering the discrete listener locations $\mathbf{r}^{(m)}$ ($m = 1, \ldots, M$), a vector notation can be introduced to describe the transfer functions between the listener and the control positions:

$$[\mathbf{H}]^{(m,\kappa)} = h(\mathbf{r}^{(m)}|\mathbf{r}_s^{(\kappa)}), \tag{5.121}$$

Equation (5.120) can then be rewritten more simply, in a vector form.

$$\mathbf{p} = \mathbf{Hq}, \tag{5.122}$$

where the column vector

$$\mathbf{p} = \left[p\left(\mathbf{r}^{(1)}\right), \ldots, p(\mathbf{r}^{(M)})\right]^T \tag{5.123}$$

represents the pressure measured at the listener positions, and

$$\mathbf{q} = \left[q^{(1)}, \ldots, q^{(K)}\right]^T \tag{5.124}$$

denotes the coefficients of the multichannel filter.

Since sound focusing deals with the acoustic potential energy that is proportional to the squared pressure, it is often necessary to consider the total squared pressure in space. Using the vector notations defined, we can rewrite the total radiated sound energy as

$$\|\mathbf{p}\|^2 = \mathbf{p}^H\mathbf{p} = M(\mathbf{q}^H\mathbf{R}\mathbf{q}). \tag{5.125}$$

Here, the Hermitian matrix **R** ($\mathbf{R} = \mathbf{R}^H$) is given by

$$\mathbf{R} = \frac{1}{M}\mathbf{H}^H\mathbf{H}, \tag{5.126}$$

for M measurement points equally sampled in space.

It should be noted that the definition of the matrix **R** is completely different, even though it uses the same symbol as the correlation matrix defined for microphone beamforming (Equation (4.6)). The correlation matrix is defined from the multiple microphone signals used for microphone beamforming, but here it represents the spatial correlation between sound fields produced by different loudspeakers. In this regard, the matrix **R** is regarded as a *spatial correlation matrix of loudspeaker transfer functions* in this chapter. One significant difference as compared to the microphone problem is that the relation in Equation (5.125) reflects the physical radiation characteristics and hence cannot be modified, while in the microphone problem, the correlation matrix can be differently defined to have better estimation of the beamforming power. For a loudspeaker array, the relation in Equation (5.125) is adjustable only when the transfer function of each loudspeaker can be changed. Therefore, in general, the only variable we can control is the source excitation **q**, which is analogous to the scan vector in the microphone beamforming technique.

Thus far, we have expressed the cause-and-effect relation between the control variable **q** and the sound field **p**, at a position of interest. The problem can then be defined as designing the coefficients of the multichannel filter **q**, so that the desired acoustic quantity can be enhanced. We next discuss a way to define the cost function and to find the best solution that maximizes/minimizes the defined function.

5.6.2 *Capon's Minimum Variance Estimator (Minimum Variance Beamformer)*

As discussed in Chapter 4, Section 4.5.1, Capon's minimum variance estimator seeks a solution that minimizes the total sound radiated to a region of interest, while the beam output in the target direction is kept constant. In this section, we apply the same principle to a loudspeaker array and investigate its similarity and difference to the microphone array problem. First, let us denote the transfer function to the target direction as a vector **h**. The formulation of Capon's approach can then be written as

$$\underset{\mathbf{q}}{\text{Minimize}}\ \mathbf{q}^H\mathbf{R}\mathbf{q}\ \text{subject to}\ \mathbf{h}^T\mathbf{q} = 1 \tag{5.127}$$

The linear constraint indicates that the array response to the target direction should have a unity gain and must be distortionless. The optimization problem with a linear constraint can be rewritten by introducing the Lagrange multiplier. Let us consider the following cost function using a Lagrange multiplier μ.

$$J = \mathbf{q}^H\mathbf{R}\mathbf{q} + \mu(1 - \mathbf{h}^T\mathbf{q}). \tag{5.128}$$

The constraints expressed in the parentheses now act as a penalty function, which increases the cost function when the constraint cannot be satisfied. A Lagrange multiplier represents how sensitively the penalty should be reflected in the cost function with respect to the total sound energy.

The optimum of J may be one of the points where the derivative of J is zero. Taking the derivative of J with respect to the variables **q** and μ (see Section A.3.3), we have

$$\mathbf{q}^H\mathbf{R} - \mu\mathbf{h}^T = 0, \tag{5.129}$$

$$\mathbf{h}^T\mathbf{q} = 1. \tag{5.130}$$

Therefore, the optimal solution that minimizes J automatically satisfies the constraint in Equation (5.130), while minimizing the total radiated energy. From Equation (5.129), we have

$$\mathbf{q} = \mu^* \mathbf{R}^{-1} \mathbf{h}^*, \tag{5.131}$$

and inserting this into Equation (5.130) gives the real-valued constant μ:

$$\mu = \mu^* = \frac{1}{\mathbf{h}^H \mathbf{R}^{-1} \mathbf{h}}. \tag{5.132}$$

Substituting Equation (5.132) into Equation (5.131) leads to the following solution:

$$\mathbf{q} = \frac{\mathbf{R}^{-1} \mathbf{h}^*}{\mathbf{h}^H \mathbf{R}^{-1} \mathbf{h}}. \tag{5.133}$$

The beam power generated by the optimal solution is given by

$$\mathbf{p}^H \mathbf{p} = \mathbf{q}^H \mathbf{R} \mathbf{q} = \frac{1}{\mathbf{h}^H \mathbf{R}^{-1} \mathbf{h}}. \tag{5.134}$$

As can be seen in Equation (5.133), the formulation is similar to that which is defined for the microphone problem (Equation (4.75)). Therefore, the minimum variance distortionless response (MVDR) technique can be directly applied to the loudspeaker problem as well. However, the definition of the correlation matrix is different and cannot be modified. Most microphone beamforming techniques that can produce a beamformer output signal can be used, but the source identification technique that requires a modified definition of beamforming power cannot be directly applied to the loudspeaker problem. For instance, the linear prediction (LP) estimator and multiple signal classification (MUSIC) beamformer that exploit a beamforming power other than that in Equation (5.125) are incompatible with the loudspeaker array.

5.6.3 *Acoustic Brightness and Contrast Control*

In many practical situations, the sound focusing problem is involved with the regional control of sound field. For example, the sound focusing is needed to be realized over a zone of finite volume, instead of a point in space. In this section, we introduce two concepts that are related to the regional control of sound field: the *acoustic brightness* and *contrast* [2, 3]. These concepts serve well to highlight what the practice of regional control of acoustic variables, such as the acoustic potential energy, can really allow us to do. As illustrated in Figure 5.35, the acoustic brightness and contrast

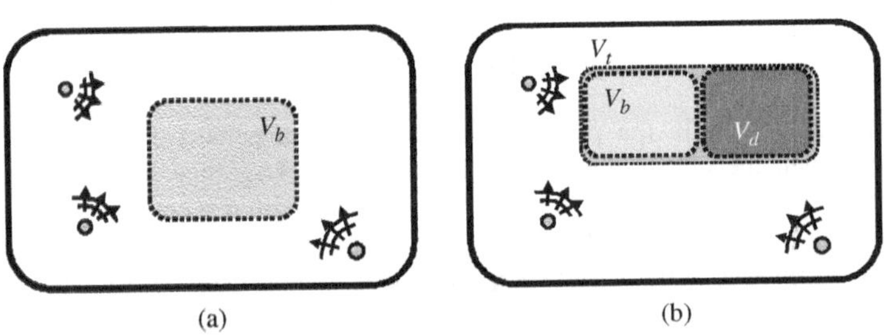

Figure 5.35 Variable definitions for enhancement of acoustic potential energy (a) acoustic brightness problem and (b) acoustic contrast problem (V_b: bright zone, V_d: dark zone, V_t: total zone of interest)

problem can enable a given listener to listen to what he or she wants, while the other region remains quiet.

To begin, let us define a measure for describing the acoustic potential energy of a zone. Since the measure or descriptor has to represent the overall acoustic potential energy of a zone, we consider a control zone V_b (Figure 5.35a) and employ the spatial average of the acoustic potential energy as the measure of the control zone's acoustic potential energy. Using the description in Equation (5.122), we can rewrite the spatial average of the acoustic potential energy as

$$\begin{aligned} e_b &= \frac{1}{V_b}\int_{V_b} |p(\mathbf{r})|^2 dV(\mathbf{r}) \\ &= \mathbf{q}^H \left[\frac{1}{M_b}\mathbf{H}_b^H \mathbf{H}_b\right]\mathbf{q}, \end{aligned} \tag{5.135}$$

where the matrix $\mathbf{H}_b$ represents the transfer function matrix $\mathbf{H}$, which is defined for the M_b measurement points in V_b.

The matrix arithmetic enclosed in brackets in Equation (5.135) can be expressed in terms of a spatial correlation matrix.

$$\mathbf{R}_b = \frac{1}{M_b}\mathbf{H}_b^H \mathbf{H}_b. \tag{5.136}$$

Since the correlation matrix is Hermitian (that is, $\mathbf{R}_b^H = \mathbf{R}_b$), the quadratic form of Equation (5.135) is positive semidefinite. In other words, the energy e_b is always greater than or equal to zero for all nonzero $\mathbf{q}$. Using the defined matrix, we can rearrange the acoustic potential energy of zone V_b in a simpler form.

$$e_b = \mathbf{q}^H \mathbf{R}_b \mathbf{q}. \tag{5.137}$$

Now, the output of the loudspeaker array system given by the spatial mean of the acoustic potential energy can be represented in a quadratic form with respect to $\mathbf{q}$.

5.6.3.1 Acoustic Brightness Problem

The first problem is to enhance the acoustic brightness of a zone, provided that the input energy is kept constant. The total input energy can be written as the squared sum of the multichannel excitations

$$e_i = h_c^2 \mathbf{q}^H \mathbf{q}, \tag{5.138}$$

where h_c is a constant that converts the dimension of the squared sum of the excitation signals into the input energy. The specific value of the normalization constant can be defined differently, depending upon the system. The input energy can be regarded as the control effort required in order to generate a focused sound in space.

Next, the *acoustic brightness* is defined as the ratio of the output and input energies of the loudspeaker system. From Equations (5.137) and (5.138), the acoustic brightness can be written as follows.

$$\alpha = \frac{e_b}{e_i} = \frac{\mathbf{q}^H \mathbf{R}_b \mathbf{q}}{h_c^2 \mathbf{q}^H \mathbf{q}}. \tag{5.139}$$

Note that the symbol α denotes the different measure from the time delay ratio of differential source (Equation (5.87)). The acoustic brightness in Equation (5.139) represents the acoustic potential energy of a zone with respect to the input energy. Therefore, we can find the most energy-efficient solution by maximizing the acoustic brightness. For a point-like bright zone, the acoustic brightness converges to the WNG in Equation (5.35).

Using an optimization technique, we can derive the optimal solution $\mathbf{q}$ that maximizes a given cost function α. First, the problem of maximizing α is equivalent to minimizing the input energy, while the output energy is constrained to a fixed value e_{bc}. Accordingly, we can define the following optimization problem:

$$\underset{\mathbf{q}}{\text{Minimize}}\,[e_i = h_c^2\mathbf{q}^H\mathbf{q}], \text{ subject to } e_{bc} - \mathbf{q}^H\mathbf{R}_b\mathbf{q} = 0. \tag{5.140}$$

Since the input energy is always greater than zero, a minimization problem with a constraint always has a global minimum. For a more straightforward derivation of the optimal solution, the quadratic optimization with a constraint can be transformed into a problem without constraint by introducing a Lagrange multiplier. For example, suppose we want to minimize the following cost function:

$$\underset{\mathbf{q}}{\text{Minimize}}\; J = h_c^2\mathbf{q}^H\mathbf{q} + a(e_{bc} - \mathbf{q}^H\mathbf{R}_b\mathbf{q}). \tag{5.141}$$

The expression within the parentheses acts as a penalty function, which increases the cost function J when the output energy $(\mathbf{q}^H\mathbf{R}_b\mathbf{q})$ is smaller than the given constraint. The variable (Lagrange multiplier) a is a kind of weighting function that determines to what degree the penalty will be reflected in the cost function. At this stage, we do not know which value of a gives the optimal solution to Equation (5.140). However, we do know that the cost function has a global minimum, so it is logical to take a derivative of Equation (5.141) with respect to the variables $\mathbf{q}$ and a, in order to examine every possible stationary point.

$$\frac{\partial J}{\partial \mathbf{q}} = h_c^2\mathbf{q} - a\mathbf{R}_b\mathbf{q} = 0, \tag{5.142}$$

$$\frac{\partial J}{\partial a} = e_{bc} - \mathbf{q}^H\mathbf{R}_b\mathbf{q} = 0. \tag{5.143}$$

From Equation (5.143), it is evident that the output power constraint in Equation (5.140) is automatically satisfied at the stationary points of J. Therefore, at the point of the global minimum in Equation (5.141), the output energy constraint can also be fulfilled. In addition, it is noteworthy that Equation (5.142) characterizes the stationary points as

$$\mathbf{R}_b\mathbf{q} = \frac{h_c^2}{a}\mathbf{q}. \tag{5.144}$$

That is, the stationary points can be obtained when $\mathbf{q}$ is one of the eigenvectors of $\mathbf{R}_b$. Once we have obtained the stationary points, we need to test whether each stationary point is a minimum, a maximum, or just a saddle point. By inserting Equations (5.144) into Equation (5.141), we find that the cost function at a stationary point can be reduced to

$$J = ae_{bc}, \tag{5.145}$$

which represents the fact that the global minimum can be obtained for the smallest value of a. Since the smallest a is attained by the maximum eigenvalue of $\mathbf{R}_b$ Equation (5.144), the optimal solution that minimizes Equation (5.141) is the eigenvector that corresponds to the maximum eigenvalue of $\mathbf{R}_b$.

The minimization problem can also be reformulated as a maximization problem. If the input energy is constrained to e_{ic} and if we want to maximize the output energy under this constraint, the problem can be described as follows:

$$\underset{\mathbf{q}}{\text{Maximize}}\,[e_b = \mathbf{q}^H\mathbf{R}_b\mathbf{q}], \text{ subject to } e_{ic} - h_c^2\mathbf{q}^H\mathbf{q} = 0. \tag{5.146}$$

The equivalent problem using a Lagrange multiplier α can then be defined as

$$\underset{\mathbf{q}}{\text{Maximize}}\ J = \mathbf{q}^H \mathbf{R}_b \mathbf{q} + \alpha(e_{ic} - h_c^2 \mathbf{q}^H \mathbf{q}). \tag{5.147}$$

It should be noted that the penalty function $e_{ic} - h_c^2 \mathbf{q}^H \mathbf{q}$ now decreases the cost function J when more input energy is consumed than the constraint e_{ic}. Taking the derivatives with respect to α and $\mathbf{q}$ leads to

$$\frac{\partial J}{\partial \mathbf{q}} = \mathbf{R}_b \mathbf{q} - \alpha h_c^2 \mathbf{q} = 0, \tag{5.148}$$

$$\frac{\partial J}{\partial \alpha} = e_{ic} - h_c^2 \mathbf{q}^H \mathbf{q} = 0, \tag{5.149}$$

and thus, the following characteristic equation can be derived:

$$\mathbf{R}_b \mathbf{q} = \alpha h_c^2 \mathbf{q}. \tag{5.150}$$

The optimal solution $\mathbf{q}_\alpha$ satisfying Equation (5.150) is the same as the solution to the minimization Equation (5.140), which is the eigenvector corresponding to the maximum eigenvalue of $\mathbf{R}_b$ ($\alpha_{\max}$).

$$\mathbf{R}_b \mathbf{q}_\alpha = \alpha_{\max} h_c^2 \mathbf{q}_\alpha \tag{5.151}$$

By deriving multiple sources using the solution of Equation (5.151), we can generate a zone of high acoustic brightness (an acoustically bright zone).

For the maximization problem, the Lagrange multiplier α is equivalent to the acoustic brightness Equation (5.139). The maximum brightness is thus the maximum eigenvalue of the correlation matrix. The ratio in Equation (5.139) is known as Rayleigh's quotient. *Rayleigh's quotient, which corresponds to a (Hermit) positive definite matrix, has stationary values in the neighborhood of the eigenvectors, where the stationary values are equal to the associated eigenvalues* (Rayleigh's principle, [4])

5.6.3.2 Acoustic Contrast Problem

To solve an acoustic contrast problem, we need to define two different control zones (Figure 5.35b). One is the acoustically bright zone V_b, where the listener is able to enjoy a high acoustic potential energy. The other zone, the acoustically dark (quiet) zone V_d, has an opposite characteristic. Then, using Equation (5.137), we can define the acoustic contrast as being the ratio between the acoustic brightness ratio of these two zones.

$$\beta = \frac{e_b}{e_d} = \frac{\mathbf{q}^H \mathbf{R}_b \mathbf{q}}{\mathbf{q}^H \mathbf{R}_d \mathbf{q}}, \tag{5.152}$$

where e_b and e_d represent the acoustic brightness of V_b and V_d respectively. As in the brightness problem, the optimal solution that maximizes this potential energy ratio can be seen as a constrained minimization or maximization problem.

$$\underset{\mathbf{q}}{\text{Minimize}}\,[e_d = \mathbf{q}^H \mathbf{R}_d \mathbf{q}], \text{ subject to } e_{bc} - \mathbf{q}^H \mathbf{R}_b \mathbf{q} = 0.$$

The matrix $\mathbf{R}_d$ is positive semi-definite, that is, it is always greater than or equal to zero for nonzero $\mathbf{q}$. Accordingly, the dark zone's energy with a constraint is bounded and therefore has the global minimum or minima. By introducing a Lagrange multiplier b, we can express the problem as a cost function.

$$\underset{\mathbf{q}}{\text{Minimize}}\ J = \mathbf{q}^H \mathbf{R}_d \mathbf{q} + b(e_{bc} - \mathbf{q}^H \mathbf{R}_b \mathbf{q}), \tag{5.153}$$

or equivalently,

$$\underset{\mathbf{q}}{\text{Maximize}}\ J = \mathbf{q}^H \mathbf{R}_b \mathbf{q} + \beta(e_{dc} - \mathbf{q}^H \mathbf{R}_d \mathbf{q}), \tag{5.154}$$

with the dark zone energy constrained to e_{dc}. At the stationary points, we have

$$\frac{\partial J}{\partial \mathbf{q}} = \mathbf{R}_b \mathbf{q} - \beta \mathbf{R}_d \mathbf{q} = 0, \tag{5.155}$$

$$\frac{\partial J}{\partial a} = e_{dc} - \mathbf{R}_d \mathbf{q} = 0, \tag{5.156}$$

and the cost function J at these points can be rewritten from Equations (5.155) and (5.156) as

$$J = \beta e_{dc}. \tag{5.157}$$

Therefore, the global maximum J is attained when $\mathbf{q}$ is given by the eigenvector that corresponds to the maximum eigenvalue ($\beta_{\max}$) of the following generalized eigenvalue problem:

$$\mathbf{R}_b \mathbf{q}_\beta = \beta_{\max} \mathbf{R}_d \mathbf{q}_\beta. \tag{5.158}$$

There are various ways to solve the generalized eigenvalue problem expressed in Equation (5.158). For nonsingular $\mathbf{R}_d$, one can convert Equation (5.158) to a standard eigenvalue problem:

$$\mathbf{R}_d^{-1} \mathbf{R}_b \mathbf{q}_\beta = \beta_{\max} \mathbf{q}_\beta. \tag{5.159}$$

However, when $\mathbf{R}_d$ is singular, factorization algorithms such as QZ decomposition [5] can be employed for the generalized eigenvalue problem expressed in Equation (5.158) in order to obtain more stable results.

5.6.3.3 Alternative Definition

Instead of the dark zone V_d, the acoustic brightness can also be defined in terms of the total zone of interest $V_t = V_b + V_d$. It can be proved that the solution to Equation (5.159) is also the solution that maximizes the following ratio.

$$\beta_t = \frac{e_b}{e_t} = \frac{\mathbf{q}^H \mathbf{R}_b \mathbf{q}}{\mathbf{q}^H \mathbf{R}_t \mathbf{q}}, \tag{5.160}$$

where e_t denotes the acoustic potential energy of V_t, and $\mathbf{R}_t$ is the spatial correlation on V_t. It is noteworthy that the total acoustic potential energy of V_t is equal to the sum of the potential energy of V_b and V_d Equation (5.160) can then be rewritten as follows

$$\beta_t = \frac{\mathbf{q}^H \mathbf{R}_b \mathbf{q}}{\frac{1}{V_t} \mathbf{q}^H (V_b \mathbf{R}_b + V_d \mathbf{R}_d) \mathbf{q}} = \frac{1}{\frac{1}{V_t}\left(V_b + V_d \frac{1}{\beta}\right)}. \tag{5.161}$$

Since volumes always have a positive value, the maximum value of β_t can be obtained only when β is a maximum. Therefore, the solution $\mathbf{q}_\beta$ in Equation (5.159) also satisfies the following relation:

$$\mathbf{R}_t^{-1} \mathbf{R}_b \mathbf{q}_\beta = \beta_t \mathbf{q}_\beta. \tag{5.162}$$

5.6.3.4 Example of Brightness and Contrast Control

The next example demonstrates the regional control characteristic of acoustic brightness and contrast control. For the sake of simplicity, the entire space was assumed to be unbounded. The control sources (monopoles) were arranged on the surface of a half-sphere with a radius $r = 20\lambda$ (λ: wavelength) (Figure 5.36). A total of 17 monopoles were distributed in the farfield from the origin, each of which was at the same intervals on the xy plane. The zone of interest was placed at the center of the xy plane and its aperture size was $L = 5\lambda$.

In Figure 5.37, the first row shows the energy distribution in a case of point focusing. The acoustically bright point is configured to be at the center. The potential energy field is normalized by the normalization constant $h_c^2 = K|i\rho_0\omega/(4\pi r)|^2$ which is the potential energy generated at the origin by perfectly correlated K monopole sources of unit input power. The magnitude of the potential energy field then indicates the amount of focused acoustic potential energy, divided by that of the perfect focusing case. In the point focusing case, the potential energy of the focused position approaches 0 dB. As the bright zone is enlarged (Figure 5.37b,c), the absolute magnitude of the potential energy field decreases because the pressure fields generated by each source are averaged out over the entire zone. In this situation, therefore, point focusing is easier than the generation of a bright zone. The cases presented here demonstrate that contrast control definitely decreases the SPL of the dark zone, whereas brightness control generates a higher SPL in the bright zone.

5.6.3.5 Hybrid Forms

One drawback of contrast control is that the acoustic brightness of the bright zone cannot always be guaranteed. This is simply because the denominator of the defined ratio can be singular. For example, consider a case in which all the control sources produce similar pressure fields in V_d. The minimum eigenvalue of $\mathbf{R}_d$ is then nearly zero, and the control sources are so excited that the control signal vector $\mathbf{q}$ belongs to the null space of $\mathbf{R}_d$, which causes the cost function β to diverge (Equation (5.152)). As a result, the maximization process tends to minimize the brightness of V_d rather than maximize the brightness of V_b. In this regard, the pressure field generated by each source must be linearly independent, but does not necessarily need to be orthogonal in the zone of interest in order to obtain adequate brightness in the bright zone. However, such a singular condition can be readily reduced by employing adequate regularization. For example, we can introduce a hybrid form of the cost functions that combines the brightness and contrast control. That is,

$$\eta = \frac{e_b}{\sigma e_i + (1-\sigma)e_d} = \frac{\mathbf{q}^H\mathbf{R}_b\mathbf{q}}{\sigma h_c^2\mathbf{q}^H\mathbf{q} + (1-\sigma)\mathbf{q}^H\mathbf{R}_d\mathbf{q}}. \tag{5.163}$$

The consideration of input power e_i prevents this singular condition, even if e_t is zero. This form is similar to the Tikhonov regularization (Chapter 6, Section 6.5.2.2 and Appendix A,

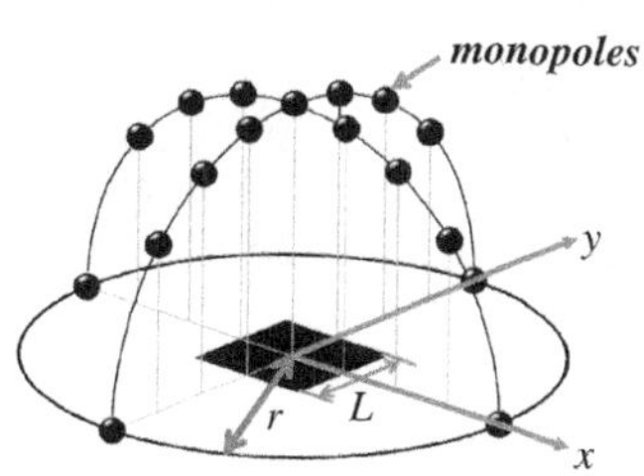

Figure 5.36 Array configuration for brightness and contrast control

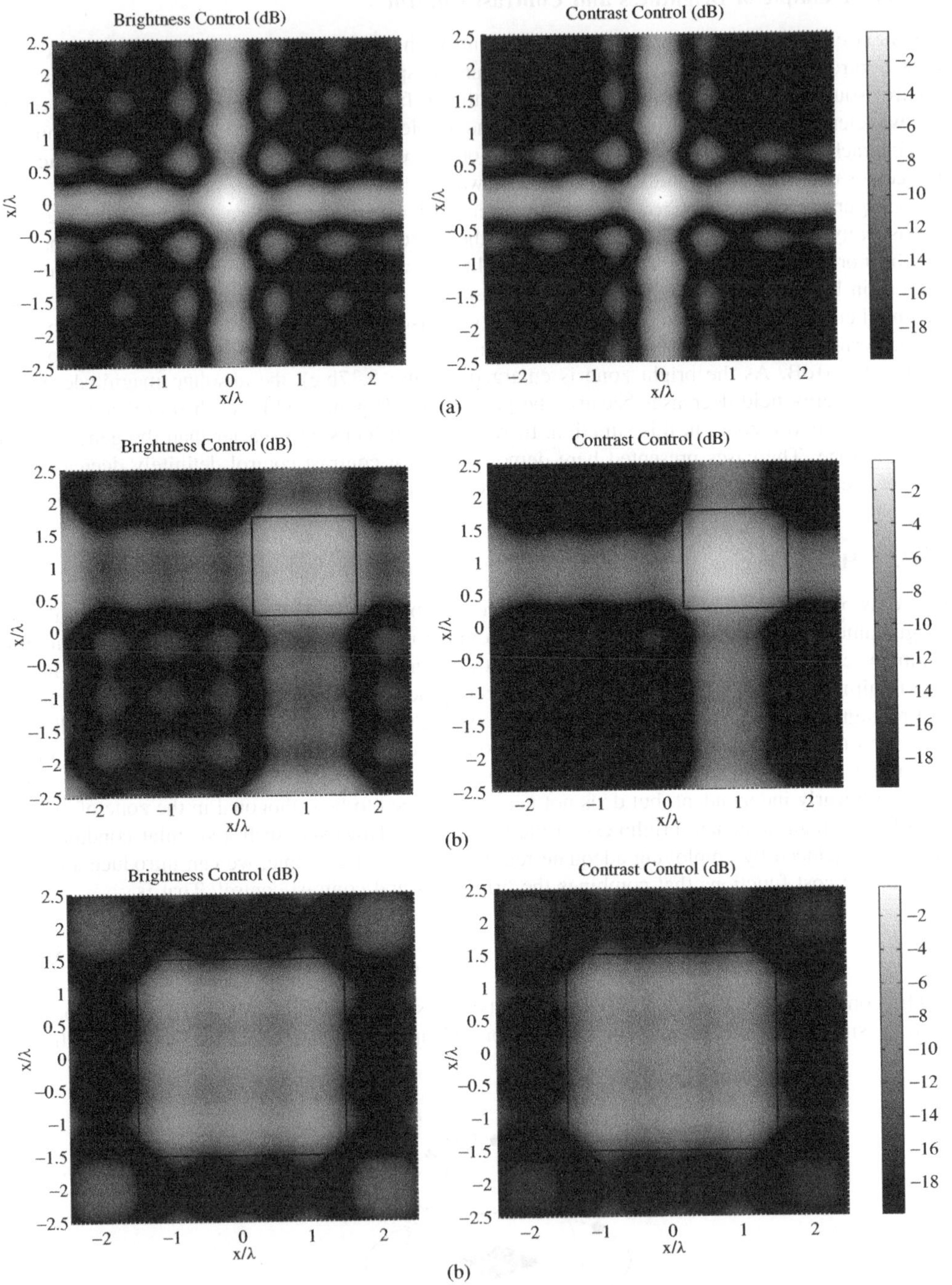

Figure 5.37 Acoustically bright and dark zones generated within illuminated region. Aperture size of bright zone: (a) point focusing, (b) $L = 1.5\lambda$, and (c) $L = 3\lambda$

Section A.4.4), in which constraint on the input power is employed. In this hybrid form, the normalization constant σ controls the priority between the brightness and contrast control. If the magnitude of σ is close to 1, the hybrid form approaches acoustic brightness, while a small σ results in contrast control.

5.6.3.6 Application to the Line Array Problem

Regional control methods can also be applied to line or planar geometries. It is interesting to see how an optimized beam pattern is connected to the basic DAS and differential patterns and how its directivity and efficiency can be analyzed from the wavenumber spectrum.

Let us again consider the farfield line array problem, as described in Section 5.2.4. To define the regional control problem for farfield radiation, the bright zone is defined as an angular region between $-\theta_b \leq \theta \leq \theta_b$, with the dark zone covering the rest of the angles (Figure 5.38).

For the line array and continuous excitation function $q(x_s)$, the total input energy e_i fed to an array is equivalent to

$$e_i = h_c^2 \int_{-\infty}^{\infty} |q(x_s)|^2 dx_s, \tag{5.164}$$

From the relation between the excitation function and the pressure field (Equation (5.22)), the normalization constant is configured as $h_c = 1/4\pi r$. The acoustic contrast can then be identified as the energy ratio of two different areas (Figure 5.39) in the wavenumber domain; it corresponds to

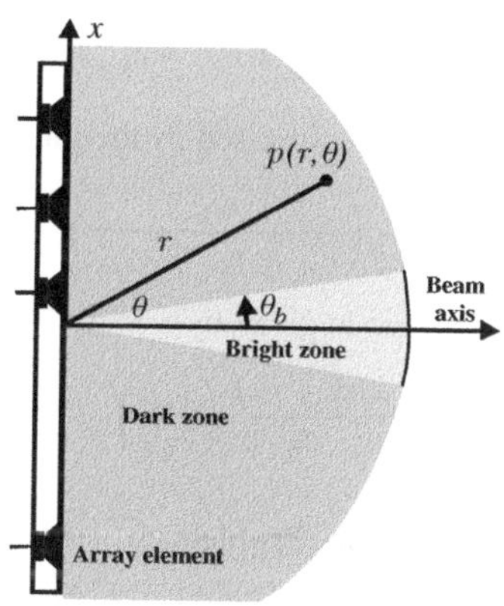

Figure 5.38 Definition of bright and dark zones for a line array problem

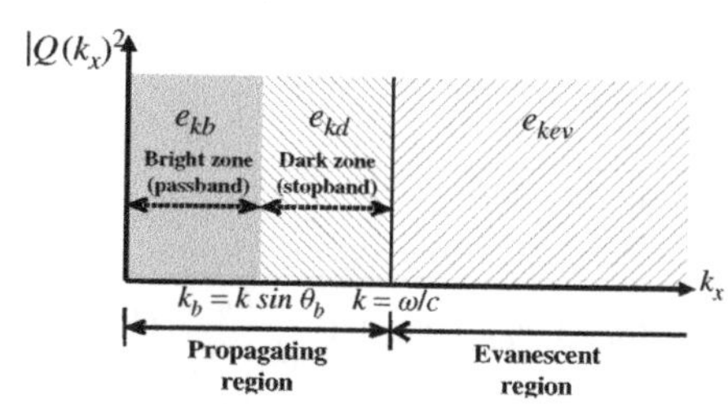

Figure 5.39 Wavenumber regions related to regional focusing problem

the passband energy divided by the stopband energy. The passband (e_{kb}) and stopband (e_{kd}) energy in the wavenumber domain can be related to the average potential energy (e_b, e_d) in space as

$$e_{kb} = \int_{-k_b}^{k_b} |Q(k_x)|^2 dk_x = 2k_b e_b / h_c^2, \tag{5.165}$$

$$e_{kd} = \left[\int_{k_b}^{k} |Q(k_x)|^2 dk_x + \int_{-k}^{-k_b} |Q(k_x)|^2 dk_x \right]$$
$$= 2(k - k_b) e_d / h_c^2, \tag{5.166}$$

where k_b is the wavenumber bound that corresponds to the angular width of a bright zone ($k_b = k \sin\theta_b$).

The wavenumber domain representation of acoustic brightness is less straightforward, since it involves two different kinds of variables: the input power and the acoustic potential energy. According to Parseval's theorem (Appendix A, Section A.1.1.5), the total sum of the squared excitation function in the space domain is equal to its total spectral energy. Applying this theorem to Equation (5.164) yields

$$\int_{-\infty}^{\infty} |q(x_s)|^2 dx_s = \frac{1}{2\pi} \int_{-\infty}^{\infty} |Q(k_x)|^2 dk_x. \tag{5.167}$$

Therefore, the total input energy is given by the total energy of the wavenumber spectrum, including the passband (e_{kb}), stopband (e_{kd}) and evanescent region (e_{kev}) (Figure 5.39):

$$e_i = \frac{h_c^2}{2\pi}(e_{kb} + e_{kd} + e_{kev}). \tag{5.168}$$

Now, the acoustic brightness can also be expressed as the ratio of the wavenumber spectrums.

$$\alpha = \frac{\pi}{k_b}\left(\frac{e_{kb}}{e_{kb} + e_{kd} + e_{kev}}\right). \tag{5.169}$$

On the other hand, the acoustic contrast is given by

$$\beta = \frac{(k - k_b)}{k_b}\left(\frac{e_{kb}}{e_{kd}}\right). \tag{5.170}$$

5.6.3.7 Ill-Conditioning of the Line Array Problem

As can be observed in Equation (5.170), the acoustic contrast deals only with the sound energy of the propagating field, and the evanescent region is completely omitted in its cost function. Correspondingly, the excitation q that gives a desired acoustic contrast can result in huge amplification of the evanescent field, and the problem is often ill-posed or ill-conditioned. The hybrid form in Equation (5.163) can circumvent this problem. If we rewrite the hybrid cost function in the wavenumber domain as

$$\eta = \frac{e_b}{\sigma e_i + (1-\sigma) e_d}$$
$$= \frac{e_{kb}}{\sigma \frac{k_b}{\pi}(e_{kb} + e_{kd} + e_{kev}) + \frac{k_b}{(k - k_b)}(1-\sigma) e_{kd}}, \tag{5.171}$$

it can be observed that the parameter σ prevents the solution from diverging by including the evanescent energy e_{kev} in the cost function.

By varying the parameter σ, we can obtain various optimal solutions that maximize the acoustic contrast, with different constraints on the acoustic brightness. To be more specific, the relationships between the efficiency (brightness) and array directivity (contrast) of the optimized solutions are presented in Figure 5.40. In this example, 14 monopoles operating at $f = 800\,\text{Hz} \sim 1.5\,\text{kHz}$ were linearly arranged at intervals of 0.07 m, and the size of the bright zone was $\theta_b = 25°$.

The brightness-contrast curve [6] shows how much the directivity of an optimal solution (contrast) can be enhanced by sacrificing the array efficiency (brightness). Higher acoustic contrast can be obtained with lower acoustic brightness, which indicates that the SPL in the bright zone will be decreased if we attempt to increase the energy ratio of the bright and dark zones.

However, because of various errors and limitations, the contrast cannot continuously increase with decreasing brightness as it does in a simulation. In a real experiment, the curve usually breaks down, and the theoretical acoustic contrast of the low efficiency solution cannot be implemented in practice. To realize a beam pattern with low efficiency in practice, the overall accuracy of the system must be enhanced. Otherwise, the target acoustic contrast must be lowered or limited to a realizable value. This realizable value of acoustic contrast can be identified from the brightness-contrast curve of an actual array setup. In addition, it should be considered that the low brightness solution also limits the maximum SPL that can be radiated by the array system. To recover the loss of SPL, loudspeaker units with higher sensitivity are required, which often leads to an increase in the overall manufacturing cost.

The optimal beam pattern corresponding to an acoustic brightness of $-15\,\text{dB}$ is shown in Figure 5.41. Because of the diffraction limit, the line array with an aperture size of 0.98 m cannot produce a sharp radiation pattern in the low-frequency region.

A typical wavenumber spectrum of hybrid control is presented in Figure 5.42. The optimal excitation calculated for the $-15\,\text{dB}$ acoustic brightness at $f = 800\,\text{Hz}$ is transformed to the wavenumber domain. As a reference, the hybrid control is compared with a conventional window-based excitation having a similar 3-dB beamwidth: a Chebyshev window with a sidelobe attenuation of $-20\,\text{dB}$.

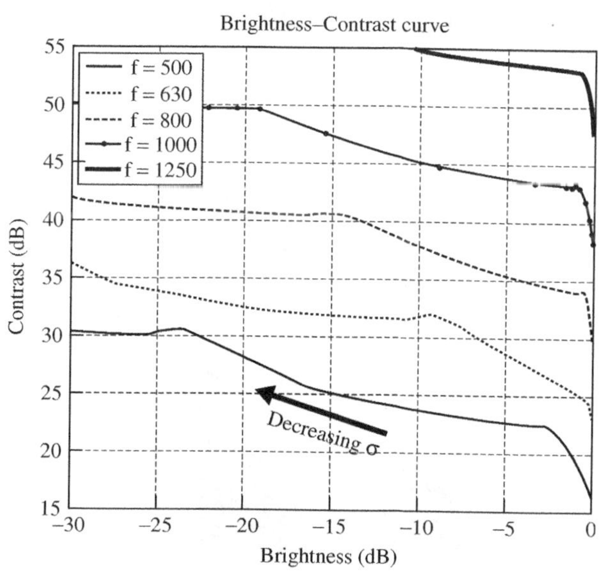

Figure 5.40 Brightness-contrast curve at different frequencies

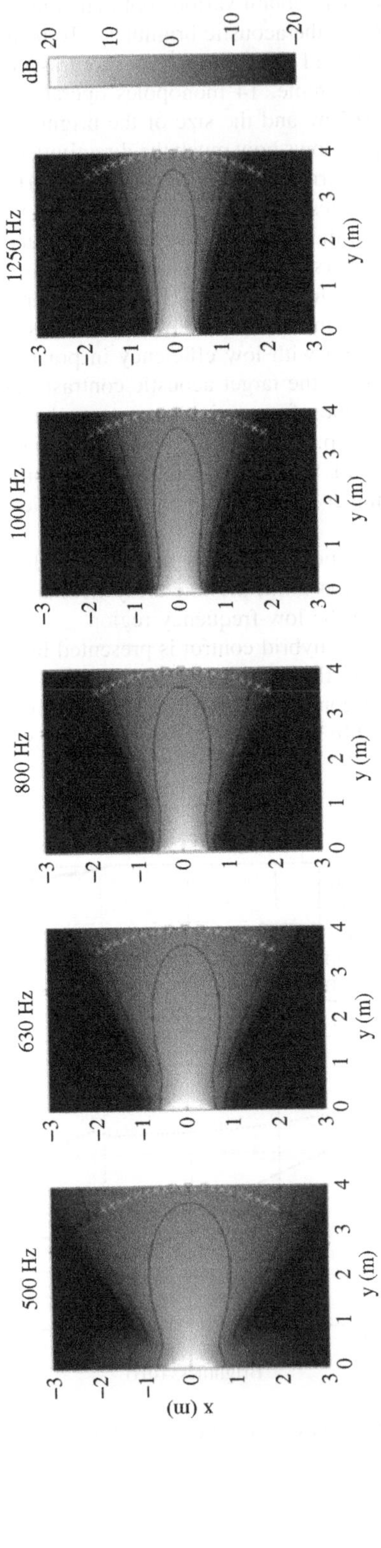

Figure 5.41 Pressure fields optimized for $10\log_{10}\alpha = -15\,\text{dB}$

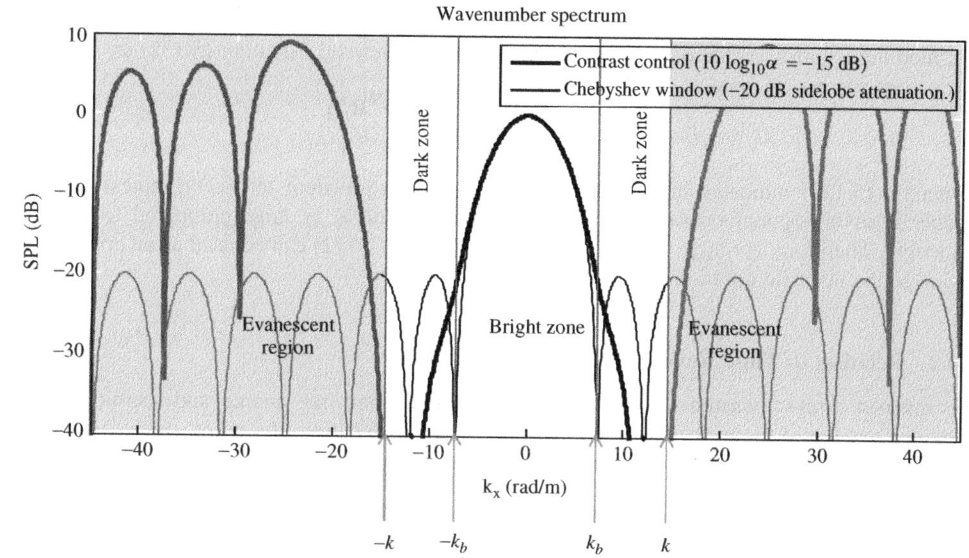

Figure 5.42 Comparison of wavenumber spectrums (solid line) of optimal excitation derived from hybrid control ($10\log_{10}\alpha = -15$ dB) (dashed line) Chebyshev window with −20 dB sidelobe attenuation. Excitation frequency: 800 Hz; number of loudspeakers: 14

One prominent difference between these two wavenumber spectrums can be found in the evanescent region and the dark zone. The spectral energy of the optimized excitation is very low within the dark zone, but this is obtained at the cost of a highly amplified evanescent region. By contrast, the Chebyshev window has constant sidelobes, even in the evanescent region. Therefore, it can be concluded that the hybrid control reduces the dark zone energy by increasing the energy in the (invisible) evanescent region. Since most of the input energy is projected to the invisible region, the acoustic brightness of the optimized beam pattern is usually lower than that of the window-based design.

The optimization procedure described here is for a pure-tone excitation. It is also somewhat simplified in order to express the acoustic contrast and brightness in the wavenumber domain. A complete procedure for implementing an FIR filter for the wideband problem is discussed in [7], and its application to various loudspeaker arrays can be found in [8, 9]. A more detailed explanation of the efficiency (acoustic brightness)-directivity (acoustic contrast) relation can also be found in [6, 10].

5.6.4 *Further Analysis of Acoustic Brightness and Contrast Control*

This section addresses how brightness and contrast control are related to well-known point focusing methods. We next explain, in detail, the relations of such control to the time-reversal arrays (TRAs; for example, [11]) and the inverse filter techniques.

5.6.4.1 Spatially Orthogonal Condition

When the sound fields of control sources are *orthogonal* to each other in V_t, the contrast problem can be treated as a brightness problem. In this case, the spatial correlations of the sound fields are zero. This means that the correlation matrix $\mathbf{R}_t$ becomes a diagonal matrix. Moreover, the diagonal elements of the matrix have a real value, since $\mathbf{R}_t$ is a positive semi-definite matrix. We can then use a modified control vector $\mathbf{q}' = \mathbf{R}_t^{\frac{1}{2}}\mathbf{q}$, which leads us to rewrite Equation (5.152) as

$$\beta' = \frac{\mathbf{q}'^H \mathbf{R}_t^{-\frac{1}{2}} \mathbf{R}_b \mathbf{R}_t^{-\frac{1}{2}} \mathbf{q}'}{\mathbf{q}'^H \mathbf{q}'} = \frac{\mathbf{q}'^H \mathbf{R}_b' \mathbf{q}'}{\mathbf{q}'^H \mathbf{q}'}. \quad (5.172)$$

Equation (5.172) indicates that the contrast problem is equivalent to the brightness problem. In general, an orthogonal condition occurs when the size of V_t is large compared to the local wavelength. Therefore, for large V_t, the solution of Equation (5.172) can be used as an approximate solution to the acoustic contrast problem.

5.6.4.2 Relation to Time Reversal

Time reversal arrays measure the signal emitted from a point-like source and reemit a phase-conjugated version of the measured signal. Let us denote the location of a point source (which also represents the focusing point) as $\mathbf{r}_f$. The sound measured by the arrays is then equal to

$$[\mathbf{h}]^{(\kappa)} = h(\mathbf{r}_f | \mathbf{r}_s^{(\kappa)}), \quad (5.173)$$

which is the Green's function between the point source and the transducers. The phase-conjugated solution emitted by the array is given by

$$\mathbf{q}_{TRA} = \mathbf{h}^H. \quad (5.174)$$

It can be shown that the brightness control gives the same solution that can be given by TRAs, particularly when the bright zone is defined as a point. When the control zone is given as a point located at $\mathbf{r}_f$, the correlation matrix becomes

$$\mathbf{R}_b = \mathbf{h}^H \mathbf{h}. \quad (5.175)$$

It is noteworthy that the following inequality holds for arbitrary matrices **A** and **B**.

$$rank\,(\mathbf{AB}) \leq \min[rank\,(\mathbf{A}), rank\,(\mathbf{B})]. \quad (5.176)$$

As regards Equations (5.175) and (5.176), we can conclude that the rank of $\mathbf{R}_b$ cannot exceed the rank of $\mathbf{h}$. Because the rank of a 1D vector is one, the correlation matrix $\mathbf{R}_b$ has a single rank and a unique eigenvector. Then, by inserting the TRAs solution of Equation (5.174) to Equation (5.150), it can be shown that Equation (5.174) is the unique eigenvector of $\mathbf{R}_0$.

$$\begin{aligned} \mathbf{R}_b \mathbf{q}_{TRA} &= (\mathbf{h}^H \mathbf{h}) \mathbf{h}^H \\ &= \mathbf{h}^H (\mathbf{h}\mathbf{h}^H) \\ &= \mathbf{h}^H (\alpha_{\max} h_c^2) \\ &= \alpha_{\max} h_c^2 \mathbf{q}_{TRA}. \end{aligned} \quad (5.177)$$

Consequently, *the solution of the TRAs is the unique solution that maximizes the acoustic brightness at a given point.*

5.6.4.3 Relation to Inverse Filtering Technique

An inverse filtering scheme has been employed to obtain sharper focusing than TRAs, particularly when the entire sound field V_t can be measured. The principle of inverse filtering is to minimize the error between a spatial delta function and the total sound field generated by multiple arrays. To explain the principle of inverse filtering, let us introduce a spatial delta-function vector $\boldsymbol{\delta}_f$, which is unity only at the focusing point $\mathbf{r}_f$.

$$\left[\boldsymbol{\delta}_f\right]^{(m)} = \delta(\mathbf{r}^{(m)} - \mathbf{r}_f). \tag{5.178}$$

The inverse filtering seeks a solution that minimizes the mean-squared error between the pressure field and the delta function. The error can be described as follows.

$$\varepsilon = \|\boldsymbol{\delta}_f - \mathbf{Hq}\|^2, \tag{5.179}$$

where $\|\|$ denotes the Euclidean norm or two-norm of a vector. We can then obtain a solution that minimizes Equation (5.179). That is,

$$\begin{aligned}\mathbf{q}_{inv} &= \mathbf{H}^{+}\boldsymbol{\delta}_f \\ &= \left[\mathbf{H}^H\mathbf{H}\right]^{-1}\mathbf{H}^H\boldsymbol{\delta}_f,\end{aligned} \tag{5.180}$$

where superscript + represents the pseudo-inverse. However, it should be noted that $\mathbf{H}^H\boldsymbol{\delta}_f = \mathbf{h}^H$, based on the definition of the delta function and Equation (5.173). As a result, the inverse solution can be simplified as follows:

$$\mathbf{q}_{inv} = \left[\mathbf{H}^H\mathbf{H}\right]^{-1}\mathbf{h}^H. \tag{5.181}$$

Then, the maximum contrast solution (Equation (5.159)) for this point focusing problem can be rewritten as

$$\mathbf{R}_t^{-1}\mathbf{R}_b\mathbf{q}_\beta = M(\mathbf{H}^H\mathbf{H})^{-1}\mathbf{h}\mathbf{h}^H\mathbf{q}_\beta = \beta'_{\max}\mathbf{q}_\beta, \tag{5.182}$$

where M is the total number of measurement positions. It should be noted that the matrix $\mathbf{R}_t^{-1}\mathbf{R}_b$ also has a single rank, because of the single-column matrix $\mathbf{h}$. Substituting the inverse solution of Equation (5.181) into Equation (5.182) gives

$$\begin{aligned}\mathbf{R}_t^{-1}\mathbf{R}_b\mathbf{q}_{inv} &= M\left[(\mathbf{H}^H\mathbf{H})^{-1}\mathbf{h}\mathbf{h}^H\right]\left[(\mathbf{H}^H\mathbf{H})^{-1}\mathbf{h}\right] \\ &= M(\mathbf{H}^H\mathbf{H})^{-1}\mathbf{h}\left[\mathbf{h}^H(\mathbf{H}^H\mathbf{H})^{-1}\mathbf{h}\right] \\ &= (\mathbf{H}^H\mathbf{H})^{-1}\mathbf{h}\beta_{inv} \\ &= \mathbf{q}_{inv}\beta_{inv}.\end{aligned} \tag{5.183}$$

The result of Equation (5.183) implies that the inverse solution $\mathbf{q}_{inv}$ is the eigenvector of $\mathbf{R}_t^{-1}\mathbf{R}_b$. Because $\mathbf{R}_t^{-1}\mathbf{R}_b$ has only a single rank (hence, a single eigenvector), the inverse solution is the only eigenvector of $\mathbf{R}_t^{-1}\mathbf{R}_b$. This states that *the maximum contrast solution converges to the inverse solution in the case of point focusing.*

The eigenvalue (acoustic contrast), $\beta_{inv} = M\mathbf{h}^H(\mathbf{H}^H\mathbf{H})^{-1}\mathbf{h} = M\mathbf{q}_{inv}^H\mathbf{h}$, is the product of the point-transfer function and the inverse solution. Therefore, the contrast of the point focusing problem is determined by the degree to which the inverse solution is similar to the point transfer function $\mathbf{h}$.

These relations indicate that the formulation of the brightness and contrast problems and their solutions are essentially a unified approach that can include both TRAs and inverse filtering techniques. Those conventional methods are among the special solutions to the brightness and contrast problem.

5.6.5 Application Examples

5.6.5.1 Personal Audio System for 17-in. Display

In this example [8, 12], acoustic contrast control was applied to develop a sound-focused personal audio system for a 17-in. monitor display. A line array system for a 17-in. monitor display (Figure 5.43a) was considered for application to a laptop computer, desktop computer, or TV. Loudspeakers with a diameter of 0.03 m were used as control sources. The experiment was performed in an anechoic chamber (3.6^W m × 3.6^L m × 2.4^H m, cut-off frequency: 100 Hz) to satisfy the free-field assumption.

Considering the scale of a 17-in. monitor display (which has a width of 0.32 m and a height of 0.27 m), the aperture of the control source array was configured to be 0.32 m. In total, $N = 9$ control sources were equally spaced on a line at intervals of 0.04 m.

The bright zone having a size of 0.4 m × 0.2 m and positioned 0.2 m away from the array (Figure 5.43b) was controlled so as to include the user's head, which is fairly large to allow for head movement. It is noteworthy that the distance between the loudspeaker and the zone of interest was not large in the low-frequency region, and the farfield approximation was not applicable. The width of the total control zone was configured to be 0.8 m. The width of each dark zone (on the left and the right sides of the bright zone) was then 0.2 m, which included the heads of two people who could be right beside the user.

The experimental setup is illustrated in Figure 5.44a. Twenty array microphones were spaced at 0.04-m intervals to prevent scattering effect among the microphones. A total of 60 steps (2 steps

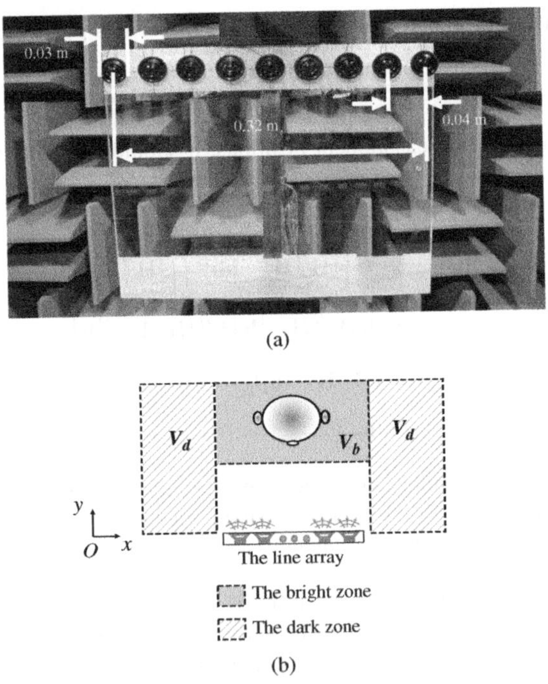

Figure 5.43 (a) Loudspeaker array configuration and (b) configuration of the bright and the dark zones: the bright zone is the region that includes the user's head in front of the array, and the dark zone is the region near the bright zone

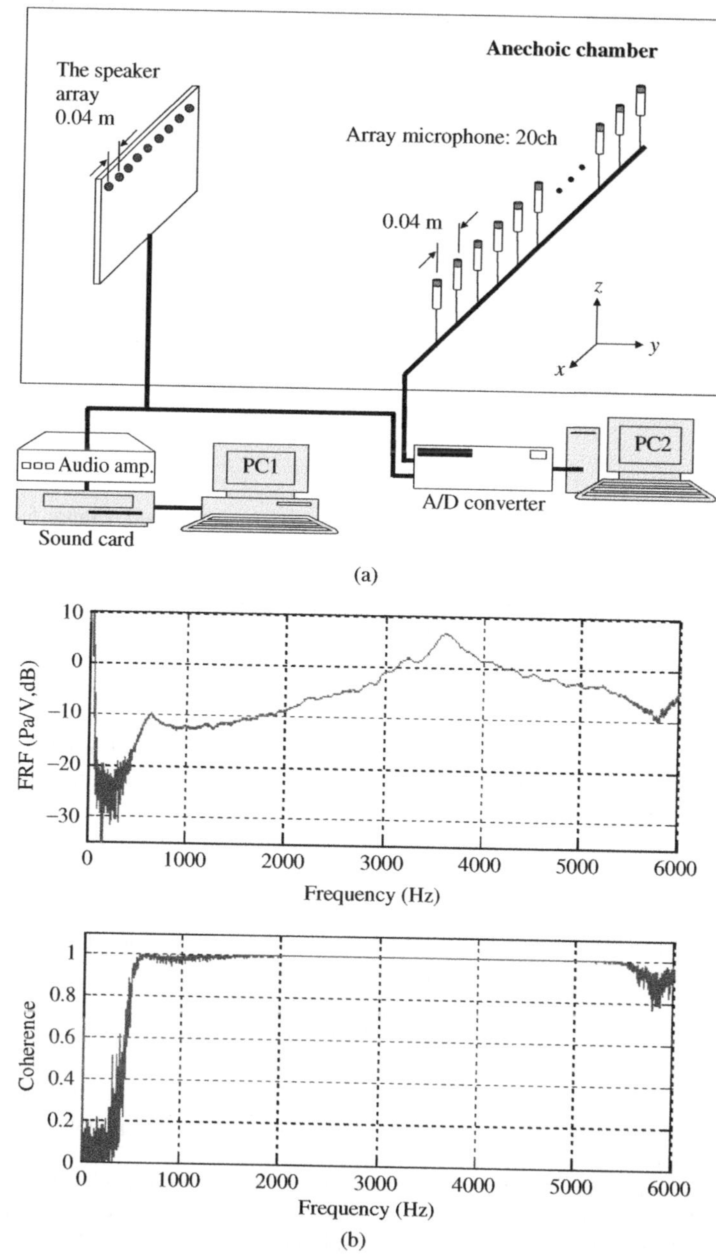

Figure 5.44 (a) Experimental set-up with array microphones: 20 channels, spacing: 0.04 m, generating part: PC, sound card, audio amp, loudspeaker array, measuring part: array microphones, A/D converter, PC. (b) Example of the measured frequency response and coherence function (800 Hz ~ 5 kHz), and (c) Magnitudes of transfer function between input voltage (V) and sound pressure (Pa) in the sound field of interest by the first loudspeaker, which is located at (−0.16, 0 m)

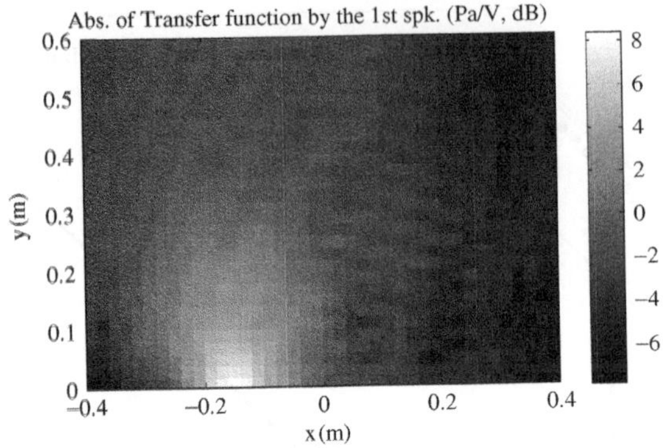

(c)

Figure 5.44 *(continued)*

in the x-axis and 30 steps in the y-axis) of measurement were performed. Figure 5.44b shows the frequency response function and the coherence between the input voltage signal into a loudspeaker unit and the output pressure (Pa) signal that is measured at a point 1 m ahead of the speaker. The available frequency range was 800 Hz to 5 kHz, which is regarded as a frequency range in which the coherence function is close to unity. The frequency range was discretized by $\Delta f = 10\,\text{Hz}$ so as to limit the duration of the impulse response to 0.1 s. This duration is sufficient for obtaining a free-field response without reflections. Figure 5.44c represents the measured magnitude response of the first loudspeaker. Using the measured transfer functions, we can construct the spatial correlation matrices $\mathbf{R}_b(\omega)$ and $\mathbf{R}_d(\omega)$, and obtain the solution of the contrast control problem.

The contrast solution is compared with two other cases: a uniform excitation case and a time-reversal case. The solution in the uniform excitation case $\mathbf{q}_u(\omega)$ is given as

$$\mathbf{q}_u(\omega) = \frac{\|\mathbf{q}_\beta(\omega)\|}{\sqrt{K}}[1 \ 1 \ \cdots \ 1]^T, \tag{5.184}$$

which is normalized to have the same power as the contrast solution (denoted by $\mathbf{q}_\beta(\omega)$). On the other hand, the solution in the TRA case $\mathbf{q}_{TRA}(\omega)$ can be written as

$$\mathbf{q}_{TRA}(\omega) = \frac{\|\mathbf{q}_c(\omega)\|}{\|\mathbf{h}(\omega)\|}\mathbf{h}(\omega)^H, \tag{5.185}$$

which is normalized so that its power is equal to that of $\mathbf{q}_c(\omega)$. The matrix $\mathbf{h}$ denotes the transfer functions from the loudspeakers to the focusing position that was configured at the center of the user's position (0.01 m, 0.4 m).

The results in Figure 5.45 show the controlled sound fields at the three selected frequencies: 800, 3150, and 5000 Hz. As can be seen in the figure, the top three sound fields in Figure 5.45a–c are shown in terms of uniform excitation, time reversal, and contrast control, respectively. The SPL is expressed by variations in color. The center rectangle indicates the bright zone, and the longer rectangles flanking it indicate the dark zone. The magnitude of the sound pressure is expressed on a dB scale when the reference value is the pressure at the center.

At all frequencies, the three sound fields at the top of Figure 5.45a–c show that the SPLs of the dark zones in the case of contrast control are lower than those in the cases of uniform excitation and a TRA. In the contrast control case, except at 800 Hz, most of the SPLs in the dark zone are lower than −20 dB.

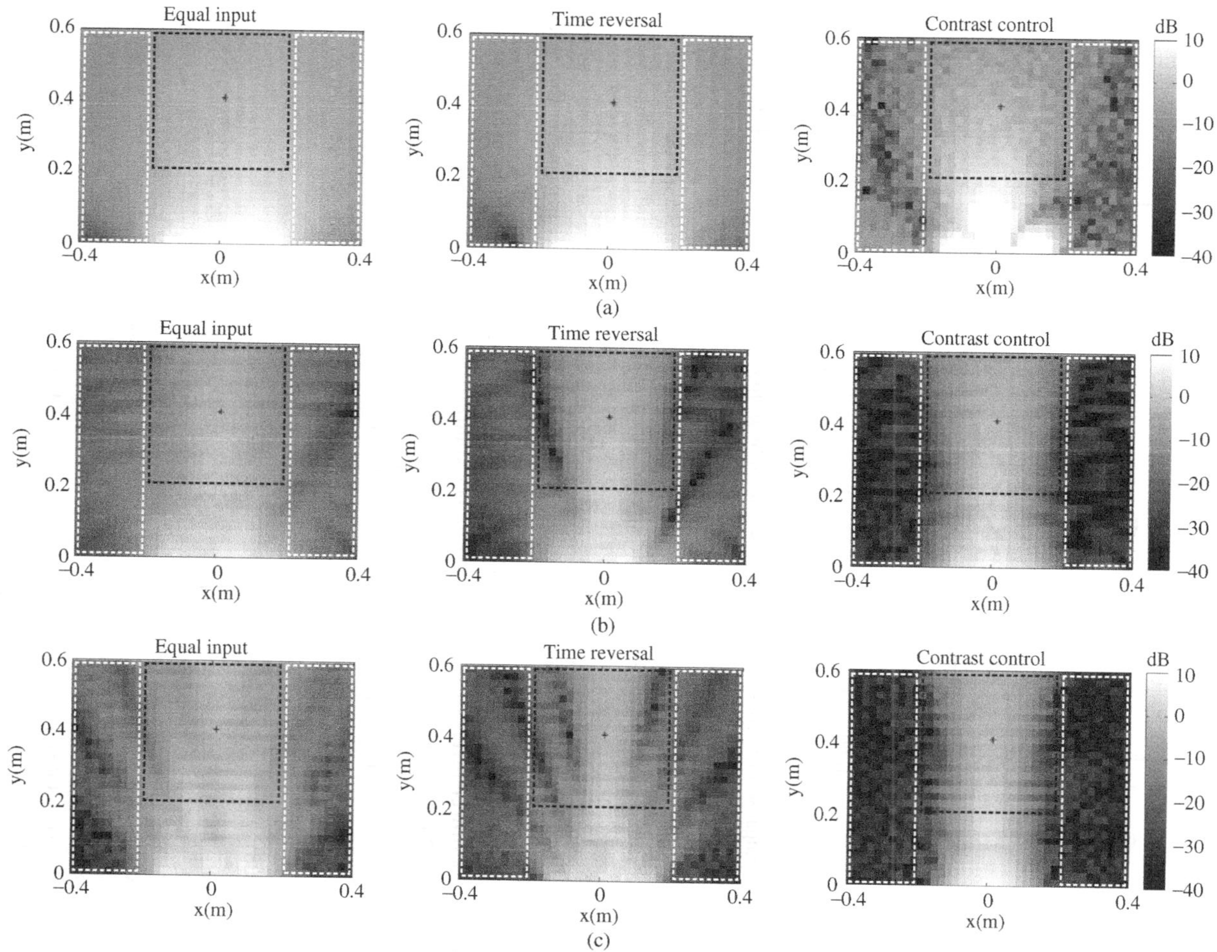

Figure 5.45 The experimental results at the selected frequencies: (a) 800 Hz, (b) 3150 Hz, and (c) 5000 Hz. The top three figures show sound fields with uniform excitation, time reversal, and contrast control, respectively. The center rectangle indicates the bright zone, and the two rectangles flanking it indicate the dark zone. The bottom figure shows the angular characteristics in the three cases

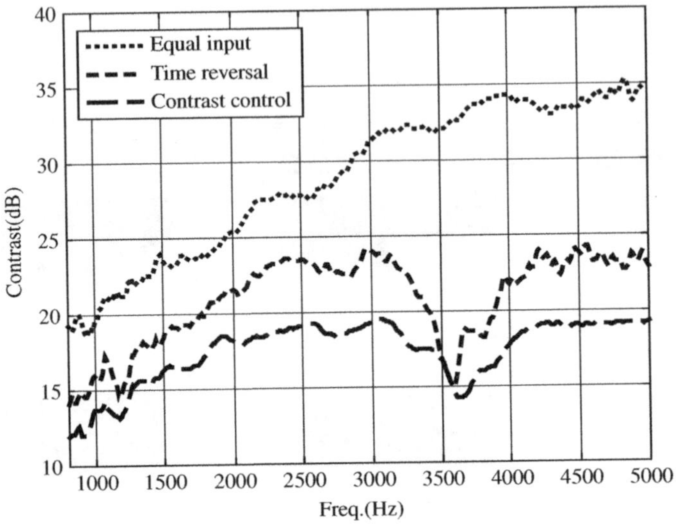

Figure 5.46 Comparison of acoustic contrast in three cases: uniform excitation, acoustic contrast control, and time-reversal array

Figure 5.46 shows the acoustic contrast in the three cases in the entire frequency region of interest, which is 800–5000 Hz. In the contrast control case, the contrasts are more than 19 dB and are higher than this in the other cases. Particularly as the frequency increases, the contrast in the contrast control case increases gradually, while the contrasts in the other cases do not; as the frequency becomes higher, the difference in the contrasts in the contrast control and in the other cases has a tendency to increase.

In summary, the experimental results show that at all frequencies, the cases of acoustic contrast control provide lower levels of sidelobes and higher contrasts than the cases of equal input and a TRA. In the contrast control case, the level of most of the sidelobes is lower than −20 dB at the selected frequencies, and the contrasts are more than 19 dB at all frequencies. In addition, as the frequency becomes higher, the difference in the contrasts in the contrast control case and in the other cases has a tendency to increase.

References

1. Olson, H.F. (1973) Gradient loudspeakers, *Journal of the Audio Engineering Society*, **21**(2), 86–93.
2. Choi, J.W. and Kim, Y.H. (2002) Generation of an acoustically bright zone within an illuminated region using multiple sources. *Journal of the Acoustical Society of America*, **111**(4), 1695–1700.
3. Choi, J.W., Kim, Y.T., Ko, S.C., and Kim, J.H. (2010) A differential approach for the implementation of superdirective loudspeaker array. Proceedings of the 128th Convention of Audio Engineering Society, Preprint 8032, London, UK, May 2010.
4. Horn, R.A. and Johnson, C.A. (1985) *Matrix Analysis*, Cambridge University Press, Cambridge, pp. 176–180.
5. Garbow, B.S. (1978) Algorithm 535: the QZ algorithm to solve the generalized eigenvalue problem for complex matrices, *Journal of the ACM Transactions on Mathematical Software*, **4**(4), 404–410.
6. Choi, J.W., Ko, S.C., Kim, J.H., and Kim, Y.T. (2008) Super-directive loudspeaker array for the generation of a personal sound zone. Proceedings of the 125th Convention of Audio Engineering Society, Preprint 7620, San Francisco, California, October 2008.

7. Choi, J.W. and Kim, Y.H. (2004) Active control for enhancement of sound field. Proceedings of Active 04, paper 042 (on CD-ROM), Williamsburg, Virginia, September 2004.
8. Chang, J.H., Lee, C.H., Park, J.Y. and Kim, Y.H. (2009) A realization of sound focused personal audio system using acoustic contrast control, *Journal of the Acoustical Society of America*, **125**(4), 2091–2097.
9. Park, J.Y., Chang, J.H. and Kim, Y.H. (2010) Generation of independent bright zones for two-channel private audio system, *Journal of the Audio Engineering Society*, **58**(5), 382–393.
10. Elliott, S.J., Cheer, J., Choi, J.W. and Kim, Y.T. (2012) Robustness and regularization of personal audio systems, *IEEE Transactions on Audio, Speech and Language Processing*, **20**(7), 2123–2133.
11. Tanter, M., Thomas, J.L. and Fink, M. (2000) Time reversal and the inverse filter, *Journal of the Acoustical Society of America*, **108**(1), 223–234.
12. Chang, J.H., Park, J.Y. and Kim, Y.H. (2009) Scattering effect on the sound focused personal audio system. *Journal of the Acoustical Society of America*, **125**(5), 3060–3066.

6

Sound Field Reproduction

6.1 Introduction

Dutch physicist Christiaan Huygens showed that the wavefront of a propagating wave at any instant can be obtained by a superposition of the waves emanating from every point on the wavefront at the prior instant of time. Mathematically, this is what the Kirchhoff–Helmholtz (KH) integral equation also explains. It says that a sound field in a selected region is determined by the pressure and velocity distribution of the closed contour of the selected region. This concept forms the basis of sound field reproduction, which attempts to reproduce a wavefront over a selected region using a loudspeaker array arranged in space. Because our goal is to find an array excitation signal that can replicate a target or desired sound field, sound field reproduction can be related to the sound visualization problem, which also deals with methods of finding the unknown source distribution that reproduces a measured sound field.

This chapter explains how the sound field reproduction problem is connected to the sound visualization problem, and how the problem of sound field reproduction can be solved by techniques used for sound visualization. Because the role of the basis function explained for sound visualization also applies to sound field reproduction, it can be said that the selection of the basis function determines the quality of sound field reproduction. As noted regarding the sound visualization problem, "the generation of the sound field depends entirely on how well the basis function mimics what happens in reality."

The choice of the basis function depends on the relationship between the listener and the loudspeakers. There are three popular approaches to sound field reproduction: field-matching, mode-matching, and integral approaches. Each uses different principles to analyze and synthesize a sound field. The method of least squares, higher-order Ambisonics, and wave field synthesis (WFS) are representative techniques using these three approaches, respectively. These techniques employ different mappings, and their reproduced fields can differ greatly with regard to the shape of the reproduction error. This chapter addresses how the basis functions of each sound field reproduction technique differ and how different basis functions affect the shape of the reproduced sound field. The early part of this chapter explains the basic principles behind the three approaches in a simple one-dimensional (1D) case. Then the general idea is extended to describe the three-dimensional (3D) sound field reproduction problem.

Regardless of the type of basis function used, sound field reproduction is an inverse problem and is sometimes ill-posed or ill-conditioned. The problem can be converted to a well-posed problem by adding extra constraints or by introducing an assumption when choosing basis functions for the selected or desired region of interest. Several regularization techniques for resolving ill-posed

Sound Visualization and Manipulation, First Edition. Yang-Hann Kim and Jung-Woo Choi.

and ill-conditioned cases and methods of converting the inverse problem into a direct problem are addressed.

6.2 Problem Statement

6.2.1 *Concept of Sound Field Reproduction*

Sound field reproduction is based on the constructive and destructive interference between multiple sound fields produced by multiple loudspeakers. To understand how the interference can be controlled to realize the desired sound field, consider two plane waves propagating in different directions whose wavenumbers are given by $(\mathbf{k}_1, \mathbf{k}_2)$:

$$p_1(\mathbf{r}, \omega) = A_0 e^{i\mathbf{k}_1 \cdot \mathbf{r}}, \tag{6.1}$$

$$p_2(\mathbf{r}, \omega) = A_0 e^{i\mathbf{k}_2 \cdot \mathbf{r}}. \tag{6.2}$$

Then the resulting sound field can be written as follows (Figure 6.1).

$$\begin{aligned} p(\mathbf{r}, \omega) &= p_1(\mathbf{r}, \omega) + p_2(\mathbf{r}, \omega) \\ &= 2A_0 e^{i\mathbf{k}_0 \cdot \mathbf{r}} \cos(\Delta\mathbf{k} \cdot \mathbf{r}), \end{aligned} \tag{6.3}$$

where $\mathbf{k}_0$ represents the mean wavenumber $(= (\mathbf{k}_1 + \mathbf{k}_2)/2)$, and $\Delta\mathbf{k}$ denotes half of the wavenumber difference $(= (\mathbf{k}_1 - \mathbf{k}_2)/2)$ between two plane waves. Within a finite zone smaller than the wavenumber difference $(\Delta\mathbf{k} \cdot \mathbf{r} \ll 1)$, the sound field described by Equation (6.3) behaves like a plane wave propagating in a new direction $\mathbf{k}_0$. This implies that the *local* sound field can be manipulated using the interference of different waves.

To manipulate the interference of different sound fields, we can control their magnitudes and phases. In this example, we manipulate the local wavenumber to $\mathbf{k}_0$, using two plane wave sources propagating in two different directions. Note, however, that the spatial variation along the direction of the wavenumber difference $(\Delta\mathbf{k})$ inevitably restricts the size of the controllable area. In sound field reproduction, the location or size of an area over which the sound field can be manipulated is limited.

6.2.2 *Objective of Sound Field Reproduction*

The objective of sound field reproduction is to replicate a desired sound field using an array of loudspeakers. Therefore, two different types of sound fields are relevant to the reproduction problem: the target field and reproduced field. The target field refers to the idealized sound field we aim to reproduce, and the reproduced field denotes the resultant field produced by loudspeakers.

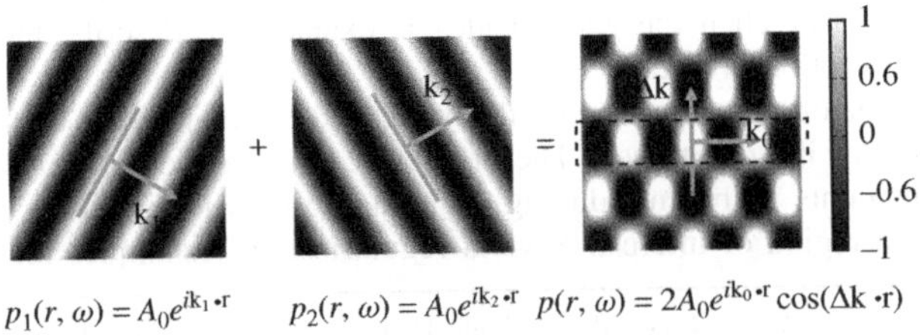

Figure 6.1 Manipulation of a local sound field by two plane waves

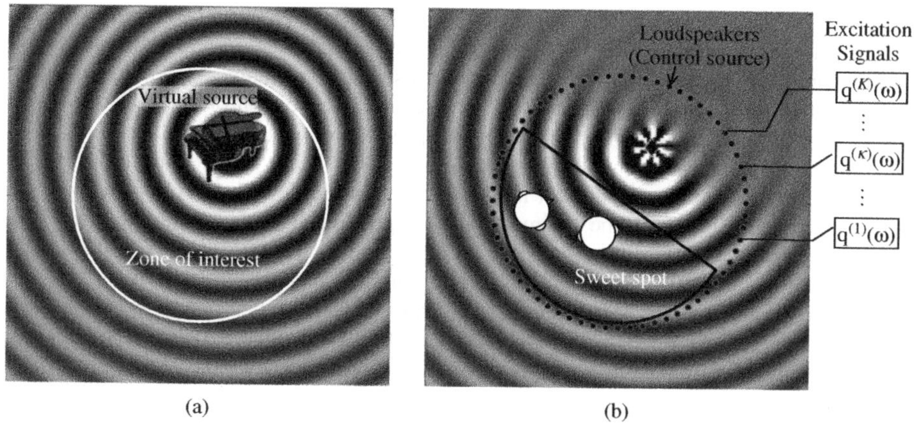

Figure 6.2 Concept of sound field reproduction (a) target sound field from a virtual source (b) reproduction with loudspeakers

The sound fields can be defined in terms of many acoustic variables depending on the purpose of reproduction; however, the sound pressure field is popular for audio applications.

To define the desired target field, the concept of a *virtual source* is often introduced. The virtual source represents an imaginary source that generates the desired target sound field (Figure 6.2a). Of course, it is not enough to describe the target field using only a virtual source. A boundary condition that produces the sound field reflected from the room boundary must also be defined. It can be said that a virtual source is a useful metaphor for describing the *direct wave* of a target field, which expresses the position and radiation pattern. When the waves reflected from the wall are also important, the reflections as well as the direct wave should be included in the definition of the target field.

Because it is necessary to match the target and reproduced pressure fields, the loudspeaker signals are controlled such that the reproduced field can mimic the target field in both space and time. The key problem to be solved here is to find the *optimal loudspeaker signals* that can reproduce the target field with minimal differences. The loudspeaker signals, called *excitation signals or functions*, are the primary unknowns of the given problem, and more complex problems can be defined if we consider the effective locations of loudspeakers or energy efficiency of an array. The loudspeakers are referred to as *control sources* because they are acoustic sources controlling a sound field.

However, as will be shown throughout this chapter, reproduction over the entire space is often impossible. The target area of a sound field to reproduce is usually limited, and we denote the limited target area as a *zone of interest* (Figure 6.2a). We can then construct an equation to derive the loudspeaker excitation signals that generate the reproduced sound field over the zone of interest. Ideally, the target field can be reproduced over the entire zone of interest, but in most cases, the actual spatial region in which the target and reproduced fields are similar is smaller than the zone of interest. The actual region with good reproduction performance is often called the *sweet spot* (Figure 6.2b).

In determining the excitation signals that produce a desired sound field, we need to use the relation between the excitation signals and the reproduced sound field. This relation is called the transfer function between the input (excitation) and output (reproduced field) of a reproduction system. By denoting the desired target field as p_d, we can construct an equation that connects the excitation signal $q^{(\kappa)}$ of the κth loudspeaker and the target sound field:

$$p_d(\mathbf{r}, \omega) = \sum_{\kappa=1}^{K} h(\mathbf{r}|\mathbf{r}_s^{(\kappa)}; \omega) q^{(\kappa)}(\omega), \tag{6.4}$$

where h denotes the transfer function.

Finding the optimal q from the given p_d is a typical inverse problem, for which we need to solve for q that gives the desired output p_d. This inverse problem may appear similar to the sound visualization or noise source identification problem discussed in Chapter 3. For the former, the predicted distribution of q, which is related to the strength and location of the noise source, is the major concern. Therefore, sound visualization and reproduction have the same type of unknowns. However, the final goal of sound field reproduction does not involve the distribution of q itself. Instead, it is to obtain the best output, that is, the sound pressure field that can most closely mimic the target sound field. Therefore, the optimal q may depend on how we define the error between the target and reproduced fields.

In the following sections, we explain three representative techniques for solving this inverse problem: the field-matching, mode-matching, and integral approaches. Each technique exploits different principles and measures to compare the target and reproduced fields. The fundamental theories from which these techniques are derived are explained for a simple 1D case.

6.3 Reproduction of One-Dimensional Sound Field

6.3.1 Field-Matching Approach

Once the target sound field is defined, the simplest reproduction method we can try is to directly match the distribution of the target and reproduced fields in space. In a field-matching approach, the measure of *similarity* is defined in terms of the error between the two sound fields.

6.3.1.1 Exterior Virtual Source

When a virtual point source radiating a pure tone of frequency ω in a free-field condition is positioned at $x = x_v$, the target sound field can be described in terms of left- and right-going waves (Figure 6.3). That is,

$$p_d(x, \omega) = \begin{cases} a_v(\omega)\, e^{ik(x-x_v)}, & x \geq x_v \\ a_v(\omega) e^{-ik(x-x_v)}, & x < x_v \end{cases} \qquad (k = \omega/c), \tag{6.5}$$

where a_v is the complex amplitude of a 1D wave at the location of a virtual source. Suppose that the sound field p_d can be measured or modeled in the listener position $x = x_a$. Then the target field is reproduced by controlling a control source, that is, a loudspeaker. In this 1D example, a loudspeaker can be located either between the listener and the virtual source or in the external region behind the virtual source.

The first case we consider is a loudspeaker between the listener and the virtual source ($x_v < x_s < x_a$), where x_s is the location of the loudspeaker. In other words, the virtual source is behind the loudspeaker; this type of problem is denoted as an *exterior virtual source* problem. If we consider a point source driven by an excitation q as the loudspeaker, its pressure field can be expressed as

$$\begin{aligned} p_r(x, \omega) &= h(x|x_s; \omega) q(x_s, \omega) \\ &= a_s(\omega) e^{ik|x-x_s|} q(x_s, \omega), \end{aligned} \tag{6.6}$$

where p_r denotes the pressure field reproduced by the loudspeaker positioned at x_s, and h expresses the transfer function between the excitation of a control source and the pressure field at a position x. The variable a_s represents the complex amplitude of the pressure field produced by a unit excitation function. For example, for a 1D Green's function, $a_s(\omega) = 1/(-2ik)$.

Now the sound field reproduction problem can be defined as finding the source excitation q that gives the sound field p_r most similar to the target field p_d. If we introduce the squared error

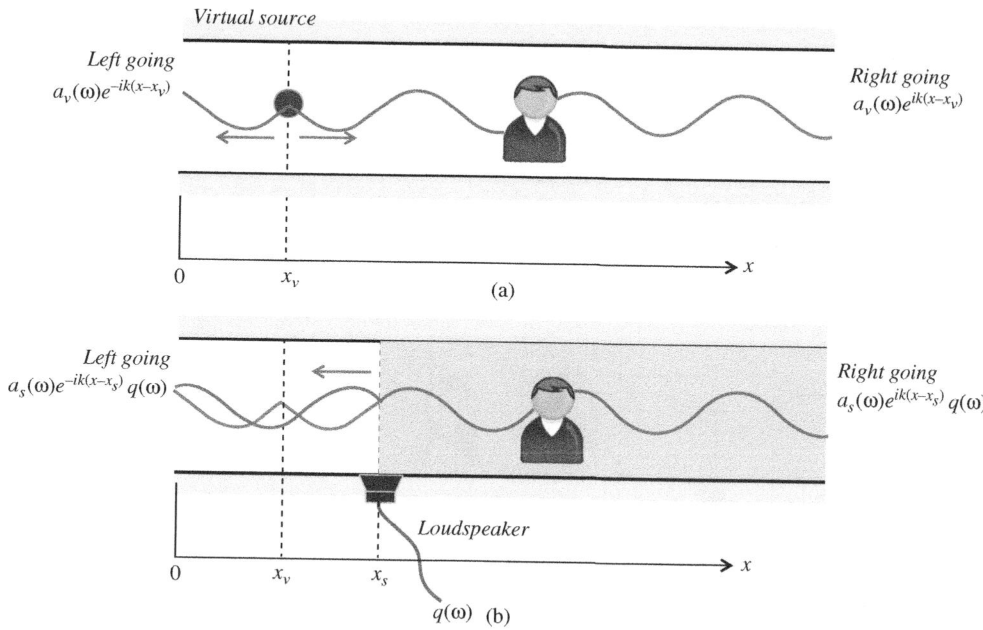

Figure 6.3 Exterior virtual source problem in 1D case (a) virtual source's field and (b) control source's field

between the target and reproduced sound fields as the measure of similarity, the error can be written as follows at the listener position x_a:

$$\varepsilon(x_a, \omega)^2 = \left|p_d\left(x_a, \omega\right) - p_r(x_a, \omega)\right|^2. \tag{6.7}$$

The excitation q that minimizes the error can be determined from Equations (6.5) and (6.6) as

$$q(\omega) = \frac{p_d(x_a, \omega)}{h(x_a|x_s; \omega)} = \frac{a_v(\omega)}{a_s(\omega)}e^{ik|x_s - x_v|} \tag{6.8}$$

for $x_v < x_s < x_a$. The exponential term, which expresses the phase change between the virtual and control sources, can be likened to the *propagator* in acoustic holography. Therefore, the excitation function compensates for the phase difference due to sound propagation from the virtual source location to the control source position.

The sound field reproduced using the excitation function of Equation (6.8) can be written as

$$p_r(x, \omega) = \begin{cases} a_v(\omega)\, e^{ik(x-x_v)} & x \geq x_s \\ a_v(\omega)e^{ik(2x_s - x - x_v)} & x < x_s \end{cases}. \tag{6.9}$$

Consequently, the target sound field p_d of Equation (6.5) can be perfectly reproduced over an area $x \geq x_s$, and the area becomes the *sweet spot* of the given reproduction problem.

6.3.1.2 Interior Virtual Source

The next case we consider is a virtual source positioned between the loudspeaker and the listener ($x_s < x_v < x_a$) (Figure 6.4). From Equations (6.5) and (6.6), the excitation that reproduces p_t is

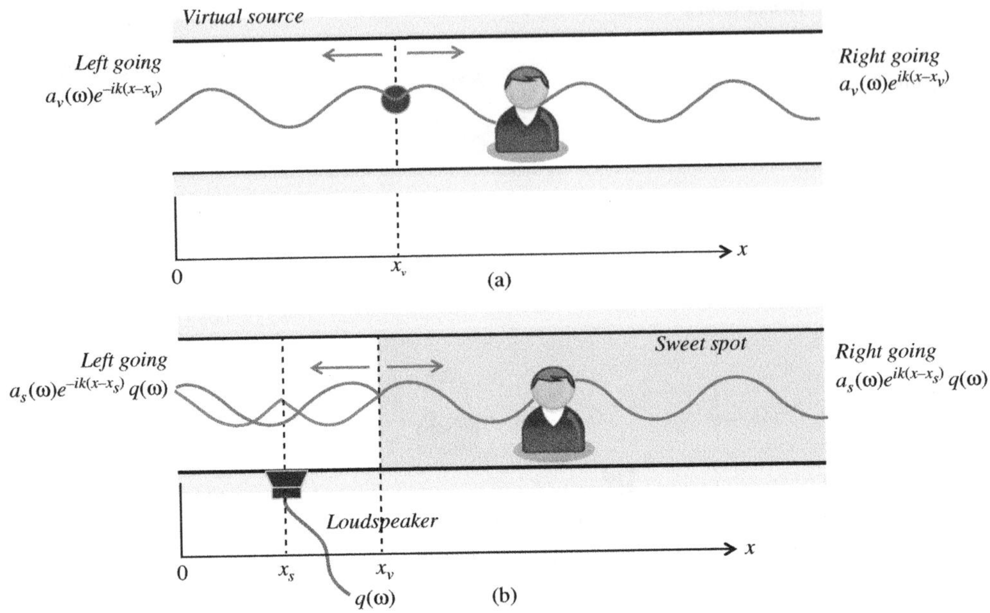

Figure 6.4 Interior virtual source problem in 1D case (a) virtual source's field and (b) control source's field

given by

$$q(\omega) = \frac{a_v(\omega)}{a_s(\omega)} \left(e^{ik|x_s - x_v|}\right)^*. \tag{6.10}$$

The reproduced sound field is given by

$$p_r(x,\omega) = \begin{cases} a_v(\omega)\, e^{ik(x-x_v)} & x \geq x_s \\ a_v(\omega) e^{ik(2x_s - x - x_v)} & x < x_s \end{cases}, \tag{6.11}$$

which looks the same as Equation (6.9). However, in this case, the virtual source is in front of the loudspeaker, so the reproduced sound field is only exact within the region $x \geq x_v$. This difference in the sweet spot tells us that *the sound field of an interior virtual source can be reproduced only beyond the location of the virtual source*.

It is noteworthy that the excitation Equation (6.10) has a complex-conjugate propagator, in contrast to that of an exterior virtual source (Equation (6.8)). Because the complex conjugation operator in the frequency domain is equivalent to time-reversal in the time domain, we can consider that the excitation q is obtained by reversing the propagator in time. Accordingly, inverse propagation is equivalent to the time-reversal operator in the 1D case.

6.3.2 Mode-Matching Approach

Field-matching solves the reproduction problem using the known pressure field at the listener position $p_d(x_a)$. On the other hand, the solution to the same problem can also be calculated if the harmonic coefficient of the left- or right-going wave is given. For instance, the excitation q that reproduces the interior or exterior virtual source (Equation (6.8) or (6.10)) can be determined from the ratio of the coefficients a_v and a_s. This is possible because we know that any 1D sound field should propagate in the form of harmonics $e^{i(\pm kx-\omega t)}$. The harmonic functions $e^{i(\pm kx-\omega t)}$ that constitute the target and control sources' sound fields are analogous to *modes*; hence, the matching of *modal coefficients* of the target and reproduced fields can reproduce the sound field.

To see how this concept operates in the 1D sound field, consider the reproduction problem for an exterior virtual source again. The 1D propagation model implies that both p_d and p_s can be expressed in terms of the right-going harmonic function e^{ikx}. That is,

$$p_d(x,\omega) = \left[a_v(\omega)\, e^{-ikx_v}\right] e^{ikx}, \quad for\, x > x_v, \tag{6.12}$$

$$p_s(x,\omega) = \left[a_s(\omega)\, e^{-ikx_s}\right] e^{ikx}, \quad for\, x > x_s. \tag{6.13}$$

The terms within brackets in Equations (6.12) and (6.13) can be regarded as modal coefficients. Accordingly, we can construct an equation to match the modal coefficient of the control source to that of the target sound field, that is,

$$\left[a_v(\omega)\, e^{-ikx_v}\right] = \left[a_s(\omega)\, e^{-ikx_s}\right] q(\omega). \tag{6.14}$$

Therefore, the solution q can also be derived from the modal coefficients without using the measured or modeled pressure fields. This type of equation involving the modal coefficients is called a *mode-matching equation*. In this 1D case, the solution of the mode-matching equation is equivalent to what we have obtained from the field-matching approach (Equation (6.8)). However, this equivalence is a peculiarity of the 1D problem. The solutions of the two different approaches are not necessarily equivalent for the two-dimensional (2D) or 3D problems, which will be explained in later sections.

The mode-matching procedure can generally be separated into two steps. The first step is to identify or extract the modal coefficient of a target sound field. For example, in the 1D problem described above, we extract the modal coefficient

$$D(\omega) = a_v(\omega)e^{-ikx_v} \tag{6.15}$$

of a right-going wave $p_d(x,\omega) = D(\omega)e^{ikx}$. This step is called *encoding* of the target sound field. We then extract the modal coefficient of the reproduced field,

$$R(\omega) = a_s(\omega)e^{-ikx_s}q(\omega), \tag{6.16}$$

and calculate the excitation function q that transforms the modal coefficient of the control source to that of the virtual source $D(\omega)$. This process is referred to as *decoding* of the sound field. The mode-matching approach hence includes the steps of encoding, decoding, and matching of the modal coefficient.

6.3.3 Integral Approach

The KH integral states that the sound field in a source-free volume can be reproduced by monopole and dipole sources distributed on the surface of the volume. Sound field reproduction in the 1D case can also be explained in terms of the KH integral.

6.3.3.1 Exterior Virtual Source

First, consider the zone of interest configured over a volume V_0 (Figure 6.5), which is bounded by two boundaries: x_{s_1} and x_{s_2}. According to the KH integral, the sound field from an exterior virtual source at $x_v < x_{s_1} < x_{s_2}$ can be reproduced by monopole and dipole sources positioned at x_{s_1}, x_{s_2}. The KH integral for the 1D case (Equation 2.101) is given by

$$p(x) = \left[p(x_s) \frac{dg(x|x_s)}{dx_s} - \frac{dp(x_s)}{dx_s} g(x|x_s) \right]_{x_s=x_{s2}}^{x_s=x_{s1}}. \tag{6.17}$$

Note that the frequency dependency ω is omitted for brevity.

To reproduce a target field p_d, the pressure field on the surface $(p(x_s))$ is replaced by the target pressure field on the boundary $(p_d(x_s))$. If we consider a virtual point source driven by excitation

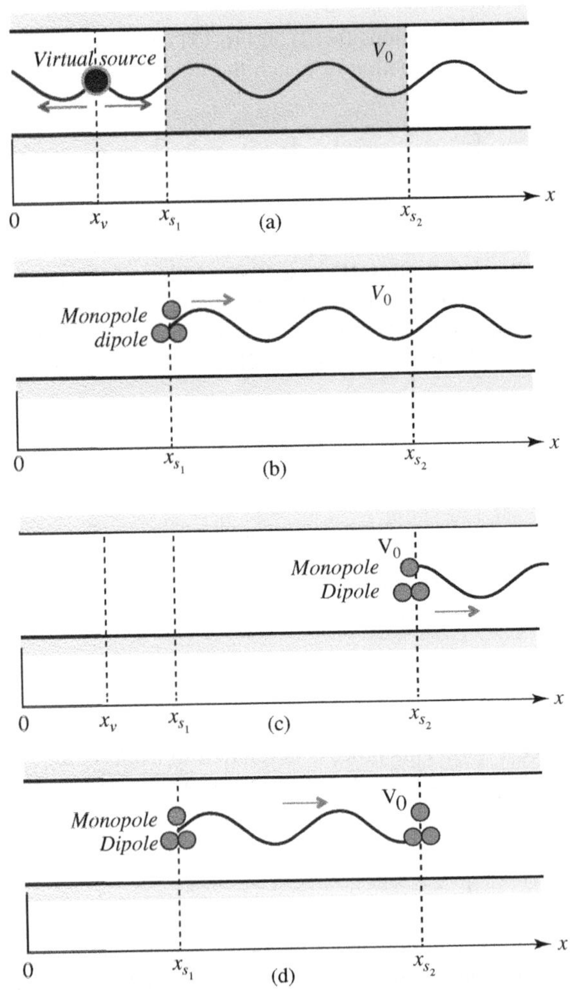

Figure 6.5 One dimensional reproduction of an exterior virtual source using the KH integral (a) target field (b) left integral (c) right integral (d) field reproduced by the complete integral

q_v, its sound field without reflections can be written in terms of the 1D free-field Green's function. That is,

$$p_d(x_s) = \begin{cases} q_v\left(\dfrac{i}{2k}e^{ik(x_s - x_v)}\right) & for\, x_s \geq x_v \\ q_v\left(\dfrac{i}{2k}e^{ik(x_v - x_s)}\right) & for\, x_s < x_v \end{cases}, \tag{6.18}$$

where the free-field Green's function and its derivative are given by

$$g_f(x|x_s) = \begin{cases} \dfrac{i}{2k}e^{ik(x-x_s)} & for\, x \geq x_s \\ \dfrac{i}{2k}e^{ik(x_s-x)} & for\, x < x_s \end{cases}, \quad \frac{dg_f(x|x_s)}{dx_s} = \begin{cases} \dfrac{1}{2}e^{ik(x-x_s)} & for\, x \geq x_s \\ -\dfrac{1}{2}e^{ik(x_s-x)} & for\, x < x_s \end{cases}. \tag{6.19}$$

Then the contributions of the monopoles and dipoles at the left boundary ($x_{s1} < x$) can be written as

$$\left[p_d(x_s)\frac{dg_f(x|x_s)}{dx_s} - \frac{dp_d(x_s)}{dx_s}g_f(x|x_s)\right]_{x_s = x_{s1}} = q_v\frac{ie^{ik(x-x_v)}}{2k} = p_d(x), \quad x > x_{s1}. \tag{6.20}$$

In contrast, for the listener on the left-hand side of x_{s1}, the integral becomes zero:

$$\left[p_d(x_s)\frac{dg_f(x|x_s)}{dx_s} - \frac{dp_d(x_s)}{dx_s}g_f(x|x_s)\right]_{x_s = x_{s1}} = 0, \quad x < x_{s1}. \tag{6.21}$$

Accordingly, the integral at the left boundary reproduces the right-going wave propagating to $x > x_{s1}$ (Figure 6.5b).

Likewise, the integral on the right boundary (x_{s2}) can be calculated as (Figure 6.5c):

$$-\left[p_d(x_s)\frac{dg_f(x|x_s)}{dx_s} - \frac{dp_d(x_s)}{dx_s}g_f(x|x_s)\right]_{x_s = x_{s2}} = \begin{cases} -p_d(x) & x > x_{s2} \\ 0 & x < x_{s2} \end{cases}. \tag{6.22}$$

By combining Equations (6.20)–(6.22), the reproduced sound field can be expressed as

$$\left[p_d(x_s)\frac{dg_f(x|x_s)}{dx_s} - \frac{dp_d(x_s)}{dx_s}g_f(x|x_s)\right]_{x_s = x_{s2}}^{x_s = x_{s1}} = \begin{cases} p_d(x) & x_{s1} < x < x_{s2} \\ 0 & x < x_{s1}, x > x_{s2} \end{cases}. \tag{6.23}$$

It can be seen that the monopole (g_f) driven by $-dp_d/dx_s$ and the dipole (dg_f/dx_s) driven by p_d can reproduce the target field from a virtual source at $x_v < x_{s1}$ over a region $x_{s1} < x < x_{s2}$ (Figure 6.5d).

The reproduced sound field is exact within $x_{s1} < x < x_{s2}$, but no sound field is produced outside. This is the *nonradiating property* of the KH integral, which arises from the combination of monopole and dipole sources. If only monopoles are used, the sound field should be symmetric with respect to the source position; hence, the nonradiating property cannot be achieved without altering the sound field within V_0.

Another aspect of the KH integral shown in Figure 6.5 is that the integral on x_{s2} (right integral) makes no contribution to V_0. If we are interested only in the region V_0, the integral on x_{s1} (left integral) can reproduce the target field exactly, without the right integral. This property is useful for reproducing a virtual source at $x_v < x_{s1}$ because we can reduce the required number of control sources by half. However, for a virtual source at $x_{s2} < x_v$, the right integral reproduces the target field, whereas the left integral makes no contribution to V_0. Therefore, depending on the possible location of a virtual source, the arrangement of control sources can differ.

6.3.3.2 Interior Virtual Source

The KH integral holds for a source-free volume V_0. If a virtual source is between the two control sources, the volume bounded by the two source locations cannot be source-free, so the KH integral cannot be directly applied.

In detail, the KH integral applied to the interior virtual source gives

$$-\left[p_d(x_s)\frac{dg_f(x|x_s)}{dx_s}-g_f(x|x_s)\frac{dp_d(x_s)}{dx_s}\right]_{x_s=x_{s2}}^{x_s=x_{s1}}=\begin{cases}0 & x_{s1}<x<x_{s2}\\ p_d(x) & x<x_{s1}\text{ or } x>x_{s2}\end{cases}, \tag{6.24}$$

which produces no sound field within V_0. Instead, the integral reproduces the external sound field outside of V_0. To reproduce the sound field from an interior virtual source, the KH integral should be modified, or a different type of integral equation is needed.

As we have seen in the field-matching problem of Equation (6.10), the reproduction of an interior virtual source is related to the time reversal operator. Accordingly, we can assume that the modified KH integral also has to do with time reversal. To derive the modified integral, we first consider *time-reversed propagation* (TRP) from the virtual source, which is defined in terms of the time-reversed Green's function:

$$\begin{aligned}p_{tr}^*(x_s)&=q_v[-g_f(x|x_v)^*]\\&=\begin{cases}q_v\dfrac{i}{2k}e^{-ik(x_s-x_v)}, & x_s\geq x_v\\ q_v\dfrac{i}{2k}e^{ik(x_s-x_v)}, & x_s<x_v\end{cases}.\end{aligned} \tag{6.25}$$

It is noteworthy that the TRP p_{tr}^* is different from the time reverse of the virtual source's sound field (p_d^*). The latter changes the phase of both the excitation q_v and the propagator $e^{ik(x_s-x_v)}$, whereas TRP inverts only the phase of the propagator g_f. Therefore, as compared to the original wavefront (Figure 6.6a) TRP changes only the propagating direction of a wavefront in space, and the wavefront converges to the location of the virtual source (Figure 6.6b).

The importance of TRP can be found from its value at the left boundary ($x_{s1}<x_v$). According to the definition of TRP, it is equivalent to the target field inversely propagated to the left boundary (Figure 6.6c). That is,

$$[p_{tr}^*(x_{s1})]=p_d(x_v)e^{ik(x_{s1}-x_v)},\quad x_{s1}<x_v. \tag{6.26}$$

In other words, the *TRP of the left-going wave extrapolates the right-going part of the target field.* The same relation holds for the TRP of the right-going wave (Figure 6.6d). Therefore, the reproduction of the left-going wave in TRP by the KH integral will generate the target field going to the right-hand side. The integral at the left boundary ($x_s=x_{s1}$) gives

$$\left[p_{tr}^*(x_s)\frac{dg_f(x|x_s)}{dx_s}-\frac{dp_{tr}^*(x_s)}{dx_s}g_f(x|x_s)\right]_{x_s=x_{s1}}=p_d(x), \tag{6.27}$$

which shows the exact reproduction of the target sound field for $x\geq x_v$ (Figure 6.6e).

On the other hand, the KH integral also produces a contribution from the right boundary. The integral at $x_s=x_{s2}$ gives

$$-\left[p_{tr}^*(x_s)\frac{dg_f(x|x_s)}{dx_s}-\frac{dp_{tr}^*(x_s)}{dx_s}g_f(x|x_s)\right]_{x_s=x_{s2}}=q_v\frac{i}{2k}e^{-ik(x-x_v)}, \tag{6.28}$$

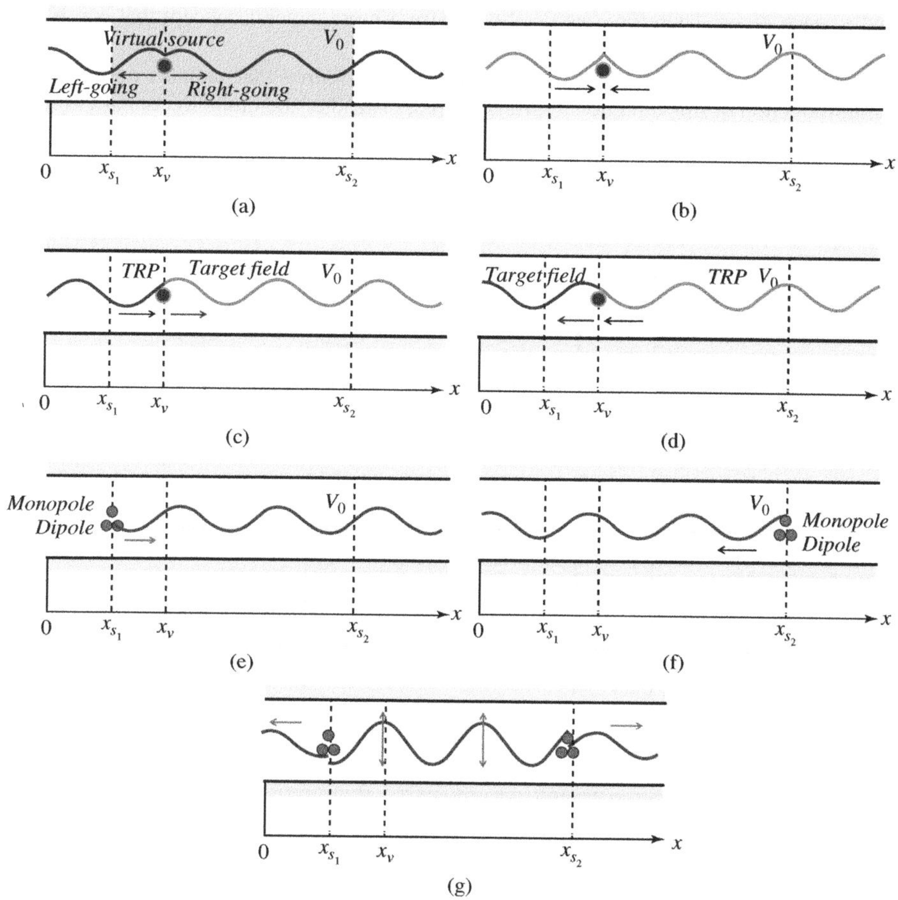

Figure 6.6 KH integral applied to an interior virtual source with time-reversed propagation (a) target field (b) time-reversed propagation (TRP) (c) TRP of the left-going wave: right-going (d) TRP of the right-going wave: left-going (e) reproduction by left integral of TRP (f) reproduction by right integral of TRP (g) total integral

which is the left-going wave of the target field extrapolated to $x > x_v$ (Figure 6.6f). The left-going wave is not what we want in the region $x_{s2} > x > x_v$, so it is regarded as an artifact. The left-going wave has the same amplitude as the right-going wave, and the combination of these two generates a standing wave within V_0 (Figure 6.6g). Therefore, the artifact produced by the right integral cannot be neglected and must be removed.

6.3.3.3 Removing the Artifact from the Reproduced Field

A straightforward method of removing this artifact is to simply turn off the monopole and dipole sources on the right boundary. However, this approach is feasible only when the right integral can be clearly separated from the left integral. Although the left and right boundaries obviously differ in this 1D case, it is not as clear for 2D or 3D problems, in which the boundaries are connected to each other. Therefore, we need to devise a more general concept that can be applied to multidimensional cases.

To derive such a formula, consider a *directional virtual source* that radiates only into $x \geq x_v$. Then the reproduction of this directional virtual source, instead of the omnidirectional virtual source (Figure 6.7a) of Equation (6.5), has the same effect as neglecting the right integral. To be more specific, let us first express the directional virtual source in 1D. From the discussion in Chapter

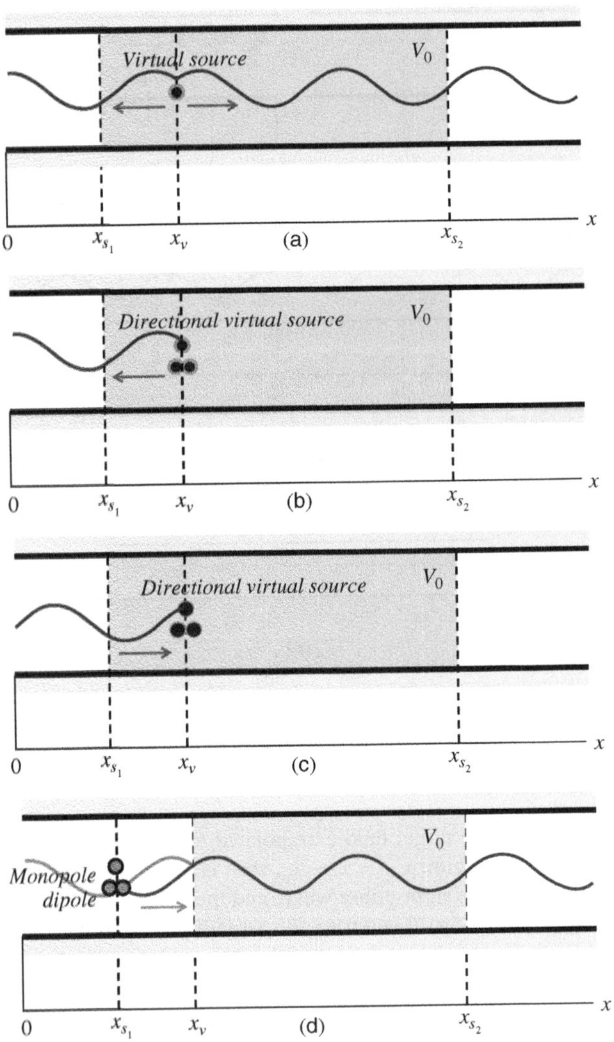

Figure 6.7 Reproduction of a directional virtual source (a) target field (b) alternative field from a directional virtual source (c) TRP of the alternative field (d) Reproduced field

5, Section 5.5.1, a directional virtual source can be made by combining a monopole and a dipole. In the 1D case, the summation of a monopole field $p_d(x)$ and a dipole field $(dp_d/dx)/ik$ of equal amplitude yields a perfect directional source propagating only in one half-space (Figure 6.7b). That is,

$$p_a(x) = \frac{1}{2}\left(p_d(x) - \frac{1}{ik}\frac{dp_d(x)}{dx}\right) = \begin{cases} 0 & x > x_v, \\ q_v \dfrac{i}{2k} e^{-ik(x-x_v)} & x \le x_v \end{cases} \tag{6.29}$$

where p_a describes the sound field from a directional virtual source and will be denoted as an *alternative sound field*.

The directional source's sound field is then time-reversed and measured at the left and right boundaries (Figure 6.7c).

$$p_{a,tr}^*(x_s) = \begin{cases} 0 & x_s > x_v \\ q_v \dfrac{i}{2k} e^{ik(x_s-x_v)} & x_s \le x_v \end{cases}. \tag{6.30}$$

Consequently, the contribution at the right boundary $(x_{s_2} > x_v)$ becomes zero, and the integral from the left boundary reproduces the same sound field as Equation (6.27) (Figure 6.7d).

So far, we have investigated the KH integral and its modified forms for the reproduction of interior and exterior virtual sources. The KH integral provides a direct solution for the exact reproduction of an exterior virtual source using monopole and dipole sources on the boundaries. In contrast, the case of an interior virtual source is less straightforward. An exact solution does not exist; however, we can find a partially exact solution by introducing time-reversal and a directional virtual source. The resultant sound field reproduced by the integral is exact only over a partial area of the interior volume. Therefore, the location and size of the sweet spot are limited for the interior virtual source.

6.3.4 Single-Layer Potential

Sound field reproduction by the KH integral requires both monopole and dipole sources. However, according to the direct field-matching discussed in Section 6.3.1, it seems that the target field can be reproduced using only monopole sources. This means that, conversely, we can derive an integral formula only for monopole control sources. This type of integral is called a *single-layer formula*, and the monopole excitation signal derived from the integral is denoted as a *single-layer potential*. Here a *potential* is a physical quantity, the spatial derivative of which produces a desired variable. For example, in fluid dynamics, the gradient of the velocity potential generates the particle velocity u. Likewise, the gradient of the pressure field (∇p) is related to the mass acceleration or inertia force $(-i\omega\rho_0 u(\omega))$ of the sound field (according to Euler equation (Equation (1.25))). From this viewpoint, the sound pressure can be regarded as the potential of the inertia force.

To see why multiple combinations of monopoles and dipoles produce the same sound field, let us compare two sound fields reproduced by the field-matching technique and integral-based formula. A simple comparison of the two fields reproduced from the exterior virtual source (Equations (6.9) and (6.23), or Figures 6.3 and 6.5d) reveals that the sound fields outside of the volume V_0 are dissimilar. The field reproduced by the monopole and dipole sources (integral approach) shows nonradiating behavior outside V_0, whereas it is nonzero in the field-matching case. Accordingly, we can see that the interior sound fields surrounded by the loudspeakers are the same, but their external radiations are not. In other words, the various monopole and dipole combinations that reproduce a given interior field within V_0 produce different external radiation. In the following sections, we will investigate various monopole-only distributions for reproducing a 3D sound field.

6.4 Reproduction of a 3D Sound Field

6.4.1 *Problem Statement and Associated Variables*

The principle for reproducing a 3D sound field is not much different from the 1D example. However, unlike the 1D control sources, the loudspeaker geometry can have various shapes in 2D or 3D spaces. In this section, the 3D reproduction problem is discussed using regular array geometries such as planar or spherical shapes.

To define the 3D reproduction problem, consider a virtual sound source located at a position $\mathbf{r}_v$ within a reproduction room (Figure 6.8). The control sources (monopoles, dipoles, or both) are distributed on a surface S and manipulate the sound field; hence, the surface S is often called a *control surface*. Further, the control surface divides the reproduction room into two regions: the interior region V and the exterior region Λ. Both regions are defined as open sets that are near the closed surface S. The exterior region is generally bounded by the physical boundary conditions of the reproduction room. However, in many reproduction problems, the direct field and its reflections from the wall are considered separately. The discussion in the preceding sections focuses mostly on the reproduction of the direct field; a free-field Green's function will be used as a transfer function between the source and receiver.

6.4.1.1 Interior and Exterior Reproduction Problems

Depending on the listener's location, the reproduction problem can be categorized into two types: interior or exterior reproduction. The reproduction problem for a listener in an enclosed volume V is called an *interior reproduction problem* (Figure 6.8), and that for a listener outside is called an *exterior reproduction problem* (Figure 6.9). The interior reproduction problem is useful when we want to provide an auditory illusion of surround sound to the listener inside, whereas exterior reproduction is useful for delivering the outward radiation characteristics of a virtual sound source such as the radiation pattern or directivity.

6.4.1.2 Interior and Exterior Virtual Sources

Depending on the locations of the virtual source and listener, we can define two different types of problems. In the first type, the virtual source is positioned within the same volume as the listener ($\mathbf{r}_v, \mathbf{r} \in V$). In this case, our aim is to reproduce the sound field from an *interior virtual source* (Figure 6.8a). In contrast, an *exterior virtual source* (Figure 6.8b) refers to a virtual source located in a different region from the listener ($\mathbf{r}_v \in \Lambda, \mathbf{r} \in V$).

Interior and exterior reproduction, or interior and exterior virtual sources, require that different problems be defined and solved. However, the basic principles for solving the defined problems can be explained by extending the technique discussed for the 1D case. In what follows, the field-matching, mode-matching, and integral approaches will be extended to 3D sound field reproduction.

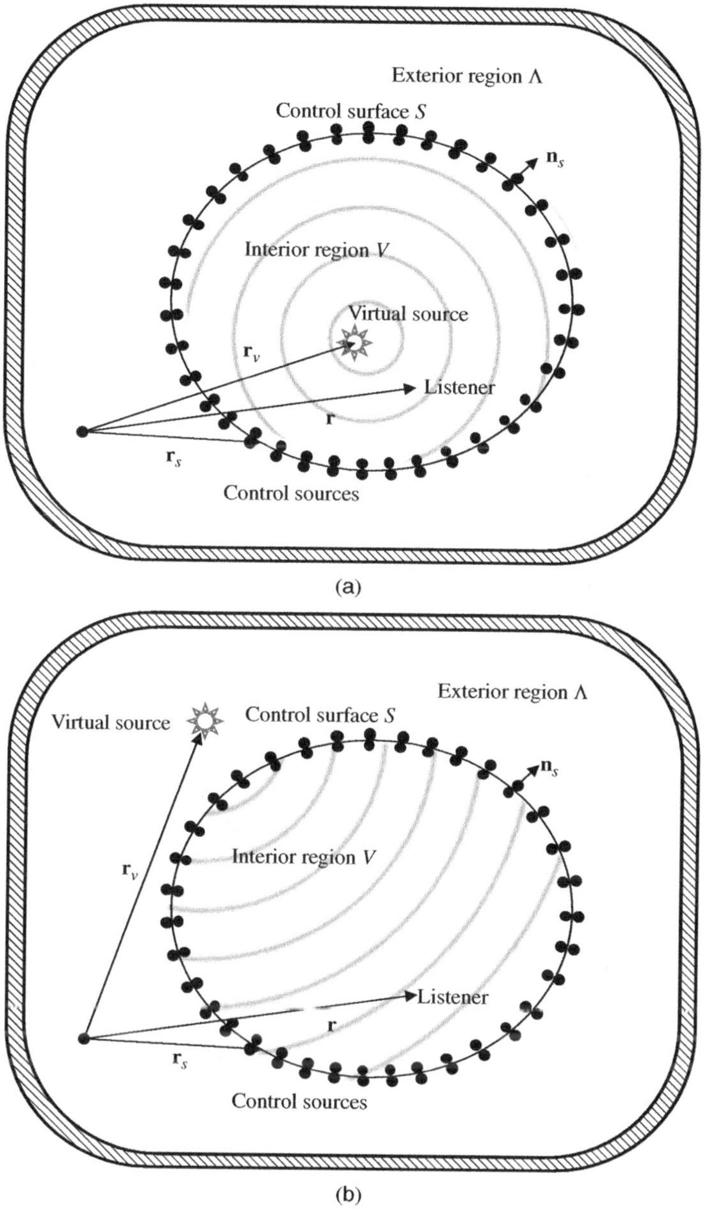

Figure 6.8 Schematic representation of interior reproduction: (a) interior virtual source and (b) exterior virtual source ($\mathbf{r}_v$: position of the virtual source, $\mathbf{r}$: listener position, $\mathbf{r}_s$: position of the control source, S: control surface, $\mathbf{n}_s$: surface normal to the exterior field, Λ: exterior region, and V: interior region)

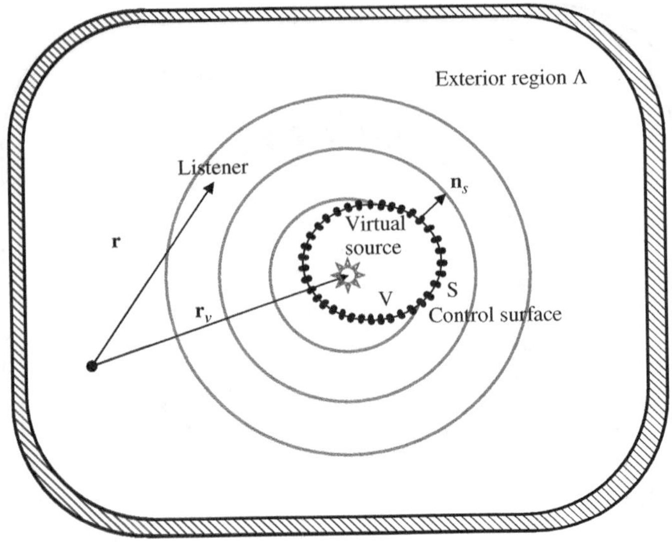

Figure 6.9 Schematic representation of exterior reproduction

6.5 Field-Matching Approach

6.5.1 Inverse Problem

The field-matching approach pursues direct reproduction of a target sound field by solving a direct inverse problem between the target field and excitation signal. For a defined zone of interest, we can define a vector consisting of sampled target pressure field:

$$\mathbf{p}_d = \left[p_d\left(\mathbf{r}^{(1)}, \omega\right), \cdots, p_d(\mathbf{r}^{(M)}, \omega)\right]^T. \tag{6.31}$$

Then the reproduced pressure field at the same positions can be described in vector form as

$$\mathbf{p}_r = \left[p_r\left(\mathbf{r}^{(1)}, \omega\right), \cdots, p_r(\mathbf{r}^{(M)}, \omega)\right]^T. \tag{6.32}$$

Using the excitation signals of the control sources (q) and the transfer functions (g), the reproduced field can be rewritten as

$$p_r(\mathbf{r}^{(m)}, \omega) = \sum_{\kappa=1}^{K} g(\mathbf{r}^{(m)}|\mathbf{r}_s^{(\kappa)}; \omega)q^{(\kappa)}(\omega),$$

$$\mathbf{p}_r = \mathbf{G}\mathbf{q}, \tag{6.33}$$

where the $M \times K$ matrix $\mathbf{G}$ denotes the transfer functions between the excitation vector $\mathbf{q}$ and the reproduced sound field. In this case, $\mathbf{q}$ is considered to be the monopole amplitude, so the Green's function is used as the transfer function.

Then the problem is to find $\mathbf{q}$ that reproduces the target field $\mathbf{p}_d$ such that

$$\mathbf{G}\mathbf{q} = \mathbf{p}_d. \tag{6.34}$$

Solving Equation (6.34) for $\mathbf{q}$ is a typical inverse problem.

The idea behind the field-matching technique is straightforward, but it also has many ambiguities. According to Hadamard [1, 2], the inverse problem of Equation (6.34) is well-posed if

1. A solution exists (existence);
2. The solution is unique (uniqueness); and
3. The solution is stable; that is, $\mathbf{G}^{-1}$ is continuous (stability),

and an inverse problem is said to be ill-posed otherwise.

The field-matching technique can also be ill-posed, for example, in the case of nonexistence, nonuniqueness, or ill-conditioned problems. The ill-posedness associated with the inverse problem can be categorized into two types: the controllability and observability issues. The former has to do with whether we have sufficient means to control a given sound field over the zone of interest. For example, we may not be able to reproduce an arbitrary target field using only one or two control sources. In this case, there may be no solution that exactly reproduces the target sound field over the zone of interest (nonexistence problem).

The latter issue – the observability – has to do with the ability to observe the target or reproduced sound field. For instance, we cannot tell that a target sound field can be reproduced over a large area by controlling only a few measurement points. The lack of observability often leads to the nonuniqueness problem in which many solutions exist. Then, how many control points are required to reproduce a desired sound field without any redundancy?

The answers to all these questions might differ depending on the method we use to solve the inverse problem, the array geometry, the target sound field, and the transfer function. However, generic tools are available that can be used to analyze the ill-posedness issues associated with the inverse problem. Singular value decomposition (SVD), which decomposes the transfer function in terms of the left and right eigenvectors, shows how the controllability and observability issues becomes problematic for a given inverse problem. SVD uses the concept of spatial modes, which provides the clue to extending the field-matching approach to the mode-matching technique, which will be discussed in Section 6.6.

6.5.1.1 Singular Value Decomposition

To investigate the inverse problem in depth, consider the least-squares approach that minimizes the error between the target and reproduced sound fields. A field-matching error can be defined as the distance between the two pressure field vectors. The Euclidean distance between two vectors can be represented using the Euclidean norm or two-norm ($\|\,\|$), that is,

$$\begin{aligned}\text{Minimize } \varepsilon^2 &= \|\mathbf{p}_d - \mathbf{p}_r\|^2 \\ &= \sum_{m=1}^{M} \left|p_d\left(\mathbf{r}^{(m)}, \omega\right) - p_r(\mathbf{r}^{(m)}, \omega)\right|^2,\end{aligned} \tag{6.35}$$

where ε^2 denotes the total squared error between the target and reproduced sound fields over a defined set of sampling points ($\mathbf{r}^{(m)}$, $m = 1, \ldots, M$). The excitation signal $\mathbf{q}$ minimizing ε^2 can be found by taking the derivative of ε^2 with respect to $\mathbf{q}$. Using the matrix derivatives of Equations (A31)–(A33), we can calculate the derivative of ε^2:

$$\begin{aligned}\frac{\partial \varepsilon^2}{\partial \mathbf{q}} &= \frac{\partial}{\partial \mathbf{q}}(\mathbf{p}_d - \mathbf{G}\mathbf{q})^H(\mathbf{p}_d - \mathbf{G}\mathbf{q}) \\ &= \frac{\partial}{\partial \mathbf{q}}(\mathbf{p}_d^H\mathbf{p}_d - \mathbf{q}^H\mathbf{G}^H\mathbf{p}_d - \mathbf{p}_d^H\mathbf{G}\mathbf{q} + \mathbf{q}^H\mathbf{G}^H\mathbf{G}\mathbf{q}) \\ &= -\mathbf{p}_d^H\mathbf{G} + \mathbf{q}^H\mathbf{G}^H\mathbf{G}.\end{aligned} \tag{6.36}$$

The optimal $\mathbf{q}$ can be found when the derivative is zero. Setting Equation (6.36) equal to zero, we obtain

$$\mathbf{G}^H\mathbf{G}\mathbf{q} = \mathbf{G}^H\mathbf{p}_d \quad \text{or} \quad \mathbf{q} = (\mathbf{G}^H\mathbf{G})^{-1}\mathbf{G}^H\mathbf{p}_d, \tag{6.37}$$

where the operator $\mathbf{G}^+_{left} = (\mathbf{G}^H\mathbf{G})^{-1}\mathbf{G}^H$ is called the *left inverse* of the matrix $\mathbf{G}$ (Equation (A40)).

The solution minimizing the reproduction error can be investigated further by using SVD of the transfer functions $\mathbf{G}$, which is given by (Appendix A, Section A.4.5)

$$\begin{aligned}\mathbf{G} &= \mathbf{U}\boldsymbol{\Sigma}\mathbf{V}^H \\ &= \sum_{i=1}^{N} \sigma_i \mathbf{u}_i \mathbf{v}_i^H,\end{aligned} \tag{6.38}$$

where N is the number of nonzero singular values and $\mathbf{u}_i$ and $\mathbf{v}_i$ are the ith columns of $\mathbf{U}$ and $\mathbf{V}$, respectively. Further, σ_i denotes singular values constituting diagonal elements of $\boldsymbol{\Sigma}$. Here, $\mathbf{v}_i$ represent orthonormal basis functions that span the vector space of $\mathbf{q}$, that is, the possible spatial distribution of source excitation signals. Thus, the basis function $\mathbf{v}_i$ can be viewed as orthonormal source modes expressing the given source excitation in space. Likewise, the columns of the left eigenvector $\mathbf{U}$ represent the orthonormal basis that expresses any pressure field produced by the source excitation $\mathbf{q}$ and given transfer function $\mathbf{G}$, so they can be denoted as pressure modes.

Let us investigate how SVD relates the pressure field and source excitation. By inserting Equation (6.38) into Equation (6.33), the reproduced sound field can be written as

$$\mathbf{p}_r = \mathbf{U}\boldsymbol{\Sigma}\mathbf{V}^H\mathbf{q}. \tag{6.39}$$

From Equation (6.39), it can be seen that the inner product of $\mathbf{q}$ and $\mathbf{V}$ extracts the contribution of each source mode. That is,

$$\mathbf{a} = \mathbf{V}^H\mathbf{q} \quad \text{or} \quad a_i = \mathbf{v}_i{}^H\mathbf{q} \tag{6.40}$$

The modal contributions $\mathbf{a}$ will be denoted as *source modal coefficients.* The elements of a diagonal matrix $\boldsymbol{\Sigma}$ then amplify the source modal coefficients, which excite different types of modes that can express the pressure field. The amplified modal coefficients, which are referred to as *pressure modal coefficients*, can be written as

$$\mathbf{b} = \boldsymbol{\Sigma}\mathbf{a}. \tag{6.41}$$

Because each column of $\mathbf{U}$ is excited by a single element of the pressure modal coefficients, there is a unique one-to-one mapping relation between the source and pressure modes.

The solution to the inverse problem in Equation (6.34) can be easily derived using SVD. By substituting Equation (6.38) into Equation (6.37), the solution can be rewritten as

$$\begin{aligned}\mathbf{q} &= \mathbf{V}\boldsymbol{\Sigma}^{-1}\mathbf{U}^H\mathbf{p}_d \\ &= \sum_{i=1}^{N} \frac{\mathbf{u}_i^H\mathbf{p}_d}{\sigma_i}\mathbf{v}_i \\ &= \sum_{i=1}^{N} a_i\mathbf{v}_i.\end{aligned} \tag{6.42}$$

SVD also shows how the target pressure field is analyzed to calculate the solution. The product of the left eigenvector and target pressure field in Equation (6.42), given by

$$\mathbf{b} = \mathbf{U}^H\mathbf{p}_d, \tag{6.43}$$

is a process for extracting the pressure modal coefficient $\mathbf{b}$ from the target sound field $\mathbf{p}_d$. Then the inverse of the singular value matrix ($\boldsymbol{\Sigma}^{-1}$) converts $\mathbf{b}$ into the source modal coefficient $\mathbf{a}$, which leads to the source excitation function $\mathbf{q}$.

6.5.1.2 Picard's Condition

In determining the existence of a solution for a given inverse problem, Picard's condition provides an important guideline. First, suppose that we have an infinite number of control sources and spatial sampling points for the target field. In this case, Equation (6.33) can be described in integral form,

$$\int_V g\left(\mathbf{r}|\mathbf{r}_s\right) q(\mathbf{r}_s)dV\left(\mathbf{r}_s\right) = p_r(\mathbf{r}), \tag{6.44}$$

and for $\mathbf{r}, \mathbf{r}_s \in V$, the integral is called the Fredholm integral of the first kind. For this continuous form, the SVD of the transfer function can be written as

$$g(\mathbf{r}|\mathbf{r}_s) = \sum_{i=1}^{\infty} \sigma_i u_i(\mathbf{r}) v_i^*(\mathbf{r}_s), \tag{6.45}$$

which is called the singular value expansion (SVE). In the same way, the solution to Equation (6.42) for an inverse problem to have $p_r = p_d$ can be rewritten as

$$q(\mathbf{r}_s) = \sum_{i=1}^{\infty} \frac{\int_V u_i^*(\mathbf{r}) p_d(\mathbf{r}) dV}{\sigma_i} v_i(\mathbf{r}_s) = \sum_{i=1}^{\infty} \frac{\langle u_i, p_d \rangle}{\sigma_i} v_i(\mathbf{r}_s). \tag{6.46}$$

Here, the linear operator $\langle f, g \rangle = \int_V (f^* g) dV$ expresses the volume integral as an inner product.

Let us consider the conditions under which the solution q can exist. To have a finite solution q, its two-norm should be finite. Because the right eigenvector v_i is orthonormal, the condition for having a finite solution q can be written as

$$\|q(\mathbf{r}_s)\|^2 = \sum_{i=1}^{\infty} |a_i|^2 = \sum_{i=1}^{\infty} \left(\frac{\langle u_i, p_d \rangle}{\sigma_i} \right)^2 < \infty, \tag{6.47}$$

where $a_i = \langle u_i, p_d \rangle / \sigma_i$. Therefore, the existence of a solution can be checked by inspecting the convergence of the source modal coefficients a_i. In other words, the pressure modal coefficients $b_i = \langle u_i, p_d \rangle$ should decay faster than the singular values σ_i; this statement is called Picard's condition (see, e.g., [3]).

In practice, we have a finite number of sources, and the number of singular values is also finite. To formulate Picard's condition for a discrete matrix, let us rewrite Equation (6.47) in terms of the matrices defined in Equation (6.42):

$$\|\mathbf{q}\|^2 = \sum_{i=1}^{N} \frac{|\mathbf{u}_i^H \mathbf{p}_d|^2}{\sigma_i^2}. \tag{6.48}$$

From Picard's condition, the pressure modal contribution $\mathbf{u}_i^H \mathbf{p}_d$ should decay on average faster than the corresponding singular values. This statement, called the *discrete Picard's condition* [3, 4], is not as clear as the original form because the solution is always bounded for a finite number of singular values, as long as they are nonzero. However, the convergence of the modal coefficient still plays an important role when we evaluate the ill-posedness of a problem or when we consider the *regularization* technique to resolve the ill-posedness of the inverse problem.

Whether an inverse problem satisfies the discrete Picard's condition is closely related to the structure of the transfer function g and the noise included in the matrix. Figure 6.11 shows one example of the decay of the pressure modal coefficient $\mathbf{b}_i = \mathbf{u}_i^H \mathbf{p}_d$ and its corresponding singular value σ_i. The Picard plot in Figure 6.11b depicts the ratio of these two values. In this example, a loudspeaker array is arranged on a plane denoted as the source plane, and a target sound field is defined over a target plane separated by 2 m from the source plane (Figure 6.10). This particular example shows that the singular values σ_i decay more rapidly for $i > 150$. Accordingly, the discrete Picard's condition may not hold as the index i increases. Another example for a higher frequency (Figure 6.12, $f = 1372\,\text{Hz}$) shows that the series is more convergent than for the low-frequency case.

6.5.1.3 Nonuniqueness and Ill-Conditioning

If any of the singular values σ_i is equal to zero, the source excitation following the shape of the corresponding eigenvector $\mathbf{v}_i$ does not radiate well as a pressure field; that is, the source excitation $\mathbf{v}_i$ is "invisible" in the reproduced pressure field $\mathbf{p}_r$. This typical observability issue due to invisible source excitation leads to the *nonuniqueness problem* in solving an inverse problem because we cannot uniquely estimate the modal contribution of $\mathbf{v}_i$ from the given target field $\mathbf{p}_d$.

In principle, we can neglect singular values of zero that do not affect the result of reproduction, and this strategy is implemented by the minimum-norm solution (Appendix A, Section A.4.2). However, in practice, noise can exist in the transfer function or target function, and small singular values produced by such noise make it impossible to identify the "invisible" eigenvectors from SVD. The inversion of any small singular value will lead to an unstable solution that is very sensitive to the noise (*ill-conditioned problem*). Without proper filtering of small singular values, their inversion can cause the solution $\mathbf{q}$ to explode. Such a problem often occurs in many practical situations.

For example, none of the cases presented in Figures 6.11 and 6.12 can reproduce the target field exactly because their small singular values are already contaminated by the machine tolerance of

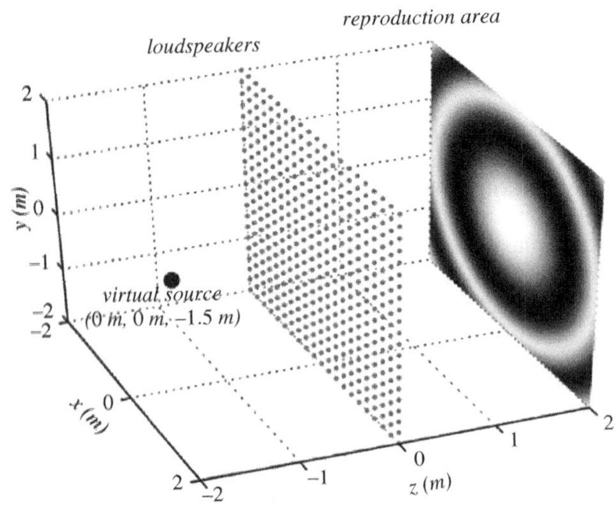

Figure 6.10 Configuration of sound field reproduction over a planar surface

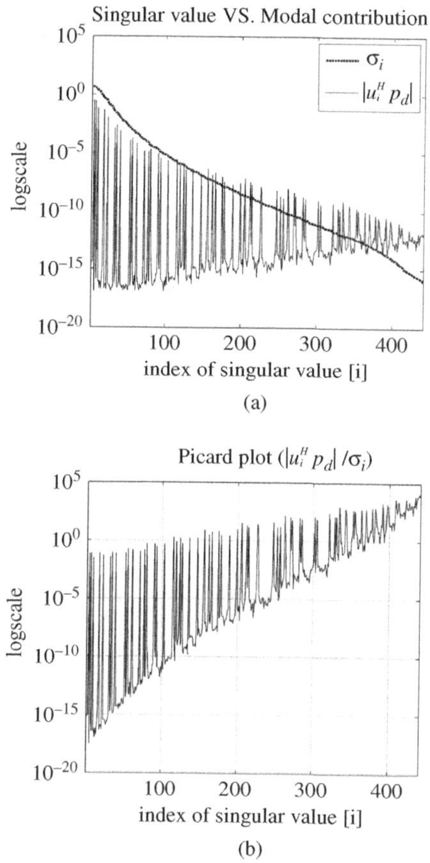

Figure 6.11 Distribution of singular values for the low-frequency ($f = 170\,\text{Hz}$) reproduction problem: (a) singular values versus modal contributions and (b) Picard plot

the computing machine. However, the amplification of singular values is much more problematic in the ill-conditioned case (Figure 6.12). Figure 6.13 compares the reproduced sound fields of the two previous cases; the reproduced sound field is clearly distorted by the amplification of noisy singular values. Such ill-conditioning can be resolved to some extent by regularization, which improves the conditioning of a matrix by modifying small singular values.

6.5.1.4 Condition Number

Sometimes the degree of ill-conditioning is represented by a condition number, which is defined as the ratio of the maximum and minimum singular values. In terms of the matrix norm, the condition number of a matrix **G** can be written as

$$\text{cond}(\mathbf{G}) = \|\mathbf{G}\|\|\mathbf{G}^{-1}\|. \tag{6.49}$$

The physical meaning of the condition number can be found by considering the noise or error in the target field $\mathbf{p}_d$. The perturbation of the solution $\Delta\mathbf{q}$ due to the target field error $\Delta\mathbf{p}_d$ can be

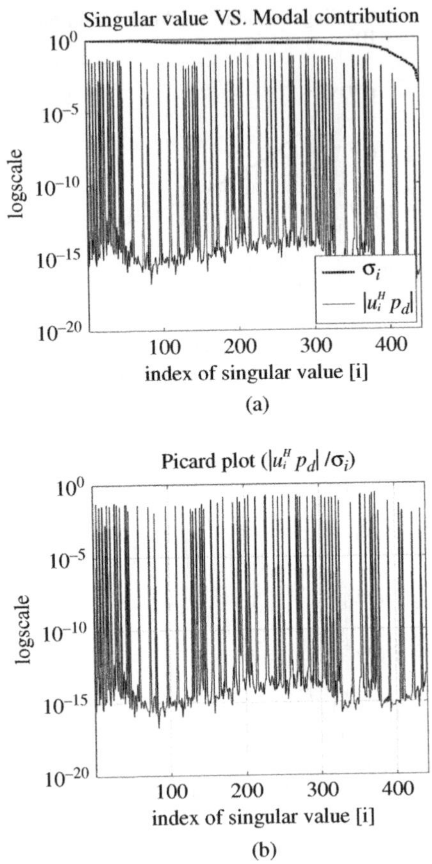

Figure 6.12 Distribution of singular values for the high-frequency ($f = 1372\,\text{Hz}$) reproduction problem: (a) singular values versus modal contributions and (b) Picard plot

written as

$$\mathbf{G}(\mathbf{q} + \Delta\mathbf{q}) = \mathbf{p}_d + \Delta\mathbf{p}_d.$$

Because $\mathbf{G}^{-1}\mathbf{p}_d = \mathbf{q}$, this yields the following inequality for $\Delta\mathbf{q}$:

$$\Delta\mathbf{q} = \mathbf{G}^{-1}\Delta\mathbf{p}_d \Rightarrow \|\Delta\mathbf{q}\| \leq \|\mathbf{G}^{-1}\|\|\Delta\mathbf{p}_d\|. \tag{6.50}$$

If we consider the relative error with respect to the norm of the original solution, the range of the relative error can be derived as

$$\begin{aligned}\frac{\|\Delta\mathbf{q}\|}{\|\mathbf{q}\|} &\leq \frac{\|\mathbf{G}^{-1}\|\|\Delta\mathbf{p}_d\|}{\|\mathbf{q}\|} \\ &\leq \frac{\|\mathbf{G}\|\|\mathbf{G}^{-1}\|\|\Delta\mathbf{p}_d\|}{\|\mathbf{G}\|\|\mathbf{q}\|} \leq \text{cond}(\mathbf{G})\frac{\|\Delta\mathbf{p}_d\|}{\|\mathbf{p}_d\|}.\end{aligned} \tag{6.51}$$

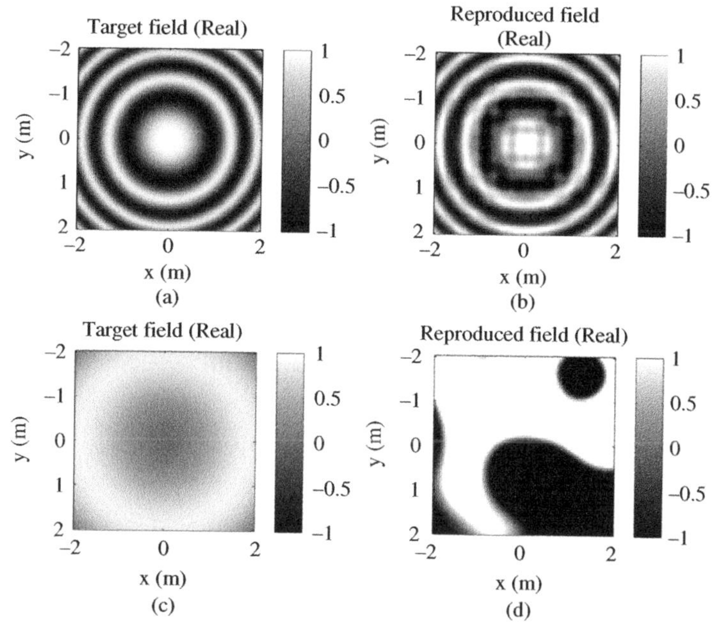

Figure 6.13 Sound field comparison of (a, b) well-conditioned and (c, d) ill-conditioned problems: real parts of (a, c) target field and (b, d) reproduced field at $f = 1372\,\text{Hz}$, $f = 170\,\text{Hz}$, respectively

Therefore, the condition number represents the maximum bound of relative perturbation in the solution $\|\mathbf{q}\|$ due to noise in the target field $\mathbf{p}_d$. A small condition number indicates less perturbation in the solution for the same amount of noise.

The condition number, however, is a very simplified measure and has to be used carefully. In particular, the condition number is determined only by the maximum and minimum singular values and thus cannot represent how the others are distributed between them. Even with the same condition number, the other singular values can have completely different decay rates. Figure 6.14 depicts an example in which the condition number is similar to that in Figure 6.11, but its singular value distribution is completely different.

6.5.2 Regularization of an Inverse Problem

6.5.2.1 Truncation of Singular Values (TSV)

There are numerous regularization algorithms for preventing the amplification of noisy singular values. A simple scheme is the truncation of singular values (TSV), for which the solution is calculated from the truncated singular values. That is,

$$\mathbf{q} = \sum_{i=1}^{N} \xi_i \frac{(\mathbf{u}_i^H \mathbf{p}_d)}{\sigma_i} \mathbf{v}_i,$$

where the singular value filter ξ_i is defined as

$$\xi_i = \begin{cases} 1 & \text{for} \quad 1 \le i \le \chi \\ 0 & \text{for} \quad \chi < i \le N \end{cases}. \tag{6.52}$$

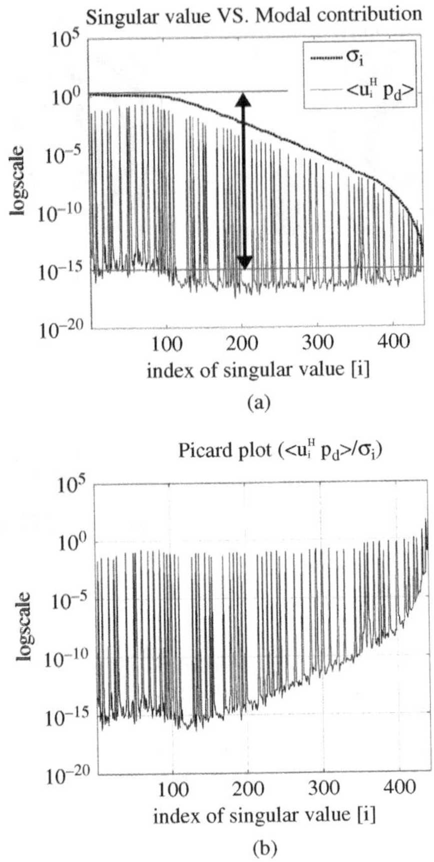

Figure 6.14 Distribution of singular values for the high-frequency ($f = 686\,\text{Hz}$) reproduction problem: (a) singular values versus modal contributions and (b) Picard plot

This type of filter simply truncates the $N - \chi$ smallest singular values from the inversion in order to neglect small singular values below a specified threshold ($\sigma_\chi = \Delta_\sigma$). In this way, the modal contributions of eigenvectors other than χ are neglected; hence, small singular values are not inverted. The TSV solution prevents the divergence of $\mathbf{q}$ by truncating the modes and can be regarded as a *modal truncation solution.* Figure 6.15 depicts the target and reproduced fields with TSV regularization. For this example, small singular values (Figure 6.12a) below a certain threshold were truncated. As the threshold of truncation increases, many eigenvectors constituting the target field are omitted.

In TSV, it is important to understand that small singular values do not necessarily make small contributions to the target field. If an eigenvector with a high modal contribution $b_i = \mathbf{u}_i^H \mathbf{p}_d$ is neglected because of its small singular value σ_i, an essential wave field component needed to express the target field can be missing (Figure 6.15c). In particular, the sound field reproduced by TSV can be written as

$$\mathbf{p}_r = \sum_{i=1}^{\chi} (\mathbf{u}_i^H \mathbf{p}_d)\mathbf{u}_i. \tag{6.53}$$

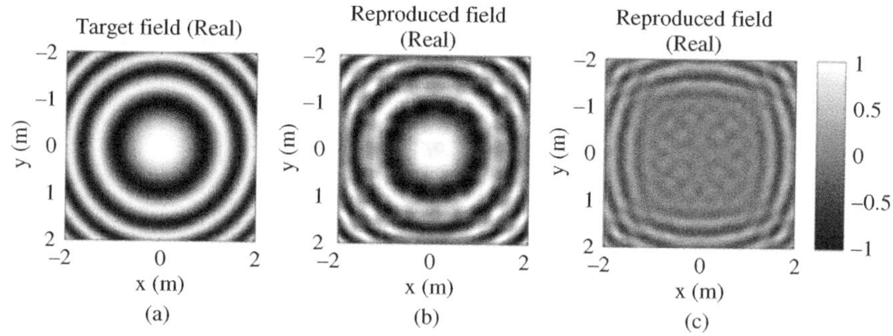

Figure 6.15 Sound fields in TSV regularization: (a) target field, (b) singular value threshold: 0.5, and (c) singular value threshold: 0.9

Hence, the reproduction error is simply given as the squared sum of the truncated modal coefficients:

$$\varepsilon^2 = \|\mathbf{p}_d - \mathbf{p}_r\|^2 = \sum_{i=\chi+1}^{N} |\mathbf{u}_i^H \mathbf{p}_d|^2.$$

Therefore, to have a small reproduction error when using TSV, the modal contribution should decay more rapidly than the small singular values, which again suggests the importance of the discrete Picard's condition.

6.5.2.2 Tikhonov Regularization

A simple yet powerful approach to preventing a diverging solution is to introduce a penalty in the size of the solution $\mathbf{q}$ ($\|\mathbf{q}\|$), as shown in Appendix A, Section A.4.4. For a regularization parameter μ, the regularized solution $\mathbf{q}_\mu$ can be rewritten as (Equation (A69))

$$\begin{aligned} \mathbf{q}_\mu &= \sum_{i=1}^{N} \left(\frac{\sigma_i^2}{\sigma_i^2 + \mu} \right) \frac{(\mathbf{u}_i^H \mathbf{p}_d)}{\sigma_i} \mathbf{v}_i \\ &= \sum_{i=1}^{N} \xi_i \frac{(\mathbf{u}_i^H \mathbf{p}_d)}{\sigma_i} \mathbf{v}_i. \end{aligned} \tag{6.54}$$

The regularized solution is expressed as the original solution multiplied by the singular value filter ξ_i, which is given by

$$\xi_i = \frac{\sigma_i^2}{\sigma_i^2 + \mu}. \tag{6.55}$$

The filter ξ_i damps out the inverse of the small singular values, so the solution is less divergent even with small singular values. Because μ determines the relative weights of the error ε and the size of the solution, a larger μ will decrease the decay of regularized singular values of Equation (6.54), given by

$$\begin{aligned} \widehat{\sigma}_i(\mu) &= \sigma_i \left(\frac{\sigma_i^2 + \mu}{\sigma_i^2} \right) \\ &= \sigma_i / \xi_i. \end{aligned} \tag{6.56}$$

From Equation (6.56), we can see that the regularization parameter affects small singular values more than large ones. As a result of regularization, the size of the solution, that is, the two-norm of $\mathbf{q}_\mu$, decreases. The "size" of a source excitation can be derived from Equation (6.54) as

$$\|\mathbf{q}_\mu\| = \sqrt{\sum_{i=1}^{N}\left(\frac{\sigma_i^2}{\sigma_i^2+\mu}\right)^2 \frac{|\mathbf{u}_i^H\mathbf{p}_d|^2}{\sigma_i^2}}, \tag{6.57}$$

which is inversely proportional to the regularization parameter μ. On the other hand, the reproduction error of Equation (6.35) is given by

$$\begin{aligned}\varepsilon &= \|\mathbf{p}_d - \mathbf{p}_r\| \\ &= \left\|\mathbf{p}_d - \sum_{i=1}^{N}\sigma_i\mathbf{u}_i\mathbf{v}_i^H\mathbf{q}_\mu\right\| \\ &= \left\|\sum_{l=N+1}^{M}(\mathbf{u}_l^H\mathbf{p}_d)\,\mathbf{u}_l\right\| + \left\|\sum_{i=1}^{N}(1-\xi_i)\,(\mathbf{u}_i^H\mathbf{p}_d)\mathbf{u}_i\right\|\end{aligned} \tag{6.58}$$

for $M > N$. The reproduction error decreases as the singular value filter ξ_i approaches one owing to the decreased regularization parameter μ. Accordingly, the reproduction error is proportional to the regularization parameter. Figure 6.16 illustrates the trade-off between the two-norm of the input $\mathbf{q}_\mu$ and that of the reproduction error ε with respect to the regularization parameter μ. The shape of the curve can vary depending on the distribution of the singular values and the pressure modal coefficient. Two different types of curves calculated from the examples in Figures 6.11 and 6.12 are presented in Figure 6.17. In the well-conditioned case (Figure 6.17b), the input size $\|\mathbf{q}\|$ does not diverge, and we can obtain a fine solution having a low reproduction error with a small regularization parameter. In contrast, the solution of an ill-conditioned problem easily diverges when we attempt to reduce the reproduction error by using a small regularization parameter.

Another practical concern is how much the regularized solution $\mathbf{q}_\mu$ differs from the ideal solution and the conditions under which the two solutions are similar or different. This question is connected

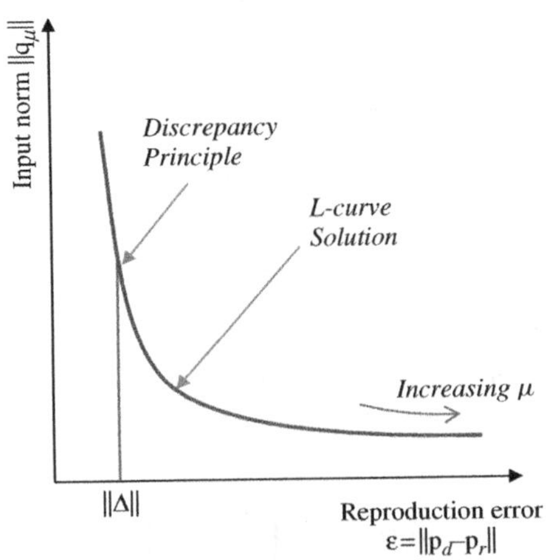

Figure 6.16 Change in the input norm and reproduction error with respect to the regularization parameter μ

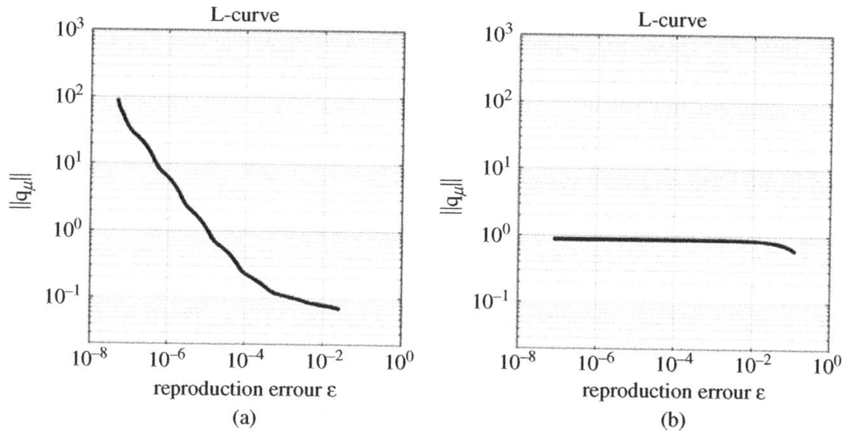

Figure 6.17 Comparison of two L-curves for (a) ill-conditioned problem ($f = 170\,\text{Hz}$) and (b) well-conditioned problem ($f = 1372\,\text{Hz}$)

with the stability of a given inverse problem, that is, whether the defined inverse problem is sensitive to the regularization parameter. If we compare the regularized solution with the ideal one of Equation (6.42), the difference can be written as

$$\begin{aligned}\|\mathbf{q} - \mathbf{q}_\mu\|^2 &= \left\| \sum_{i=1}^{N} \left(\frac{\mu}{\sigma_i^2 + \mu} \right) \frac{(\mathbf{u}_i^H \mathbf{p}_d)}{\sigma_i} \mathbf{v}_i \right\|^2 \\ &= \sum_{i=1}^{N} \left(\frac{\mu}{\sigma_i^2 + \mu} \right)^2 \frac{|\mathbf{u}_i^H \mathbf{p}_d|^2}{\sigma_i^2} \\ &= \sum_{i=1}^{N} \left(\frac{\mu}{\sigma_i^2 + \mu} \right)^2 |a_i|^2 \end{aligned} \tag{6.59}$$

for any real-valued, positive μ. The term in parentheses of Equation (6.59) is always less than one, and the convergence of the difference depends on both the regularization parameter and the modal contribution a_i. Interestingly, even for the same regularization parameter μ, the regularized solution will quickly converge to the ideal solution when the amplified modal contribution $|\mathbf{u}_i^H \mathbf{p}_d|$ decays more rapidly than the singular values σ_i. Therefore, the discrete Picard's condition can be understood as a criterion for the regularized solution to converge more rapidly to the ideal solution.

6.5.3 *Selection of the Regularization Parameter*

Another practically important issue in Tikhonov regularization is how to determine an optimal regularization parameter μ. There is clearly no golden rule that can be applied to every inverse problem, but by introducing an extra criterion to the problem, we can set up a reasonable rule for determining the regularization parameter.

One popular criterion is limiting the *increase of error* due to regularization. Regularization damps out the small singular values required to reproduce the target field, and inevitably increases the error between the target and reproduced sound fields as compared to the ideal solution of Equation (6.42).

The other criterion we can use is the *norm or size of the solution* $\mathbf{q}_\mu$. Because regularization keeps the solution from diverging, the solution's norm decreases monotonically as the regularization parameter increases. By configuring an upper bound to the allowable size of the solution, we can determine the optimal parameter μ.

6.5.3.1 Morozov's Discrepancy Principle

The criterion based on the total amount of error is called Morozov's discrepancy principle [5]. The principle is based on the observation that there is an intrinsic error due to the noise in the target field or in the transfer function and that a certain amount of error can thus be allowed for the regularized inverse. For example, suppose that there is noise $\Delta\mathbf{p}_d$ in the target field $\mathbf{p}_d$, and the amount of noise $\|\Delta\mathbf{p}_d\|$ is known a priori. The observed reproduction error between the target and reproduced fields is given by

$$\varepsilon = \|(\mathbf{p}_d + \Delta\mathbf{p}_d) - \mathbf{G}\mathbf{q}\|. \tag{6.60}$$

It is noteworthy that, even if the solution is ideal ($\mathbf{p}_d = \mathbf{G}\mathbf{q}$), we have an error observation of

$$\varepsilon = \|\Delta\mathbf{p}_d\|. \tag{6.61}$$

Therefore, it is reasonable to determine the regularization parameter such that the reproduction error is within the bound of Equation (6.61). In practice, the bound can be expressed as a form of the range of acceptance,

$$\tau_1\|\Delta\mathbf{p}_d\| \le \varepsilon \le \tau_2\|\Delta\mathbf{p}_d\|, \tag{6.62}$$

for given constants $1 \le \tau_1 < \tau_2$. The optimal μ is then obtained by repeating the calculation of $\mathbf{q}_\mu$ for different values of μ until criterion Equation (6.62) is satisfied.

Another case of practical importance for sound field reproduction is when noise $\Delta\mathbf{G}$ exists in the transfer function $\mathbf{G}$. Assuming that the matrix norm of the noise or its upper bound is known ($\|\Delta\mathbf{G}\|$), the Tikhonov regularization method can be written as

$$\text{Minimize}\, J(\mathbf{q}, \mu) = \|(\mathbf{G} + \Delta\mathbf{G})\mathbf{q} - \mathbf{p}_d\|^2 + \mu\|\mathbf{q}\|^2. \tag{6.63}$$

Again, even for the perfect solution $\mathbf{q}$ ($\mathbf{p}_d = \mathbf{G}\mathbf{q}$), the reproduction error ε is given by

$$\varepsilon = \|(\mathbf{G} + \Delta\mathbf{G})\mathbf{q} - \mathbf{p}_d\| = \|\Delta\mathbf{G}\mathbf{q}\|. \tag{6.64}$$

In this case, the bound of the error depends on both the noise and the solution. However, using the inequality

$$\varepsilon = \|\Delta\mathbf{G}\mathbf{q}\| \le \|\Delta\mathbf{G}\|\,\|\mathbf{q}\|, \tag{6.65}$$

the discrepancy principle can be expressed as

$$\varepsilon \le \|\Delta\mathbf{G}\|\,\|\mathbf{q}\|. \tag{6.66}$$

The regularized solution $\mathbf{q}_\mu$ is then searched for using different values of μ until criterion Equation (6.66) is satisfied over some range of acceptance (Figure 6.16).

6.5.3.2 L-Curve Method

The discrepancy principle requires information on the amount of noise. In many practical situations, however, little information is available. The L-curve method (e.g., [6]) attempts to determine the optimal regularization parameter without knowledge of the noise by considering both the input size and the reproduction error. The principle of this method is to select the regularization parameter that gives the best compromise between the input power $\|\mathbf{q}_\mu\|$ and the reproduction error ε. As illustrated in Figure 6.16, the input norm and reproduction error are inversely proportional to each other as the regularization parameter changes. Therefore, the regularization parameter that performs well in both worlds would be that at the corner of the curve. The name L-curve arises from the typical L-shape of the curve drawn for the input norm and reproduction error. In particular, the input power and reproduction error calculated from a certain μ are presented in logarithmic scale on the x and y axes, respectively. By scanning the curve with a discrete set of μ, we can obtain the L-curve. The regularization parameter giving the point near the corner of the curve is then selected as the optimal one (Figure 6.16). However, the curve shape can be distorted in many practical situations, and some heuristic selection can be made to determine the optimal regularization parameter.

6.5.3.3 Summary

The field-matching approach is basically an inverse problem that can be approached by inverting a known transfer function between the target field and source excitations. The rank deficiency of the transfer matrix often leads to nonuniqueness of the solution, and in this case we need some added constraint to determine a unique solution. Regularization is a process for introducing additional constraints to the problem, and the regularization parameter represents the relative weights of the reproduction error and additional constraints. Tikhonov regularization is a popular regularization technique; however, the process explained in this chapter is somewhat simplified for ease of description compared to its general form [7, 8]. The selection of an appropriate regularization parameter is another important issue, which may depend on the problem to be solved. Unlike methods that use the input norm and reproduction error, there are numerous techniques for determining the regularization parameter, such as generalized cross validation [9], which attempts to minimize the effect of transducer failure. However, the best regularization parameter can be determined only when we define a rule for treating the constraint and reproduction error.

6.6 Mode-Matching Approach

6.6.1 *Encoding and Decoding of Sound Field*

Although direct field-matching can determine the source excitation that produces a sound field that matches the target sound field well, there is always a matching error. The mean-squared error is a physical error, which is in fact not directly related to what we want to achieve. For example, can the pressure error at the center of Figure 6.15c be treated as equal in importance to the error at the edge of a zone of interest? Alternatively, is the error in the high-wavenumber component as important as that in the low-wavenumber component?

In fact, there can be countless types of errors depending on the desired outcome. The error must be minimized in such a way that the most essential component in the sound field can be reproduced. One possible way to realize this goal is to transform the sound field into another domain, in which we can discriminate the desired and unwanted wave components. Then, by matching only the desired wave components, we can minimize the error in the desired component without wasting control effort on the unwanted components.

This type of approach is called *mode-matching*, The mode-matching approach decomposes a sound field into a sum of orthogonal modes (Figure 6.18). SVD of the field-matching approach can also be viewed as a mode-matching approach because it uses eigenvectors as the modes. However, there are generally numerous choices for the modes. To be specific, let us denote the modal function

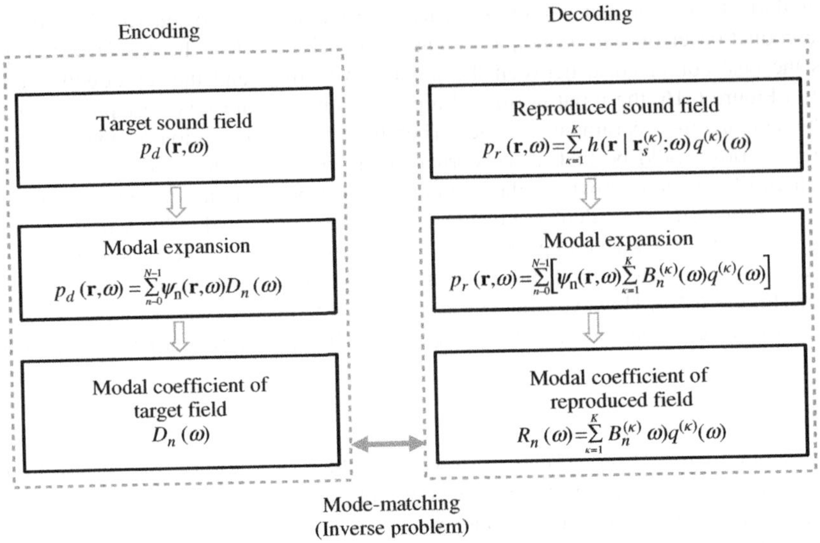

Figure 6.18 Brief description of mode-matching approach

as ψ_n, where n is the index of a mode ($n = 0, \cdots, N-1$). We can then express the target pressure field as a sum of modal functions, each of which is weighted by the modal contribution D_n:

$$p_d(\mathbf{r}, \omega) = \sum_{n=0}^{\infty} D_n(\omega)\psi_n(\mathbf{r}, \omega). \tag{6.67}$$

Then the contribution of each mode is matched by using the sound fields generated by the control sources. The reproduced sound field p_r is the sum of the sound field produced by the control sources:

$$p_r(\mathbf{r}, \omega) = \sum_{\kappa=1}^{K} h(\mathbf{r}|\mathbf{r}_s^{(\kappa)}; \omega)q^{(\kappa)}(\omega). \tag{6.68}$$

The transfer function h can also be expressed in terms of modal functions. Denoting the modal contribution of the κth transfer function to the nth mode as $B_n^{(\kappa)}$, we have

$$h(\mathbf{r}|\mathbf{r}_s^{(\kappa)}; \omega) = \sum_{n=0}^{\infty} B_n^{(\kappa)}(\omega)\psi_n(\mathbf{r}, \omega). \tag{6.69}$$

By inserting Equation (6.69) into Equation (6.68), the reproduced field can be expressed as

$$p_r(\mathbf{r}, \omega) = \sum_{n=0}^{\infty}[R_n(\omega)\psi_n(\mathbf{r}, \omega)], \tag{6.70}$$

$$R_n(\omega) = \sum_{\kappa=1}^{K} B_n^{(\kappa)}(\omega)q^{(\kappa)}(\omega). \tag{6.71}$$

The coefficient R_n represents the modal contribution of the nth mode for expressing the reproduced sound field. Equation (6.71) plays an important role in finding an excitation function q that reproduces the target sound field.

To reproduce the target sound field, its modal coefficients (D_n) should be equal to those of the reproduced sound field ($R_n(\omega)$). Accordingly, we can construct an equation using Equation (6.71),

$$D_n(\omega) = \sum_{\kappa=1}^{K} B_n^{(\kappa)}(\omega) q^{(\kappa)}(\omega), \tag{6.72}$$

which is called the *mode-matching equation*.

In deriving Equations (6.67) and (6.70), we considered an infinite number of modes. In practice, however, only a finite number of modes (N) can be considered; hence, there will be a mode-matching error in the reproduced sound field. Nevertheless, in the mode-matching approach, one can select the modes to reproduce. The distribution of error hence depends on the shape and characteristics of the modes.

Therefore, this is again the elephant and greyhound problem (Chapter 3, Section 3.1). Various types of modes can express a sound field, and depending on the selection of modes, one can control the distribution of the error in space.

Plane waves and spherical harmonics are popular modes. The former is advantageous in that the Fourier transform can be used as a transform operator. The latter has unique radial modes, which allow the size of the controllable region to be related to the number of modes.

Unlike the direct field-matching technique, the mode-matching approach first transforms the sound field to identify the contribution of each mode (D_n). This step is called *encoding* of the sound field. The encoded sound field is the group of modal coefficients extracted from the target or measured sound field to reproduce.

The modal contribution is then delivered to the reproduction system. The reproduction system analyzes the received modal contribution and translates it to calculate the excitation signal of the loudspeakers (q). This process is called *decoding* of the sound field. Because various types of loudspeaker arrangements can be used, the decoding process should be different for dissimilar loudspeaker layouts.

In the following sections, we review several mode-matching approaches using different types of modal functions. The advantage of each modal function and its artifacts, along with the use of a finite number of modes, are discussed.

6.6.2 *Mode-Matching with Plane Waves*

6.6.2.1 Encoding of Sound Field

The use of plane waves as modes is beneficial when we use a planar control source (loudspeaker) or microphone distribution. To formulate the planar geometry of the microphones or loudspeakers, we consider an infinite planar surface S_a at $z = z_a$ that divides the locations of the virtual source and listener (Figure 6.19). Assuming that the virtual source is on the left-hand side of S_a, an arbitrary target sound field produced on the right-hand side of the virtual source ($z > z_a$) can be expressed as an infinite sum of plane waves propagating to the right-hand side. That is,

$$\begin{aligned} p_d(\mathbf{r}, \omega) &= \int_{-\infty}^{\infty}\int_{-\infty}^{\infty} D(k_x, k_y, \omega) e^{i(k_x x + k_y y + k_z z)} dk_x dk_y \\ &= \mathcal{F}_s^{-1}[D(k_x, k_y, \omega) e^{ik_z z}], \end{aligned} \tag{6.73}$$

where $\mathcal{F}_s^{-1}[]$ denotes the inverse spatial Fourier transform with respect to k_x, k_y. The integration kernel $e^{i(k_x x + k_y y + k_z z)}$ is a plane wave propagating in the direction $\mathbf{e}_k = \mathbf{k}/k = (k_x, k_y, k_z)/k$, and

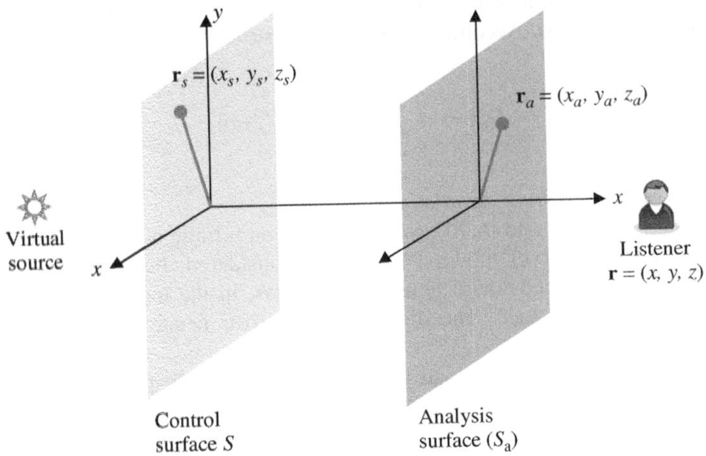

Figure 6.19 Configuration for mode-matching problem using plane waves

$D(k_x, k_y, \omega)$ represents its complex amplitude. Accordingly, the plane wave is the modal function of Equation (6.67),

$$\psi_n(x, y, z) = e^{i(k_x x + k_y y + k_z z)},$$

and D becomes the modal coefficient.

From the wave equation, the wavenumbers (k_x, k_y, k_z) are coupled to each other ($k_z^2 = \sqrt{k^2 - k_x^2 - k_y^2}$ Equation 2.52); hence, only two of the three are independent. In addition, we are considering the plane waves propagating to the right-hand side, so the wavenumber in the z direction (k_z) is always positive ($k_z \geq 0$).

To extract the modal coefficient D, we consider a continuous distribution of microphones arranged on the planar surface S_a. The contribution of each plane wave can then be identified by applying a spatial Fourier transform ($\mathcal{F}_s[\,]$) over the surface. From Equation (6.73), the contribution can be extracted as

$$\begin{aligned} D(k_x, k_y, \omega) e^{ik_z z_a} &= \mathcal{F}_s[p_d(\mathbf{r}_a, \omega)] \\ &= \int_{-\infty}^{\infty} \int_{-\infty}^{\infty} p_d(\mathbf{r}_a, \omega) e^{-i(k_x x_a + k_y y_a)} dx_a dy_a, \end{aligned} \tag{6.74}$$

where $\mathbf{r}_a = (x_a, y_a, z_a)$ denotes a position on S_a. Because the modal coefficient D is identified from the measurement on S_a, we denote the plane S_a as an *analysis surface*.

The contribution of each plane wave identified from the 2D Fourier transform includes an extra term $e^{ik_z z_a}$ that depends on the measurement distance z_a. The distance-independent modal contribution is calculated by dividing the result of the spatial Fourier transform by $e^{ik_z z_a}$:

$$D(k_x, k_y, \omega) = \frac{1}{e^{ik_z z_a}} \mathcal{F}_s[p_d(\mathbf{r}_a, \omega)]. \tag{6.75}$$

Equation (6.75) completely describes the encoding process for planar microphone arrays. The coefficient D is then recorded or transferred to the reproduction system to calculate the excitation signals for multiple loudspeakers.

6.6.2.2 Decoding of Sound Field

Next, consider a continuous loudspeaker array configured on an infinite planar surface S at $z_s < z_a$. Any sound field p_r produced by the loudspeaker array on the control surface S can also be expanded in terms of plane waves as

$$p_r(\mathbf{r}, \omega) = \mathcal{F}_S^{-1}\left[R\left(k_x, k_y, \omega\right) e^{ik_z z}\right], \tag{6.76}$$

where the coefficient $R(k_x, k_y; \omega)$ represents the modal contribution of each plane wave. According to Equation (6.75), the modal contribution can also be identified from the spatial Fourier transform of p_r,

$$R(k_x, k_y, \omega)e^{ik_z z} = \mathcal{F}_s\left[p_r(\mathbf{r}, \omega)\right]. \tag{6.77}$$

The modal coefficient of the reproduced sound field now must be related to the excitation signal in order to find the excitation that can best reproduce the desired modal coefficients. The pressure field reproduced by the planar loudspeaker array can also be described as a superposition of the sound fields from each control source:

$$\begin{aligned} p_r(\mathbf{r}, \omega) &= \int_S \left[g\left(\mathbf{r}|\mathbf{r}_s; \omega\right) q(\mathbf{r}_s, \omega)\right] dS(\mathbf{r}_s) \\ &= \int_{-\infty}^{\infty}\int_{-\infty}^{\infty} \left[g\left(x, y, z|x_s, y_s, z_s; \omega\right) q(x_s, y_s, \omega)\right] dx_s dy_s, \end{aligned} \tag{6.78}$$

The modal coefficient of the reproduced field can then be obtained by taking the 2D Fourier transform of Equation (6.78). That is,

$$R(k_x, k_y, \omega) = \int_{-\infty}^{\infty}\int_{-\infty}^{\infty} B(k_x, k_y|x_s, y_s, z_s; \omega)q(x_s, y_s, \omega)dx_s dy_s, \tag{6.79}$$

where B is the modal coefficient of the Green's function g:

$$B(k_x, k_y|x_s, y_s, z_s; \omega)e^{ik_z z} = \mathcal{F}_s\left[g\left(x, y, z|x_s, y_s, z_s; \omega\right)\right]. \tag{6.80}$$

Note that Equation (6.79) is equivalent to Equation (6.71).

The equation can be further simplified if a Green's function satisfies the shift invariance, that is, if the Green's function depends on the relative distance between the source and listener. For instance, for a monopole control source in a free-field condition, $g(\mathbf{r}|\mathbf{r}_s)$ becomes the free-field Green's function $g_f(\mathbf{r} - \mathbf{r}_s)$ having shift invariance in the space domain. Accordingly, the integral Equation (6.78) can be written as a spatial convolution. That is,

$$\begin{aligned} p_r(\mathbf{r}, \omega) &= \int_S \left[g_f\left(\mathbf{r} - \mathbf{r}_s; \omega\right) q(\mathbf{r}_s, \omega)\right] dS(\mathbf{r}_s), \\ p_r(x, y, z, \omega) &= \int_{-\infty}^{\infty}\int_{-\infty}^{\infty} \left[g_f\left(x - x_s, y - y_s, z - z_s; \omega\right) q(x_s, y_s, \omega)\right] dx_s dy_s. \end{aligned} \tag{6.81}$$

Consider the spatial Fourier transform of the excitation q and free-field Green's function $g_f(x, y, z, \omega)$, defined as

$$Q(k_x, k_y, \omega) = \mathcal{F}_s\left[q\left(x, y, \omega\right)\right], \tag{6.82}$$

$$G_f(k_x, k_y, z; \omega) = \mathcal{F}_s\left[g_f\left(x, y, z; \omega\right)\right]. \tag{6.83}$$

The spatial convolution of Equation (6.81) is equivalent to multiplication in the wavenumber domain. Therefore, using Equation (6.77), we can rewrite Equation (6.81) as

$$R(k_x, k_y, \omega)e^{ik_z z} = G_f(k_x, k_y, z - z_s; \omega)Q(k_x, k_y, \omega). \tag{6.84}$$

From Weyl's identity (Appendix B, Section B.2.1), the spatial Fourier transform of the free-field Green's function G_f is given by

$$G_f(k_x, k_y, z; \omega) = i\frac{e^{ik_z z}}{2k_z}, \quad \text{for } z > 0. \tag{6.85}$$

Substituting Equation (6.85) into Equation (6.84) yields the simplified relation between the modal coefficient of the reproduced field and the source excitation:

$$R(k_x, k_y, \omega) = \frac{ie^{-ik_z z_s}}{2k_z} Q(k_x, k_y, \omega). \tag{6.86}$$

In fact, this is nothing more than an expression for the plane wave propagating from the control surface. The pressure field at the control surface $z = z_s$ is given by $R(k_x, k_y, \omega)e^{ik_z z_s}$. From the Euler equation, we can derive a result equivalent to Equation (6.86):

$$\begin{aligned} Q(k_x, k_y, \omega) &= -2i\omega\rho_0 U(k_x, k_y, \omega) \\ &= -2ik_z\left[R\left(k_x, k_y, \omega\right)e^{ik_z z_s}\right], \end{aligned} \tag{6.87}$$

where $U(k_x, k_y, \omega)$ denotes the spatial Fourier transform of the particle velocity field at the control surface.

Because our aim is to reproduce the modal coefficient $D(k_x, k_y, \omega)$ using a loudspeaker array, we construct an equation for matching the modal coefficient of the reproduced field ($R(k_x, k_y, \omega)$) with that of the target field ($D(k_x, k_y, \omega)$). From Equation (6.86) and $D = R$, we have

$$Q(k_x, k_y, \omega) = -2ik_z e^{ik_z z_s} D(k_x, k_y, \omega). \tag{6.88}$$

The mode-matching approach using plane waves is similar to the planar acoustic holography discussed in Chapter 3, Section 3.3.3. The only difference is that we separate the prediction operators for the encoding and decoding processes. This allows us to extract and deliver a distance-independent property, that is, the complex amplitude $D(k_x, k_y, \omega)$ of the plane waves, and enables the recording and reproduction of the sound field using loudspeaker arrays installed at an arbitrary distance z_s.

Finally, we can obtain the loudspeaker excitation signal that reproduces the encoded sound field. Taking the inverse spatial Fourier transform, we have

$$\begin{aligned} q(x_s, y_s, \omega) &= \frac{1}{4\pi^2}\int_{-\infty}^{\infty}\int_{-\infty}^{\infty} Q(k_x, k_y, \omega)e^{i(k_x x_s + k_y y_s)}dk_x dk_y \\ &= \frac{1}{2\pi^2}\int_{-\infty}^{\infty}\int_{-\infty}^{\infty}\left[-ik_z e^{ik_z z_s} D\left(k_x, k_y, \omega\right)\right]e^{i(k_x x_s + k_y y_s)}dk_x dk_y, \end{aligned} \tag{6.89}$$

with $k_z^2 = \sqrt{k^2 - k_x^2 - k_y^2}$.

6.6.2.3 Ill-Conditioned Problem

There is also an ill-conditioned problem associated with the encoding and decoding of a sound field. The evanescent wave, which is characterized by an imaginary wavenumber in the z direction $(k_z = i\sqrt{k_x^2 + k_y^2 - k^2} = i\zeta)$, is produced when the wavenumber in the x and y directions is higher than k $(k_x^2 + k_y^2 > k^2)$. The evanescent wave shows exponential decay for propagation in the positive z direction $(e^{ik_z z} = e^{-\zeta z})$.

Therefore, at the encoding stage of Equation (6.74), the microphone array cannot measure the evanescent waves with sufficient sensitivity. To extract the distance-independent modal coefficient $D(k_x, k_y, \omega)$ using Equation (6.75), we need a huge amplification $(e^{\zeta z})$ of the spatial Fourier transform. However, if the listener is in the farfield with respect to the virtual source, the evanescent wave would not propagate to the listener's location, so neglecting it in the encoding stage does not cause a serious problem.

6.6.2.4 Effect of Discrete Arrays

So far, we have assumed a continuous distribution of microphones and loudspeakers, which is practically impossible. A discrete distribution of either microphones or loudspeakers produces several artifacts compared to a continuous distribution. Because the spatial Fourier transform is involved in the encoding and decoding in planar geometries, the artifacts can be likened to the aliasing phenomenon, which is quite well-known in the sampling theory of discrete time signal processing. Because of the analogy with temporal sampling, the artifacts induced by spatial sampling of the pressure field are called *spatial aliasing*.

First, let us describe the sound field measured by microphones at discrete positions: $x_a = n\Delta x_a$ and $y_a = n\Delta y_a$. Then the continuous integral of Equation (6.74) can be reduced to the following summations:

$$D_D(k_x, k_y, \omega)e^{ik_z z_a} = \Delta x_a \Delta y_a \sum_{n=-\infty}^{\infty} \sum_{m=-\infty}^{\infty} p_d(m\Delta x_a, n\Delta y_a, z_a, \omega)e^{-i(mk_x \Delta x_a + nk_y \Delta y_a)}. \tag{6.90}$$

Here, D_D represents the modal coefficient of the target field estimated from the discrete measurement. This summation is equivalent to the spatial Fourier transform of the target field multiplied by the periodic impulse trains. That is,

$$D_D(k_x, k_y, \omega)e^{ik_z z_a} = \int_{-\infty}^{\infty} \int_{-\infty}^{\infty} p_{d,D}(x_a, y_a, z_a, \omega)e^{-i(k_x x_a + k_y y_a)} dx_a dy_a, \tag{6.91}$$

where $p_{d,D}$ is defined as

$$p_{d,D}(x_a, y_a, z_a, \omega) = p_d(x_a, y_a, z_a, \omega)(\Delta x_a \Delta y_a) \sum_{n=-\infty}^{\infty} \sum_{m=-\infty}^{\infty} \delta(x_a - m\Delta x_a)\delta(y_a - n\Delta y_a). \tag{6.92}$$

Because the multiplication of two functions in the space domain is represented as a convolution in the wavenumber domain (Appendix A, Equation A10), the modal coefficient D_D can be written as

$$D_D(k_x, k_y, \omega)e^{ik_z z_a} = \left[D\left(k_x, k_y, \omega\right) e^{ik_z z_a}\right] *_{k_x} *_{k_y} \left[\sum_{n=-\infty}^{\infty} \sum_{m=-\infty}^{\infty} \delta\left(k_x - m\Delta k_x\right) \delta(k_y - n\Delta k_y)\right], \tag{6.93}$$

where $\Delta k_x = 2\pi/\Delta x_a$ and $\Delta k_y = 2\pi/\Delta y_a$. The asterisks, $*_{k_x}$ and $*_{k_y}$, represent the convolution operators with respect to k_x and k_y, respectively. Therefore, the true modal coefficient is corrupted by the convolution by impulse trains at $m\Delta k_x$ and $n\Delta k_y$.

The spatial aliasing occurring at the encoding stage can be reduced by using a directional microphone. Specifically, a directional microphone whose beam axis is in the z direction can narrow the bandwidth of the measured sound field and hence can prevent overlapping due to the convolution of periodic impulses. If an ideal directional microphone successfully truncates the spatial bandwidth above $\pm\Delta k_x/2$, $\pm\Delta k_y/2$, we can ensure that the modal coefficient within Δk_x, Δk_y can be free from aliasing artifacts.

The discrete distribution of control sources can also be problematic. A discrete distribution of loudspeakers can be formulated as a spatially sampled excitation function,

$$q_D(x_s, y_s, \omega) = q(x_s, y_s, \omega)(\Delta x_s \Delta y_s)\left[\sum_{n=-\infty}^{\infty}\sum_{m=-\infty}^{\infty} \delta\left(k_x - m\Delta x_s\right)\delta(k_y - n\Delta y_s)\right], \tag{6.94}$$

where Δx_s and Δy_s are the source intervals in the x and y directions, respectively.

The wavenumber spectrum of the sampled excitation function is the convolution of the original wavenumber spectrum and a periodic delta function,

$$Q_D(k_x, k_y, \omega) = Q(k_x, k_y, \omega) *_{k_x} *_{k_y} \left(\sum_{n=-\infty}^{\infty}\sum_{m=-\infty}^{\infty} \delta\left(k_x - m\Delta k_{xs}\right)\delta(k_y - n\Delta k_{ys})\right), \tag{6.95}$$

with $\Delta k_{xs} = 2\pi/\Delta x_s$ and $\Delta k_{ys} = 2\pi/\Delta y_s$. Accordingly, Q_D also becomes periodic function. The modal coefficient, that is, the wavenumber spectrum of the reproduced sound field (R_D), is then given by the product of the Green's function and Q_D:

$$R_D(k_x, k_y, \omega)e^{ik_z z} = G_f(k_x, k_y, z - z_s; \omega) Q_D(k_x, k_y, \omega). \tag{6.96}$$

Therefore, the reproduced spectrum R_D is a periodic function in k_x and k_y weighted by the wavenumber spectrum of the Green's function. The periods of repetition are Δk_{xs} and Δk_{ys}; hence, we can control the spectrum R_D only within $|k_x| \le {}^{\Delta k_{xs}}/_2$, $|k_y| \le {}^{\Delta k_{ys}}/_2$. The periodic repetition of Q_D is a physical phenomenon, so it cannot be removed by signal processing techniques. One can attempt to limit the bandwidth of the original wavenumber spectrum Q to avoid possible overlapping due to the convolution, but the repetition itself is unavoidable. If we consider a band-limited wavenumber spectrum of Q (Q_{LP}) within $|k_x| \le {}^{\Delta k_{xs}}/_2$ and $|k_y| \le {}^{\Delta k_{ys}}/_2$, then Equations (6.95) and (6.96) can be rewritten as

$$\begin{aligned} R_D(k_x, k_y, \omega)e^{ik_z z} &= G_f(k_x, k_y, z - z_s; \omega) Q_{LP}(k_x, k_y, \omega) \\ &\quad + G_f(k_x, k_y, z - z_s; \omega)\sum_{n\neq 0}\sum_{m\neq 0} Q_{LP}(k_x - m\Delta k_{xs}, k_y - n\Delta k_{ys}, \omega) \\ &= R_{LP}(k_x, k_y, \omega)e^{ik_z z} + \varepsilon_{HP}(k_x, k_y, \omega), \end{aligned} \tag{6.97}$$

which implies that we can match the band-limited spectrum R_{LP} within $|k_x| \le {}^{\Delta k_{xs}}/_2$ and $|k_y| \le {}^{\Delta k_{ys}}/_2$, and the aliasing artifact (ε_{HP}) begins to appear at wavenumbers above $\pm\Delta k_{xs}/2$.

The prominent difference between the sampling problem for the decoding stage and that for the encoding stage is in the Green's function g_f. Because the Green's function follows the wave equation, most of its wavenumber spectrum is bounded within $\sqrt{k_x^2 + k_y^2} \le k$. The evanescent wave outside of the radiation circle will be quickly attenuated at some distance from the control sources. Accordingly, the Green's function acts as an anti-aliasing filter that reduces the effect of ε_{HP}.

To use this anti-aliasing characteristic of the Green's function, the wavenumber periods $\pm\Delta k_{xs}/2$ and $\pm\Delta k_{ys}/2$ of Equation (6.97) should be higher than k, which means that the distance between adjacent loudspeakers should be

$$\Delta x_s, \Delta y_s \leq \frac{\lambda}{2}. \tag{6.98}$$

This rule requires dense spatial sampling at higher frequencies.

Another way to decrease ε_{HP} is to modify the Green's function. A loudspeaker with broadside directivity can limit the wavenumber bandwidth to less than k, as explained in Chapter 5, Section 5.4.1. Therefore, the use of directional loudspeakers that filter out the wavenumber components above $\pm\Delta k_{xs}/2$ can suppress the aliasing artifact ε_{HP}.

6.6.2.5 Effect of Finite Aperture

The finite aperture of the loudspeaker or microphone array limits the encoding resolution of the modal functions. The finite aperture can be mathematically represented as truncation of a measured sound field by a finite window function $w(x, y)$. By using the rectangular window function (Equation A5)

$$w(x, y) = \text{rect}(x/L_x)\text{rect}(y/L_y), \tag{6.99}$$

the estimated wavenumber spectrum can be written as

$$\tilde{D}(k_x, k_y, \omega)e^{ik_z z_a} = \int_{-\infty}^{\infty}\int_{-\infty}^{\infty} \left[w\left(x_a, y_a\right) p_d(x_a, y_a, z_a, \omega)\right] e^{-i(k_x x_a + k_y y_a)} dx_a dy_a. \tag{6.100}$$

Multiplication in the space domain is given in convolution form in the wavenumber domain,

$$\tilde{D}(k_x, k_y, \omega) = \frac{1}{4\pi^2} W(k_x, k_y) *_{k_x} *_{k_y} D(k_x, k_y, \omega). \tag{6.101}$$

According to the Fourier transform of the rectangular window function (Appendix A, Section A.1.1), $W(k_x, k_y)$ is given as a 2D sinc function,

$$W(k_x, k_y) = (L_x L_y)\text{sinc}(k_x L_x/2)\text{sinc}(k_y L_y/2). \tag{6.102}$$

The effect of convolution with a 2D sinc function is that the true wavenumber spectrum $D(k_x, k_y, \omega)$ at (k_x, k_y) is mixed with other spectra, so it becomes difficult to separate adjacent plane waves. Consequently, the resolution of encoding in wavenumber domain is degraded as the aperture size L decreases.

Another artifact can arise when a discrete Fourier transform (DFT) is applied to solve the mode-matching equation. It is well known that a wrap-around error is produced when two finite sequences are convolved in terms of a DFT. In the convolution process, which involves the multiplication of two wavenumber spectra, the wrap-around error can be avoided by applying zero-padding to the time-domain sequences. However, the mode-matching equation is a deconvolution process, which requires division of two wavenumber spectra. In the deconvolution process, the original spectrum cannot be completely resolved by zero-padding, and additional constraints are required to derive a unique solution (e.g., [10]). The deconvolution technique has been thoroughly studied; however, for it to be applied to the mode-matching problem, the constraint imposed on the technique has to be compatible with complex data, and the stability of a solution should be verified.

6.6.3 Mode-Matching with Spherical Harmonics

6.6.3.1 Encoding of Sound Field (Interior Problem)

As with the mode-matching in Cartesian coordinates, exactly the same sort of technique can be derived for spherical coordinates. Recall that a spherical harmonic expansion can describe an arbitrary sound field in spherical coordinates ($\mathbf{r} = (r, \theta, \phi)$) in terms of the spherical basis functions. That is,

$$p_d(\mathbf{r}, \omega) = \sum_{n=0}^{\infty} \sum_{m=-n}^{n} D_{nm}(\omega) j_n(kr) Y_n^m(\theta, \phi) \tag{6.103}$$

for a listener in a source-free region V ($\mathbf{r} \in V$). The coefficients of spherical harmonic expansion D_{nm} express the contributions of spherical harmonic functions constituting the target field p_d. The harmonic coefficients D_{nm} include all the information on the spatial sound distribution. In this regard, the spherical harmonic expansion can be regarded as a *spatial encoder* of a sound field in spherical coordinates.

A generic procedure for extracting the spherical harmonic coefficient of a given sound field is to take a measurement over an analysis surface S_a surrounding a source-free volume V. A convenient choice of analysis surface is a spherical surface of radius r_a (Figure 6.20). The volume V then becomes a source-free region if $r_v > r_a$ for any virtual source at $\mathbf{r}_v = (r_v, \theta_v, \phi_v)$. The spherical harmonic coefficient multiplied by the range dependency $j_n(kr)$ can be extracted by integrating the target sound field with a complex conjugate of a spherical harmonic of a specific order and degree (Equation (B137)). That is,

$$\begin{aligned} j_n(kr_a) D_{nm}(\omega) &= \frac{1}{r_a^2} \int_{S_a} p_d(\mathbf{r}_a, \omega) Y_n^m(\theta_a, \phi_a)^* dS_a \\ &= \int_{\Omega} p_d(\mathbf{r}_a, \omega) Y_n^m(\theta_a, \phi_a)^* d\Omega_a, \end{aligned} \tag{6.104}$$

where $\mathbf{r}_a = (r_a, \theta_a, \phi_a)$ is a position vector with respect to the analysis surface, and $\int_{\Omega} d\Omega_a = \int_0^{2\pi} \int_0^{\pi} \sin\theta_a d\theta_a d\phi_a$. Because the extracted coefficient depends on the measurement geometry $j_n(kr_a)$, measurement-independent encoding $D_{nm}(\omega)$ can be obtained by dividing Equation (6.104) by $j_n(kr_a)$.

For example, consider a virtual monopole source at $\mathbf{r}_v$. Using the representation of the Green's function in spherical coordinates (Equation (B174)), we have

$$\begin{aligned} p_d(\mathbf{r}_a, \omega) &= g_f(r_a, \theta_a, \phi_a | r_v, \theta_v, \phi_v; \omega) \\ &= ik \sum_{n=0}^{\infty} j_n(kr_a) h_n^{(1)}(kr_v) \sum_{m=-n}^{n} Y_n^m(\theta_v, \phi_v)^* Y_n^m(\theta_a, \phi_a) \quad \textit{for } r_a \le r_v, \end{aligned} \tag{6.105}$$

From Equations (6.103) and (6.105), spherical harmonic expansion of the sound field p_d will yield the following harmonic coefficients $D_{nm}(\omega)$:

$$\begin{aligned} D_{nm}(\omega)_{mono} &= \frac{1}{j_n(kr_a)} \int_{\Omega} p_d(\mathbf{r}_a, \omega) Y_n^m(\theta_a, \phi_a)^* d\Omega_a \\ &= ikh_n^{(1)}(kr_v) Y_n^m(\theta_v, \phi_v)^*. \end{aligned} \tag{6.106}$$

As another example, if a virtual source produces a plane wave, the harmonic coefficient is given by

$$D_{nm}(\omega)_{plane} = 4\pi i^n Y_n^m(\theta_v, \phi_v)^*. \tag{6.107}$$

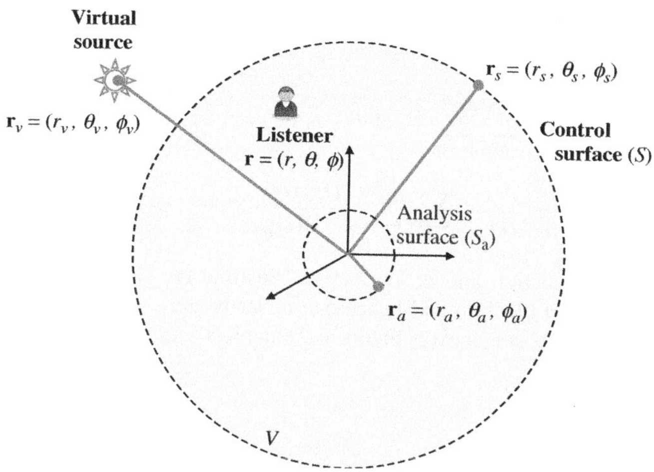

Figure 6.20 Encoding and decoding problem for interior reproduction in spherical coordinates

The encoded data D_{nm} are then transferred to a reproduction system in such a way that the spatial characteristics of the sound field can be reproduced by multiple loudspeakers.

6.6.3.2 Ill-Posed and Ill-Conditioned Problems in Encoding

The encoding procedure can suffer from the nonuniqueness problem at certain frequencies, that is, when $j_n(\omega r_a/c) = 0$ at a given frequency ω. These frequencies, called *forbidden frequencies*, can occur irrespective of the shape of the interior volume. The physical meaning of this phenomenon is that we cannot identify the contribution of a certain harmonic function if it can be zero over the entire surface. For instance, when $j_n(kr_a) = 0$ in the above example, we cannot observe the spherical harmonic of order n even if it readily exists in p_d. This is a typical observability problem, which can arise when the interior volume surrounded by the analysis surface has a resonance at those frequencies. Specifically, consider a (fictitious) pressure-release boundary condition imposed on the analysis surface. At the resonance frequency of the interior volume surrounded by the surface, a sound field of which the pressure distribution is zero over the entire surface can exist.

To circumvent the nonuniqueness problem, a rigid microphone array is a popular option. If a rigid boundary condition is imposed on a spherical surface, the incident wave p_d is scattered from the surface. Because the scattered wave p_{sc} is described by an outward radiating function, its spherical harmonic expansion can be described in terms of the spherical Hankel function:

$$p_{sc}(\mathbf{r}, \omega) = \sum_{n=0}^{\infty} \sum_{m=-n}^{n} S_{nm}(\omega) h_n^{(1)}(kr) Y_n^m(\theta, \phi). \tag{6.108}$$

To satisfy the rigid boundary condition at $r = r_a$, the radial derivative of the pressure field should be zero.

$$\left.\frac{\partial \left(p_d(\mathbf{r}, \omega) + p_{sc}(\mathbf{r}, \omega)\right)}{\partial r}\right|_{r=r_a} = k \sum_{n=0}^{\infty} \sum_{m=-n}^{n} (D_{nm}(\omega) j_n'(kr_a) + S_{nm}(\omega) h_n'^{(1)}(kr_a)) Y_n^m(\theta_a, \phi_a) = 0. \tag{6.109}$$

From Equation (6.109) and the Wronskian relation (Equation (B164)), the total field measured at the spherical surface is given by

$$p_d(\mathbf{r}_a,\omega)+p_{sc}(\mathbf{r}_a,\omega)=\sum_{n=0}^{\infty}\sum_{m=-n}^{n}D_{nm}(\omega)\left(j_n(kr_a)-\frac{j_n'(kr_a)}{h_n'^{(1)}(kr_a)}h_n^{(1)}(kr_a)\right)Y_n^m(\theta_a,\phi_a)$$

$$=\frac{i}{(kr_a)^2}\sum_{n=0}^{\infty}\sum_{m=-n}^{n}\frac{D_{nm}(\omega)}{h_n'^{(1)}(kr_a)}Y_n^m(\theta_a,\phi_a), \qquad (6.110)$$

The spherical Hankel function and its derivative illustrate the radial propagation of spherical harmonics in the unbounded medium and hence can never be zero. Therefore, the coefficient D_{nm} can be extracted without any singularity. Figure 6.21 depicts a commercial spherical microphone with a rigid spherical surface.

Figure 6.21 Example of spherical microphone array: em32 Eigenmike® microphone array produced by mh acoustics

The sensitivity of the array is an important performance measure for a microphone. The amplification of a single spherical harmonics of order (Equation (6.110) at the surface can be approximated as

$$\frac{1}{(kr_a)^2}\frac{1}{h_n'^{(1)}(kr_a)} \approx \frac{(kr_a)^n}{i(n+1)(2n-1)!!} \tag{6.111}$$

for an acoustically small sphere (using the small argument approximation of Equation (B159)). Therefore, the sensitivity of a small spherical array is low, particularly at low frequencies and high-order harmonics (Figure 6.22). The higher-order harmonics are more vulnerable to the low-sensitivity problem; thus, the spherical microphone array should be designed by considering the trade-off between compactness and low sensitivity. For high frequencies, the large argument approximation (Equation (B161)) yields

$$\frac{1}{(kr_a)^2}\frac{1}{h_n'^{(1)}(kr_a)} \approx i^n \frac{e^{-ikr_a}}{kr_a}, \tag{6.112}$$

which is inversely proportional to $\omega = k/c$. The high-frequency sensitivity drops with increasing frequency; however, the slope is moderate compared to the low-frequency roll-off.

6.6.3.3 Decoding of Sound Field (Interior Problem)

Once the sound field is encoded, the reproduction system (such as Figure 6.23) analyzes and calculates the loudspeaker excitation signals using the encoded data. This process is referred to as the decoding of sound field. Because the loudspeaker array geometry can vary, the decoded signal is a function of the loudspeaker arrangement. However, in terms of spherical harmonics, the

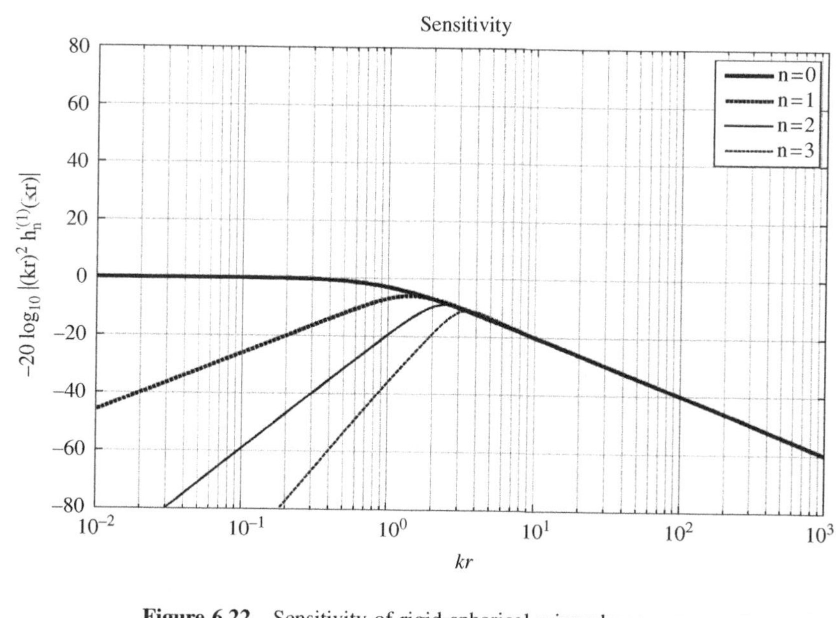

Figure 6.22 Sensitivity of rigid spherical microphone array

(a) (b)

Figure 6.23 (a,b) Spherical loudspeaker array for interior reproduction: 50 channel Lebedev sphere. Center for Noise and Vibration Control (NOVIC), Korea Advanced Institute of Science and Technology (KAIST), 2012

arbitrary sound field generated by exterior loudspeakers ($r_s > r$) can be written as

$$p_r(\mathbf{r}, \omega) = \sum_{n=0}^{\infty} \sum_{m=-n}^{n} R_{nm}(\omega) j_n(kr) Y_n^m(\theta, \phi). \tag{6.113}$$

From analogy with Equation (6.106), the spherical harmonic coefficient of the reproduced field is given by

$$R_{nm}(\omega) = \frac{1}{j_n(kr)} \int_{4\pi} p_r(\mathbf{r}, \omega) Y_n^m(\theta, \phi)^* d\Omega. \tag{6.114}$$

As a simple example, consider a spherical loudspeaker array for which monopole sources are distributed over a spherical surface of radius $r_s > r$. Then the pressure field reproduced by control sources with the excitation signal $q(\mathbf{r}_s)$ may be written as

$$\begin{aligned} p_r(\mathbf{r}, \omega) &= \int_S [g_f(\mathbf{r}|\mathbf{r}_s; \omega) q(\mathbf{r}_s, \omega)] dS(\mathbf{r}_s) \\ &= r_s^2 \int_{4\pi} [g_f(r, \theta, \phi | r_s, \theta_s, \phi_s; \omega) q(r_s, \theta_s, \phi_s, \omega)] d\Omega_s. \end{aligned} \tag{6.115}$$

Using spherical harmonic expansion of a monopole sound field (Equation (B174) and (6.114)), the reproduced field can be encoded as

$$R_{nm}(\omega) = ikr_s^2 h_n^{(1)}(kr_s) \int_{4\pi} \left[Y_n^m(\theta_s, \phi_s)^* q(r_s, \theta_s, \phi_s, \omega) \right] d\Omega_s. \tag{6.116}$$

The sound field can then be reproduced by equating the spherical harmonics of the reproduced and target fields. However, to do that, we need to know how to translate the source excitation q into the desired spherical harmonic coefficients R_{nm}. The integral of Equation (6.116) over the entire solid angle $\Omega = 4\pi$ is equivalent to the spherical harmonic coefficient of the excitation q. In an analogous manner to Equation (6.82), we can denote the coefficient as

$$Q_{nm}(\omega) = \int_{4\pi} Y_n^m(\theta_s, \phi_s)^* q(r_s, \theta_s, \phi_s, \omega) d\Omega_s, \tag{6.117}$$

or conversely the excitation q can be written as

$$q(\mathbf{r}_s, \omega) = \sum_{n=0}^{\infty} \sum_{m=-n}^{n} Q_{nm}(\omega) Y_n^m(\theta_s, \phi_s). \tag{6.118}$$

From Equations (6.116) and (6.117), we have

$$R_{nm}(\omega) = ikr_s^2 h_n^{(1)}(kr_s) Q_{nm}(\omega). \tag{6.119}$$

Substituting Equation (6.119) into Equation (6.113) gives

$$p_r(\mathbf{r}, \omega) = ikr_s^2 \sum_{n=0}^{\infty} j_n(kr) h_n^{(1)}(kr_s) \sum_{m=-n}^{n} Q_{nm}(\omega) Y_n^m(\theta, \phi). \tag{6.120}$$

Equations (6.119) and (6.118) express the relationship between the source excitation and the spherical harmonic coefficients of the reproduced field. Interestingly, the spherical harmonic expansion of the spatial convolution (Equation (6.115)) is equivalent to the product of the spherical harmonics of the Green's function ($ikr_s^2 h_n^{(1)}$) and the excitation function ($Q_{nm}(\omega)$). This can be seen as a product theorem for convolution in spherical coordinates.

Because the spherical harmonics of the target field should be equal to that of the reproduced field, we can construct a mode-matching equation that relates the spherical harmonic coefficients of the target and excitation signals:

$$D_{nm}(\omega) = R_{nm}(\omega) = ikr_s^2 h_n^{(1)}(kr_s) Q_{nm}(\omega). \tag{6.121}$$

The decoding is then completed by calculating the excitation signal q using the encoded data D_{nm}.

$$\begin{aligned} q(\mathbf{r}_s, \omega) &= \sum_{n=0}^{\infty} \sum_{m=-n}^{n} Q_{nm}(\omega) Y_n^m(\theta_s, \phi_s)^* \\ &= \sum_{n=0}^{\infty} \sum_{m=-n}^{n} \frac{D_{nm}(\omega)}{ikr_s^2 h_n^{(1)}(kr_s)} Y_n^m(\theta_s, \phi_s)^*. \end{aligned} \tag{6.122}$$

For a virtual monopole source, the excitation function can be described directly in terms of the locations of the virtual and control sources. Inserting Equation (6.106) into Equation (6.122), we have

$$q(\mathbf{r}_s, \omega) = \frac{1}{r_s^2} \sum_{n=0}^{\infty} \sum_{m=-n}^{n} \frac{h_n^{(1)}(kr_v)}{h_n^{(1)}(kr_s)} Y_n^m(\theta_s, \phi_s)^* Y_n^m(\theta_v, \phi_v). \tag{6.123}$$

The loudspeaker excitation for a virtual point source is often denoted as a *panning function.* For an exterior virtual source, the angular dependency of the panning function is determined by the Legendre polynomial. On the other hand, the radial dependency is related to the spherical Hankel function of the first kind. Because the amplitude of the spherical Hankel function is a monotonically decreasing function with an increasing argument, the ratio of two spherical Hankel functions depends on the locations of the virtual and control sources. For example, if both r_v and r_s are much greater than the wavelength, the large argument approximation of the spherical Hankel function (Equation (B161)) yields

$$q(\mathbf{r}_s, \omega) \approx \frac{1}{r_s r_v} \sum_{n=0}^{\infty} Y_n^m(\theta_s, \phi_s)^* Y_n^m(\theta_v, \phi_v). \tag{6.124}$$

This simplified panning function implies that the radial dependency (spherical Hankel functions) can be independent of the harmonic order for the farfield virtual source and control sources. Ideally, the infinite summation of this series converges to a delta function, which implies that only the secondary source whose direction coincides with that of the virtual source is activated:

$$q(\mathbf{r}_s, \omega) \approx \frac{1}{r_s r_v} \delta(\cos\theta_s - \cos\theta_v)\delta(\phi_s - \phi_v). \tag{6.125}$$

6.6.3.4 Encoding and Decoding for Exterior Reproduction Problem

For the exterior reproduction problem, the virtual sound source is located within a volume V, and its sound field propagates outward. The radiating field is then measured using microphone arrays surrounding the sound source (Figures 6.24 and 6.25a). The measured data can be expressed in terms of the spherical Hankel function as

$$p_d(\mathbf{r}_a, \omega) = \sum_{n=0}^{\infty} \sum_{m=-n}^{n} D_{nm}(\omega) \left[h_n^{(1)}(kr_a) Y_n^m(\theta_a, \phi_a) \right] \tag{6.126}$$

at the analysis surface $\mathbf{r}_a \in S_a$. Using the same procedure as for the interior reproduction problem, the coefficient of the spherical harmonics can be encoded as

$$D_{nm}(\omega) = \frac{1}{h_n^{(1)}(kr_a)} \int_{\Omega} p_d(\mathbf{r}_a, \omega) Y_n^m(\theta_a, \phi_a)^* d\Omega_a \tag{6.127}$$

for a spherical analysis surface of radius r_a.

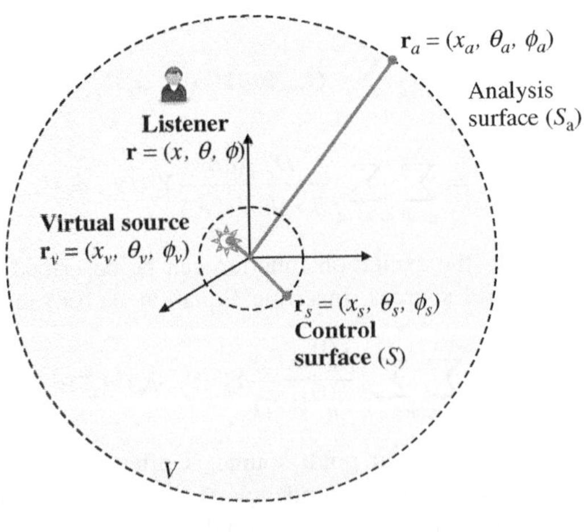

Figure 6.24 Encoding and decoding problem for exterior reproduction in spherical coordinates

(a) (b)

Figure 6.25 (a) Spherical microphone and (b) loudspeaker array for exterior reproduction problem. (IEM, Graz; photograph courtesy of Franz Zotter and Hannes Pomberger.)

Note that the spherical Hankel function of Equation (6.127) does not cause the nonuniqueness or singularity problems, unlike the spherical Bessel function of Equation (6.104). Accordingly, no special treatment is required to extract the coefficient D_{mn} in the encoding step. Furthermore, the large-argument approximation of the spherical Hankel function converges to $1/(kr_a)$, so the sensitivity of the measured pressure at the surface S_a is nearly constant irrespective of the order of harmonics. The change in sensitivity with respect to the frequency change is given as a first-order high-pass filter, the magnitude of which is inversely proportional to the frequency.

The sound field is then decoded by an interior loudspeaker array facing outward. The sound field radiating outward can be expanded to

$$p_r(\mathbf{r}, \omega) = \sum_{n=0}^{\infty} \sum_{m=-n}^{n} R_{nm}(\omega) h_n^{(1)}(kr) Y_n^m(\theta, \phi), \tag{6.128}$$

$$R_{nm}(\omega) = \frac{1}{h_n^{(1)}(kr_a)} \int_{4\pi} p_r(\mathbf{r}_a, \omega) Y_n^m(\theta_a, \phi_a)^* d\Omega_a. \tag{6.129}$$

If we again consider a simple spherical loudspeaker array, the reproduced field can be expressed as

$$\begin{aligned} R_{nm}(\omega) &= ikr_s^2 j_n(kr_s) \int_{4\pi} \left[Y_n^m(\theta_s, \phi_s)^* q(r_s, \theta_s, \phi_s, \omega) \right] d\Omega_s \\ &= ikr_s^2 j_n(kr_s) Q_{nm}(\omega) \end{aligned} \tag{6.130}$$

using the spherical harmonic representation of the monopole and the surface integral of Equation (6.115). Note that the boundary S of the sphere is acoustically transparent and does not induce any scattering.

We can then construct a mode-matching equation for the encoded data D_{nm} and the harmonic coefficient of the reproduced field.

$$D_{nm}(\omega) = R_{nm}(\omega) = ikr_s^2 j_n(kr_s) Q_{nm}(\omega). \tag{6.131}$$

The excitation signal q for reproducing the encoded data is given by

$$\begin{aligned} q(\mathbf{r}_s, \omega) &= \sum_{n=0}^{\infty} \sum_{m=-n}^{n} Q_{nm}(\omega) Y_n^m(\theta_s, \phi_s)^* \\ &= \sum_{n=0}^{\infty} \sum_{m=-n}^{n} \frac{D_{nm}(\omega)}{ikr_s^2 j_n(kr_s)} Y_n^m(\theta_s, \phi_s)^*. \end{aligned} \tag{6.132}$$

For a monopole virtual source at $\mathbf{r}_v$, the excitation function can be directly described as

$$q(\mathbf{r}_s, \omega) = \frac{1}{r_s^2} \sum_{n=0}^{\infty} \sum_{m=-n}^{n} \frac{j_n(kr_v)}{j_n(kr_s)} Y_n^m(\theta_s, \phi_s)^* Y_n^m(\theta_v, \phi_v). \tag{6.133}$$

The decoding Equation (6.133) contains the spherical Bessel function in its denominator and can be singular when $j_n(kr_s) = 0$. This problem resembles the nonuniqueness problem of interior reproduction, but the associated physical situation is completely different. The singularity occurs because the loudspeaker array at the same radius cannot generate the related radial modes, as shown in Equation (6.130). The excitation signal of Equation (6.133) therefore goes to infinity with $j_n(kr_s) = 0$ in order to amplify the small output produced by the loudspeaker array.

The singularity can also be circumvented by introducing a rigid sphere (Figure 6.25b). By imposing a rigid boundary condition on the loudspeaker array, the particle velocity distribution on the surface can be expanded to

$$u(\mathbf{r}_s, \omega) = \sum_{n=0}^{\infty} \sum_{m=-n}^{n} U_{nm}(\omega) h_n^{(1)}(kr_s) Y_n^m(\theta_s, \phi_s). \tag{6.134}$$

From Euler equation, the velocity field is related to the pressure field as

$$i\rho_0 \omega [U_{nm}(\omega) h_n^{(1)}(kr_s)] = k h_n'^{(1)}(kr_s) R_{nm}(\omega). \tag{6.135}$$

Therefore, the surface velocity distribution for reproducing the target field ($R_{nm} = D_{nm}$) is given by

$$u(\mathbf{r}_s, \omega) = \frac{1}{i\rho_0 c} \sum_{n=0}^{\infty} \sum_{m=-n}^{n} D_{nm}(\omega) h_n'^{(1)}(kr_s) Y_n^m(\theta_s, \phi_s)^*. \tag{6.136}$$

The particle velocity can then be converted into the source excitation (inertial force, Equation (1.67))

$$\begin{aligned} q(\mathbf{r}_s, \omega) &= -i\rho_0 \omega u(\mathbf{r}_s, \omega) \\ &= -k \sum_{n=0}^{\infty} \sum_{m=-n}^{n} D_{nm}(\omega) h_n'^{(1)}(kr_s) Y_n^m(\theta_s, \phi_s)^*. \end{aligned} \tag{6.137}$$

6.6.3.5 Sensitivity of Spherical Loudspeaker Array

The sensitivity of a loudspeaker array is also affected by its size. From Equations (6.128) and (6.135), the general relation between the surface velocity distribution and reproduced field can be written as

$$p_r(\mathbf{r};\omega) = i\rho_0 c \sum_{n=0}^{\infty}\sum_{m=-n}^{n}\left[U_{nm}(\omega)\frac{h_n^{(1)}(kr_s)}{h_n'^{(1)}(kr_s)}\right]h_n^{(1)}(kr)Y_n^m(\theta,\phi). \tag{6.138}$$

The modal weight consisting of spherical Hankel functions $(h_n^{(1)}(kr_s)/h_n'^{(1)}(kr_s))$ is related to the sensitivity of the loudspeaker array or the radiation efficiency of each harmonic. If the weight is small, the corresponding spherical harmonic coefficient is attenuated, and the array cannot radiate efficiently.

For an acoustically small array ($kr_s \ll 1$), small argument approximation of the spherical Hankel functions (Equation (B159)) would give

$$\frac{h_n^{(1)}(kr_s)}{h_n'^{(1)}(kr_s)} \approx -\frac{kr_s}{(n+1)}, \tag{6.139}$$

which is a first-order high-pass filter proportional to $\omega = ck$. Therefore, more excitation amplitude on the spherical surface is required at lower frequencies, and the response is biased by the harmonic order $(n+1)$.

At high frequencies or for an acoustically large array ($kr_s \gg 1$), the large argument approximation applies, and the sensitivity yields

$$\frac{h_n^{(1)}(kr_s)}{h_n'^{(1)}(kr_s)} \approx -i, \tag{6.140}$$

which converges to a constant value across all frequencies and harmonic orders (Figure 6.26).

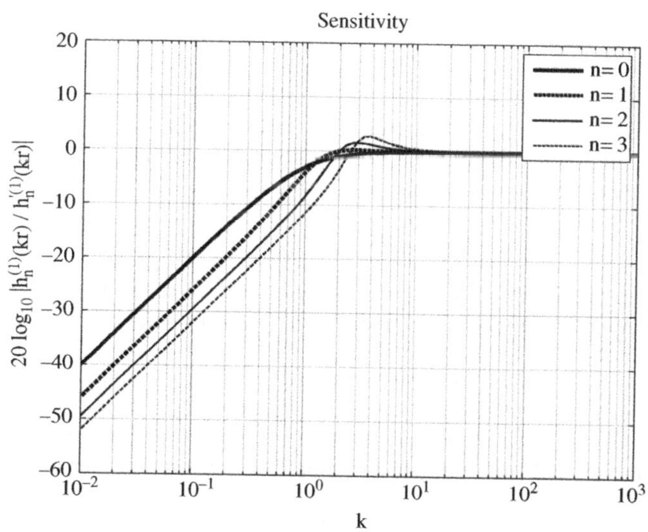

Figure 6.26 Sensitivity curve of a spherical loudspeaker array

6.6.3.6 Encoding with a Finite Number of Microphones

A continuous distribution of microphones can rarely be implemented in practice. In practical applications, measurement is implemented with a discrete distribution of microphones. As for the linear microphone array, the discrete distribution can be treated as spatial sampling. In this section, we discuss methods of achieving efficient sampling on a sphere.

The encoding of a sound field is all about spherical harmonic expansion; therefore, we must investigate spatial sampling in relation to spherical harmonic expansion. The discrete form of the spherical harmonic expansion of Equation (6.104) can generally be described as

$$j_n(kr_a)D_{nm}(\omega) = \sum_{l=1}^{L} p_d(\mathbf{r}_a^{(l)})Y_n^m(\theta_a^{(l)}, \phi_a^{(l)})^* \Delta\Omega^{(l)} \tag{6.141}$$

for the discrete positions $\mathbf{r}_a^{(l)} = [r_a, \theta_a^{(l)}, \phi_a^{(l)}]$ over a sphere of radius r_a. The factor $\Delta\Omega^{(l)}$ is analogous to the infinitesimal solid angle $d\Omega$ of Equation (6.104); hence, it represents the weighting of each discrete position to realize a continuous integral.

If we consider the spherical harmonic expansion to degree N, a total of $(N+1)^2$ $(= \sum_{n=0}^{N}\sum_{m=-n}^{n} 1)$ harmonic coefficients are required to express a given sound field. The discrete summation of Equation (6.141) can be rewritten in matrix form as

$$\mathbf{d} = \frac{1}{j_n(kr_a)}\left[\mathbf{\Psi}^H\mathbf{\Omega}\right]\mathbf{p}_d, \tag{6.142}$$

where $\mathbf{d}$ is an $(N+1)^2 \times 1$ vector containing the spherical harmonic coefficients D_{nm} of the target field, and vector $\mathbf{p}_d$ represents the target pressure field sampled at given microphone positions. The $(N+1)^2 \times L$ spherical harmonic matrix $\mathbf{\Psi}$ is defined as

$$\mathbf{\Psi} = \begin{bmatrix} Y_0^0\left(\theta_a^{(1)}, \phi_a^{(1)}\right) & Y_1^{-1}(\theta_a^{(1)}, \phi_a^{(1)}) & \cdots & Y_N^N(\theta_a^{(1)}, \phi_a^{(1)}) \\ Y_0^0(\theta_a^{(2)}, \phi_a^{(2)}) & Y_1^{-1}(\theta_a^{(2)}, \phi_a^{(2)}) & \cdots & Y_N^N(\theta_a^{(2)}, \phi_a^{(2)}) \\ \vdots & \vdots & \ddots & \vdots \\ Y_0^0(\theta_a^{(L)}, \phi_a^{(L)}) & Y_1^{-1}(\theta_a^{(L)}, \phi_a^{(L)}) & & Y_N^N(\theta_a^{(L)}, \phi_a^{(L)}) \end{bmatrix}, \tag{6.143}$$

and the weighting matrix $\mathbf{\Omega}$ has $\Delta\Omega^{(l)}$ as its diagonal elements.

The quality of the discrete spherical harmonic expansion depends on the sampling positions and relevant weightings. One necessary property of the discrete sampling scheme is the orthogonality of the spherical harmonics; that is, the spherical harmonic matrix should satisfy

$$\left[\mathbf{\Psi}^H\mathbf{\Omega}\right]\mathbf{\Psi} = \mathbf{I} \tag{6.144}$$

to realize continuous expansion without error. If the orthogonality condition cannot be met, discrete spherical harmonic expansion of a single spherical harmonic Y_n^m would produce multiple harmonic coefficients.

Numerous studies have been conducted to identify efficient sampling positions and weightings. The simplest method is to design the sampling points in the θ and ϕ directions separately. The spherical harmonics is the harmonic function in the ϕ direction, so regular sampling at equal angular distances yields the orthogonality relation. For sampling in the θ direction, we can use Gaussian–Legendre quadrature (e.g., [11]) that exactly integrates the Legendre polynomials. However, this two-way approach is somewhat inefficient because it requires $(N+1)^2/2$ sampling points ($(N+1)/2$ points for the θ direction and $N+1$ points for the ϕ direction) to integrate the spherical harmonics up to degree N.

One of the more efficient sampling schemes, known as Lebedev quadrature [12], has been popular for obtaining the surface integral over a sphere. A Lebedev grid is constructed so as to have invariance with respect to octahedral rotation with inversion; it is known to exactly integrate all the spherical harmonics up to degree N. However, the nodes cannot be calculated for an arbitrary degree and are currently solved up to 131th degree. The number of sampling points required to integrate harmonics up to the N th degree is approximately $(N+1)^2/3$, which is 3/2 times more efficient than Gaussian–Legendre quadrature. Figure 6.27 illustrates Lebedev quadrature of different degrees.

Even if a given sampling grid is designed to satisfy the orthogonality up to degree N, the higher-order harmonics ($n > N$) included in the target sound field would produce an artifact. This can be likened to the spatial aliasing of a linear microphone array. As we apply an anti-aliasing filter to reduce the aliasing components before sampling, the higher-order harmonics should be reduced before they are sampled by the microphone array.

Anti-aliasing can be realized by modifying the microphone array hardware such that the higher-order harmonics over the array surface are minimized. For example, the rigid spherical array considered in Equation (6.110) naturally decreases the higher-order contribution at low kr_a. Therefore, the measured higher-order spherical harmonic coefficient is naturally attenuated by Equation (6.111), so the small microphone array acts as a natural low-pass filter. Then we can adjust the radial compensation factor at the reconstruction stage. Specifically, the sound field on the surface of a microphone array is given by

$$p_d(\mathbf{r}_a, \omega) + p_{sc}(\mathbf{r}_a, \omega) = \frac{i}{(kr_a)^2} \sum_{n=0}^{\infty} \sum_{m=-n}^{n} \frac{D_{nm}(\omega)}{h_n'^{(1)}(kr_a)} Y_n^m(\theta_a, \phi_a). \tag{6.145}$$

If we apply a discrete spherical harmonic expansion, the estimated harmonic coefficients can be written as

$$\begin{aligned}
\frac{i}{(kr_a)^2 h_{n'}'^{(1)}(kr_a)}[D_{Dn'm'}(\omega)] &= \sum_{l=1}^{L} [(p_d(\mathbf{r}_a^{(l)}, \omega) + p_{sc}(\mathbf{r}_a^{(l)}, \omega)) Y_{n'}^{m'}(\theta_a^{(l)}, \phi_a^{(l)})^* \Delta\Omega^{(l)}] \\
&= \sum_{n=0}^{\infty} \sum_{m=-n}^{n} \left[\frac{D_{nm}(\omega)}{h_n'^{(1)}(kr_a)} \sum_{l=1}^{L} (Y_n^m(\theta_a^{(l)}, \phi_a^{(l)}) Y_{n'}^{m'}(\theta_a^{(l)}, \phi_a^{(l)})^* \Delta\Omega^{(l)}) \right] \\
&= \frac{i}{(kr_a)^2} \sum_{n=0}^{\infty} \sum_{m=-n}^{n} \frac{D_{nm}(\omega)}{h_n'^{(1)}(kr_a)} (\delta_{nn'}\delta_{mm'} + \varepsilon_{nn'mm'}),
\end{aligned} \tag{6.146}$$

where $\varepsilon_{nn'mm'}$ represents the discrete harmonic expansion of the higher-order spherical harmonics ($n > n'$). Compensating for the radial dependency $i/(kr_a)^2 h_{n'}'^{(1)}(kr_a)$, the estimated coefficient is given by

$$D_{Dn'm'}(\omega) = D_{n'm'}(\omega) + \sum_{n=0}^{\infty} \sum_{m=-n}^{n} \frac{h_{n'}'^{(1)}(kr_a)}{h_n'^{(1)}(kr_a)} D_{nm}(\omega) \varepsilon_{nn'mm'}. \tag{6.147}$$

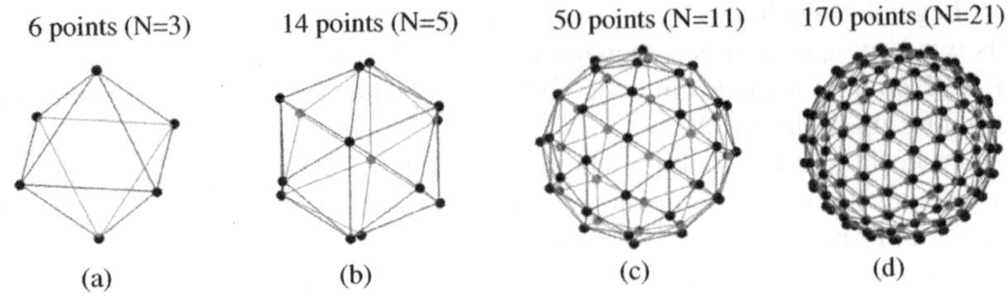

Figure 6.27 (a–d) Lebedev sphere with different degrees (number of points $\approx (N+1)^2/3$)

Therefore, the aliased component $\varepsilon_{nn'mm'}D_{nm}(\omega)$ is attenuated by $h'^{(1)}_{n'}(kr_a)/h'^{(1)}_{n}(kr_a)$. If the rigid sphere is smaller than the wavelength ($kr_a << 1$), we have

$$\frac{h'^{(1)}_{n'}(kr_a)}{h'^{(1)}_{n}(kr_a)} \approx \frac{(n'+1)(2n'-1)!!}{(n+1)(2n-1)!!}(kr_a)^{(n-n')}, \tag{6.148}$$

which indicates that the aliased component can be suppressed by decreasing the size of the microphone array. Because the sensitivity of an array is inversely proportional to kr_a, there is a trade-off between aliasing suppression and sensitivity.

Another approach to reducing the aliasing artifact is to increase the diaphragm area of each microphone sensor or, equivalently, to use directional microphones. According to the convolution theorem of spherical harmonics [13], a directional microphones can efficiently reduce the bandwidth of the measured spherical harmonic coefficients. A detailed derivation and application of this technique can be found in [14].

These anti-aliasing techniques reduce the higher-order harmonics, so the encoded data are limited or truncated in order. If we record a finite number of modes, the details of the sound field that can be expressed by higher-order harmonics would be omitted. Thus, a mode-truncation error will occur when mode-truncated encoding is reproduced by a loudspeaker array. However, as the order of the spherical harmonics increases, the higher-order spherical Hankel or Bessel functions contribute most to the farfield away from the center; consequently, the mode-truncation error tends to disappear at the center of the spherical coordinates (the center of the microphone array). Figure 6.28 shows the monopole sound fields truncated at different orders. Monopole truncation at $N = 15$ shows a much larger sweet spot from the origin of the coordinate (0 m, 0 m) than the sound field truncated at low order ($N = 6$).

On the other hand, the truncation error reveals another important advantage of mode-matching by spherical harmonics: the center position is less disturbed by modal truncation. This property enables scalable encoding and decoding of the sound field; low-order encoding can be transmitted when the bandwidth of the transmission channel or the number of loudspeakers is limited. Even with low-order encoding and decoding, the reproduction quality can be ensured at the center position.

6.6.3.7 Decoding with a Finite Number of Loudspeakers

The reproduced sound field can generally be expressed as

$$p_r(\mathbf{r},\omega) = \sum_{\kappa=1}^{K} g(\mathbf{r}|\mathbf{r}_s^{(\kappa)};\omega)q^{(\kappa)}(\omega)\Delta S^{(\kappa)},$$

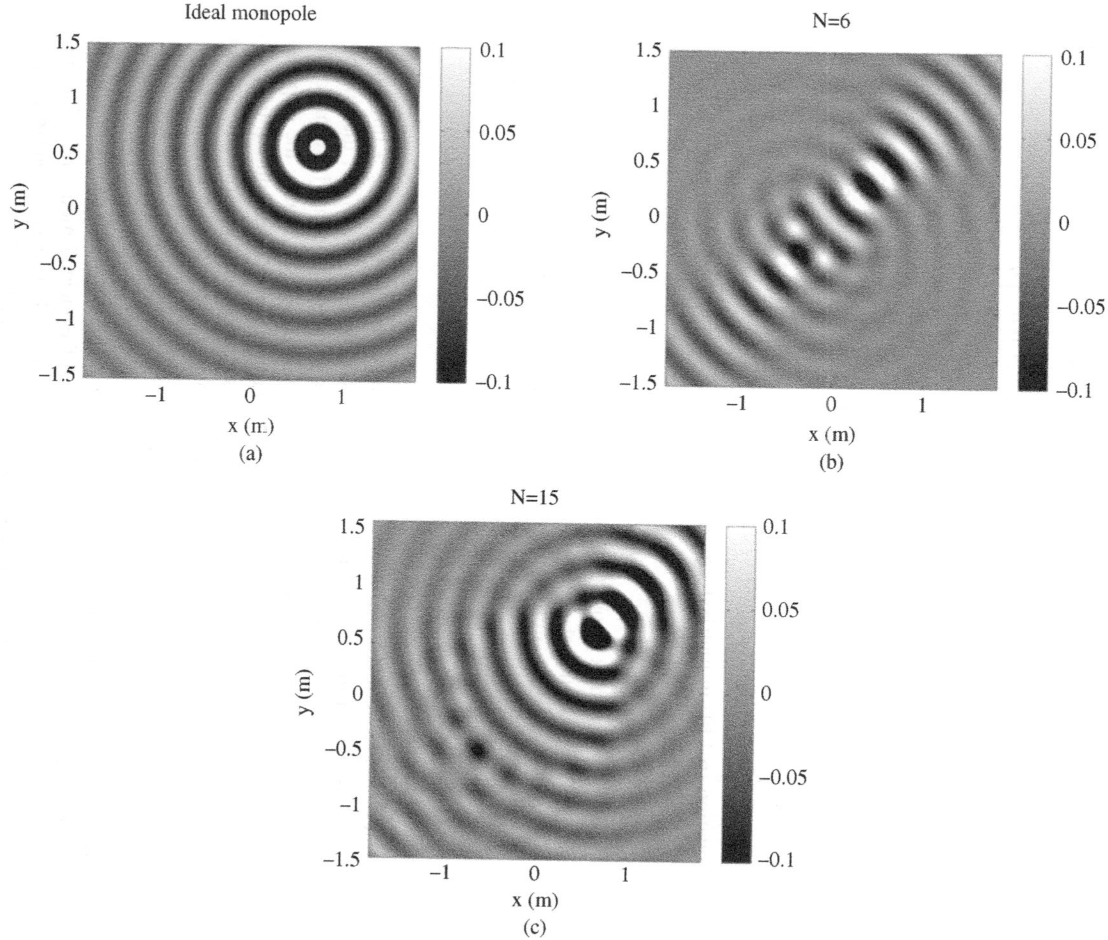

Figure 6.28 Effect of mode-truncation. Real parts of (a) ideal monopole sound field at 1000 Hz, (b) mode-truncated sound field up to $N = 6$, and (c) mode-truncated sound field up to $N = 15$

where $\mathbf{r}_s^{(\kappa)}$ indicates the location of each loudspeaker, and $\Delta S^{(\kappa)}$ is the weight representing the sectorial area around the κth loudspeaker position. For an interior reproduction problem (exterior loudspeaker array), the spherical harmonic expansion of the reproduced sound field with respect to $\mathbf{r}$ may be written as

$$\begin{aligned} R_{nm}(\omega) &= \frac{1}{h_n^{(1)}(kr_a)} \int_{4\pi} p_r(\mathbf{r}_a, \omega) Y_n^m(\theta_a, \phi_a)^* d\Omega_a \\ &= \sum_{\kappa=1}^{K} \left[\frac{1}{h_n^{(1)}(kr_a)} \int_{4\pi} g(\mathbf{r}_a|\mathbf{r}_s^{(\kappa)}; \omega) Y_n^m(\theta_a, \phi_a)^* d\Omega_a \right] q^{(\kappa)}(\omega)\Delta S^{(\kappa)} \\ &= \sum_{\kappa=1}^{K} G_{nm}(\mathbf{r}_s^{(\kappa)}, \omega) q^{(\kappa)}(\omega)\Delta S^{(\kappa)}, \end{aligned} \tag{6.149}$$

where $G_{nm}(\mathbf{r}_s^{(\kappa)}, \omega)$ represents the spherical harmonic expansion of the κ th control source's sound field. That is,

$$G_{nm}(\mathbf{r}_s^{(\kappa)}, \omega) = \frac{1}{h_n^{(1)}(kr)} \int_{\Omega} g(r, \theta, \phi|\mathbf{r}_s^{(\kappa)}; \omega) Y_n^m(\theta, \phi)^* d\Omega. \tag{6.150}$$

Rewriting Equation (6.149) in matrix form, we have

$$\boldsymbol{\rho} = \boldsymbol{\Gamma}\mathbf{S}\mathbf{q}, \tag{6.151}$$

where $\boldsymbol{\Gamma} \in \mathbb{C}^{(N+1)^2 \times K}$ is a matrix consisting of $G_{nm}(\mathbf{r}_s^{(\kappa)}, \omega)$, and $\boldsymbol{\rho} \in \mathbb{C}^{(N+1)^2 \times 1}$ is a vector of the coefficients R_{nm}. The elements of vector $\mathbf{q}$ correspond to the excitation signal $q(\mathbf{r}_s^{(\kappa)})$, and the diagonal matrix $\mathbf{S}$ consists of the weighting factor $\Delta S^{(\kappa)}$ of each loudspeaker.

Using the spherical harmonic coefficients of the target field $\mathbf{d}$, the mode-matching equation ($\boldsymbol{\rho} = \mathbf{d}$) for the target and reproduced fields can be written as

$$\mathbf{d} = \boldsymbol{\Gamma}\mathbf{S}\mathbf{q}. \tag{6.152}$$

To approximate a target field consisting of an infinite number of modes using a finite number of loudspeakers, we have to truncate the number of considered modes. Therefore, the number of spherical harmonics, that is, the size of vector $\mathbf{d}$, is inevitably truncated in the matrix formula. When the number of considered modes is equal to the number of loudspeakers, we have the opportunity for a one-to-one mapping between the excitation signal $\mathbf{q}$ and a given $\mathbf{d}$. The solution of a mode-matching equation obtained from this condition is sometimes called the mode-truncation solution. The mode-matching equation can be solved by taking the inverse of the matrix $\boldsymbol{\Gamma}\mathbf{S}$:

$$\mathbf{q} = (\boldsymbol{\Gamma}\mathbf{S})^{-1}\mathbf{d}. \tag{6.153}$$

The mode-truncation solution neglects the higher-order harmonics of the encoded data, so mode truncation will cause an error. Figure 6.29 shows an example of truncated mode matching. In this example, a monopole at a distance of 2.5 m from the center of a 3D spherical loudspeaker array of radius 2 m is reproduced by the mode-matching equation up to degree $N = 10$. The reproduced sound field (Figure 6.29b) shows a distinct difference from the original monopole field (Figure 6.29a), especially at a distance of more than 1 m from the center. The relative error between the ideal and reproduced monopole fields, normalized by the magnitude of the ideal monopole field, illustrates that the sweet spot (dark area) is generated mainly near the center.

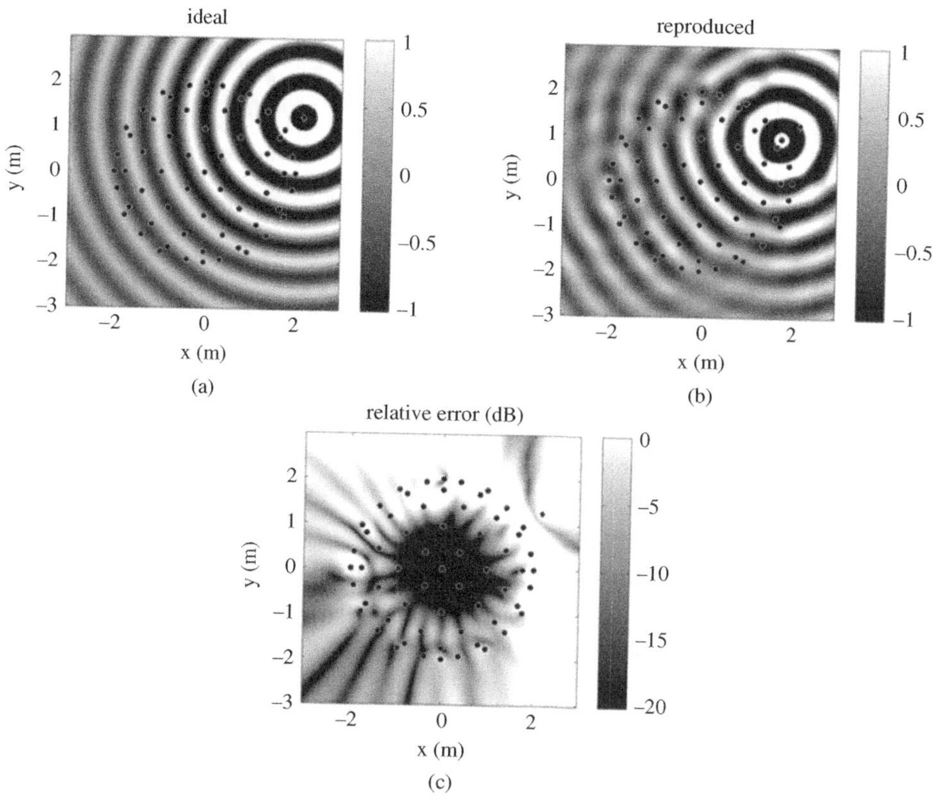

Figure 6.29 Interior reproduction of an exterior monopole source: (a) ideal monopole field (b) field reproduced by 110 secondary sources on Lebedev quadrature points ($r_s = 2$ m), and (c) relative reproduction error (dots: positions of control sources)

However, the reproduction error is not due only to truncation of the harmonics. Control sources themselves inevitably generate higher-order harmonics that lead to reproduction artifacts. If we denote the higher-order components of each control source as $\mathbf{\Gamma}'$, the higher-order modes excited by the mode-truncation solution generate the following higher-order contribution to the reproduced sound field:

$$\boldsymbol{\rho}' = \mathbf{\Gamma}'\mathbf{S}(\mathbf{\Gamma}\mathbf{S})^{-1}\mathbf{d}, \tag{6.154}$$

where $\boldsymbol{\rho}'$ represents the modal coefficients of higher-order modes ($n > N$) in the reproduced sound field.

The uncontrolled higher-order harmonics due to the finite number of control sources can be likened to the spatial aliasing artifact of a planar array (Equation (6.97)). However, the spherical harmonics are multiplied by the radial functions – the spherical Bessel or Hankel function – to be transformed into a spatial pressure distribution. Because radial functions of low order typically describe the sound field around the center of the sphere, the aliasing artifacts due to the higher-order harmonics appear at locations away from the center. Figure 6.30 shows the reproduced sound field of the controlled spherical harmonics (up to $N = 10$) and the sound field of the higher-order

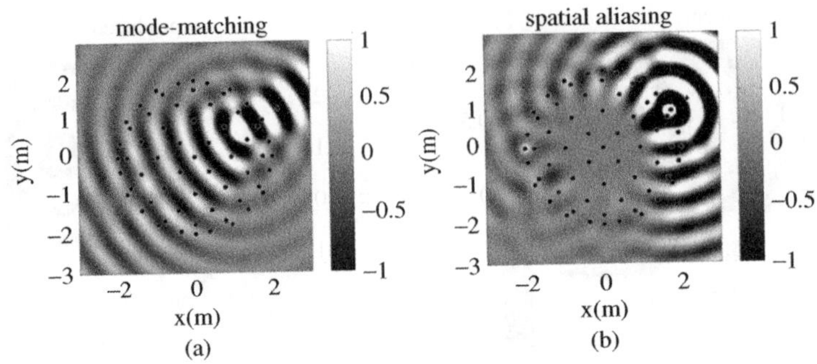

Figure 6.30 (a) Sound field of Figure 6.29b with truncated order and (b) sound field of higher-order harmonics (spatial aliasing)

modes. The low-order modes reproduce the sound field in the central area (Figure 6.30a), whereas the aliasing artifacts are distributed mainly in the farfield (Figure 6.30b).

6.6.3.8 Mode-Matching Equation for Higher-Order Harmonics

A mode-matching equation can be constructed to deal with a situation where the number of spherical harmonics is not equal to the number of loudspeakers. If the number of loudspeakers is smaller than the considered number of modes ($K < (N+1)^2$), we have more equations than unknowns, and the problem becomes overdetermined. In this case, the error can never be zero, but the solution can be determined so as to minimize the overall mode-matching error for both lower- and higher-order harmonics. The mode-matching error can be defined in terms of the difference between the modal coefficients,

$$\varepsilon^2 = \sum_{n=0}^{N} \sum_{m=-n}^{n} |D_{nm} - R_{nm}|^2 = \|\mathbf{d} - \boldsymbol{\rho}\|^2. \tag{6.155}$$

The solution that minimizes the defined error can be obtained from a pseudo-inverse of the matrix $\boldsymbol{\Gamma}\mathbf{S}$.

$$\begin{aligned} \mathbf{q} &= (\boldsymbol{\Gamma}\mathbf{S})^{+}\mathbf{d} \\ &= [(\boldsymbol{\Gamma}\mathbf{S})^{H}(\boldsymbol{\Gamma}\mathbf{S})]^{-1}(\boldsymbol{\Gamma}\mathbf{S})^{H}\mathbf{d}. \end{aligned} \tag{6.156}$$

When the integral weight $\mathbf{S}$ is well-conditioned, the excitation signal can also be calculated from

$$\begin{aligned} \mathbf{q} &= \mathbf{S}^{-1}(\boldsymbol{\Gamma})^{+}\mathbf{d} \\ &= \mathbf{S}^{-1}[\boldsymbol{\Gamma}^{H}\boldsymbol{\Gamma}]^{-1}\boldsymbol{\Gamma}^{H}\mathbf{d}. \end{aligned} \tag{6.157}$$

The mode-matching equation using the pseudo-inverse can provide a unique answer if the transfer function matrix $\boldsymbol{\Gamma}$ is well-conditioned. However, if the number of modes is greater than the number of loudspeakers, most of the control efforts may be wasted on manipulation of uncontrollable higher-order harmonics.

On the other hand, if the number of harmonics is less than the number of loudspeakers ($(N+1)^2 < K$), the problem can be underdetermined. In this case, we can have multiple solutions that minimize the error defined in Equation (6.155). This ill-posedness can be resolved by adding an extra constraint on the error-minimization problem; for example, a minimum-norm solution

(Equation (A42)) that also minimizes the control effort $\|\mathbf{q}\|$ can be computed by the right inverse of the matrix $\mathbf{\Gamma S}$.

$$\mathbf{q} = (\mathbf{\Gamma S})^H \left[(\mathbf{\Gamma S})(\mathbf{\Gamma S})^H\right]^{-1} \mathbf{d}. \tag{6.158}$$

In the simulation shown in Figure 6.29, the number of spherical harmonics ($(N+1)^2 = 121$) was greater than the number of loudspeakers ($K = 110$). Reproduction using more or fewer modes can be attempted; however, care must be taken to obtain acceptable results. The regularization method for the field-matching technique (Section 6.5.2) can also be applied to the mode-matching problem to search for more stable solutions.

6.7 Surface Integral Equations

The field-matching or mode-matching approach requires that an inverse problem be solved. Unlike these two approaches, the integral approach is based on an integral equation that can directly describe a target sound field in terms of the contributions of control sources distributed on a surface. For example, the KH integral equation expresses the interior or exterior sound fields as a superposition of monopoles and dipoles. Accordingly, the sound fields of monopoles and dipoles are the basis functions expressing a given sound field.

The use of both monopole and dipole sound fields allows greater flexibility in solving the reproduction problem. The integral equation is a direct form for which we need not solve an inverse problem. The KH integral directly yields the excitation signals on the surface to produce the desired sound field, provided that we can manipulate both monopoles and dipoles. Physically, the monopoles and dipoles distributed on the surface act like variable-directivity sources whose surface normal directivity can be changed according to the relative weight between monopoles and dipoles (remember the differential sources discussed in Chapter 5, Section 5.5.1).

The KH integral equation provides a theoretical background for sound field reproduction over a volume using control sources distributed on the surface. However, the KH integral has many different forms depending on the locations of the listener and virtual source. Some of them are useful for sound field reproduction, but some are inadequate. In this section, we introduce four integral equations for different source and listener locations (Figure 6.31). We will show that in some cases the KH integral cannot be used directly for sound field reproduction.

6.7.1 Source Inside, Listener Inside ($V_0 \subset V$, $\mathbf{r} \in V$)

We first consider an interior reproduction problem in which the listener is located inside of a volume V ($\mathbf{r} \in V$). In addition, it is also assumed that virtual sources are positioned within a smaller volume $V_0 \subset V$. The target sound field p_d at a certain point $\mathbf{r}_0 \in V$, generated by sound sources of excitation $q(\mathbf{r}_0)$, should satisfy the inhomogeneous wave equation in V. That is,

$$\left[\nabla_0^2 + k^2\right] p_d(\mathbf{r}_0) = -q(\mathbf{r}_0), \tag{6.159}$$

where the Laplacian ∇_0^2 is taken with respect to the position $\mathbf{r}_0$. The pressure field at a listener point $\mathbf{r}$ can be described as an integral form by using the Green's function

$$p_d(\mathbf{r}) = \int_{V_0} g_d(\mathbf{r}|\mathbf{r}_0) q(\mathbf{r}_0) dV(\mathbf{r}_0), \tag{6.160}$$

where g_d denotes the Green's function following the boundary condition of the room where the target pressure field p_d is obtained.

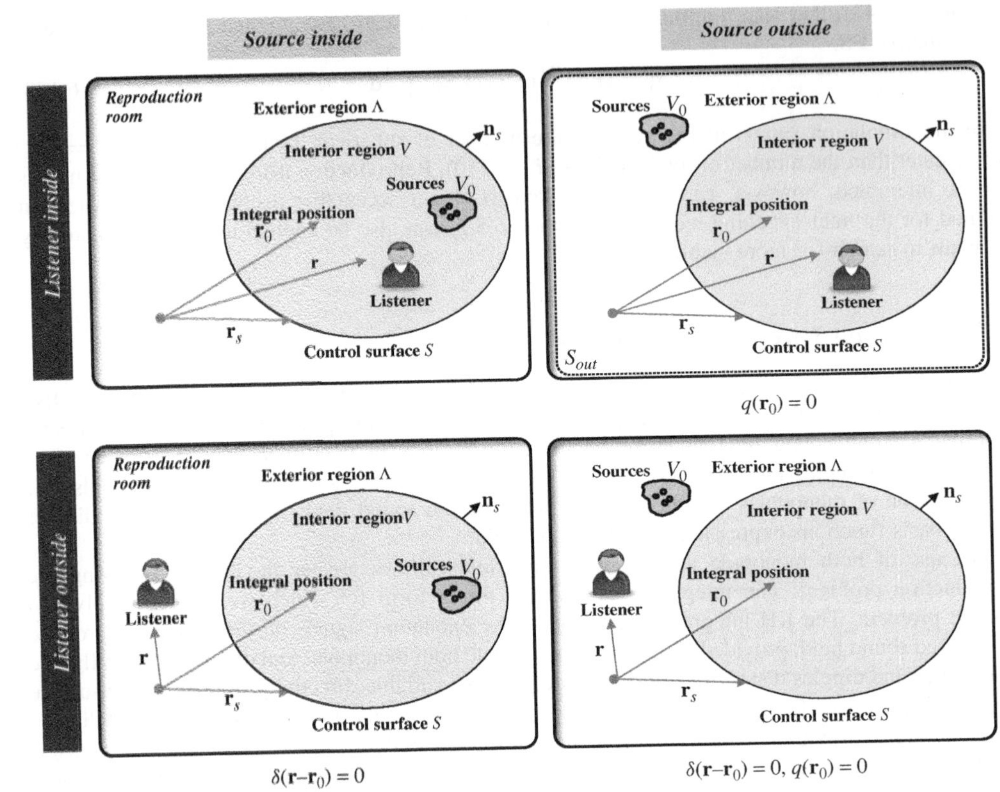

Figure 6.31 Four different combinations of listener and source locations

Then we consider another wave equation describing the behavior of the Green's function.

$$\left[\nabla_0^2 + k^2\right] g(\mathbf{r}|\mathbf{r}_0) = -\delta(\mathbf{r} - \mathbf{r}_0). \tag{6.161}$$

Equation (6.161) expresses the overall relationship between the listener position $\mathbf{r}$ and the other position $\mathbf{r}_0$. The wave equation of Equation (6.161) can be homogeneous or inhomogeneous depending on the location of the listener. For example, if the listener lies outside of the volume V, the delta function of Equation (6.161) is always zero because the listener position $\mathbf{r}$ is never equal to $\mathbf{r}_0 \in V$. However, it is nonzero for this listener-inside case.

Note also that the Green's function g of Equation (6.161) can be different from g_d in Equation (6.160). Because the differential equation describes only the behavior of the medium, it does not say anything about the boundary condition. Therefore, any type of Green's function satisfying an arbitrary boundary condition outside V satisfies the differential equation of Equation (6.161).

Using the arbitrary Green's function g, we attempt to express the given sound field p_d. In other words, we are going to identify a given sound field as a superposition of arbitrary point sources' sound fields. The sound fields of these point sources are the basis function for expressing the target sound field p_d.

The expression can be derived as follows. Multiplying both sides of Equations (6.159) and (6.161) by g and p, respectively, and subtracting one from the other yields

$$g(\mathbf{r}|\mathbf{r}_0)\nabla_0^2 p_d(\mathbf{r}_0) - p_d(\mathbf{r}_0)\nabla_0^2 g(\mathbf{r}|\mathbf{r}_0) = p_d(\mathbf{r}_0)\delta(\mathbf{r}-\mathbf{r}_0) - q(\mathbf{r}_v)g(\mathbf{r}|\mathbf{r}_v). \tag{6.162}$$

Then, an integral equation can be obtained by integrating Equation (6.162) with respect to the *integral position* $\mathbf{r}_0 \in V$. That is,

$$\int_V [g(\mathbf{r}|\mathbf{r}_0)\nabla_0^2 p_d(\mathbf{r}_0) - p_d(\mathbf{r}_0)\nabla_0^2 g(\mathbf{r}|\mathbf{r}_0)]dV(\mathbf{r}_0)$$
$$= \int_V [p_d(\mathbf{r}_0)\delta(\mathbf{r}-\mathbf{r}_0) - q(\mathbf{r}_0)g(\mathbf{r}|\mathbf{r}_0)]dV(\mathbf{r}_0). \tag{6.163}$$

Applying the divergence theorem yields the integral equation for the interior problem.

$$p_d(\mathbf{r}) = \int_{V_0} g(\mathbf{r}|\mathbf{r}_0)q(\mathbf{r}_0)dV + \int_S \left[\frac{\partial p_d(\mathbf{r}_s)}{\partial n_s} g(\mathbf{r}|\mathbf{r}_s) - p_d(\mathbf{r}_s)\frac{\partial g(\mathbf{r}|\mathbf{r}_s)}{\partial n_s}\right] dS, \tag{6.164}$$

where $\mathbf{r}_s \in S$ is the integral point on the control surface and $\partial/\partial n_s$ represents the partial derivative with respect to the surface outward normal $\mathbf{n}_s$.

The pressure field at the listener position is a superposition of two integrals. The volume integral and surface integral express the sound field from the interior virtual source. To understand the meaning of Equation (6.164), the characteristics of the Green's function g should be emphasized here. For the sound field reproduction problem, g represents the transfer function of a monopole source installed in the reproduction room. Therefore, g follows the boundary condition of the reproduction room.

Remember that Equation (6.164) holds for any type of Green's function that satisfies the differential equation of Equation (6.161). Accordingly, various types of Green's functions can be used for Equation (6.164), which means that the equation can be applied to any type of reproduction room. For example, if we use the free-field Green's function g_f for g, Equation (6.164) can be reduced to

$$p_d(\mathbf{r}) = \int_{V_0} g_f(\mathbf{r}|\mathbf{r}_0)q(\mathbf{r}_0)dV + \int_S \left[\frac{\partial p_d(\mathbf{r}_s)}{\partial n_s} g_f(\mathbf{r}|\mathbf{r}_s) - p_d(\mathbf{r}_s)\frac{\partial g_f(\mathbf{r}|\mathbf{r}_s)}{\partial n_s}\right] dS. \tag{6.165}$$

In this case, the first volume integral represents sound radiation without reflections, that is, the *direct field*,

$$p_{d,direct}(\mathbf{r}) = \int_{V_0} g_f(\mathbf{r}|\mathbf{r}_0)q(\mathbf{r}_0)dV. \tag{6.166}$$

The surface integral of Equation (6.165), which is now equivalent to the total field subtracted from the direct field, becomes the *reflections from the boundary*:

$$p_{d,refl}(\mathbf{r}) = \int_S \left[\frac{\partial p_d(\mathbf{r}_s)}{\partial n_s} g_f(\mathbf{r}|\mathbf{r}_s) - p_d(\mathbf{r}_s)\frac{\partial g_f(\mathbf{r}|\mathbf{r}_s)}{\partial n_s}\right] dS. \tag{6.167}$$

With the free-field Green's function, the two integrals represent the direct and reflected fields of p_d. In typical circumstances where the arbitrary Green's function g is used, *the volume integral represents the contribution of the sound field satisfying the boundary condition imposed on* g, *and the surface integral expresses correction terms to fulfill the actual boundary condition imposed on* p_d. Therefore, if we attempt to reproduce the direct wave of p_d, the surface integral of Equation (6.167) becomes zero because no reflected waves exists in the target field.

$$\int_S \left[\frac{\partial p_{d,direct}(\mathbf{r}_s)}{\partial n_s} g_f(\mathbf{r}|\mathbf{r}_s) - p_{d,direct}(\mathbf{r}_s) \frac{\partial g_f(\mathbf{r}|\mathbf{r}_s)}{\partial n_s} \right] dS = 0. \tag{6.168}$$

A very special case occurs when the Green's function g_d is used for g:

$$p_d(\mathbf{r}) = \int_{V_0} g_d(\mathbf{r}|\mathbf{r}_0) q(\mathbf{r}_0) dV + \int_S \left[\frac{\partial p_d(\mathbf{r}_s)}{\partial n_s} g_d(\mathbf{r}|\mathbf{r}_s) - p_d(\mathbf{r}_s) \frac{\partial g_d(\mathbf{r}|\mathbf{r}_s)}{\partial n_s} \right] dS. \tag{6.169}$$

From Equation (6.160), the volume integral can completely describe the pressure field p_d, so the contribution of the surface integral becomes zero.

$$\int_S \left[\frac{\partial p_d(\mathbf{r}_s)}{\partial n_s} g_d(\mathbf{r}|\mathbf{r}_s) - p_d(\mathbf{r}_s) \frac{\partial g_d(\mathbf{r}|\mathbf{r}_s)}{\partial n_s} \right] dS = 0. \tag{6.170}$$

In this case, the Green's function $g = g_d$ exactly expresses the sound field p_d, so no correction from the control surface is necessary. Therefore, the surface integral makes no contribution to the interior sound field.

6.7.2 *Source Inside, Listener Outside* ($V_0 \subset V$, $\mathbf{r} \in \Lambda$)

The integral derived in Equation (6.164) expresses the interior pressure field. If the listener is outside of the integral surface *S*, the differential equation for the Green's function differs from Equation (6.161). The Dirac delta function on the right-hand side of Equation (6.161) is always zero because the integral position ($\mathbf{r}_0 \in V$) can never be equal to the listener position ($\mathbf{r} \in \Lambda$). Accordingly, the wave equation can now be described as

$$[\nabla_0^2 + k^2] g(\mathbf{r}|\mathbf{r}_0) = 0. \tag{6.171}$$

Following the procedure we used for Equations (6.162) and (6.163), the integral equation can be rearranged to give

$$\int_{V_0} g(\mathbf{r}|\mathbf{r}_0) q(\mathbf{r}_0) dV = -\int_S \left[\frac{\partial p_d(\mathbf{r}_s)}{\partial n_s} g(\mathbf{r}|\mathbf{r}_s) - p_d(\mathbf{r}_s) \frac{\partial g(\mathbf{r}|\mathbf{r}_s)}{\partial n_s} \right] dS. \tag{6.172}$$

The integral equation implies that the surface integral on the right-hand side of Equation (6.172) can be used to produce the sound field of the volume integral on the left-hand side. The volume integral simply represents the sound field radiated by the virtual sources in the reproduction room. However, the volume integral differs from the target field of Equation (6.160) because its Green's function follows the boundary condition of the reproduction room. Accordingly, the reflected wave of the reproduced field differs from that of the target field.

For example, if we use the free-field Green's function g_f for g, the surface integral is given by

$$\begin{aligned} \int_{V_0} g_f(\mathbf{r}|\mathbf{r}_0) q(\mathbf{r}_0) dV &= p_{d,direct}(\mathbf{r}) \\ &= -\int_S \left[\frac{\partial p_d(\mathbf{r}_s)}{\partial n_s} g_f(\mathbf{r}|\mathbf{r}_s) - p_d(\mathbf{r}_s) \frac{\partial g_f(\mathbf{r}|\mathbf{r}_s)}{\partial n_s} \right] dS, \end{aligned} \tag{6.173}$$

which shows that we can reproduce the direct wave of p_d. Likewise, the total sound field can be reconstructed by using the exact Green's function that satisfies the boundary condition for the target field.

$$p_d(\mathbf{r}) = \int_{V_0} g_d(\mathbf{r}|\mathbf{r}_0)q(\mathbf{r}_0)dV$$
$$= -\int_S \left[\frac{\partial p_d(\mathbf{r}_s)}{\partial n_s} g_d(\mathbf{r}|\mathbf{r}_s) - p_d(\mathbf{r}_s)\frac{\partial g_d(\mathbf{r}|\mathbf{r}_s)}{\partial n_s} \right] dS, \tag{6.174}$$

Equation (6.174) reveals that perfect exterior reproduction is possible if we excite the monopole and dipole sources in the same room where the pressure signal p_d and its derivative $\partial p_d/\partial n_s$ were recorded.

6.7.3 *Source Outside, Listener Outside* ($V_0 \subset \Lambda$, $\mathbf{r} \in \Lambda$)

When the virtual sources are also outside of the integral volume V ($q(\mathbf{r}_0) = 0$ for $\mathbf{r}_0 \in V$), the differential equation for the pressure field p_d (Equation (6.159)) becomes homogeneous.

$$\left[\nabla_0^2 + k^2\right] p_d(\mathbf{r}_0) = 0 \quad \text{for } \mathbf{r}_0 \in V. \tag{6.175}$$

Because both differential equations, Equations (6.161) and (6.159), are homogeneous, the surface integral simply becomes zero irrespective of the type of Green's function:

$$\int_S \left[\frac{\partial p_d(\mathbf{r}_s)}{\partial n_s} g(\mathbf{r}|\mathbf{r}_s) - p_d(\mathbf{r}_s)\frac{\partial g(\mathbf{r}|\mathbf{r}_s)}{\partial n_s} \right] dS = 0. \tag{6.176}$$

The integral over the control surface makes *no contribution* to the listener outside. Therefore, the surface integral of the sound field satisfying the homogeneous wave equation produces no sound for the listener outside of the integration volume.

The result is similar to the source-inside and listener-inside case, but Equation (6.176) is always zero, whereas Equation (6.169) becomes zero only when the special Green's function is used (Equation (6.170)). This difference arises from the geometric configuration of the integral volume. The integral surface S is the only surface surrounding V_0 for the source-inside, listener-inside case. In contrast, for the source-outside, listener-outside case, there are two surfaces surrounding the source and listener (S and S_{out}). If we integrate over the exterior volume Λ, the integral equation should be the same as Equation (6.164) because both the listener and the source are inside of the integration volume. The major difference is, however, that we have two surface integrals for both S and S_{out}. That is,

$$p_d(\mathbf{r}) = \int_\Lambda g(\mathbf{r}|\mathbf{r}_0)q(\mathbf{r}_0)dV - \int_{S+S_{out}} \left[\frac{\partial p_d(\mathbf{r}_s)}{\partial n_s} g(\mathbf{r}|\mathbf{r}_s) - p_d(\mathbf{r}_s)\frac{\partial g(\mathbf{r}|\mathbf{r}_s)}{\partial n_s} \right] dS. \tag{6.177}$$

Note that the sign of the surface integral is inverted because the surface normal vector n_s points in the inward direction with respect to the integration volume. The surface integral for S is zero from Equation (6.176), so the equation can be rewritten as

$$p_d(\mathbf{r}) = \int_\Lambda g(\mathbf{r}|\mathbf{r}_0)q(\mathbf{r}_0)dV - \int_{S_{out}} \left[\frac{\partial p_d(\mathbf{r}_s)}{\partial n_s} g(\mathbf{r}|\mathbf{r}_s) - p_d(\mathbf{r}_s)\frac{\partial g(\mathbf{r}|\mathbf{r}_s)}{\partial n_s} \right] dS. \tag{6.178}$$

Consequently, the pressure field inside the volume Λ can also be expressed as a superposition of the volume integral and surface integral, as shown in Equation (6.164).

6.7.4 Source Outside, Listener Inside ($V_0 \subset \Lambda$, $\mathbf{r} \in V$)

For the case of an outside source and an inside listener, we can use the differential equations (Equations (6.175) and (6.161)). This gives the integral equation

$$p_d(\mathbf{r}) = \int_S \left[\frac{\partial p_d(\mathbf{r}_s)}{\partial n_s} g(\mathbf{r}|\mathbf{r}_s) - p_d(\mathbf{r}_s) \frac{\partial g(\mathbf{r}|\mathbf{r}_s)}{\partial n_s} \right] dS, \tag{6.179}$$

which indicates that the pressure field from the exterior source can be reconstructed by the surface integral. In this case, any Green's function can be used to reproduce the target sound field; hence, the reflected waves, as well as the direct wave of p_d, will be reproduced regardless of the type of Green's function. This surface integral is a popular form of the KH integral and can be obtained when the virtual source and the listener are not located within the same volume.

Note that the source-inside, listener-outside case of Equation (6.172) is similar in form to Equation (6.179). However, for the source-inside, listener-outside case, the reflected waves differ from that of the target field p_d.

6.7.5 Listener on the Control Surface

As the listener approaches the control surface ($\mathbf{r} \to \mathbf{r}_s$), the surface integral will be singular owing to the singularity of the Green's function $g(\mathbf{r}|\mathbf{r}_s)$. To derive an integral equation for the listener on the control surface S, we modify the shape of S such that a small internal bump (S_ϵ) surrounds the listener position. As a result, the integral volume surrounded by the bump (S_ϵ) and the rest (S_{re}) excludes the listener position (Figure 6.32). For a smooth surface S, the small bump S_ϵ can be considered to be a hemisphere of radius ϵ and solid angle $\Omega = 2\pi$. By shrinking the radius to zero ($\epsilon \to 0$), we can formulate the integral equation at the listener location on the surface.

First, consider a virtual source in V. Because the listener is outside of the modified integral volume, we can use the integral Equation (6.172):

$$\int_{V_0} g(\mathbf{r}|\mathbf{r}_0) q(\mathbf{r}_0) dV = -\lim_{\epsilon \to 0} \int_{S_{re}+S_\epsilon} \left[\frac{\partial p_d(\mathbf{r}_s)}{\partial n_s} g(\mathbf{r}|\mathbf{r}_s) - p_d(\mathbf{r}_s) \frac{\partial g(\mathbf{r}|\mathbf{r}_s)}{\partial n_s} \right] dS. \tag{6.180}$$

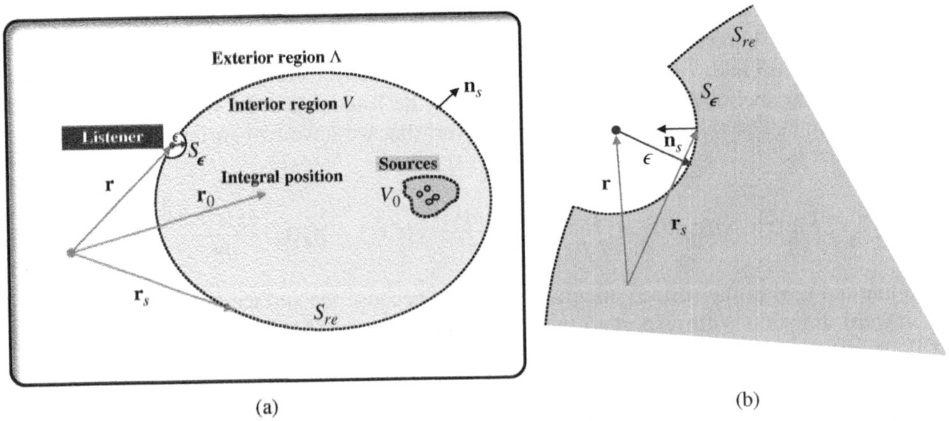

Figure 6.32 Integral volume modified for the listener on the surface: (a) exclusion of the listener from the integration volume and (b) hemispherical surface S_ϵ at the listener

Before inspecting this integral, we separate the contributions of the direct and reflected waves in the Green's function g. The direct wave is equal to the free-field Green's function g_f, and the reflected wave can be denoted as

$$g_{refl}(\mathbf{r}|\mathbf{r}_s) = g(\mathbf{r}|\mathbf{r}_s) - g_f(\mathbf{r}|\mathbf{r}_s). \tag{6.181}$$

Because the reflected wave of the Green's function does not have any singularity on the surface, the integral on the modified surface $S_{re} + S_\varepsilon$ is no different from an ordinary integral on S. Therefore, we have

$$\lim_{\varepsilon \to 0} \int_{S_{re}+S_\varepsilon} [\,]dS = \int_S [\,]dS \tag{6.182}$$

for $g_{refl}(\mathbf{r}|\mathbf{r}_s)$.

In contrast, the free-field Green's function g_f becomes singular near $\mathbf{r} = \mathbf{r}_s$. The distance between the surface and the listener ($|\mathbf{r} - \mathbf{r}_s|$) is equal to ϵ on the small hemispherical surface S_ϵ, and the surface normal derivative can be approximated by

$$\frac{\partial}{\partial n_s} \approx -\frac{\partial}{\partial \epsilon}. \tag{6.183}$$

The free-field Green's function and its derivative can then be written as

$$g_f(\mathbf{r}|\mathbf{r}_s) = \frac{e^{ik\epsilon}}{4\pi\epsilon}, \quad \frac{\partial g_f(\mathbf{r}|\mathbf{r}_s)}{\partial n_s} = -\frac{\partial}{\partial \epsilon}\left(\frac{e^{ik\epsilon}}{4\pi}\right) = -\left(ik - \frac{1}{\epsilon}\right)\frac{e^{ik\epsilon}}{4\pi\epsilon}. \tag{6.184}$$

The small integral surface dS can be regarded as a solid angle multiplied by the square of the radius ($\varepsilon^2 d\Omega$). Hence, the integral over S_ϵ is given by

$$-\int_{S_\epsilon} [\,]dS = \int_\Omega \left[\frac{\partial p_0(\mathbf{r}_s)}{\partial \epsilon}\frac{e^{ik\epsilon}}{4\pi\epsilon} - \left(ik - \frac{1}{\epsilon}\right)\frac{e^{ik\epsilon}}{4\pi\epsilon} p_0(\mathbf{r}_s)\right]\epsilon^2 d\Omega, \tag{6.185}$$

where the solid angle Ω is 2π for the hemisphere. By taking the limit as $\epsilon \to 0$, we have

$$-\lim_{\varepsilon \to \infty} \int_{S_\varepsilon} [\,]dS = \frac{1}{2} p_0(\mathbf{r}). \tag{6.186}$$

The rest of the integral over S_{re} converges to the integral for S; thus, we have

$$\frac{1}{2} p_d(\mathbf{r}) = \int_{V_0} g(\mathbf{r}|\mathbf{r}_0) q(\mathbf{r}_0) dV + \int_S \left[\frac{\partial p_d(\mathbf{r}_s)}{\partial n_s} g(\mathbf{r}|\mathbf{r}_s) - p_d(\mathbf{r}_s)\frac{\partial g(\mathbf{r}|\mathbf{r}_s)}{\partial n_s}\right] dS, \quad \text{for } \mathbf{r} \in S \tag{6.187}$$

for the interior virtual source.

In the same manner, the integral for the exterior virtual source can be derived from Equation (6.179). That is,

$$\frac{1}{2} p_d(\mathbf{r}) = \int_S \left[\frac{\partial p_d(\mathbf{r}_s)}{\partial n_s} g(\mathbf{r}|\mathbf{r}_s) - p_d(\mathbf{r}_s)\frac{\partial g(\mathbf{r}|\mathbf{r}_s)}{\partial n_s}\right] dS, \quad \textit{for } \mathbf{r} \in S \tag{6.188}$$

6.7.6 Summary of Integral Equations

The integral equations we have derived so far can be summarized as follows: for the interior virtual source in V, the integral equation is given by

$$\alpha p_d(\mathbf{r}) = \int_{V_0} g(\mathbf{r}|\mathbf{r}_0)q(\mathbf{r}_0)dV + \int_S \left[\frac{\partial p_d(\mathbf{r}_s)}{\partial n_s} g(\mathbf{r}|\mathbf{r}_s) - p_d(\mathbf{r}_s)\frac{\partial g(\mathbf{r}|\mathbf{r}_s)}{\partial n_s}\right] dS, \quad \text{for } V_0 \subset V. \tag{6.189}$$

where the listener-dependent parameter α is

$$\alpha = \begin{cases} 1 & \text{for } \mathbf{r} \in V \\ 1/2 & \text{for } \mathbf{r} \in S \\ 0 & \text{for } \mathbf{r} \in \Lambda. \end{cases}$$

For the exterior virtual source in Λ, we have

$$\alpha p_d(\mathbf{r}) = \int_S \left[\frac{\partial p_d(\mathbf{r}_s)}{\partial n_s} g(\mathbf{r}|\mathbf{r}_s) - p_d(\mathbf{r}_s)\frac{\partial g(\mathbf{r}|\mathbf{r}_s)}{\partial n_s}\right] dS. \tag{6.190}$$

Table 6.1 summarizes the various integrals and their physical meanings. Each integral equation has its own characteristics and applications. For example, the integral equations marked as *perfect reconstruction* can be used to predict the sound field from the measured data on the surface. Likewise, if we have monopole and dipole sources on the surface S and control them, the sound field p_d can be *reproduced* over the region V.

6.7.7 Nonradiating Sound Field and Nonuniqueness Problem

Two particular cases in Table 6.1 exhibit special behavior: the nonradiating sound field. That is, the integral of the monopole and dipole sources on the surface makes no contribution at the listener's position. This happens when the virtual source and listener are within the same volume (either V or Λ). For the listener outside, the integral is always zero, and it is zero with $g = g_d$ for the listener inside.

This nonradiating sound field can be regarded as a *homogeneous solution* of the reconstruction problem, and the homogeneous solution causes a nonuniqueness problem. The nonuniqueness

Table 6.1 Four integral equations for different source and listener locations $\left(\int_S [\,]dS = \int_S \left[\frac{\partial p_d}{\partial n_s} g - p_d \frac{\partial g}{\partial n_s}\right] dS\right)$

	Source inside	Source outside
Listener inside	$\int_S [\,]dS = p_d(\mathbf{r}) - \int_{V_0} g(\mathbf{r}|\mathbf{r}_0)q(\mathbf{r}_0)dV$ ($g = g_f$) Scattering field reconstruction ($g = g_d$) Nonradiating	$\int_S [\,]dS = p_d(\mathbf{r})$ Perfect reconstruction
Listener outside	$-\int_S [\,]dS = \int_{V_0} g(\mathbf{r}|\mathbf{r}_0)q(\mathbf{r}_0)dV$ ($g = g_f$) Direct field reconstruction ($g = g_d$) Perfect reconstruction	$\int_S [\,]dS = 0$ Nonradiating

problem arises from the following simple question: *can we uniquely determine the surface pressure and its derivative on the surface that gives a desired sound field in* Λ? To answer this question, suppose that there is a particular solution for the surface pressure and derivative. Then we can always add the homogeneous solution, that is, the surface pressure and velocity that produce the nonradiating sound field in Λ (that is, the sound field from the source outside). The addition of this homogenous solution does not change the sound field in Λ, but the surface data are not the same as before. Therefore, in general, *we cannot uniquely determine the surface pressure and its derivative from the known sound field within* Λ.

This type of nonuniqueness problem arises because we do not know whether the sound is impinging from the exterior or interior of the surface. However, if we apply an additional constraint that the sound should come from the interior only, a unique solution is possible.

The nonuniqueness problem is not bad at all for sound field reproduction because it means that we have many candidates that can reproduce a given sound field in Λ. Then what are the differences between the many solutions? The answer is that they differ in the interior field. Although the exterior fields are the same, the interior fields are not. To be more specific, suppose that a particular solution p_c and $\partial p_c/\partial n_s$ generates a desired sound field p_Λ in Λ and p_V in V.

$$\int_S \left[\frac{\partial p_c(\mathbf{r}_s)}{\partial n_s} g(\mathbf{r}|\mathbf{r}_s) - p_c(\mathbf{r}_s)\frac{\partial g(\mathbf{r}|\mathbf{r}_s)}{\partial n_s}\right] dS = \begin{cases} p_V(\mathbf{r}) & \mathbf{r}\in V \\ p_\Lambda(\mathbf{r}) & \mathbf{r}\in \Lambda \end{cases}. \tag{6.191}$$

Then, from Equation (6.179), the homogeneous solution p_h satisfies the following integral equation:

$$\int_S \left[\frac{\partial p_h(\mathbf{r}_s)}{\partial n_s} g(\mathbf{r}|\mathbf{r}_s) - p_h(\mathbf{r}_s)\frac{\partial g(\mathbf{r}|\mathbf{r}_s)}{\partial n_s}\right] dS = \begin{cases} -\int_{V_0} g(\mathbf{r}|\mathbf{r}_0)\, q(\mathbf{r}_0)dV & \mathbf{r}\in V \\ 0 & \mathbf{r}\in \Lambda. \end{cases} \tag{6.192}$$

By adding the homogeneous solution scaled by an arbitrary factor α to the particular solution, the integral becomes

$$\int_S \left[\frac{\partial (p_c+\alpha p_h)}{\partial n_s} g - (p_c+\alpha p_h)\frac{\partial g}{\partial n_s}\right] dS = \begin{cases} p_V(\mathbf{r}) - \alpha\int_{V_0}(gq)dV & \mathbf{r}\in V \\ p_\Lambda(\mathbf{r}) & \mathbf{r}\in \Lambda, \end{cases} \tag{6.193}$$

which means that the new surface pressure $p_c+\alpha p_h$ and its spatial derivative do not alter the exterior sound field, whereas the interior sound field is different. This indicates that, conversely, *we can generate various types of interior field while reproducing the same exterior field, provided that both monopole and dipole sources can be used on the surface S*. Therefore, the use of monopole and dipole sources gives us more flexibility in controlling the interior and exterior fields.

The nonuniqueness discussed above can also be applied to the source-inside case if the Green's function g_d is used. In this case, there can be many different exterior fields for a given interior sound field in V. One positive application of this phenomenon is to find a nonradiating sound field in which the exterior radiation is suppressed without inducing any change in the interior field. The nonradiating field is useful in relation to a personal sound system that reproduces audio within a finite space without disturbing other listeners outside.

Another famous use of nonuniqueness is the *single-layer formula*. Because we have multiple candidates for the surface source distributions that reproduce a target sound field within V, we can also find a surface distribution that does not contain any dipole contribution. Conversely, it is possible to derive a double-layer formula that includes only the dipole expression. These formulas are useful for sound field reproduction problems because working with both monopole and dipole sources is less practical. The directivity of the conventional loudspeaker is fixed, and at low frequencies many loudspeakers are nearly omnidirectional, so they can be viewed as producing monopole radiation. The derivation of the single-layer formula is discussed in more detail in the following section.

6.8 Single-layer Formula

When only monopole sources are available, or when the radiation pattern of control sources is unchangeable, we need to set up another problem to find the excitation signals. The problem of converting the KH integral to a single-layer formula is a typical inverse problem. There are generally two ways to resolve an inverse problem. One is to approximate the integral equation into a direct form, and the other is to solve it using a numerical or analytic method. The direct form is less exact than solving the inverse problem, but its computation is simple. In solving the inverse problem, the uniqueness of the solution must be ensured, and several ill-conditioned problems can arise. However, once the problem is well-conditioned, its solution can be accurate and exact.

Apart from ill-conditioned problems, there are situations in which a solution cannot exist. The nonexistence problem typically arises when we attempt to reproduce a virtual source inside a region of interest [15]. In this section, direct integral formulas are derived for the cases of interior and exterior virtual sources.

6.8.1 Single-layer Formula for Exterior Virtual Source

6.8.1.1 Single-layer Formula

We return to the discussion on the integral equations for interior and exterior virtual sources. If we confine our interest to the direct waves of both the target and reproduced fields, the integral equation for the exterior virtual source (Equation (6.190)) can be rewritten as

$$\int_S \left[\frac{\partial p_d(\mathbf{r}_s)}{\partial n_s} g_f(\mathbf{r}|\mathbf{r}_s) - p_d(\mathbf{r}_s) \frac{\partial g_f(\mathbf{r}|\mathbf{r}_s)}{\partial n_s} \right] dS = \begin{cases} p_d(\mathbf{r}) & \mathit{for}\ \mathbf{r} \in V \\ \dfrac{p_d(\mathbf{r})}{2} & \mathit{for}\ \mathbf{r} \in S. \\ 0 & \mathit{for}\ \mathbf{r} \in \Lambda \end{cases} \tag{6.194}$$

The integral equation itself is a direct formula; it tells us directly how we can reproduce a desired target field p_d. That is, the monopole contribution $\partial p_d / \partial n_s$ and dipole contribution p_d can certainly reproduce the target field p_d. The single-layer formula for an exterior virtual source is obtained by converting the integral Equation (6.194) such that it can be expressed in terms of monopoles only.

One useful property we can use here is that the monopole and dipole distributions reproducing the given interior sound field within V are not unique. We have two different integral equations depending on whether the virtual source is inside or outside of the volume V.

First, consider an arbitrary sound field p_{in} propagating from the interior volume V. From Equation (6.189) and using

$$p_d(\mathbf{r}) = \int_{V_0} g_f(\mathbf{r}|\mathbf{r}_0) q(\mathbf{r}_0) dV \quad \text{for direct waves,} \tag{6.195}$$

the sound field p_{in} can be written as

$$\int_S \left[\frac{\partial p_{in}(\mathbf{r}_s)}{\partial n_s} g_f(\mathbf{r}|\mathbf{r}_s) - p_{in}(\mathbf{r}_s) \frac{\partial g_f(\mathbf{r}|\mathbf{r}_s)}{\partial n_s} \right] dS = \begin{cases} 0 & \mathit{for}\, \mathbf{r} \in V \\ -\dfrac{p_{in}(\mathbf{r})}{2} & \mathit{for}\ \mathbf{r} \in S\ . \\ -p_{in}(\mathbf{r}) & \mathit{for}\ \mathbf{r} \in \Lambda \end{cases} \tag{6.196}$$

This second equation, which produces no sound field inside V, can be used to alter the monopole and dipole contributions on S without changing the interior sound field. Subtracting Equation (6.196) from Equation (6.194) yields another integral equation:

$$\int_S \left[\left(\frac{\partial p_d(\mathbf{r}_s)}{\partial n_s} - \frac{\partial p_{in}(\mathbf{r}_s)}{\partial n_s} \right) g_f(\mathbf{r}|\mathbf{r}_s) - (p_d(\mathbf{r}_s) - p_{in}(\mathbf{r}_s)) \frac{\partial g_f(\mathbf{r}|\mathbf{r}_s)}{\partial n_s} \right] dS$$

$$= \begin{cases} p_d(\mathbf{r}) & for\ \mathbf{r} \in V \\ (p_d(\mathbf{r}) + p_{in}(\mathbf{r}))/2 & for\ \mathbf{r} \in S \\ p_{in}(\mathbf{r}) & for\ \mathbf{r} \in \Lambda \end{cases}. \tag{6.197}$$

According to Equation (6.197), the dipole contribution can be eliminated by finding a pressure field p_{in} that is equal to the target field on the control surface. That is,

$$p_d(\mathbf{r}_s) = p_{in}(\mathbf{r}_s) \quad \text{for } \mathbf{r}_s \in S. \tag{6.198}$$

If such a p_{in} exists, the derivatives of p_d and p_{in} are not the same at S because p_{in} propagates outward from the interior sound source, whereas p_d propagates from the exterior region. The difference between the two sound fields becomes discontinuous across the boundary.

Here we can ask whether this sound field p_{in} from the interior source always exists, and if it exists, how we can calculate it. The existence of such a sound field can be simply shown: consider a volume with a pressure-release boundary condition imposed on its surface. Then, the target field incident on the sphere is scattered to satisfy the boundary condition. To satisfy the pressure release condition, the incident pressure field on the surface ($p_d(\mathbf{r}_s)$) should be equal to the scattered field with a negative sign. The summation of the incident and scattered fields should be zero on the surface, but the sum of the surface normal derivative is not. Accordingly, we can conclude that the field scattered from a pressure-release boundary is $p_{in}(\mathbf{r}_s)$.

With the scattering field $p_{in}(\mathbf{r}_s)$, we can derive the single-layer formula

$$\int_S q(\mathbf{r}_s) g_f(\mathbf{r}|\mathbf{r}_s) dS = \begin{cases} p_d(\mathbf{r}) & for\ \mathbf{r} \in V \\ (p_d(\mathbf{r}) + p_{in}(\mathbf{r}))/2 & for\ \mathbf{r} \in S(p_d = p_{in}) \\ p_{in}(\mathbf{r}) & for\ \mathbf{r} \in \Lambda \end{cases} \tag{6.199}$$

with the monopole excitation

$$q(\mathbf{r}_s) = \frac{\partial p_d(\mathbf{r}_s)}{\partial n_s} - \frac{\partial p_{in}(\mathbf{r}_s)}{\partial n_s}. \tag{6.200}$$

Therefore, finding the source excitation q for a single-layer formula is equivalent to solving a scattering problem over the control surface S.

The source excitation can be calculated from the integral formula (Equation (6.199)). If we take the derivative of the pressure field with respect to the surface normal $\mathbf{n}$, the integral equation at the surface S becomes

$$\int_S q(\mathbf{r}_s) \frac{\partial g_f(\mathbf{r}|\mathbf{r}_s)}{\partial n} dS = \begin{cases} \dfrac{\partial p_d(\mathbf{r})}{\partial n} & for\ \mathbf{r} \in V \\ \dfrac{\partial (p_d(\mathbf{r}) + p_{in}(\mathbf{r}))}{2\partial n} & for\ \mathbf{r} \in S \\ \dfrac{\partial p_{in}(\mathbf{r})}{\partial n} & for\ \mathbf{r} \in \Lambda \end{cases}. \tag{6.201}$$

Note that the subscript s is omitted in the surface normal vector $\mathbf{n}$ and the derivative $\partial/\partial n$ because the derivative is for the listener's position $\mathbf{r}$. From Equations (6.200) and (6.201), we can derive an equation for the source excitation q:

$$\int_S q(\mathbf{r}_s)\frac{\partial g_f(\mathbf{r}|\mathbf{r}_s)}{\partial n_s}dS = \frac{\partial p_d(\mathbf{r})}{\partial n} - \frac{1}{2}q(\mathbf{r}), \quad \mathit{for}\ \mathbf{r} \in S. \tag{6.202}$$

Solving this equation for q completes the single-layer formula.

The discontinuity of the pressure derivative on the boundary S, called the *jump relation*, characterizes the sound field reproduced by the single-layer formula. The discontinuity in the pressure derivative stems from the singularity imposed on the boundary; any scattering boundary or the surface source distribution sets singularities along the surface. On the other hand, the pressure field is continuous across S, but the exterior field given in Equation (6.199) is dissimilar to the original nonradiating field of Equation (6.194). This is because the monopole source radiates symmetrically for the interior and exterior regions.

6.8.1.2 Boundary-Element-Method-Based Approach

One way to solve Equation (6.202) for q is to use the boundary element method (BEM) on the surface. To that end, we discretize the integral Equation (6.202) on the surface S, which leads to

$$d\mathbf{G}_f d\mathbf{S}\mathbf{q} = d\mathbf{p}_d - \mathbf{q}, \tag{6.203}$$

where $d\mathbf{G}_f$ and $d\mathbf{p}_d$ denote matrices consisting of the surface normal derivatives of the Green's function and target pressure field, respectively.

$$[d\mathbf{G}_f]^{(m,\kappa)} = \frac{\partial g_f(\mathbf{r}^{(m)}|\mathbf{r}_s^{(\kappa)})}{\partial n}, \quad [d\mathbf{p}_d]^{(m,1)} = \frac{\partial p_d(\mathbf{r}^{(m)})}{\partial n} \quad (m = 1, \cdots, M;\ \ \kappa = 1, \cdots, K).$$

The diagonal matrix $d\mathbf{S}$ contains the area of the surface patch dS at the location of the given integral position $\mathbf{r}_s^{(\kappa)}$. We assume that the number of control sources K is the same as the number of measurement positions M on the surface S. Then the source excitation vector $\mathbf{q} = [q(\mathbf{r}_s^{(1)}), \cdots, q(\mathbf{r}_s^{(M)})]^T$ satisfying Equation (6.203) is given by

$$\mathbf{q} = (d\mathbf{G}_f d\mathbf{S} + \mathbf{I})^{-1} d\mathbf{p}_d. \tag{6.204}$$

This is a very simplified form of the direct BEM approach. Unlike the KH integral, the single-layer formula requires that an inverse problem be solved. There are particular conditions under which the inverse of Equation (6.204) becomes singular, similar to the forbidden frequency of the mode-matching approach (Section 6.6.3). Moreover, its computational complexity can be very high depending on the frequency of interest and the size of the control surface, for example, at high frequencies in which the control surface S should be sampled at fine spatial intervals smaller than the wavelength.

There is another integral technique for deriving the single-layer formula at high frequency. This is possible by approximating the KH integral using some assumptions suitable at high frequencies.

6.8.1.3 Farfield and Stationary Phase Approximations

A simpler derivation is possible when a virtual point source and listener are in the farfield away from the control source. For a virtual point source radiating under free-field conditions, the target sound field on the control surface is given by

$$p_d(\mathbf{r}_s) = \frac{e^{ikR_v}}{4\pi R_v}, \quad R_v = |\mathbf{r}_v - \mathbf{r}_s|, \tag{6.205}$$

where $\mathbf{r}_v$ is the position of the virtual source. By considering a free-field Green's function $g_f(\mathbf{r}|\mathbf{r}_s) = e^{ikR}/4\pi R$ ($R = |\mathbf{r} - \mathbf{r}_s|$), Equation (6.194) can be written as

$$p_d(\mathbf{r}) = \frac{1}{(4\pi)^2}\int_S \left\{ \left[\frac{\partial R_v}{\partial n_s}\left(ik - \frac{1}{R_v}\right) - \frac{\partial R}{\partial n_s}\left(ik - \frac{1}{R}\right)\right] \frac{e^{ik(R_v+R)}}{RR_v} \right\} dS \tag{6.206}$$

for an exterior virtual source.

If the virtual source and the listener are at the farfield away from the control surface ($kR \gg 1$, $kR_v \gg 1$), we can approximate Equation (6.206) as

$$p_d(\mathbf{r}) \approx \frac{ik}{(4\pi)^2}\int_S \left(\underbrace{\frac{\partial R_v}{\partial n_s}}_{\substack{\textit{monopole}\\\textit{contribution}}} - \underbrace{\frac{\partial R}{\partial n_s}}_{\substack{\textit{dipole}\\\textit{contribution}}} \right) \frac{e^{ik(R_v+R)}}{RR_v} dS, \tag{6.207}$$

where the derivatives of R_v and R are related to the contributions of the monopole and dipole sources, respectively. The change in the distances R_v and R in the surface normal direction $\mathbf{n}_s$ can be expressed in terms of the angle between $-\mathbf{n}_s$ and the listener (φ), and that between $-\mathbf{n}_s$ and the virtual source (φ_v), respectively (Figure 6.33). That is,

$$p_d(\mathbf{r}) \approx \frac{ik}{(4\pi)^2}\int_S \left[(\cos\varphi_v - \cos\varphi) \frac{e^{ik(R_v+R)}}{RR_v} \right] dS. \tag{6.208}$$

Another important approximation we introduce here is the stationary phase approximation (Section B.2.2). Suppose that the phase change of the integrand ($e^{ik(R_v+R)}$) in Equation (6.208) varies rapidly along the control surface S, except at the point of the stationary phase, and that the magnitude varies more slowly than the phase change. Under these conditions, most of the integral is canceled out by the rapidly varying phase, and only the integral near the point of the stationary phase contributes to the reproduced field.

This assumption is valid when the wavenumber k and the associated path change are large along the control surface. Note that the phase of the integrand is related to the total length of the acoustic

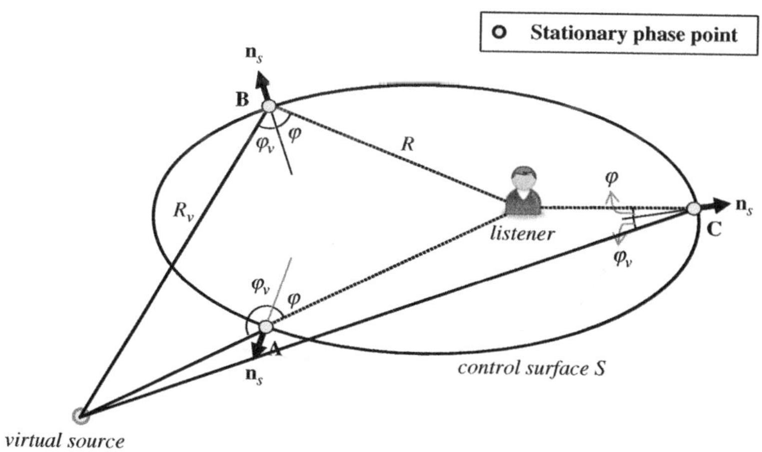

Figure 6.33 Three stationary phase points for an elliptical control surface

transfer path ($R_v + R$) with respect to the wavelength. From the triangle inequality, the global minimum of the total path is given as a line directly connecting the virtual source and the listener (Figure 6.33). Therefore, the global minimum of the phase is obtained at position A in Figure 6.33.

However, multiple stationary phase points can exist for a closed control surface. In the example shown in Figure 6.33, there are three stationary phase points on the elliptic control surface. The global phase minimum occurs at position A, but stationary values can be obtained at the other two positions as well. Nevertheless, there is an important difference between position A and the others. When observed from the virtual source location, the surface near position A is convex, but it is concave from the listener's position. At B and C, the surface is always concave from both the virtual source and listener locations.

This difference results in dissimilar contributions of monopoles and dipoles. At position A, we have

$$\cos\varphi_v = -\cos\varphi. \tag{6.209}$$

In contrast, at B and C, the contributions of monopoles and dipoles are inverted in sign ($\cos\varphi = \cos\varphi_v$), which results in the zero integrand of Equation (6.207). Because the sum of the contributions is zero at these points, point A contributes the most to the given listener position. Moreover, from Equations (6.208) and (6.209), the contributions of monopole and dipole control sources are similar near the stationary phase point A. Therefore, we can reproduce the target sound field just by doubling the contribution of the monopole sources, which yields an *approximated single-layer formula*. For the single-layer formula, the monopole source at the concave part of the control surface (points B and C) should be turned off. Otherwise, the monopole-only integral produces a nonzero sound field that differs from that with both monopole and dipole sources.

The stationary phase point A is a function of the virtual source and listener. Therefore, for a given listener's position, there is a region of control sources that effectively contribute to the listener (Figure 6.34). If the listener moves, the effective region also changes. Therefore, multiple regions on the control surface should be activated to reproduce the target field over multiple listener positions at the same time.

From all these observations, the integration of the doubled monopole contribution of Equation (6.207) over the convex part of the control surface (S' in Figure 6.34) can reproduce the sound

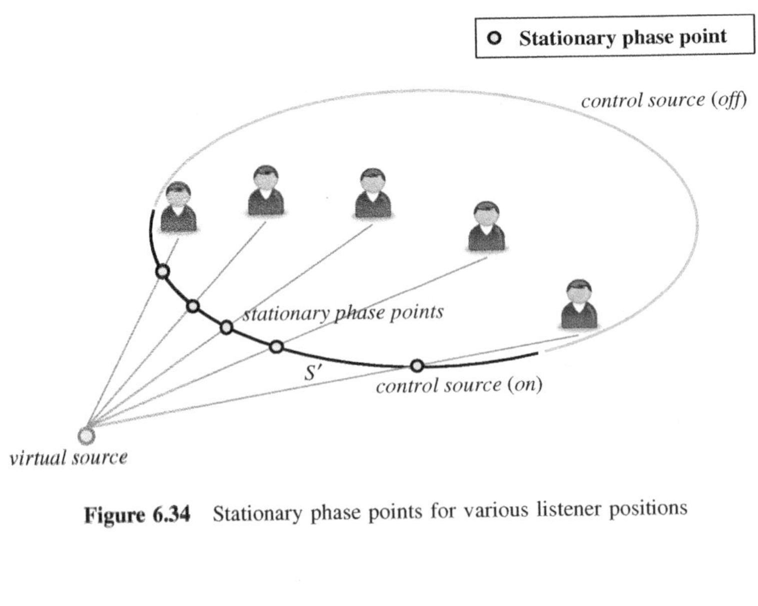

Figure 6.34 Stationary phase points for various listener positions

field over a wide area:

$$p_d(\mathbf{r}) \approx 2ik \int_{S'} \left[\left(\frac{\partial R_v}{\partial n_s} \frac{e^{ikR_v}}{4\pi R_v} \right) \frac{e^{ikR}}{4\pi R} \right] dS$$

$$\approx \int_{S'} \left(q_{mono}(\mathbf{r}_s) \frac{e^{ikR}}{4\pi R} \right) dS, \quad \text{where } q_{mono}(\mathbf{r}_s) = 2ik \frac{\partial R_v}{\partial n_s} \frac{e^{ikR_v}}{4\pi R_v}. \tag{6.210}$$

Now the KH integral is converted into a single-layer formula that incorporates only monopole control sources. The deriving function q_{mono} can also be viewed as a doubled particle velocity in the surface normal direction produced by a virtual monopole source in the farfield.

$$q_{mono}(\mathbf{r}_s) \approx 2ik \frac{\partial}{\partial n_s} \left(\frac{e^{ikR_v}}{4\pi R_v} \right), \quad (kR_v \gg 1). \tag{6.211}$$

6.8.1.4 Wave Field Synthesis

In practice, it is not practical to construct loudspeaker arrays covering the entire control surface. Because the stationary phase point on the control surface depends on the listener position, we can reduce the control surface by limiting the allowed listener positions (sweet spot).

For example, let us limit the listener's position to a single lateral plane. Because the stationary phase point exists on a cross point of the control surface and a line connecting the listener and the virtual source, the control sources near a contour C defined by the intersection of the horizontal plane and control surface (Figure 6.35) contribute most to the integral. To evaluate the integral near the stationary phase point, we can use the method for the stationary phase described in Appendix B, Section B.2.2. First, consider a small fraction of the control surface in the vertical direction ($y_s \in [-\Delta/2, \Delta/2]$). According to the stationary phase approximation, the integral over a strip defined by the open contour C and Δ can replace the complete integral. That is,

$$p_d(\mathbf{r}) \approx 2ik \int_C \left\{ \int_{-\Delta/2}^{\Delta/2} \left[\left(\frac{\partial R_v}{\partial n_s} \frac{e^{ikR_v}}{4\pi R_v} \right) \frac{e^{ikR}}{4\pi R} \right] dy_s \right\} dl, \tag{6.212}$$

where dl indicates an infinitesimal element of the contour C. The phase-related term can be expressed as

$$\zeta(y_s) = k(R_v + R) = k(\sqrt{R_{v\perp}^2 + y_s^2} + \sqrt{R_\perp^2 + y_s^2}), \tag{6.213}$$

where $R_{v\perp}$ and $R_\perp$ are projections of R_v and R, respectively, to the lateral plane. The phase change with respect to y_s can be evaluated as

$$\zeta'(y_s) = ky_s \left(\frac{1}{R_v} + \frac{1}{R} \right), \tag{6.214}$$

which is zero for $y_s = 0$. Using the relation

$$\zeta(0) = k(R_{v\perp} + R_\perp), \zeta''(0) = k \left(\frac{1}{R_{v\perp}} + \frac{1}{R_\perp} \right) \tag{6.215}$$

and the method for the stationary phase (Appendix B, Section B.2.2), we can reduce the surface integral Equation (6.212) to a contour integral along a curve C defined on the plane $y_s = 0$

$$p_d(\mathbf{r}) \approx \sqrt{-8\pi ik} \int_C \left[\left(\sqrt{\frac{RR_v}{(R + R_v)}} \frac{\partial R_v}{\partial n_s} \frac{e^{ikR_v}}{4\pi R_v} \right) \frac{e^{ikR}}{4\pi R} \right] dl. \tag{6.216}$$

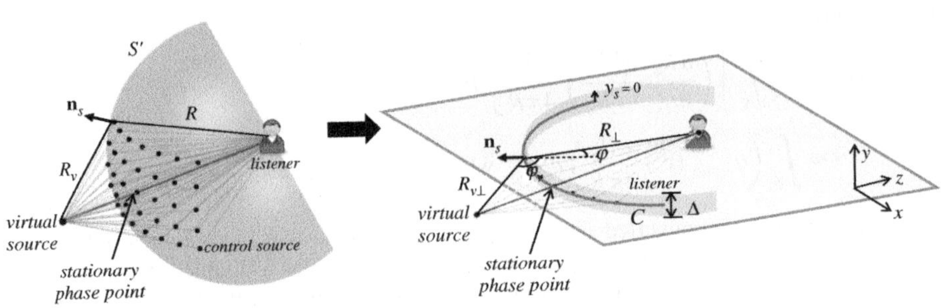

Figure 6.35 Approximation of the surface integral to the contour integral for a virtual source and listener on the same lateral plane

The contour integral implies that the target sound field can be reproduced over the $y = 0$ plane using a linear or curved array of loudspeakers. The excitation signal within the parenthesis of Equation (6.216), however, depends on the listener's position R. It is somewhat impractical to identify the listener location and update the excitation signal, so we usually assign a representative listener distance R_{ref} called a *reference line* or *position*. For a fixed listener at the reference position ($R = R_{ref}$), the integral equation yields

$$p_d(\mathbf{r}) \approx \sqrt{-8\pi ik} \int_C \left[\left(\sqrt{\frac{R_{ref} R_v}{\left(R_{ref} + R_v\right)}} \frac{\partial R_v}{\partial n_s} \frac{e^{ikR_v}}{4\pi R_v} \right) \frac{e^{ikR}}{4\pi R} \right] dl$$

$$\approx \int_C \left(q_{2.5D}(\mathbf{r}_s) \frac{e^{ikR}}{4\pi R} \right) dl, \text{ where } q_{2.5D}(\mathbf{r}_s) = \sqrt{\frac{-ik}{2\pi}} \sqrt{\frac{R_{ref}}{(R_{ref} + R_v)}} \cos\varphi_v \frac{e^{ikR_v}}{\sqrt{R_v}}. \quad (6.217)$$

The approximated integral, commonly referred to as a 2.5D integral, shows a greater reproduction error as the listener moves away from the reference position R_{ref}. Nevertheless, the error due to the use of the reference position appears mainly as a magnitude error. If we remember that the control sources near the stationary phase point contribute most to the integral, the sound field produced by the contour integral equation (Equation (6.216)) can be also determined by the integrand near the stationary phase points. According to Equation (B115), the approximated contour integral has the form

$$p_d(\mathbf{r}) \approx \sqrt{\frac{2\pi i}{\zeta''(\mathbf{l}_0)}} f(\mathbf{l}_0) e^{i\zeta(\mathbf{l}_0)}, \quad (6.218)$$

where $\mathbf{l}_0$ is the stationary phase point for a listener at R_{ref}, and $f(\mathbf{l}_0)$ is a magnitude-related term in the integrand in Equation (6.216). The phase-related term $\zeta = k(R_v + R)$ does not depend on R_{ref}, so it can be rendered properly irrespective of the listener position. In contrast, the magnitude dependencies $f(\mathbf{l}_0)$ calculated by Equations (6.216) and (6.217) differ when the listener is not at the prescribed position R_{ref}. Accordingly, the error in $f(\mathbf{l}_0)$ appears mainly as a magnitude error and does not distort the phase of the reproduced field.

Sound field reproduction techniques using the 2.5D integral have been extensively developed in WFS. There are various issues related to the number of loudspeakers, finite size of the control surface, loudspeaker directivity, and other factors. For further details on WFS, please consult [16, 17].

6.8.1.5 Relationship between the KH Integral and Mode-Matching Approach

When the control surface is given by a spherical surface, the KH integral can then be simplified by means of spherical harmonics. Consider the KH integral for an exterior virtual source and interior listener.

$$p_r(\mathbf{r}) = \int_S \left[\frac{\partial p_r(\mathbf{r}_s)}{\partial n_s} g_f(\mathbf{r}|\mathbf{r}_s) - p_r(\mathbf{r}_s) \frac{\partial g_f(\mathbf{r}|\mathbf{r}_s)}{\partial n_s} \right] dS. \tag{6.219}$$

For a spherical control surface of radius r_s, the surface normal is equal to the radial derivative, and the surface integral can be replaced with the angular integral over solid angle Ω.

$$p_r(\mathbf{r}) = r_s^2 \int_\Omega \left[\frac{\partial p_r(\mathbf{r}_s)}{\partial r_s} g_f(\mathbf{r}|\mathbf{r}_s) - p_r(\mathbf{r}_s) \frac{\partial g_f(\mathbf{r}|\mathbf{r}_s)}{\partial r_s} \right] d\Omega. \tag{6.220}$$

Using spherical harmonic expansion of the surface pressure distribution

$$p_r(\mathbf{r}_s) = \sum_{n=0}^{\infty} \sum_{m=-n}^{n} Q_{nm} j_n(kr_s) Y_n^m(\theta_s, \phi_s) \tag{6.221}$$

and the free-field Green's function (Equation (B173) and (B174)), the integral can be rewritten as

$$p_r(r,\theta,\phi) = i(kr_s)^2 \sum_{n=0}^{\infty} \sum_{m=-n}^{n} j_n(kr) Y_n^m(\theta,\phi) Q_{nm}(\omega) [j_n(kr_s)' h_n^{(1)}(kr_s) - j_n(kr_s) h_n^{(1)}(kr_s)']. \tag{6.222}$$

The operation in the brackets in Equation (6.222) is identical to the Wronskian relation (Equation (B164)); hence, the KH integral leads to

$$p_r(r,\theta,\phi) = \sum_{n=0}^{\infty} \sum_{m=-n}^{n} Q_{nm}(\omega) j_n(kr) Y_n^m(\theta,\phi). \tag{6.223}$$

Consequently, the KH integral over a spherical surface is simply equivalent to spherical harmonic expansion.

From this result, we can also determine the surface pressure distribution that reproduces a target sound field and its derivative. The surface pressure of the KH integral can be calculated by equating the target and reproduced fields (a *field-matching equation*).

$$p_d(r,\theta,\phi) = \sum_{n=0}^{\infty} \sum_{m=-n}^{n} Q_{nm}(\omega) j_n(kr) Y_n^m(\theta,\phi). \tag{6.224}$$

If the radial dependency of a target sound field is expressed in terms of the spherical Bessel function, we can expand the left-hand side of Equation (6.224) as

$$\sum_{n=0}^{\infty} \sum_{m=-n}^{n} D_{nm}(\omega) j_n(kr) Y_n^m(\theta,\phi) = \sum_{n=0}^{\infty} \sum_{m=-n}^{n} Q_{nm}(\omega) j_n(kr) Y_n^m(\theta,\phi); \tag{6.225}$$

hence, it can be seen that the KH integral is equivalent to the mode-matching equation

$$Q_{nm}(\omega) = D_{nm}(\omega). \tag{6.226}$$

Accordingly, the surface pressure and its normal derivative can be derived from

$$p_r(r_s, \theta_s, \phi_s) = \sum_{n=0}^{\infty} \sum_{m=-n}^{n} D_{nm}(\omega) j_n(kr_s) Y_n^m(\theta_s, \phi_s),$$

$$\partial p_r(r_s, \theta_s, \phi_s)/\partial r_s = k \sum_{n=0}^{\infty} \sum_{m=-n}^{n} D_{nm}(\omega) j_n(kr_s)' Y_n^m(\theta_s, \phi_s). \tag{6.227}$$

The physical meaning of the expansion using a spherical Bessel function is that a sound field is propagating from outside with respect to the spherical surface of radius r. If this is not the case, that is, if the radial function is expanded in terms of the radiating function $h_n^{(1)}(kr)$, the field-matching equation is given by

$$\sum_{n=0}^{\infty} \sum_{m=-n}^{n} D_{nm}(\omega) h_n^{(1)}(kr) Y_n^m(\theta, \phi) = \sum_{n=0}^{\infty} \sum_{m=-n}^{n} Q_{nm}(\omega) j_n(kr) Y_n^m(\theta, \phi), \tag{6.228}$$

which can never be equal because of the discrepancy between the spherical Hankel and Bessel functions.

An interesting example of this behavior is the case in which the target field is produced by an interior monopole source ($r_v < r_s$), and the listener is inside the control surface ($r < r_s$). From the spherical harmonic expansion of the free-field Green's function, the target field can be written as

$$p_d(\mathbf{r}) = g_f(\mathbf{r}|\mathbf{r}_v)$$
$$= \begin{cases} ik \sum_{n=0}^{\infty} \left[h_n^{(1)}(kr) j_n(kr_v) \sum_{m=-n}^{n} Y_n^m(\theta, \phi) Y_n^m(\theta_v, \phi_v)^* \right] & r > r_v \\ ik \sum_{n=0}^{\infty} \left[j_n(kr) h_n^{(1)}(kr_v) \sum_{m=-n}^{n} Y_n^m(\theta, \phi) Y_n^m(\theta_v, \phi_v)^* \right] & r < r_v \end{cases}. \tag{6.229}$$

Thus, there are two types of expansion depending on the listener's location. If the listener's distance from the center is larger than that of the virtual source, the radial dependency of the target sound field is given by the superposition of spherical Hankel functions. This corresponds to the case of Equation (6.228), which cannot be solved by the mode-matching equation. On the other hand, for a listener closer to the center than the virtual source, the harmonic expansion of the sound field is a function of the spherical Bessel function and can be reproduced by solving the mode-matching equation of Equation (6.226). The harmonic coefficients of the target field are given by

$$D_{nm} = ikh_n^{(1)}(kr_v) Y_n^m(\theta_v, \phi_v)^*, \tag{6.230}$$

and the corresponding surface pressure and its normal derivative can be derived from Equation (6.227).

The sound field reproduced from the surface monopoles and dipoles is given by

$$p_r(r, \theta, \phi) = ik \sum_{n=0}^{\infty} \left[j_n(kr) h_n^{(1)}(kr_v) \sum_{m=-n}^{n} Y_n^m(\theta, \phi) Y_n^m(\theta_v, \phi_v)^* \right] \tag{6.231}$$

for both listener positions, $r > r_v$ and $r \leq r_v$. The reproduced sound field is identical for a listener at $r \leq r_v$ but completely different for $r > r_v$. This implies that the singularity of the virtual source cannot be reproduced by the exterior source. Therefore, sound field reproduction of an interior virtual source requires a different type of integral equation.

6.8.2 *Integral Formulas for Interior Virtual Source*

In this section, we consider a specific interior reproduction problem in which the virtual source is located inside the loudspeaker array (Figure 6.8a). The goal of sound field reproduction from an interior virtual source is to provide the illusion of sound sources that are closer to the listener than the array location; this is commonly referred to as *focused source reproduction* in WFS studies [17].

6.8.2.1 Nonexistence Problem of the Interior Virtual Source

The interior virtual source problem is interesting in many respects. First, the KH integral cannot be used directly. For free-field radiation p_d, the KH integral makes no contribution to the interior sound field (Equation (6.168)), which indicates that the contribution of the surface monopoles and dipoles is zero for an interior virtual source. Therefore, the interior virtual source problem cannot be handled using the conventional KH integral and requires a different approach.

The main difficulty of the source-inside case is that a solution that reproduces the interior source over an entire region V does not exist. Specifically, consider a volume V enclosed by a distribution of secondary sources on S (Figure 6.8a). The target field generated by a virtual point source positioned at $\mathbf{r}_v \in V$ satisfies the following inhomogeneous differential equation within the region V:

$$\nabla^2 g(\mathbf{r}|\mathbf{r}_v) + k^2 g(\mathbf{r}|\mathbf{r}_v) = -\delta(\mathbf{r} - \mathbf{r}_v), \quad \mathbf{r}, \mathbf{r}_v \in V. \tag{6.232}$$

The sound field induced by the interior point source has a singularity on $\mathbf{r} = \mathbf{r}_v$. However, the control sources used for the reproduction are outside of the volume V, and their sound fields follow the homogeneous wave equation.

$$\nabla^2 g(\mathbf{r}|\mathbf{r}_s) + k^2 g(\mathbf{r}|\mathbf{r}_s) = 0, \quad \mathbf{r} \in V, \mathbf{r}_s \in S. \tag{6.233}$$

Therefore, it is impossible to generate a sound field having a singularity within V in terms of the control sources satisfying the homogeneous differential equation. This is a typical *nonexistence problem*, in which the solution cannot exist because of certain physical limitations.

One possible way to resolve this problem is to replace the target field with an *alternative field* that satisfies the homogeneous differential equation within V but still resembles the ideal field except at the singular point $\mathbf{r}_v$. If we can find such an alternative field, we can reproduce it using the integral equation without solving any inverse problems. One candidate for the alternative field can be obtained by subtracting the target field from its time inverse. That is,

$$p(\mathbf{r}) = g(\mathbf{r}|\mathbf{r}_v) - g(\mathbf{r}|\mathbf{r}_v)^*. \tag{6.234}$$

The conjugation operator * in the frequency domain represents time reversal in the time domain. Because the time reversal of the wave equation also has the same singularity on $\mathbf{r}_v$,

$$\nabla^2 g(\mathbf{r}|\mathbf{r}_v)^* + k^2 g(\mathbf{r}|\mathbf{r}_v)^* = -\delta(\mathbf{r} - \mathbf{r}_v), \quad \mathbf{r}, \mathbf{r}_v \in V, \tag{6.235}$$

the subtracted field in Equation (6.234) satisfies the homogeneous wave equation and has no singularity within the region V.

$$\nabla^2 p(\mathbf{r}) + k^2 p(\mathbf{r}) = 0, \quad \mathbf{r} \in V. \tag{6.236}$$

Because it satisfies the homogeneous wave equation, the sound field in Equation (6.234) can be reproduced by the ordinary KH integral in Equation (6.190). That is,

$$g(\mathbf{r}|\mathbf{r}_v) - g(\mathbf{r}|\mathbf{r}_v)^* = \int_S \left[\frac{\partial \left(g\left(\mathbf{r}_s|\mathbf{r}_v\right) - g(\mathbf{r}_s|\mathbf{r}_v)^*\right)}{\partial n_s} g(\mathbf{r}|\mathbf{r}_s) - (g(\mathbf{r}_s|\mathbf{r}_v) - g(\mathbf{r}_s|\mathbf{r}_v)^*) \frac{\partial g(\mathbf{r}|\mathbf{r}_s)}{\partial n_s} \right] dS. \tag{6.237}$$

However, from Equation (6.170), the surface integral of the sound field $g(\mathbf{r}_s|\mathbf{r}_v)$ is zero. As a result, we can obtain the following integral to reproduce Equation (6.234):

$$g(\mathbf{r}|\mathbf{r}_v) - g(\mathbf{r}|\mathbf{r}_v)^* = \int_S \left[g(\mathbf{r}_v|\mathbf{r}_s)^* \frac{\partial g(\mathbf{r}|\mathbf{r}_s)}{\partial n_s} - \frac{\partial g(\mathbf{r}_v|\mathbf{r}_s)^*}{\partial n_s} g(\mathbf{r}|\mathbf{r}_s) \right] dS(\mathbf{r}_s), \quad \mathbf{r}_v, \mathbf{r} \in V. \tag{6.238}$$

Equation (6.238) implies that the time reversal of the surface pressure and its derivative at the surface S generate two types of sound fields: a converging wave field $-g(\mathbf{r}|\mathbf{r}_v)^*$ and a diverging wave field $G(\mathbf{r}|\mathbf{r}_v)$. Because the time-reversed field can exist for $t < 0$, the alternative field first converges toward the focal point $\mathbf{r}_v$ and then begins to diverge. From the viewpoint of sound field reproduction, the diverging wavefront is what we want to reproduce, and the converging wavefront is an artifact to be removed or suppressed. The integral of Equation (6.238) does not require any inversion to find the excitation function of the secondary sources at $\mathbf{r}_s$. However, the converging wavefront is the time-reverse of the omnidirectional radiation and as such is incident from every direction; its sound is audible in every location between the control surface and the virtual source.

6.8.2.2 Integral Equation for a Multipole Virtual Source[1]

The basic concept for reducing the effect of the converging wavefront is to employ multipole virtual sources instead of a virtual point source [18]. First, consider a group of point sources (i.e., a

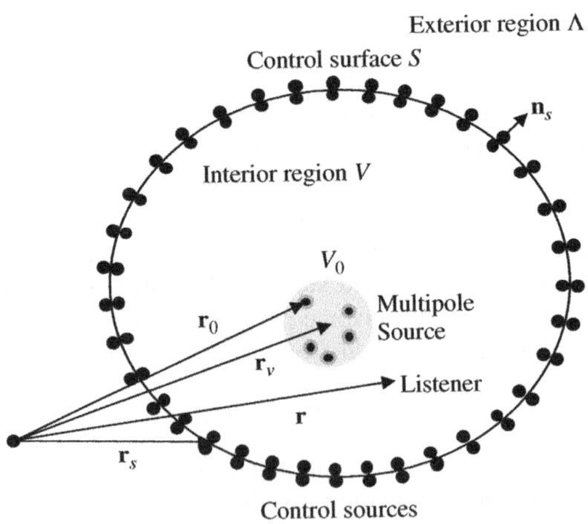

Figure 6.36 Replacement of a virtual source with multipole sources

[1] © [2012] IEEE. Reprinted, with permission, from [Jung-Woo Choi and Yang-Hann Kim, Integral Approach for Reproduction of Virtual Sound Source Surrounded by Loudspeaker Array, Audio, Speech, and Language Processing, IEEE Transactions on, Sept. 2012].

multipole source) compactly packed within the volume V_0 (Figure 6.36). When the multipole source is driven by an *excitation function* $q(\mathbf{r}_0)$, the target field $p_d(\mathbf{r})$ is given by

$$p_d(\mathbf{r}) = \int_{V_0} g(\mathbf{r}|\mathbf{r}_0)q(\mathbf{r}_0)dV(\mathbf{r}_0), \mathbf{r} \in V, \mathbf{r}_0 \in V_0 \subset V, \tag{6.239}$$

where the target field satisfies the wave equation

$$\nabla^2 p_d(\mathbf{r}) + k^2 p_d(\mathbf{r}) = -q(\mathbf{r}), \mathbf{r} \in V. \tag{6.240}$$

Although the desired excitation q has not been specified yet, we first assume that there exists an excitation pattern that is suitable for the reproduction.

Next, we consider another sound field p_{tr}^* that can be obtained by using the time-reversed Green's function $g(\mathbf{r}|\mathbf{r}_0)^*$. The excitation signal $q(\mathbf{r})$, however, remains unchanged.

$$p_{tr}^*(\mathbf{r}) = \int_{V_0} g(\mathbf{r}|\mathbf{r}_0)^* q(\mathbf{r}_0)dV(\mathbf{r}_0). \tag{6.241}$$

The sound field p_{tr}^* is also driven by the same excitation function q; hence, it satisfies the following wave equation:

$$\nabla^2 p_{tr}(\mathbf{r})^* + k^2 p_{tr}(\mathbf{r})^* = -q(\mathbf{r}), \quad \mathbf{r} \in V. \tag{6.242}$$

The sound field produced by the time-reversed Green's function differs from the time-reversed field p_d^* because the wavefront of p_d^* is generated by the time-reversed excitation q^*, whereas p_{tr}^* is driven by q. Because of its practical importance, we denote the sound field p_{tr}^* as a *time-reversed propagation* TRP field. Both the target field and its TRP satisfy the same differential equations, Equations (6.240) and (6.242); therefore, we can reproduce the subtracted field $p_d - p_{tr}^*$.

Because the TRP satisfies the inhomogeneous wave equation, we can use the integral for the source-inside, listener-inside case, (Equation (6.189)). Replacing p_d in Equation (6.189) with p_{tr}^*, and using the free-field Green's function g_f for g yields

$$p_{tr}(\mathbf{r}_s)^* = \int_{V_0} g_f(\mathbf{r}|\mathbf{r}_0)q(\mathbf{r}_0)dV + \int_S \left[\frac{\partial p_{tr}(\mathbf{r}_s)^*}{\partial n_s} g_f(\mathbf{r}|\mathbf{r}_s) - p_{tr}(\mathbf{r}_s)^* \frac{\partial g_f(\mathbf{r}|\mathbf{r}_s)}{\partial n_s}\right] dS, \mathbf{r} \in V. \tag{6.243}$$

From the definition of the multipole field, Equation (6.239), the volume integral of Equation (6.243) becomes the target field p_d. Therefore, the integral equation can be arranged as

$$\begin{aligned} p_r(\mathbf{r}) &= p_d(\mathbf{r}) - p_{tr}(\mathbf{r})^* \\ &= \int_S \left[p_{tr}(\mathbf{r}_s)^* \frac{\partial g(\mathbf{r}|\mathbf{r}_s)}{\partial n_s} - \frac{\partial p_{tr}(\mathbf{r}_s)^*}{\partial n_s} g(\mathbf{r}|\mathbf{r}_s)\right] dS(\mathbf{r}_s). \quad \mathbf{r} \in V \end{aligned} \tag{6.244}$$

Equation (6.244) shows that both the target field and its TRP are reproduced by driving the monopole and dipole sources using the TRP and its derivative (p_{tr}^* and $\partial p_{tr}^*/\partial n_s$), respectively.

We want to reproduce the target field, so the TRP field $-p_{tr}^*$ is the only artifact we have. This is similar to the single-monopole case in Equation (6.238), but in this multipole case, the radiation patterns of the target and TRP can be dissimilar. Accordingly, we can design the multipole excitation q that can minimize the artifacts due to the reproduction of TRP.

To see how the TRP field differs from the target field, consider a simple case in which the multipole consists of two monopoles close to each other. Configuring the polar coordinates $\mathbf{r} = (r, \theta)$, the origin of which is at the center of the multipole ($\mathbf{r}_v$), the two monopole's locations

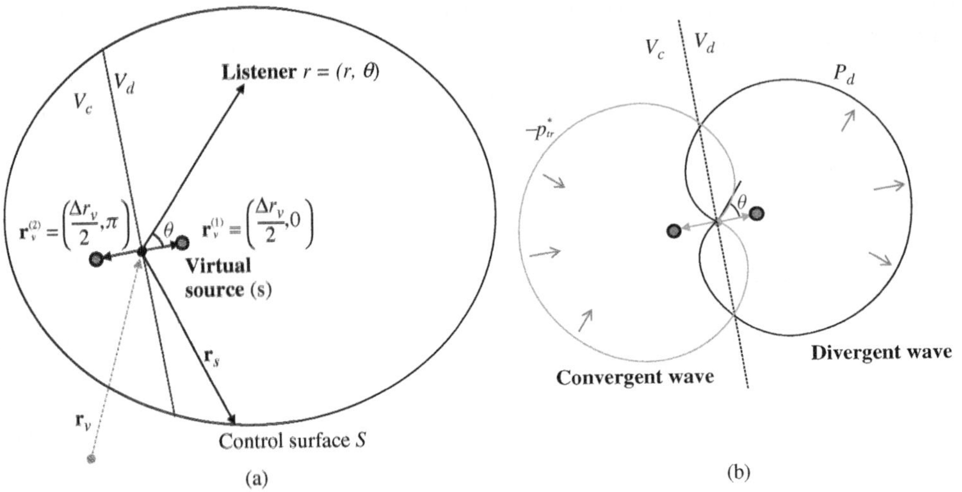

Figure 6.37 (a,b) Cardioid virtual source configuration

can be denoted as $\mathbf{r}_v^{(1)} = (\Delta r_v/2, 0)$ and $\mathbf{r}_v^{(2)} = (\Delta r_v/2, \pi)$, as shown in Figure 6.37a. If we excite the two monopoles by the differential excitation function (Equation (5.85)) with an extra phase change $\alpha k \Delta r_v$,

$$q(\mathbf{r}_0) = \delta(\mathbf{r}_0 - \mathbf{r}_v^{(1)}) - \delta(\mathbf{r}_0 - \mathbf{r}_v^{(2)})e^{i\alpha k \Delta r_v}, \tag{6.245}$$

then, according to Equation (6.239), the target field can be expressed as

$$p_d(\mathbf{r}) = g_f(\mathbf{r}|\mathbf{r}_v^{(1)}) - g_f(\mathbf{r}|\mathbf{r}_v^{(2)})e^{i\alpha k \Delta r_v}. \tag{6.246}$$

Because the separation distance is smaller than the wavelength ($k\Delta r_v \ll 1$), the target sound field from the two monopoles can be approximated as (Equation 5.88)

$$p_d(\mathbf{r}) \approx -i(k\Delta r_v) g_f(\mathbf{r}|\mathbf{r}_v)(\alpha + \cos\theta), \tag{6.247}$$

where the angle of the listener observed from the center of the virtual source is indicated by θ (Figure 6.37a). According to Equation (6.247), the radiation pattern generated by the first-order differential excitation is given by $b(\theta) \approx -ik\Delta r_v(\alpha + \cos\theta)$. In contrast, the TRP is given by

$$\begin{aligned} p_{tr}(\mathbf{r})^* &= g_f^*(\mathbf{r}|\mathbf{r}_v^{(1)}) - g_f^*(\mathbf{r}|\mathbf{r}_v^{(2)})e^{i\alpha k \Delta r_v} \\ &\approx -i(k\Delta r_v) g(\mathbf{r}|\mathbf{r}_v)^*(\alpha - \cos\theta). \end{aligned} \tag{6.248}$$

Comparing Equations (6.247) and (6.248), we observe a notable difference between the target and TRP fields. The radiation patterns of the two fields are in opposite directions to each other. For example, when the differential sources are configured to be cardioid ($\alpha = 1$), the contribution of the TRP is zero in the $\theta = 0$ direction, but the magnitude of the target field is maximum (Figure 6.37b). The contributions of the target and TRP fields are reversed with respect to each other. These symmetric radiation patterns can be used to separate the contributions of the converging and diverging wavefronts in space. For example, the virtual cardioid can generate two regions (V_c and V_d) in which either the TRP or the target field dominates (Figure 6.37b).

This two-monopole case is just one of the candidates for the alternative target field. To derive a general formula for the alternative field, we need to analyze the radiation patterns of the convergent and divergent waves from multiple sources.

6.8.2.3 Two Direct Fields of Symmetric Radiation Patterns Generated by a Multipole Source

Consider L multiple sources compactly located around a position $\mathbf{r}_v$ (a multipole source). We assume that the multipole source is radiating in the free-field condition. Then, we consider spherical coordinates centered at $\mathbf{r}_v$ (Figure 6.38) and denote the location of the lth virtual monopole in the coordinates as $\mathbf{r}_v^{(l)} = (\Delta r_v^{(l)}, \theta_v^{(l)}, \phi_v^{(l)})$. The sound field of a monopole measured at the listener position $\mathbf{r} = (r, \theta, \phi)$ is given by Equation (B173)

$$g_f(\mathbf{r}|\mathbf{r}_v^{(l)}) = ik\sum_{n=0}^{\infty} j_n(k\Delta r_v^{(l)})h_n^{(1)}(kr)\sum_{m=-n}^{n} Y_n^m(\theta,\phi)Y_n^m(\theta_v^{(l)},\phi_v^{(l)})^*. \tag{6.249}$$

We can express the target field from the multipole sources as

$$p_d(\mathbf{r}) = \sum_{l=1}^{L} g_f(\mathbf{r}|\mathbf{r}_v^{(l)})q(\mathbf{r}_v^{(l)})$$
$$= ik\sum_{n=0}^{\infty} h_n^{(1)}(kr)a_n(\theta,\phi), \tag{6.250}$$

where $a_n(\theta,\phi)$ represents the complex radiation pattern of order n generated by the multipoles. That is,

$$a_n(\theta,\phi) = \sum_{m=-n}^{n} Y_n^m(\theta,\phi)\sum_{l=1}^{L} j_n(k\Delta r_v^{(l)})Y_n^m(\theta_v^{(l)},\phi_v^{(l)})^* q^{(l)}, \tag{6.251}$$

and $q^{(l)}$ is an abbreviation for $q(\mathbf{r}_v^{(l)})$. On the other hand, for the TRP, we have

$$p_{tr}(\mathbf{r})^* = \sum_{l=1}^{L} g_f(\mathbf{r}|\mathbf{r}_v^{(l)})^* q(\mathbf{r}_v^{(l)}) = -ik\sum_{n=0}^{\infty} h_n^{(2)}(kr)a_n(\theta,\phi). \tag{6.252}$$

The farfield behavior of the multipoles can be analyzed using these expressions. Because the radial decay of the multipole sound field should resemble the radiation from a virtual point source,

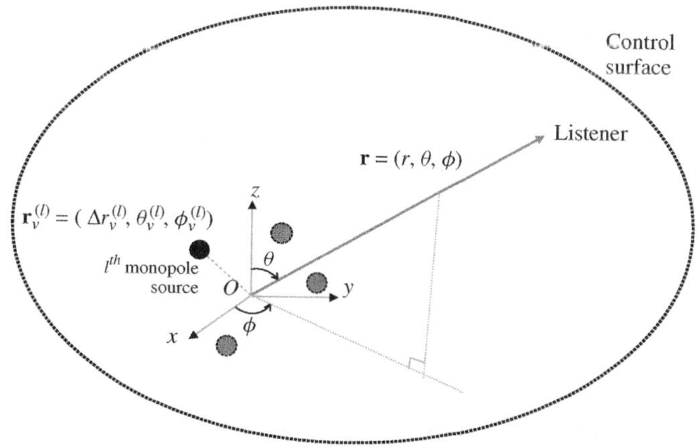

Figure 6.38 Spherical coordinates defined at the center of the virtual multipole ($\mathbf{r}_v$). ($\mathbf{r}_v^{(l)}$: Location of the lth virtual source, $\mathbf{r}$: location of the listener)

the multipole region should be sufficiently compact (small $k\Delta r_v^{(l)}$). The compact source assumption decreases the contribution of the high-order harmonics to the radiation pattern a_n because of the small contributions of the high-order Bessel functions in Equation (6.252). In this case, we can use a large argument approximation (Equation (B161)), from which the first and second kinds of spherical Hankel functions can be reduced to

$$h_n^{(1)}(kr) \approx (-i)^{n+1}\frac{e^{ikr}}{kr},\ h_n^{(2)}(kr) \approx (i)^{n+1}\frac{e^{-ikr}}{kr}. \tag{6.253}$$

for the condition

$$kr \gg n(n+1)/2. \tag{6.254}$$

The large argument approximation implies that the listener should be at the farfield with respect to the wavelength and the order of harmonics used for the radiation pattern a_n. With this approximation, we can rewrite Equation (6.250) as

$$\begin{aligned} p_d(\mathbf{r}) &\approx \frac{e^{ikr}}{r}\left(\sum_{n=0}^{\infty}(-i)^n a_n(\theta,\phi)\right) \\ &\approx b(\theta,\phi)\frac{e^{ikr}}{4\pi r}, \end{aligned} \tag{6.255}$$

where the radiation pattern $b(\theta,\phi)$ is defined as

$$b(\theta,\phi) \equiv 4\pi\sum_{n=0}^{\infty}(-i)^n a_n(\theta,\phi). \tag{6.256}$$

Equation (6.255) expresses the farfield radiation from the multipole as a product of the radiation pattern and monopole radiation. Using the same approximation and parity of the spherical harmonics ($(-1)^n Y_n^m(\theta,\phi) = Y_n^m(\pi-\theta,\phi+\pi)$: Equation (B139)), the farfield of the TRP field can be reduced to

$$\begin{aligned} p_{tr}(\mathbf{r})^* &\approx \frac{e^{-ikr}}{r}\left(\sum_{n=0}^{\infty}(-i)^n a_n(\pi-\theta,\phi+\pi)\right) \\ &\approx b(\pi-\theta,\phi+\pi)\frac{e^{-ikr}}{4\pi r}. \end{aligned} \tag{6.257}$$

The TRP field also consists of monopole radiation multiplied by a radiation pattern. However, its directional radiation pattern is an exact mirror image ($b(\pi-\theta,\phi+\pi)$) of that of the target field ($b(\theta,\phi)$). Consequently, the direct TRP field converges toward the center position of the multipoles with a reversed radiation pattern, as long as the listener is at the farfield of the virtual source.

Substituting Equations (6.255) and (6.257) into Equation (6.244), we can derive the integral equation for the farfield case as follows:

$$\begin{aligned} p_r(\mathbf{r}) &\approx g_f(\mathbf{r}|\mathbf{r}_v)b(\theta,\phi) - g_f(\mathbf{r}|\mathbf{r}_v)^* b(\pi-\theta,\phi+\pi) \\ &\approx \int_S \left[p_{tr}^*(\mathbf{r}_s)\frac{\partial g_f(\mathbf{r}|\mathbf{r}_s)}{\partial n_s} - \frac{\partial p_{tr}^*(\mathbf{r}_s)}{\partial n_s} g_f(\mathbf{r}|\mathbf{r}_s)\right] dS(\mathbf{r}_s), \end{aligned} \tag{6.258}$$

where $p_{tr}^*(\mathbf{r}_s) = b(\pi-\theta_s,\phi_s+\pi)g_f(\mathbf{r}_s|\mathbf{r}_v)^*$. Therefore, if we design an appropriate directional radiation pattern $b(\theta,\phi)$, the target field and its TRP can be separated in space, as in the cardioid example.

The nearfield behavior of the reproduced sound field also needs to be investigated. According to Equations (6.250) and (6.252), the reproduced field can be rewritten as

$$p_d(\mathbf{r}) - p_{tr}(\mathbf{r})^* = 2ik\sum_{n=0}^{\infty} j_n(kr)a_n(\theta, \phi). \tag{6.259}$$

The contribution of high-order harmonics to a_n usually increases the nearfield components. As mentioned, the compact source assumption (small $k\Delta r_v^{(l)}$) imposed on the multipole source suppresses the nearfield component. Nevertheless, if the nearfield components are amplified by the excitation q, the nearfield components of the target and TRP fields do not have opposite radiation patterns to each other, and it is difficult to separate the target field from the TRP field. Therefore, we cannot use an arbitrary radiation pattern for the target field. If relatively high-order harmonics are used extensively for the radiation pattern $b(\theta, \phi)$, the distance from the multipole to the listener may violate the condition of Equation (6.254), and the target and TRP fields can no longer be separated.

6.8.2.4 Examples and Analysis

The general form of Equation (6.244) enables us to reproduce an interior virtual source by using the TRP from a directional multipole source. The radiation characteristics of the target and TRP fields shown in Equation (6.258) allows us to predict the behavior of the convergent wave in space; it also provides a guideline for designing the radiation pattern for separating the converging and diverging wavefronts. To understand the requirement for the design of the directivity in more detail, we examine certain cases having two different radiation patterns.

Figure 6.39 shows an example of reproduction of an ideal monopole source (which will be denoted as p_i) in 2D. The target sound field consists of two monopole sources with a cardioid radiation pattern ($\alpha = 1$). For the control sources, 64 monopoles and dipoles were arranged on a circle of radius 4λ (λ: wavelength) in order to reduce the artifacts from discrete sampling of the control surface. Then, two virtual point sources were located at $\mathbf{r}_v = (r_v, \theta_v) = (2\lambda, \pi/3)$ from the center of the sphere with a separation distance $\Delta r_v = 0.01\lambda$. The first three panels (Figure 6.39a–c) depict the imaginary part of the ideal monopole field, the target cardioid field from the virtual multipole, and the reproduced sound field, respectively. The main difference between the reproduced field and the ideal monopole is the backward radiation (V_c in Figure 6.37b) in the interior region. Because the radiation of the TRP field is time reversed, the phase of the backward radiation is opposite to that of the ideal monopole source. On the other hand, the exterior radiation of the reproduced field corresponds to that of the target field.

To determine the size of the sweet spot, the relative error of the reproduced field with respect to the ideal monopole field is calculated for both interior and exterior regions. We can define the relative error in the ideal field as

$$\varepsilon^2(\mathbf{r}) = |(p_i(\mathbf{r}) - p_r(\mathbf{r}))/p_i(\mathbf{r})|^2. \tag{6.260}$$

The relative error is depicted in Figure 6.39d. The completely dark area represents the region where the relative error is less than −20 dB, which is equivalent to a 1% error. The error map shows that even low-order directivity such as a cardioid can generate a reasonably sized sweet spot. Further, the sweet spot coincides with the angular region where the amplitude of the cardioid pattern is high.

To examine the nearfield effect of the high-order harmonics, we considered another radiation pattern with a sharp discontinuity: an angular step function. The step function is expected to secure a wider sweet spot than the cardioid because it restricts the converging wavefront within one half

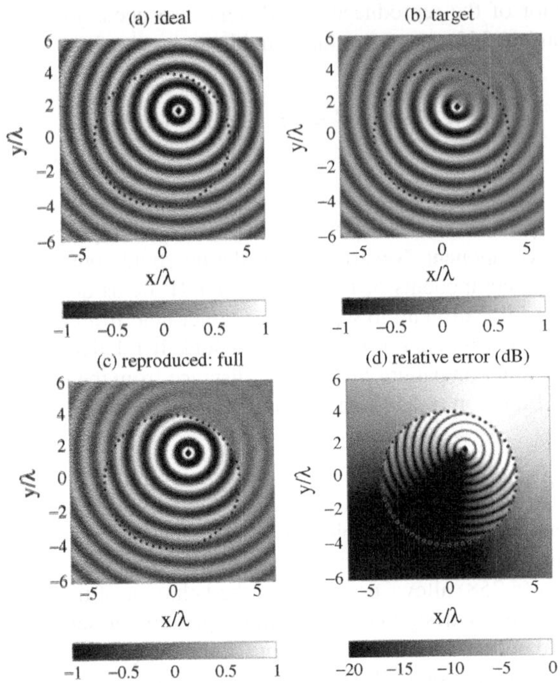

Figure 6.39 Sound field reproduction of an interior cardioid virtual source in 2D. Imaginary parts of the (a) ideal monopole field, (b) target cardioid field, (c) reproduced field, and (d) reproduction error on a decibel scale

of the space and reproduces the diverging wavefront in the other. Regarding Equations (6.255) and (6.257), the farfield response should have the form

$$p_d(\mathbf{r}) - p_{tr}(\mathbf{r})^* \approx \frac{e^{ikr}}{4\pi r} w(\theta) - \frac{e^{-ikr}}{4\pi r} w(\pi - \theta), \tag{6.261}$$

where $w(\theta)$ is a step function defined as

$$w(\theta) = \begin{cases} 0 & 0 \le \theta < \pi/2 \\ 1/2 & \theta = \pi/2 \\ 1 & \pi/2 \le \theta < \pi \end{cases}. \tag{6.262}$$

Note that we can arbitrarily define the origin of the coordinates such that the plane $\theta = \pi/2$ divides the two half-spaces for the convergent region V_c and divergent region V_d in Figure 6.37.

The farfield characteristics of the sound field generated by the angular step function seems ideal, but its nearfield behavior becomes problematic. Because the step function has a sharp discontinuity at the boundary between the two half-spaces, the contribution of the high-order harmonics is greater in this case than in the cardioid case. Figure 6.40 shows an example of reproduction with an angular step function. For this example, 20 virtual sources were arranged on a circle centered at $\mathbf{r}_v = (r_v, \theta_v) = (2\lambda, \pi/3)$ from the center of a control surface with radius $\Delta r_v = 0.01\lambda$. Circular harmonics up to the ninth order were controlled to mimic the angular step function. The other

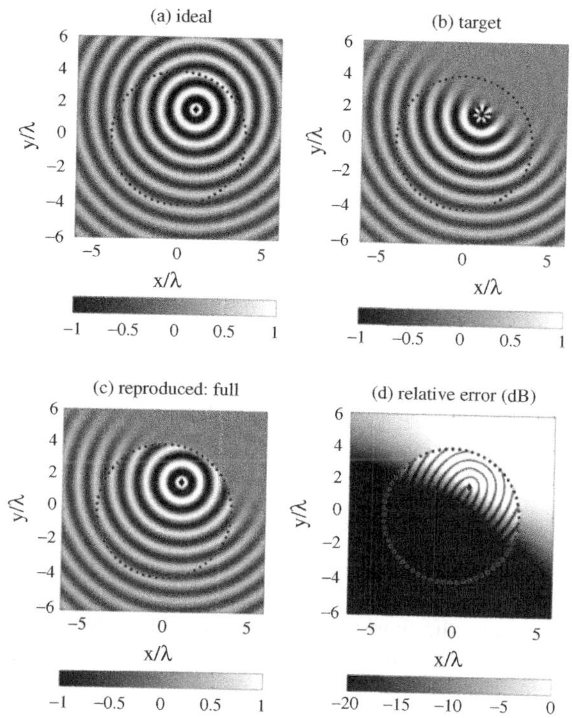

Figure 6.40 Two-dimensional reproduction using a multipole virtual source ($L = 20$). ($\mathbf{r}_v = (r_v, \theta_v) = (2\lambda, \pi/3)$, number of control sources: 64 monopoles and dipoles). Imaginary parts of the (a) ideal monopole field p_i, (b) target multipole field p_d, and (c) reproduced field p_r, and magnitude of the (d) relative error $\varepsilon^2(\mathbf{r}) = |(p_i(\mathbf{r}) - p_r(\mathbf{r}))/p_i(\mathbf{r})|^2$ on a decibel scale. Magnitudes are scaled such that $|p_i|(x = 0, y = 0) = 1$

parameters, such as the location of the virtual source, were the same as in the cardioid example. Compared to Figure 6.39, in which the sweet spot is close to the virtual source, the sweet spot in the step function case is retarded away from the location of the virtual source.

Consequently, there is a trade-off between the angular width of the sweet spot and the nearest distance of the sweet spot from the virtual source. If the nearest distance between the sweet spot and the virtual source is important, the high-order harmonics constituting the radiation pattern should be suppressed. This implies that the overall radiation pattern $b(\theta, \phi)$ should be synthesized only with low-order modes. The consequence of high-order suppression is the unclear separation between the converging and diverging wavefronts and the narrowed angular width of the sweet spot. The previous cardioid example employs only the zero- and first-order harmonics, and thus can be regarded as an extreme case of low-order reproduction.

Next, a 3D case was tested with a spherical loudspeaker array (Figure 6.41). For this example, a sphere with radius 2λ was sampled according to the rule of Lebedev quadrature. In all, 170 loudspeakers were employed so that the effect of spatial sampling could be avoided ($K \geq (kr_s + 1)^2$ [19], r_s: radius of sphere) over the entire interior volume V. The resultant field in the 3D case (Figure 6.41c) shows the target cardioid field reproduced within a finite angular region, with the external field identical to the target field.

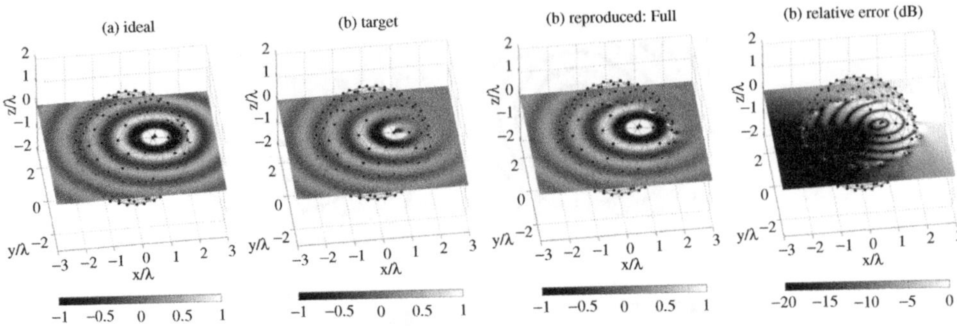

Figure 6.41 Three-dimensional reproduction using a cardioid virtual source ($\mathbf{r}_v = (r_v, \theta_v, \phi_v) = (\lambda, 0, \pi/6)$, number of control sources: 170 monopoles and dipoles). Imaginary parts of the (a) ideal monopole field p_i, (b) target cardioid field p_d, and (c) reproduced field p_r, and magnitude of the (d) relative error $\varepsilon^2(\mathbf{r}) = |(p_i(\mathbf{r}) - p_r(\mathbf{r}))/p_i(\mathbf{r})|^2$ on a decibel scale

6.8.2.5 Discussion of the Integral Equation

The procedure for reproducing the sound field using Equation (6.258) is summarized in Figure 6.42. First, we design the multipole excitation signal on the basis of the desired radiation pattern and the source signal, for example, the music signal that we want to reproduce. Then, the excitation signal is time reversed and fed to the multipole source so as to have reversed radiation compared to the virtual source location. The radiated sound is recorded at the control surface and time-reversed again to drive the monopole and dipole sources. This two-step time reversal gives the surface distribution of the TRP field.

$$p_{tr}(\mathbf{r}_s)^* = \left[\int_{V_0} g_f(\mathbf{r}_s|\mathbf{r}_0)\, q(\mathbf{r}_0)^* dV \right]^*. \tag{6.263}$$

Instead of two-step time reversal, a direct calculation of the excitation function is also possible from Equation (6.257). Once we design a radiation pattern $b(\theta, \phi)$, the surface excitation is simply determined by

$$p_{tr}(\mathbf{r}_s)^* \approx b(\pi - \theta_s, \phi_s + \pi) g_f(\mathbf{r}_s|\mathbf{r}_v)^*. \tag{6.264}$$

Equation (6.264) also shows that tapering the speaker weights or turning off the array behind the virtual source has an effect similar to assigning directivity to the virtual source. However, beyond such similarity, the multipole formula is advantageous in that the behavior of the converging wavefront is highly predictable. Because the converging wavefront has a radiation pattern exactly opposite to that of the diverging source, the converging and diverging wavefronts are well separated in space. In principle, these characteristics are not restricted by the shape of the control surface. The only requirement for obtaining Equation (6.258) is that the distance between the center of the virtual multipoles and the listener position should be large compared to the wavelength and the order of the harmonics employed to generate the directivity.

The advantage of using a virtual multipole source is the simplicity of the KH integral equation: what we need to do is to design a target field from a point source with directivity. The remaining required work is the calculation of the TRP of the target field at the control surface. Furthermore, the separation of the converging and diverging wavefronts can be conducted in a scalable manner. For instance, we can simply use cardioid or other first-order directivity when a precise separation is not required. If a clearer separation is needed, we can design a more accurate radiation pattern by considering a greater number of virtual sources.

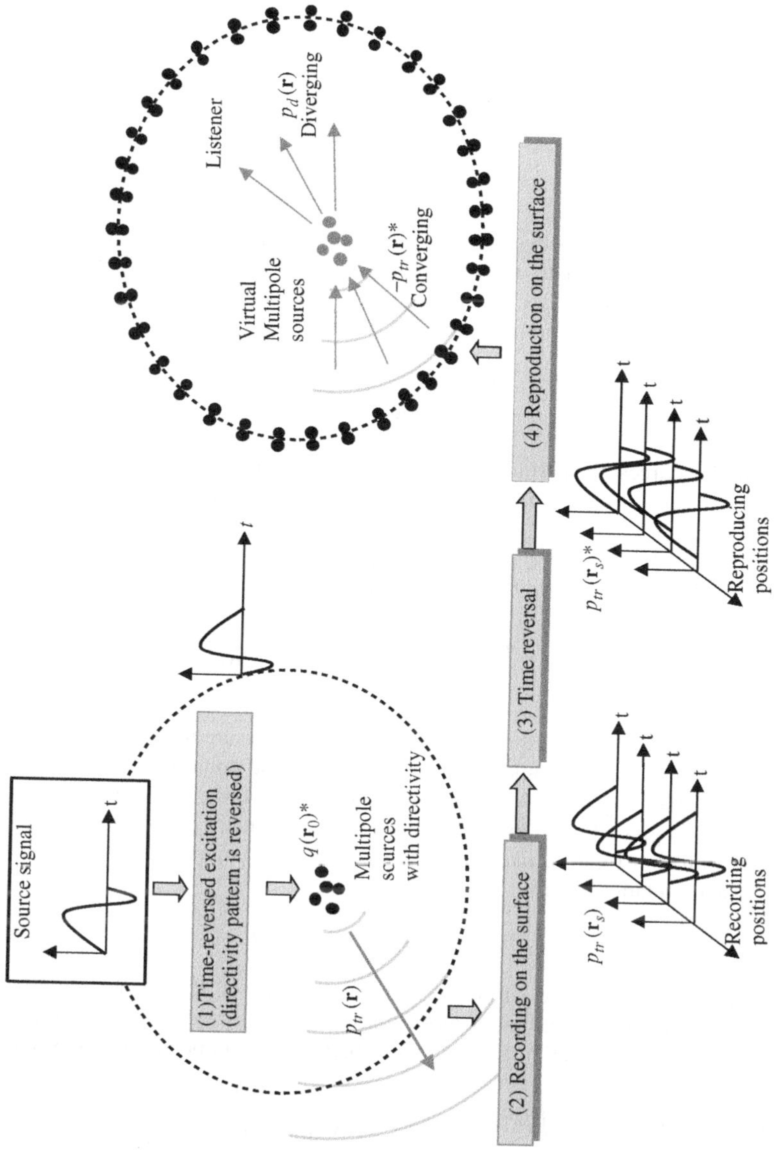

Figure 6.42 Procedure for reproducing directional multipole sources. (1) The source signal is time-reversed; (2) the radiation from the time-reversed excitation is calculated on the control surface; (3) the calculated signals are time-reversed; and (4) the signals excite the control sources to reproduce the virtual source inside

6.8.2.6 Single-layer Formula

Because we derived the KH-type integral for an interior virtual source, the single-layer formula can also be derived using the technique applied to an exterior virtual source. The equivalent scattering problem (Equation (6.199)) is applicable to the control surface S, or the stationary phase method can be used at high frequencies. In this section, we examine the stationary phase method, which can be simply derived without solving an inverse problem.

Denoting the distance from the control surface to the listener as $R = |\mathbf{r} - \mathbf{r}_s|$, we can rewrite the multipole Equation (6.244) as

$$p_d(\mathbf{r}) - p_{tr}(\mathbf{r})^* = \int_S \left\{ \left[p_{tr}(\mathbf{r}_s)^* \frac{\partial R}{\partial n_s}(ik - 1/R) - \frac{\partial p_{tr}(\mathbf{r}_s)^*}{\partial n_s} \right] \frac{e^{ikR}}{4\pi R} \right\} dS(\mathbf{r}_s), \quad \mathbf{r} \in V \qquad (6.265)$$

with $e^{ikR}/(4\pi R) = g_f(\mathbf{r}|\mathbf{r}_s)$.

Let us then denote the distance between the multipole source and the control surface as $R_0 = |\mathbf{r}_0 - \mathbf{r}_s|$. When both the listener and the multipole source are located in the farfield ($kR_s, kR_0 \gg 1$) from the control surface with respect to the wavelength, a combination of Equations (6.241) and (6.265) can be approximated as

$$p_d(\mathbf{r}) - p_{tr}(\mathbf{r})^* \approx \int_{V_0} \left\{ \int_S \left[ik \left(\frac{\partial R}{\partial n_s} + \frac{\partial R_0}{\partial n_s} \right) \frac{e^{-ikR_0}}{4\pi R_0} \right] \frac{e^{ikR}}{4\pi R} dS(\mathbf{r}_s) q(\mathbf{r}_0) \right\} dV(\mathbf{r}_0), \quad \mathbf{r} \in V. \qquad (6.266)$$

For compact multipole sources, the partial derivative $\partial R_0/\partial n_s$ is nearly equal to all the multipole source positions $\mathbf{r}_0$ ($\partial R_0/\partial n_s \approx \partial R_v/\partial n_s$). From Equation (6.264), the integral can be expressed as

$$\begin{aligned} p_d(\mathbf{r}) - p_{tr}(\mathbf{r})^* &\approx ik \int_S \left[\left(\frac{\partial R}{\partial n_s} + \frac{\partial R_v}{\partial n_s} \right) p_{tr}^*(\mathbf{r}_s) \frac{e^{ikR}}{4\pi R} \right] dS(\mathbf{r}_s) \\ &\approx \frac{ik}{16\pi^2} \int_S \left[b(\pi - \theta_s, \phi_s + \pi) \left(\frac{\partial R}{\partial n_s} + \frac{\partial R_v}{\partial n_s} \right) \frac{e^{ik(R - R_v)}}{R_v R} \right] dS(\mathbf{r}_s), \quad \mathbf{r} \in V. \end{aligned} \qquad (6.267)$$

The stationary phase approximation requires a slowly varying pressure magnitude over the integral surface S, and the phase change should be rapid except near the point of the stationary phase. If we assume that the radiation pattern b is a smooth function on the surface, the phase-related term in Equation (6.267) is given by the difference between two transfer paths ($\exp[ik(R - R_v)]$). At a high frequency, Equation (6.267), like Equation (6.209), has stationary phase points, and at these points the following relation holds:

$$\partial(R - R_v)/\partial n_s = \cos\varphi_s - \cos\varphi_v = 0. \qquad (6.268)$$

For a convex control surface S, the surface normal derivatives of R and R_v have the same sign, which indicates that the contributions of monopoles and dipoles are always the same at the stationary phase point. Using this relation and Equation (6.267), we can obtain the 3D single-layer formula.

$$p_d(\mathbf{r}) - p_{tr}(\mathbf{r})^* \approx \frac{ik}{8\pi^2} \int_S b(\pi - \theta_s, \phi_s + \pi) \left(\frac{\partial R_v}{\partial n_s} \frac{e^{ik(R - R_v)}}{R_v R} \right) dS(\mathbf{r}_s). \qquad (6.269)$$

Moreover, if the listener position is limited to a lateral plane (A) at the same height as the virtual source, the 3D integral can be reduced to a 2.5D integral. In this case, we can approximate the integral in Equation (6.267) as

$$p_d(\mathbf{r}) - p_{tr}(\mathbf{r})^* \approx \frac{1}{4\pi}\int_C \left(b\left(\pi - \theta_s, \phi_s + \pi\right) \frac{\partial R_v}{\partial n_s} \sqrt{\frac{ik}{2\pi} \frac{R_v R_{ref}}{R_{ref} - R_v}} \frac{e^{-ikR_v}}{R_v} \right) \frac{e^{ikR}}{R} dl\left(\mathbf{r}_s\right), \quad \mathbf{r} \in A, \tag{6.270}$$

where C denotes the contour consisting of the stationary phase points. However, $b(\pi - \theta_s, \phi_s + \pi)$ should consist of low-order harmonics in order to reduce the nearfield error and satisfy the stationary phase approximation. If this condition can be met, even the nearfield of the virtual source can be reproduced.

6.8.2.7 Performance Evaluation

In the following example, the single-layer formula is tested with the same 2D configuration used for the monopole and dipole control sources. The only difference from the previous examples is that the dipole control sources are neglected. Figure 6.43 shows the sound field reproduced by a single-layer potential. The error field depicted in Figure 6.43b shows that the reproduced result is similar to the double-layer case (or reproduction with monopole and dipoles sources) in Figure 6.40. In this simulation, the distance from the virtual source to the nearest control surface is just two wavelengths.

In contrast to the interior radiation, the exterior radiation behind the virtual source differs considerably from the ideal or target field. This difference is the main drawback of omitting the dipole control sources. Because the monopole control sources are omnidirectional, it is not possible to control the exterior and interior radiation simultaneously. Nevertheless, the magnitude of the external radiation does not diverge and is still comparable to the ideal case. Therefore, the single-layer formula for monopole control sources is useful when the external radiation to the rear of the virtual source does not disturb the listener greatly.

Next, the frequency dependency of the reproduction error is tested with 128 control sources that are evenly distributed on a circle of radius 1.5 m. The virtual source is positioned at $\mathbf{r}_v = (r_v, \theta_v) = (0.75\,\text{m}, \pi/3)$ from the center of the circle. To represent the total reproduction

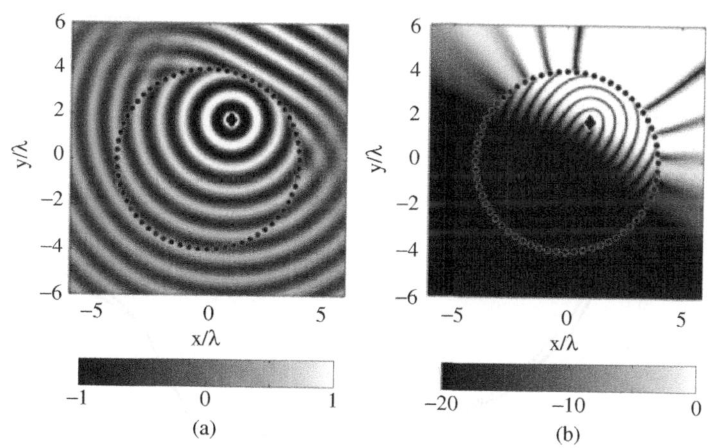

Figure 6.43 Two-dimensional reproduction using the single-layer formula: multipole virtual source ($L = 20$) ($\mathbf{r}_v = (r_v, \theta_v) = (2\lambda, \pi/3)$, number of control sources: 64 monopoles). (a) Imaginary part of the sound field reproduced by the single-layer potential and (b) relative error of the reproduced sound field

error at each frequency, the spatially averaged error is defined as

$$\overline{\varepsilon}^2 = \frac{\frac{1}{V'}\int_{V'} |p_r(\mathbf{r}) - p_i(\mathbf{r})|^2 dV(\mathbf{r})}{\frac{1}{V'}\int_{V'} |p_i(\mathbf{r})|^2 dV(\mathbf{r})}, \tag{6.271}$$

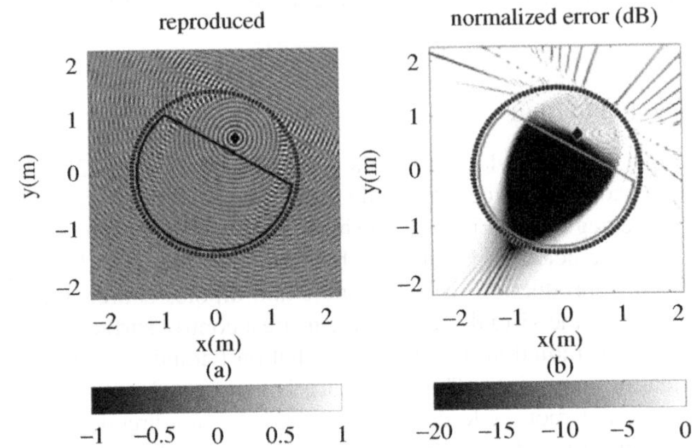

Figure 6.44 Reproduced sound field with spatial aliasing ($f = 4\,\text{kHz}$). (a) Imaginary part of the pressure field and (b) normalized error distribution (in decibel). Truncated circular area represents the volume for which the spatially averaged error is calculated

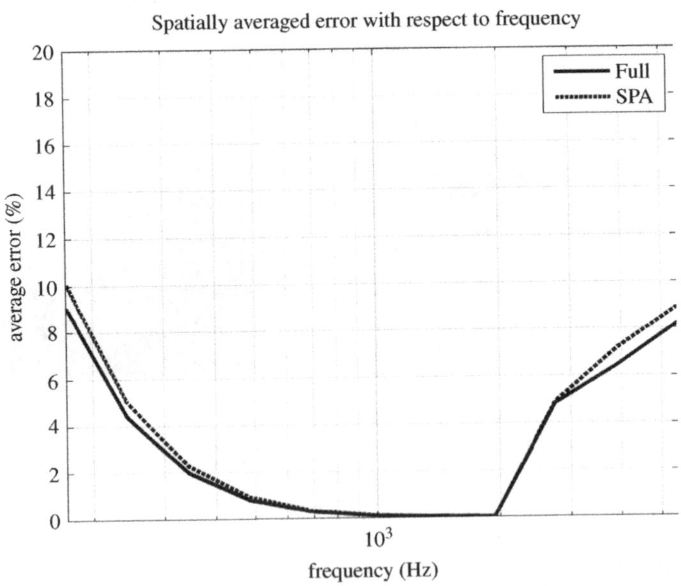

Figure 6.45 Spatially averaged error of single-layer formula. Full: integral with monopole and dipole sources, SPA: single-layer potential. (Two-dimensional simulation with 128 control sources equally distributed on a circle of radius 1.5 m. Location of the virtual source: $\mathbf{r}_v = (r_v, \theta_v) = (0.75\,\text{m}, \pi/3)$)

where the averaged volume (area) V' is set as the partial interior volume separated by 25 cm from the virtual source (truncated circular area in Figure 6.44a). Figure 6.45 shows the reproduction error across various frequencies. As expected, the error increases in the low-frequency region because the farfield approximation of Equation (6.254) does not hold. Nevertheless, all the simple source formulas have modest reproduction errors in the mid-frequency range up to 3 kHz. The high-frequency error above 3 kHz is due mainly to the discrete sampling of the control surface. Aliasing at high frequencies has been identified as the source of a pre-echo artifact [20, 21], which causes audible coloration or even shifts the location of the perceived virtual source. An additional spatial filtering of the excitation function [16] or the use of directional loudspeakers [17, 22] can be helpful for reducing this aliasing artifact in the high-frequency region. However, even at high frequencies, the artifact is minimal at locations close to the virtual source (Figure 6.44b). Therefore, even if the spatial aliasing artifact decreases the size of the sweet spot, the sound field within the sweet spot can still be free of the artifact for sound with a limited bandwidth.

References

1. Hadamard, J. (1902) Sur les problèmes aux dérivées partielles et leur signification physique. *Princeton University Bulletin*, **13**, 49–52.
2. Hadamard, J. (1923) *Lectures on the Cauchy Problem in Linear Partial Differential Equations*. New Haven Yale University Press, New Haven, CO.
3. Hansen, P.C. (1990) The discrete Picard condition for discrete ill-posed problems, *BIT Numerical Mathematics*, **30**(4), 658–672.
4. Varah, J.M. (1979) A practical examination of some numerical methods for linear discrete ill-posed problems, *SIAM Review*, **21**(1), 100–111.
5. Morozov, V.A. (1962) The choice of parameter when solving functional equations by regularization, *Doklady Akademii Nauk SSSR*, **175**, 1225–1228.
6. Hansen, P.C. and O'Leary, D.P. (1993) The use of the L-curve in the regularization of discrete ill-posed problems, *SIAM Journal on Scientific Computing*, **14**(6), 1487–1503.
7. Tikhonov, A.N. (1963) Solution of incorrectly formulated problems and the regularization method, *Soviet Mathematics Doklady*, **4**, 1036–1038.
8. Tikhonov, A.N. (1963) Regularization of incorrectly posed problems, *Soviet Mathematics Doklady*, **4**, 1624–1627.
9. Golub, G.H., Heath, M. and Wahba, G. (1979) Generalized cross-validation as a method for choosing a good ridge parameter, *Technometrics*, **21**(2), 215–223.
10. Press, W.H., Flannery, B.P., Teukolsky, S.A., Vetterling, W.T. (1992) *Numerical Recipes in Fortran 77: The Art of Scientific Computing*, Chapter 13, Cambridge University Press, New York.
11. Abramowitz, M. and Stegun, I.A. (1972) *Integration, Handbook of Mathematical Functions with Formulas, Graphs, and Mathematical Tables*, Dover Publications New York.
12. Lebedev, V.I. (1975) Values of the nodes and weights of ninth to seventeenth order Gauss-Markov quadrature formulae invariant under the octahedron group with inversion. *USSR Computational Mathematics and Mathematical Physics*, **15**(1), 44–51.
13. Driscoll, J.R. and Healy, D.M. (1994) Computing Fourier transforms and convolutions on the 2-sphere. *Advances in Applied Mathematics*, **15**(2), 202–250.
14. Rafaely, B., Weiss, B. and Bachmat, E. (2007) Spatial aliasing in spherical microphone arrays, *IEEE Transactions on Signal Processing*, **55**(3), 1003–1010.
15. Fazi, F.M. (2010) Sound field reproduction. PhD thesis. Institute of Sound and Vibration Research, University of Southampton, Southampton, UK.
16. Start, E. (1997) Direct sound enhancement by wave field synthesis. PhD thesis. Delft University of Technology, Delft, the Netherlands.
17. Verheijen, E.N.G. (1998) Sound reproduction by wave field synthesis. PhD thesis. Delft University of Technology, Delft, the Netherlands.
18. Choi, J.W. and Kim, Y.H. (2012) Integral approach for the reproduction of a virtual sound source surrounded by loudspeaker array, *IEEE Transactions on Audio, Speech and Language Processing*, **20**(7), 1976–1989.

19. Ward, D.B. and Abhayapala, T.D. (2001) Reproduction of a plane-wave sound field using an array of loudspeakers. *IEEE Transactions on Speech and Audio Processing*, **9**(6), 697–707.
20. Spors, S., Wierstorf, H., Geier, M., and Ahrens, J. (2009) Physical and perceptual properties of focused sources in wave field synthesis. Proceedings of the 127th Convention of Audio Engineering Society, New York, USA, preprint 7914.
21. Geier, M., Wierstorf, H., Ahrens, J. *et al.* (2010) Perceptual evaluation of focused sources in wave field synthesis. Proceedings of the 128th Convention of Audio Engineering Society, London, UK, preprint 8069.
22. Ahrens, J. and Spors, S. (2011) On the anti-aliasing loudspeaker for sound field synthesis employing linear and circular distributions of secondary sources. Proceedings of the 129th Convention of Audio Engineering Society, San Francisco, California, 2010, preprint 8246.

Appendix A

Useful Formulas

A.1 Fourier Transform

A.1.1 Fourier Transform Table

The definition of Fourier transform depends on the sign convention of acoustic phase. For $e^{i(k_x x-\omega t)}$ notation, the spatial Fourier transform $\mathcal{F}_s[\,]$ is defined as $\int_{-\infty}^{\infty} f(x)e^{-ik_x x}dx$, but it is $\int_{-\infty}^{\infty} f(x)e^{ik_x x}dx$ for $e^{i(\omega t-k_x x)}$. The followings are defined for the $e^{i(k_x x-\omega t)}$ convention.

A.1.1.1 Temporal Fourier Transform

$$F(\omega) = \mathcal{F}_t[f(t)] = \int_{-\infty}^{\infty} f(t)e^{i\omega t}dt \tag{A1}$$

$$f(t) = \mathcal{F}_t^{-1}[F(\omega)] = \frac{1}{2\pi}\int_{-\infty}^{\infty} F(\omega)e^{-i\omega t}d\omega \tag{A2}$$

A.1.1.2 Spatial Fourier Transform

$$F(k_x) = \mathcal{F}_s[f(x)] = \int_{-\infty}^{\infty} f(x)e^{-ik_x x}dx \tag{A3}$$

$$f(x) = \mathcal{F}_s^{-1}[F(k_x)] = \frac{1}{2\pi}\int_{-\infty}^{\infty} F(k_x)e^{ik_x x}dk_x. \tag{A4}$$

Useful formulas for the spatial Fourier transform are shown in Table A.1.

A.1.1.3 Time-Reversal Operator

The time-reversal operator is a convenient tool for the sound field prediction and reproduction problems. Consider a signal $f(t)$ which is a function of time, then its frequency spectrum is given by

$$F(\omega) = \int_{-\infty}^{\infty} f(t)e^{i\omega t}dt. \tag{A5}$$

Sound Visualization and Manipulation, First Edition. Yang-Hann Kim and Jung-Woo Choi.

Table A.1 Useful spatial Fourier transforms

$f(x)$	Fourier transform $F(k_x) = \int_{-\infty}^{\infty} f(x)e^{-ik_x x}dx$	$F(k_x)$	Inverse Fourier transform $f(x) = \frac{1}{2\pi}\int_{-\infty}^{\infty} F(k_x)e^{-ik_x x}dk_x$
$e^{ik_0 x}$	$2\pi\delta(k_x - k_0)$	$e^{-ik_x x_0}$	$\delta(x - x_0)$
$f(ax)$	$\frac{1}{a}F\left(\frac{k_x}{a}\right)$	$F(k_x a)$	$\frac{1}{\|a\|}f\left(\frac{x}{a}\right)$
$\mathrm{rect}(x/a)^*$	$\|a\|\mathrm{sinc}\left(\frac{k_x a}{2}\right)^\dagger$	$\mathrm{rect}(k_x/a)$	$\frac{\|a\|}{2\pi}\mathrm{sinc}\left(\frac{ax}{2}\right)$
e^{iax^2}	$\sqrt{\frac{\pi i}{a}}e^{i\frac{k_x^2}{4a}}$	$e^{-i\frac{k_x^2}{4a}}$	$\sqrt{\frac{a}{\pi i}}e^{iax^2}$
$\sum_{n=-\infty}^{\infty}\delta(x - n\Delta x)$	$\frac{2\pi}{\Delta x}\sum_{m=-\infty}^{\infty}\delta\left(k_x - \frac{2\pi}{\Delta x}m\right)$	$\sum_{n=-\infty}^{\infty}\delta(k_x - \Delta k_x n)$	$\frac{1}{\Delta k}\sum_{n=-\infty}^{\infty}\delta\left(x - \frac{2\pi}{\Delta k}n\right)$
$f(x)*_x g(x)^\ddagger$	$F(k_x)G(k_x)$	$F(k_x)*_{k_x}G(k_x)^\S$	$2\pi f(x)g(x)$

*Rectangular function:

$$\mathrm{rect}\left(\frac{x}{L}\right) = \begin{cases} 1 & \text{for } |x| < L/2 \\ 1/2 & \text{for } |x| = L/2 \\ 0 & \text{for } |x| > L/2 \end{cases}. \quad \text{(A6)}$$

†$\mathrm{sinc}(x) = \sin(x)/x$

‡Spatial convolution $f(x)*_x g(x) = \int_{-\infty}^{\infty} f(x - x')g(x')dx'$

§Spectral convolution $F(k_x)*_{k_x}G(k_x) = \int_{-\infty}^{\infty} F(k_x - k_x')G(k_x')dk_x'$

For temporal Fourier transform, the result of transform is complex conjugated $f(x) \Rightarrow f(t)$, $F(k_x)^* \Rightarrow F(\omega)$.

The time-reverse of the function $f(t)$ is written as $f(-t)$ in time domain. The frequency spectrum of the time reversed signal F_{trm} can be related to $F(\omega)$ as follows:

$$\begin{aligned} F_{trm}(\omega) &= \int_{-\infty}^{\infty} f(-t)e^{i\omega t}dt \\ &= \int_{-\infty}^{\infty} f(t')e^{-i\omega t'}dt' \quad (t' = -t). \\ &= F(\omega)^* \end{aligned} \quad \text{(A7)}$$

That is, the frequency spectrums of the original and the time-reversed signal are complex conjugate pair to each other.

A.1.1.4 Convolution Theorem

For two functions $f(x)$ and $g(x)$, the convolution is defined as

$$\begin{aligned} h(x) &= f(x)*_x g(x) \\ &= \int_{-\infty}^{\infty} f(x')g(x - x')dx' = \int_{-\infty}^{\infty} f(x - x')g(x')dx'. \end{aligned} \quad \text{(A8)}$$

Fourier transform of the convolved two sequences is equivalent to the multiplication of their wavenumber spectrum. That is,

$$\begin{aligned}\mathcal{F}_s[f(x)*_x g(x)] &= \int_{-\infty}^{\infty}\int_{-\infty}^{\infty} f(x-x')g(x)dx' e^{-ik_x x}dx \\ &= \int_{-\infty}^{\infty}\left[\int_{-\infty}^{\infty} f\left(x-x'\right)e^{-ik_x(x-x')}dx\right] g(x')e^{-ik_x x'}dx' \\ &= F(k_x)G(k_x).\end{aligned} \tag{A9}$$

In an analogous manner, the multiplication in space domain can be written in the wavenumber domain as

$$\begin{aligned}\mathcal{F}_s[f(x)g(x)] &= \int_{-\infty}^{\infty}\left(\frac{1}{2\pi}\int_{-\infty}^{\infty} F\left(k_x'\right)e^{ik_x'x}dk_x'\right)\left(\frac{1}{2\pi}\int_{-\infty}^{\infty} G\left(k_x''\right)e^{ik_x''x}dk_x''\right)e^{-ik_x x}dx \\ &= \frac{1}{4\pi^2}\int_{-\infty}^{\infty}\int_{-\infty}^{\infty}\left[F\left(k_x'\right)G(k_x'')\left(\int e^{i(k_x'+k_x''-k_x)x}dx\right)\right]dk_x'dk_x'' \\ &= \frac{1}{2\pi}F(k_x)*_{k_x}G(k_x).\end{aligned} \tag{A10}$$

For two-dimensional Fourier transform, we have

$$\mathcal{F}_s[f(x,y)g(x,y)] = \frac{1}{4\pi^2}F(k_x,k_y)*_{k_x}*_{k_y}G(k_x,k_y). \tag{A11}$$

A.1.1.5 Parseval's Theorem

The total energy of a signal in space domain, given by

$$e_x = \int_{-\infty}^{\infty} f(x)f(x)^*dx, \tag{A12}$$

is equivalent to the energy in wavenumber domain:

$$e_{k_x} = \frac{1}{2\pi}\int_{-\infty}^{\infty} F(k_x)F(k_x)^*dk_x. \tag{A13}$$

(proof) Substituting $g(x) = f(x)^*$ into Equation (A10) gives

$$\begin{aligned}\mathcal{F}_s\left[f(x)f(x)^*\right] &= \frac{1}{2\pi}F(k_x)*_{k_x}F(-k_x). \\ \int_{-\infty}^{\infty} f(x)f(x)^*e^{-ik_x x}dx &= \frac{1}{2\pi}\int_{-\infty}^{\infty} F(k_x')F(k_x+k_x')dk_x.\end{aligned} \tag{A14}$$

The total energy of Equation (A12) is equivalent to the Fourier transform of Equation (A14) at $k_x = 0$. Therefore, we have the relation

$$\int_{-\infty}^{\infty}|f(x)|^2dx = \frac{1}{2\pi}\int_{-\infty}^{\infty}|F(k_x)|^2dk_x. \tag{A15}$$

A.2 Dirac Delta Function

The Dirac delta function refers to a function that is zero everywhere except at zero. It is a generalized function, which is indirectly defined by what it does to the other function. Consider a sequence $\delta_a(x)$ satisfying the following relation:

$$\lim_{a\to\infty} \int \delta_a(x) f(x) dx = f(0).$$

This type of sequence, called delta sequence, becomes the Dirac delta function $\delta(x)$ in the limit as $a \to \infty$.

$$\lim_{a\to\infty} \delta_a(x) \equiv \delta(x).$$

Representative examples of the delta sequence are

$$\frac{a}{\pi}\text{sinc}(ax), \quad \frac{1}{2\pi}\frac{\sin[(a+1/2)x]}{\sin(x/2)}, \quad \frac{a}{\sqrt{\pi}}e^{-a^2x^2}.$$

The delta sequence has a shape of spike that goes to infinity as $a \to \infty$.

$$\delta(0) = \infty.$$

The total area under the spike, however, is preserved, and hence the Dirac delta function should satisfy the identity

$$\int_{-\infty}^{\infty} \delta(x) dx = 1.$$

Some useful definitions and properties of the Dirac delta function are:

- $\lim_{a\to 0}\left[\frac{1}{a\sqrt{\pi}}e^{-x^2/a^2}\right] = \delta(x)$: Gaussian function
- $\frac{1}{2\pi}\int_{-\infty}^{\infty} e^{i\zeta x} d\zeta = \delta(x)$: Fourier transform
- $\int_{-\infty}^{\infty} \delta(ax) dx = \frac{1}{a}$: Scaling property
- $\delta(\delta(x)) = \sum_n \frac{\delta(x - x_n)}{|g'(x_n)|}$: Delta function of a function
- $\delta(-x) = \delta(x)$: Even function
- $\delta(a\mathbf{r}) = \frac{\delta(\mathbf{r})}{a^n}$: For a N-dimensional vector $\mathbf{r}$ (such as $\mathbf{r} = (x, y, z)$)
- $\sum_{n=-\infty}^{\infty} e^{inx} = \sum_{m=-\infty}^{\infty} \delta(x + 2\pi m)$: Periodic impulse

A.3 Derivative of Matrices

A.3.1 Derivative of Real-Valued Matrix

Vector derivative of matrices can be derived from the derivative of the summation formula. For example, consider two real-valued column vectors $\mathbf{f}$ and $\mathbf{q}$ ($\mathbf{f}, \mathbf{q} \in \mathbb{R}^N$). Then consider the derivative of a vector product $\mathbf{f}^T\mathbf{q}$ with respect to the single element of $\mathbf{q}$ ($q^{(j)}$):

$$\frac{\partial \sum_{n=1}^{N} f^{(n)} q^{(n)}}{\partial q^{(j)}} = f^{(j)}. \tag{A16}$$

Since the vector $\mathbf{q}$ consists of N independent variables, its derivative is nonzero only when the index n equals to j. Calculating the derivative with respect to all elements of $\mathbf{q}$, we can derive the following matrix derivative:

$$\frac{\partial \mathbf{f}^T\mathbf{q}}{\partial \mathbf{q}} = \frac{\partial \mathbf{q}^T\mathbf{f}}{\partial \mathbf{q}} = \mathbf{f}^T. \tag{A17}$$

The derivative Equation (A17) can be defined as either $\mathbf{f}^T$ or $\mathbf{f}$; however, in this book we will follow the Jacobian formulation($\mathbf{f}^T$). From Equation (A16), it is easy to see that a matrix $\mathbf{A} \in \mathbb{R}^{M\times N}$ follows

$$\frac{\partial \mathbf{A}\mathbf{q}}{\partial \mathbf{q}} = \mathbf{A} \text{ or } \frac{\partial \mathbf{q}^T\mathbf{A}}{\partial \mathbf{q}} = \mathbf{A}^T. \tag{A18}$$

As a more complex case, the derivative of a quadratic form $\mathbf{q}^T\mathbf{q}$ can be derived as

$$\frac{\partial \sum_{n=1}^{N} (q^{(n)})^2}{\partial q^{(j)}} = 2q^{(j)} \quad \Rightarrow \quad \frac{\partial \mathbf{q}^T\mathbf{q}}{\partial \mathbf{q}} = 2\mathbf{q}^T. \tag{A19}$$

A quadratic form $\mathbf{q}^T\mathbf{B}\mathbf{q}$, involving a matrix $\mathbf{B} \in \mathbb{R}^{N\times N}$, satisfy the following derivative relation:

$$\frac{\partial \sum_{m=1}^{N}\sum_{n=1}^{N}\left(q^{(m)}B^{(m,n)}q^{(n)}\right)}{\partial q^{(j)}} = \sum_{m=1}^{N}\left(q^{(m)}B^{(m,j)}\right) + \sum_{n=1}^{N}\left(B^{(j,n)}q^{(n)}\right), \tag{A20}$$

which can be generalized to

$$\frac{\partial \mathbf{q}^T\mathbf{B}\mathbf{q}}{\partial \mathbf{q}} = \mathbf{q}^T\mathbf{B} + (\mathbf{B}\mathbf{q})^T = \mathbf{q}^T(\mathbf{B}+\mathbf{B}^T) \tag{A21}$$

or

$$\frac{\partial \mathbf{q}^T\mathbf{B}\mathbf{q}}{\partial \mathbf{q}^T} = \mathbf{B}\mathbf{q} + (\mathbf{q}^T\mathbf{B})^T = (\mathbf{B}+\mathbf{B}^T)\mathbf{q}. \tag{A22}$$

For a symmetric matrix $\mathbf{B} = \mathbf{B}^T$, the derivative becomes $2\mathbf{B}\mathbf{q}$.

A.3.2 Derivative of Complex-Valued Function

For a complex scalar variable z and its conjugate z^*, the following relations hold:

$$z = x + iy, \quad z^* = x - iy \tag{A23}$$

$$x = \frac{z+z^*}{2}, \quad y = \frac{z-z^*}{2i}, \tag{A24}$$

where x and y are the real and imaginary parts of z, respectively. The complex variable consists of two independent variables (x, y), and hence, its complex conjugate can be considered to be independent of z. Therefore, the generalized complex differential of a scalar function $f(z, z^*)$: $\mathbb{C}\times\mathbb{C}\to\mathbb{C}$ can be written as

$$df = \frac{\partial f}{\partial z}dz + \frac{\partial f}{\partial z^*}dz^* \tag{A25}$$

or

$$df = \frac{\partial f}{\partial x}dx + \frac{\partial f}{\partial y}dy. \tag{A26}$$

From the relations of Equation (A24), we have

$$dx = \frac{dz + idz^*}{2}, \quad dy = \frac{dz - idz^*}{2i}. \tag{A27}$$

Inserting Equation (A27) to Equation (A26) gives

$$\begin{aligned} df &= \frac{\partial f}{\partial x}\left(\frac{dz + idz^*}{2}\right) + \frac{\partial f}{\partial y}\left(\frac{dz - idz^*}{2i}\right) \\ &= \frac{1}{2}\left(\frac{\partial f}{\partial x} - i\frac{\partial f}{\partial y}\right)dz + \frac{1}{2}\left(\frac{\partial f}{\partial x} + i\frac{\partial f}{\partial y}\right)dz^*. \end{aligned} \tag{A28}$$

Comparing this result with Equation (A25), it can be seen that the partial derivatives with respect to z and z^* are given by

$$\frac{\partial f}{\partial z} = \frac{1}{2}\left(\frac{\partial f}{\partial x} - i\frac{\partial f}{\partial y}\right), \quad \frac{\partial f}{\partial z^*} = \frac{1}{2}\left(\frac{\partial f}{\partial x} + i\frac{\partial f}{\partial y}\right). \tag{A29}$$

A.3.3 Derivative of Complex Matrix

The derivative for a single complex variable z can be extended for the matrices. We consider again the differential $f(\mathbf{z})$ with respect to a arbitrary complex vector $\mathbf{z} \in \mathbb{C}^{N\times 1}$.

$$\frac{\partial f(\mathbf{z})}{\partial \mathbf{z}} = \frac{1}{2}\left(\frac{\partial f(\mathbf{z})}{\partial \mathbf{x}} - i\frac{\partial f(\mathbf{z})}{\partial \mathbf{y}}\right), \quad \frac{\partial f(\mathbf{z})}{\partial \mathbf{z}^*} = \frac{1}{2}\left(\frac{\partial f(\mathbf{z})}{\partial \mathbf{x}} + i\frac{\partial f(\mathbf{z})}{\partial \mathbf{y}}\right). \tag{A30}$$

From this relation we can rewrite Equation (A17) for the complex vector $\mathbf{z}$:

$$\frac{\partial \mathbf{f}^H\mathbf{z}}{\partial \mathbf{z}} = \frac{1}{2}\left(\frac{\partial \mathbf{f}^H(\mathbf{x} + i\mathbf{y})}{\partial \mathbf{x}} - i\frac{\partial \mathbf{f}^H(\mathbf{x} + i\mathbf{y})}{\partial \mathbf{y}}\right) = \mathbf{f}^H \tag{A31}$$

$$\frac{\partial \mathbf{z}^H\mathbf{f}}{\partial \mathbf{z}} = \frac{1}{2}\left(\frac{\partial (\mathbf{x} - i\mathbf{y})^T\mathbf{f}}{\partial \mathbf{x}} - i\frac{\partial (\mathbf{x} - i\mathbf{y})^T\mathbf{f}}{\partial \mathbf{y}}\right) = 0, \tag{A32}$$

where superscript H denotes the Hermitian operator (complex conjugate transpose).

Using Equation (A30), the derivative of quadratic form can be derived as

$$\begin{aligned} \frac{\partial(\mathbf{z}^H\mathbf{B}\mathbf{z})}{\partial \mathbf{z}} &= \frac{1}{2}\left(\frac{\partial(\mathbf{x} + i\mathbf{y})^H\mathbf{B}(\mathbf{x} + i\mathbf{y})}{\partial \mathbf{x}} - i\frac{\partial(\mathbf{x} + i\mathbf{y})^H\mathbf{B}(\mathbf{x} + i\mathbf{y})}{\partial \mathbf{y}}\right). \\ &= \frac{1}{2}\left[\mathbf{z}^H\left(\mathbf{B} + \mathbf{B}^T\right) + \mathbf{z}^H\mathbf{B} - \mathbf{z}^H\mathbf{B}^T\right] = \mathbf{z}^H\mathbf{B} \end{aligned} \tag{A33}$$

A.4 Inverse Problem

In this section, we construct a linear algebraic problem defined by

$$\mathbf{A}\mathbf{x} = \mathbf{b} \tag{A34}$$

in which the matrix $\mathbf{A} \in \mathbb{C}^{M\times N}$ indicates the input ($\mathbf{x} \in \mathbb{C}^{M\times 1}$) and output ($\mathbf{b} \in \mathbb{C}^{N\times 1}$) relation of a given system. The objective of this problem is to find an optimal solution $\mathbf{x}$ that produces a desired output $\mathbf{b}$. The number of equations M of this problem represents the number of constraint we have, and the number of elements in $\mathbf{x}$ (N) represents the number of variables we can control.

The least square problem attempts to solve Equation (A34) by minimizing the error between the desired and actually produced outputs. Since an error of two vectors can be defined by a Euclidean norm, we can construct an optimization problem defined as

$$\text{Minimize } \|\Delta\mathbf{b}\|, \text{ subject to } \mathbf{Ax} = \mathbf{b} + \Delta\mathbf{b} \tag{A35}$$

or, simply

$$\text{Minimize } \varepsilon = \|\mathbf{b} - \mathbf{Ax}\|,$$

where $\|\mathbf{x}\| = \sqrt{\mathbf{x}^H\mathbf{x}}$ denotes the Euclidean norm (two-norm) of a vector $\mathbf{x}$. The error ε of this case is called a *cost function* of the optimization problem.

For a square matrix $\mathbf{A}$ ($M = N$) consisting of linearly independent column or row vectors, the solution to Equation (A34) can be obtained by calculating an inverse of matrix $\mathbf{A}$:

$$\mathbf{x} = \mathbf{A}^{-1}\mathbf{b}. \tag{A36}$$

The inverse of a matrix $\mathbf{A}$ is a matrix that satisfies $\mathbf{AA}^{-1} = \mathbf{A}^{-1}\mathbf{A} = \mathbf{I}$. Some useful expressions involving the matrix inverse are

$$\begin{aligned}(\mathbf{A} + \mathbf{BCD})^{-1} &= \mathbf{A}^{-1} - \mathbf{A}^{-1}\mathbf{BC}(\mathbf{I} + \mathbf{DA}^{-1}\mathbf{BC})^{-1}\mathbf{DA}^{-1} \\ &= \mathbf{A}^{-1} - \mathbf{A}^{-1}\mathbf{B}(\mathbf{I} + \mathbf{CDA}^{-1}\mathbf{B})^{-1}\mathbf{CDA}^{-1},\end{aligned} \tag{A37}$$

which hold for the invertible matrix $\mathbf{A}$.

A.4.1 Overdetermined Linear Equations and Least Squares (LS) Solution

If the number of linearly independent column vectors of $\mathbf{A}$ is less than the number of independent constraints, then we can find a unique solution that minimizes ε. For example, a full rank and skinny matrix $\mathbf{A}$ ($M > N$) has more constraints than the control variables. It is obvious that the point of minimum ε can be found where the derivative of ε^2 is zero. That is,

$$\frac{\partial\varepsilon^2}{\partial\mathbf{x}} = \frac{\partial(\mathbf{b} - \mathbf{Ax})^H(\mathbf{b} - \mathbf{Ax})}{\partial\mathbf{x}} = 0.$$

Using the complex derivative Equations (A31) and (A33), The optimal point can be found at

$$\frac{\partial(\mathbf{b} - \mathbf{Ax})^H(\mathbf{b} - \mathbf{Ax})}{\partial\mathbf{x}} = \mathbf{x}^H(\mathbf{A}^H\mathbf{A}) - \mathbf{b}^H\mathbf{A} = 0. \tag{A38}$$

Simplifying Equation (A38) with respect to $\mathbf{x}$ gives

$$\mathbf{x} = (\mathbf{A}^H\mathbf{A})^{-1}\mathbf{A}^H\mathbf{b}. \tag{A39}$$

The solution of Equation (A39) is referred to as a *least squares solution* of a given linear equation. The matrix operator of Equation (A39) given by

$$\mathbf{A}^{+}{}_{\text{left}} = (\mathbf{A}^H\mathbf{A})^{-1}\mathbf{A}^H \tag{A40}$$

is denoted as a *left inverse* of $\mathbf{A}$ ($\mathbf{A}^{+}{}_{\text{left}}\mathbf{A} = \mathbf{I}$).

A.4.1.1 Nonexistence Problem

If the vector **b** can be expressed as a weighted sum of column vectors of **A**, then Equation (A39) will give an exact solution that results in $\varepsilon = 0$. Nevertheless, in general the Euclidean distance cannot be zero, since the number of controllable column vectors are less than the number of equations. Consequently, we encounter a nonexistence problem for which the solution cannot exist. The least squares approach tries to find the nearest solution in two-norm sense under the nonexistence situation.

A.4.2 Underdetermined Linear Equations and Minimum-Norm Problem

A.4.2.1 Nonuniqueness and Ill-Posed Problem

When the number of independent constraints is less than the number of variables, the linear equation becomes underdetermined. A typical case is when **A** is fat ($M < N$). Since the number of constraints (equations) is not enough, we cannot determine a unique solution and there can be multiple solutions. This problem is referred to as the *uniqueness problem*, and the linear equations are being said *ill-posed*.

The underdetermined condition is equivalent to having a nontrivial solution $\mathbf{x}_h$ of the problem $\mathbf{A}\mathbf{x}_h = \mathbf{0}$. This is because any two different solutions satisfying $\mathbf{A}\mathbf{x}_1 = \mathbf{b}$ and $\mathbf{A}\mathbf{x}_2 = \mathbf{b}$ will give $\mathbf{A}(\mathbf{x}_1 - \mathbf{x}_2) = \mathbf{0}$. Such a solution $\mathbf{x}_h$ is called *homogeneous solution*. Conversely, we can say that there will be the uniqueness problem when the homogeneous solution $\mathbf{x}_h$ exists.

The underdetermined situation can be resolved by adding extra constraints to the problem. In principle, the additional constrains should reflect the practical condition involved with the problem; however, one of popularly used constraints is the *minimum-norm* of the input **x**. That is,

$$\text{Minimize } \|\mathbf{x}\|, \text{ subject to } \mathbf{A}\mathbf{x} = \mathbf{b}. \tag{A41}$$

The physical implication of the minimum-norm constraint is that the total input power or control effort required to produce the desired output **b** should be minimized. Therefore, the minimum-norm solution is beneficial when we want to save the control effort of a system.

The solution to Equation (A41) is known as

$$\mathbf{x}_{RH} = \mathbf{A}^H(\mathbf{A}\mathbf{A}^H)^{-1}\mathbf{b}, \tag{A42}$$

and the inverse operator

$$\mathbf{A}^+{}_{\text{right}} = \mathbf{A}^H(\mathbf{A}\mathbf{A}^H)^{-1} \tag{A43}$$

is referred to as the *right inverse* of a matrix **A**. The proof is as follows: first, the least-norm solution Equation (A42) satisfies $\mathbf{A}\mathbf{x}_{RH} = \mathbf{b}$. Accordingly, if we consider an arbitrary solution **x** satisfying $\mathbf{A}\mathbf{x} = \mathbf{b}$, then it leads to a homogeneous solution given by

$$\mathbf{A}(\mathbf{x} - \mathbf{x}_{RH}) = \mathbf{A}\mathbf{x}_h = \mathbf{0}. \tag{A44}$$

Now check the orthogonality of the two solution vectors: $\mathbf{x}_h$ and $\mathbf{x}_{RH}$. From Equation (A44), we have

$$\begin{aligned} \mathbf{x}_h{}^H\mathbf{x}_{RH} &= \mathbf{x}_h{}^H\mathbf{A}^H(\mathbf{A}\mathbf{A}^H)^{-1}\mathbf{b} \\ &= (\mathbf{A}\mathbf{x}_h)^H(\mathbf{A}\mathbf{A}^H)^{-1}\mathbf{b} \\ &= \mathbf{0} \end{aligned} \tag{A45}$$

which means that $\mathbf{x}_h$ and $\mathbf{x}_{RH}$ are orthogonal to each other. Since two vectors are orthogonal to each other, we can derive the following inequality:

$$\|\mathbf{x}\|^2 = \|\mathbf{x}_{RH} + \mathbf{x} - \mathbf{x}_{RH}\|^2$$
$$= \|\mathbf{x}_{RH}\|^2 + \|\mathbf{x}_h\|^2 \geq \|\mathbf{x}_{RH}\|^2 \quad \text{(A46)}$$

which shows that the norm of $\mathbf{x}_{RH}$ is the smallest among any solution.

The physical meaning lying behind of this proof is that, any homogeneous solution $\mathbf{x}_h = \mathbf{x} - \mathbf{x}_{RH}$ is orthogonal to $\mathbf{x}_{RH}$, and its norm is always greater than zero. This conversely states that any solution other than $\mathbf{x}_{RH}$ is not the minimum-norm solution, and hence the minimum-norm solution is unique.

A.4.2.2 Pseudo Inverse

The pseudo inverse or Moore-Penrose inverse of a matrix $\mathbf{A}$ is defined as a matrix $\mathbf{A}^+$ that fulfills all of the following properties:

$$\mathbf{A}\mathbf{A}^+\mathbf{A} = \mathbf{A} \quad \text{(A47)}$$
$$\mathbf{A}^+\mathbf{A}\mathbf{A}^+ = \mathbf{A}^+ \quad \text{(A48)}$$
$$\mathbf{A}\mathbf{A}^+ = (\mathbf{A}\mathbf{A}^+)^H \quad \text{(A49)}$$
$$\mathbf{A}^+\mathbf{A} = (\mathbf{A}^+\mathbf{A})^H. \quad \text{(A50)}$$

The right and left inverse $\mathbf{A}^+{}_{\text{left}}$, $\mathbf{A}^+{}_{\text{right}}$ of Equations (A40) and (A43) are the pseudo inverses of skinny and fat matrices, respectively. As shown Equations (A44)–(A46), the pseudo inverse of a matrix is unique.

The pseudo inverse should be differentiated from the generalized inverse that has a much loose definition. A generalized inverse $\mathbf{A}^-$ is any matrix that only satisfies Equation (A47). That is,

$$\mathbf{A}\mathbf{A}^-\mathbf{A} = \mathbf{A}. \quad \text{(A51)}$$

A.4.3 *Method of Lagrange Multiplier*

Method of Lagrange multiplier is useful when solving a constrained optimization problem such as Equation (A41). It converts the constrained optimization problem to an unconstrained problem, by including the constraint in the cost function. For example, a new cost function for the Equation (A41) can be written as follows:

$$\text{Minimize } J(\mathbf{x}, \mu) = \mathbf{x}^H\mathbf{x} + \mu(\mathbf{A}\mathbf{x} - \mathbf{b}) \quad \text{(A52)}$$

with μ being the Lagrange multiplier. The optimal solution of Equation (A52) can be derived by seeking the zero derivatives point of J with respect to $\mathbf{x}$ and μ

$$\frac{\partial J}{\partial \mathbf{x}} = \mathbf{x}^H + \mu\mathbf{A} = 0 \quad \text{(A53)}$$
$$\frac{\partial J}{\partial \mu} = \mathbf{A}\mathbf{x} - \mathbf{b} = 0. \quad \text{(A54)}$$

It can be seen that the Equation (A54) is equivalent to the constraint of Equation (A41). From Equation (A53), we have

$$\mathbf{x} = -\mathbf{A}^H \mu^*, \tag{A55}$$

and substitute this result into Equation (A54) to obtain

$$\mu^* = -(\mathbf{A}\mathbf{A}^H)^{-1}\mathbf{b}. \tag{A56}$$

Consequently, the optimal solution Equation (A55) is given by

$$\mathbf{x} = \mathbf{A}^H(\mathbf{A}\mathbf{A}^H)^{-1}\mathbf{b}, \tag{A57}$$

which is the same result as Equation (A42).

A.4.4 Regularized Least Squares

The constrained optimization of Section A.4.3 is a kind of step-by-step approach, which first collects solutions satisfying given constraints and then finds a unique solution minimizing the cost function. However, sometimes a solution slightly violating the constraint can be better in regard to the cost function. To find such a solution, we can setup a new cost function defined by the weighted sum of the constraint and the original cost function.

$$J = \|\mathbf{A}\mathbf{x} - \mathbf{b}\|^2 + \mu\|\mathbf{x}\|^2, \tag{A58}$$

where a real-valued constant μ is a kind of tuning factor that controls relative weight between the fitness of the constraint and the original cost function. Therefore, using Equation (A58), we attempt to minimize both the squared error and the input power. If $\mu \to 0$, then the cost function is equivalent to Equation (A35).

The optimal solution minimizing Equation (A58) can be found from the derivative

$$\frac{\partial J}{\partial \mathbf{x}} = \mathbf{x}^H\mathbf{A}^H\mathbf{A} - \mathbf{b}^H\mathbf{A} + \mu\mathbf{x}^H = 0, \tag{A59}$$

which leads to

$$\mathbf{x} = (\mathbf{A}^H\mathbf{A} + \mu\mathbf{I})^{-1}\mathbf{A}^H\mathbf{b}. \tag{A60}$$

This technique is referred to as a regularized least-squares, since the inverse of skinny matrix $\mathbf{A}$ is regularized by the extra factor μ. It can be seen that the regularized solution Equation (A60) converges to the least squares solution Equation (A39) as $\mu \to 0$.

A.4.5 Singular Value Decomposition

There can be various basis functions or vectors spanning a given vector space. The eigenvalue analysis is one of the useful tools for finding the orthonormal basis function. The eigenvalue analysis searches the orthonormal basis function on the basis of the output power produced by a system. Let us go back to the input–output relation defined by the linear equation (Equation (A34)). Consider a problem of maximizing the output power with input power constrained to one.

$$\text{Maximize } \mathbf{b}^H\mathbf{b}, \text{ subject to } \mathbf{x}^H\mathbf{x} = 1$$

The solution can be solved by introducing real-valued Lagrange multiplier μ_j:

$$J = \mathbf{b}^H\mathbf{b} - \mu_j(\mathbf{x}^H\mathbf{x} - 1), \tag{A61}$$

and its solution is given by

$$\frac{\partial J}{\partial \mathbf{x}} = \mathbf{x}^H\mathbf{A}^H\mathbf{A} - \mu_j\mathbf{x}^H = 0 \Rightarrow \ \mathbf{A}^H\mathbf{A}\mathbf{x} = \mu_j\mathbf{x} \tag{A62}$$

$$\frac{\partial J}{\partial \mu} = \mathbf{x}^H\mathbf{x} - 1 = 0 \ \ \Rightarrow \mathbf{x}^H\mathbf{x} = 1.$$

The relation given by Equation (A62) is called characteristic equation and can have multiple solutions with different μ_js. The vector $\mathbf{u}_j$ is a *right eigenvector* of $\mathbf{B} = \mathbf{A}^H\mathbf{A}$ if it satisfies the characteristic equation

$$\mathbf{B}\mathbf{u}_j = \mu_j\mathbf{u}_j, \tag{A63}$$

and corresponding μ_j becomes the eigenvalue. Likewise, a vector $\mathbf{v}_j$ is called a *left eigenvector* of $\mathbf{B}$ if it satisfies $\mathbf{v}_j^H\mathbf{B} = \mu_j\mathbf{v}_j^H$.

The first eigenvector $\mathbf{u}_1$ corresponding to maximum eigenvalue μ_1 maximizes the output power of the system with a given input power. The second eigenvector is orthogonal to $\mathbf{u}_2$ and thus it represents the maximum energy solution among the vectors that are orthogonal to $\mathbf{u}_1$. Therefore, the eigenvalue analysis finds orthogonal basis functions based on the maximum energy principle.

The singular value decomposition (SVD) is a factorization that decomposes any matrix $\mathbf{A}$ into a product of three matrices:

$$\mathbf{A} = \mathbf{U}\Sigma\mathbf{V}^H \tag{A64}$$

such that $\mathbf{U}$ and $\mathbf{V}$ are unitary matrices whose columns are the right and left eigenvectors, respectively. From the orthogonality of eigenvectors, we have

$$\mathbf{U}^H\mathbf{U} = \mathbf{I}, \ \ \mathbf{V}^H\mathbf{V} = \mathbf{I}. \tag{A65}$$

The $M \times N$ matrix $\mathbf{\Sigma}$ is a diagonal matrix having singular values σ_j as its diagonal entries. If the rank of the matrix $\mathbf{A}$ is r, then $\mathbf{\Sigma}$ has r nonzero entries:

$$\mathbf{\Sigma} = \mathrm{diag}(\sigma_1, \cdots, \sigma_r, 0, \cdots, 0). \tag{A66}$$

The columns of $\mathbf{U}$ and $\mathbf{V}$ are given by eigenvectors of $\mathbf{A}\mathbf{A}^H$ and $\mathbf{A}^H\mathbf{A}$, therefore, they are called left and right singular vectors (or eigenvectors) respectively. From Equation (A63), the r singular values are related to eigenvalues as $\sigma_j^2 = \mu_j$.

The SVD is useful for solving the least squares problem, since the left and right singular vectors do not change the Euclidean norm of the vector $\mathbf{b}$. That is,

$$\|\mathbf{U}\mathbf{\Sigma}\mathbf{V}^H\mathbf{x} - \mathbf{b}\| = \|\mathbf{\Sigma}\mathbf{V}^H\mathbf{x} - \mathbf{U}^H\mathbf{b}\|. \tag{A67}$$

This can be proved from the fact that the pseudo inverse of the matrix $\mathbf{A}$ is given by

$$\mathbf{A}^- = \mathbf{V}\mathbf{\Sigma}^-\mathbf{U}^H \tag{A68}$$

where the $\mathbf{\Sigma}^-$ represents the inverse of $\mathbf{\Sigma}$, which has the inverse of nonzero diagonal elements of $\mathbf{\Sigma}$ as its diagonal entries ($\mathbf{\Sigma}^- = \mathrm{diag}(\sigma_1^{-1}, \cdots, \sigma_r^{-1}, 0, \cdots, 0)$). The matrix $\mathbf{A}^-$ satisfies the properties of pseudo inverse given by Equations (A47)–(A50)).

Using SVD, we can express the solution of the regularized least squares problem as well. Inserting Equation (A64) to Equation (A60) gives

$$\mathbf{x} = \mathbf{V}[(\mathbf{\Sigma}^2 + \mu\mathbf{I})^{-1}\mathbf{\Sigma}]\mathbf{U}^H\mathbf{b}. \tag{A69}$$

A.4.6 Total Least Squares (TLS)

In solving the following linear equations:

$$\mathbf{AX} = \mathbf{B}, \tag{A70}$$

the TLS considers errors in both the transfer matrix $\mathbf{A} \in \mathbb{C}^{M \times N}$ and target $\mathbf{B} \in \mathbb{C}^{M \times K}$. Denoting the errors in $\mathbf{A}$ and $\mathbf{B}$ as $\Delta\mathbf{A}$ and $\Delta\mathbf{B}$, the TLS problem can be defined as

$$\text{Minimize} \left\| \left[\Delta\mathbf{A} \quad \Delta\mathbf{B}\right] \right\|_F \text{ subject to } (\mathbf{A} + \Delta\mathbf{A})\mathbf{X} = \mathbf{B} + \Delta\mathbf{B}. \tag{A71}$$

This is in contrast with the objective of least-squares (LS) approach given by

$$\text{Minimize} \left\| \Delta\mathbf{B} \right\|_F \text{ subject to } \mathbf{AX} = \mathbf{B} + \Delta\mathbf{B}$$

which only considers error in the target matrix $\mathbf{B}$. Therefore, TLS aims to reduce both errors in the transfer matrix and target, under the constraint that the perturbed target $\mathbf{B} + \Delta\mathbf{B}$ can be expressed in terms of (in the range of) $\mathbf{A} + \Delta\mathbf{A}$, where $\mathbf{X} \in \mathbb{C}^{N \times K}$.

The constraint of Equation (A71) can be written as

$$\text{Minimize } \|\Delta\mathbf{C}\|_F \text{ subject to } (\mathbf{C} + \Delta\mathbf{C}) \begin{bmatrix} \mathbf{X} \\ -\mathbf{I} \end{bmatrix} = \mathbf{0} \tag{A72}$$

with $\mathbf{C} = [\mathbf{A}\ \mathbf{B}] \in \mathbb{C}^{M \times (N+K)}$ and $\Delta\mathbf{C} = [\Delta\mathbf{A}\ \Delta\mathbf{B}]$.

The constraint of Equation (A72) implies that the solution $\mathbf{X}$ partitioned with an identity matrix $\mathbf{I} \in \mathbb{R}^{K \times K}$ belongs to the null space of $\mathbf{C} + \Delta\mathbf{C}$. Due to the identity matrix, the K columns of $[\mathbf{X}^T - \mathbf{I}^T]^T$ are linearly independent to each other. Therefore, if the constraint of Equation (A72) holds, then the null-space of $\mathbf{C} + \Delta\mathbf{C}$ should be of dimension K at least (the opposite is not always true). In other words, the rank of a matrix $\mathbf{C} + \Delta\mathbf{C}$ should be reduced by K.

The solution to the Equation (A72) can be obtained from the SVD of the matrix $\mathbf{C}$. Denote the SVD of $\mathbf{C}$ as a partitioned form:

$$\begin{aligned} \mathbf{C} &= \mathbf{U}\boldsymbol{\Sigma}\mathbf{V}^H \\ &= \mathbf{U} \begin{bmatrix} \boldsymbol{\Sigma}_{AA} & 0 \\ 0 & \boldsymbol{\Sigma}_{BB} \end{bmatrix} \begin{bmatrix} \mathbf{V}_{AA}^H & \mathbf{V}_{BA}^H \\ \mathbf{V}_{AB}^H & \mathbf{V}_{BB}^H \end{bmatrix} \end{aligned} \tag{A73}$$

such that

$$\mathbf{A} = \mathbf{U} \begin{bmatrix} \boldsymbol{\Sigma}_{AA}\mathbf{V}_{AA}^H \\ \boldsymbol{\Sigma}_{BB}\mathbf{V}_{AB}^H \end{bmatrix}, \quad \mathbf{B} = \mathbf{U} \begin{bmatrix} \boldsymbol{\Sigma}_{AA}\mathbf{V}_{BA}^H \\ \boldsymbol{\Sigma}_{BB}\mathbf{V}_{BB}^H \end{bmatrix}.$$

The singular value matrix $\boldsymbol{\Sigma}_{BB}$, consisting of K arbitrary singular values of $\mathbf{C}$, has a dimension of $K \times K$. Therefore, if the perturbation matrix $\Delta\mathbf{C}$ is defined such that

$$\Delta\mathbf{C} = -\mathbf{U} \begin{bmatrix} 0 & 0 \\ 0 & \boldsymbol{\Sigma}_{BB} \end{bmatrix} \begin{bmatrix} \mathbf{V}_{AA}^H & \mathbf{V}_{BA}^H \\ \mathbf{V}_{AB}^H & \mathbf{V}_{BB}^H \end{bmatrix}, \tag{A74}$$

then the perturbed matrix $\mathbf{C} + \Delta\mathbf{C}$ can be written as

$$\mathbf{C} + \Delta\mathbf{C} = \mathbf{U} \begin{bmatrix} \boldsymbol{\Sigma}_{AA} & 0 \\ 0 & 0 \end{bmatrix} \begin{bmatrix} \mathbf{V}_{AA}^H & \mathbf{V}_{BA}^H \\ \mathbf{V}_{AB}^H & \mathbf{V}_{BB}^H \end{bmatrix}. \tag{A75}$$

Equation (A75) means that the rank of the matrix $\mathbf{C} + \Delta\mathbf{C}$ will be always reduced by K, and thus the matrix has the null space of dimension K. The vectors spanning the null space of $\mathbf{C} + \Delta\mathbf{C}$ can be found by using the orthogonal property of the columns of $\mathbf{V}$. Multiplying both sides of Equation (A75) by $\begin{bmatrix} \mathbf{V}_{AB}^T & \mathbf{V}_{BB}^T \end{bmatrix}^T$ gives

$$\mathbf{C} + \Delta\mathbf{C} \begin{bmatrix} \mathbf{V}_{AB} \\ \mathbf{V}_{BB} \end{bmatrix} = \mathbf{U} \begin{bmatrix} \mathbf{\Sigma}_{AA} & 0 \\ 0 & 0 \end{bmatrix} \begin{bmatrix} 0 \\ \mathbf{I} \end{bmatrix} = 0. \tag{A76}$$

Accordingly, the right eigenvectors of $\mathbf{C}$ corresponding to the K singular values span the null space of $\mathbf{C} + \Delta\mathbf{C}$. Now let us transform the solution to have the form of Equation (A72). By multiplying $-\mathbf{V}_{BB}^{-1}$ to the solution of Equation (A76), we have

$$-\begin{bmatrix} \mathbf{V}_{AB}\mathbf{V}_{BB}^{-1} \\ \mathbf{V}_{BB}\mathbf{V}_{BB}^{-1} \end{bmatrix} = \begin{bmatrix} -\mathbf{V}_{AB}\mathbf{V}_{BB}^{-1} \\ -\mathbf{I} \end{bmatrix}. \tag{A77}$$

From the comparison of Equations (A77) and (A72), we can see that the solution $\mathbf{X}$ satisfying the constraint of Equation (A72) can be written as

$$\mathbf{X} = -\mathbf{V}_{AB}\mathbf{V}_{BB}^{-1}. \tag{A78}$$

However, the solution of Equation (A78) is not unique, because any K vectors of $\mathbf{V}$ can be used to calculate $\begin{bmatrix} \mathbf{V}_{AB}^T & \mathbf{V}_{BB}^T \end{bmatrix}^T$. Among them, we need to select the solution minimizing the Frobenius norm of $\Delta\mathbf{C}$. The Frobenius norm of Equation (A74) is given by

$$\|\Delta\mathbf{C}\|_F = \|\mathbf{\Sigma}_{BB}\|_F, \tag{A79}$$

and hence the optimal solution is given by Equation (A78) calculated from the eigenvectors of the K smallest singular values of $\mathbf{C}$.

Appendix B

Description of Sound Field

B.1 Three-Dimensional Acoustic Wave Equation

B.1.1 Conservation of Mass

Consider the conservation of mass in a small volume located in an arbitrary three-dimensional space. According to the conservation of mass, a net increase of mass per unit time is equal to a net mass flux per unit time, which enters and exits through the surface of small volume. For convenience, mass per unit time that enters and exits through $dy\,dz$ faces, namely two faces orthogonal to the x coordinate, can be written as Figure B.1:

$$\rho u\,dy\,dz - \left(\rho u + \frac{\partial \rho u}{\partial x}dx\right)dy\,dz = -\frac{\partial \rho u}{\partial x}dV. \tag{B1}$$

In the same manner, increased mass that enters and exits through $dz\,dx$, $dx\,dy$ faces in the volumes $dx\,dy\,dz$ are equal to $-\frac{\partial \rho v}{\partial y}dV$ and $-\frac{\partial \rho w}{\partial z}dV$, respectively.

In addition, increased mass in the volumes per unit time $dx\,dy\,dz$ due to fluid compression can be written as:

$$\frac{\partial}{\partial t}(\rho\,dx\,dy\,dz) = \frac{\partial}{\partial t}(\rho dV) = dV\frac{\partial \rho}{\partial t}. \tag{B2}$$

By considering Equations (B1) and (B2) and the mass increase rates through the surfaces in the y, z directions, we can write

$$\frac{\partial \rho}{\partial t}dV = \left(\frac{\partial \rho u}{\partial x} + \frac{\partial \rho v}{\partial y} + \frac{\partial \rho w}{\partial z}\right)dV = \nabla\cdot\rho\mathbf{u}dV, \tag{B3}$$

where $\mathbf{u} = (u, v, w)$. By applying the conservation of mass, we obtain

$$\frac{\partial \rho}{\partial t} = -\nabla\cdot\rho\mathbf{u}. \tag{B4}$$

The assumption employed here is that the fluid is a continuum and compressible.

B.1.2 Conservation of Momentum

As shown in Figure B.2, Newton's second law dictates that the force applied to a small volume and the change of momentum should be balanced. We represent this law with selected coordinate

Sound Visualization and Manipulation, First Edition. Yang-Hann Kim and Jung-Woo Choi.

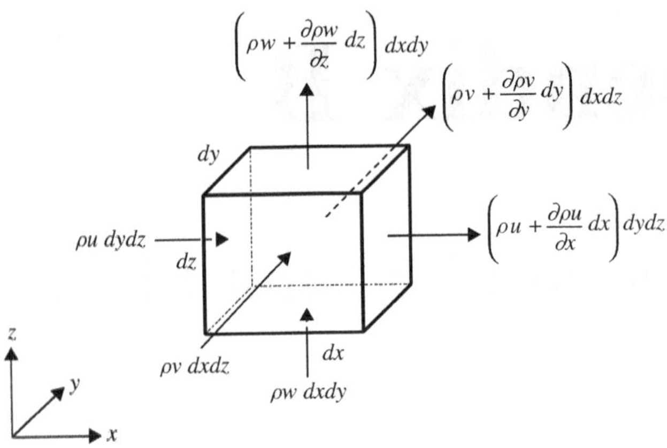

Figure B.1 Conservation of mass for a small volume located in three-dimensional space

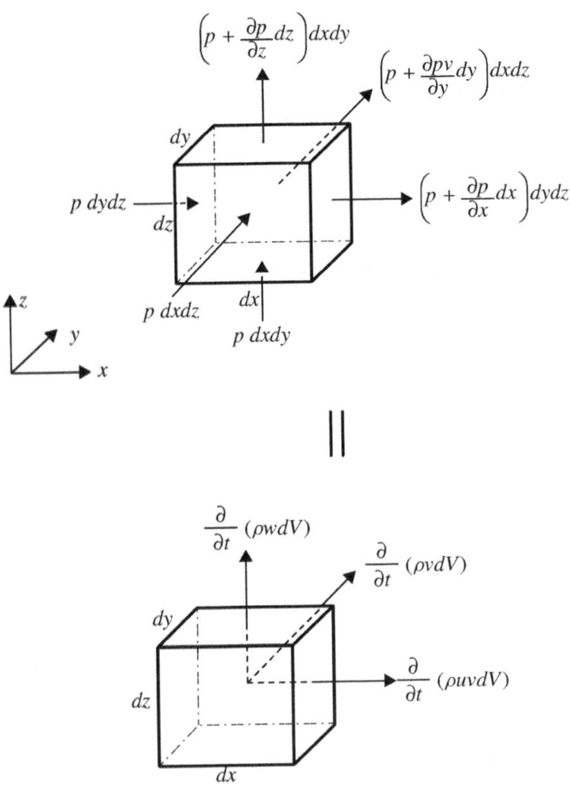

Figure B.2 Relationship between force and motion for a small volume located in three-dimensional space

systems and physical measures including velocity and mass. First, the balance between the change of momentum and force can be specified as

$$\begin{aligned}&\text{Change of momentum through the surface of small volume}\\&+\text{Change of momentum within the body of small volume}\\&=\text{Sum of forces (surface or body) applied to small volume}\end{aligned}\tag{B5}$$

Here, the overall change of momentum can be written as:[1]

$$\frac{\partial}{\partial x}(u\rho\mathbf{u})dx\,dy\,dz+\frac{\partial}{\partial y}(v\rho\mathbf{u})dy\,dz\,dx+\frac{\partial}{\partial z}(w\rho\mathbf{u})dz\,dx\,dy+\frac{\partial}{\partial t}(\rho\mathbf{u}dV)\tag{B6}$$

There are two types of force that act on the surface of a small volume. One is the surface normal force that works normal to the surface and the other is shear force, which is a surface tangential force that works parallel to the surface. Of these, the forces that work parallel to the surface bring about an angular displacement change of a small amount of fluid in a hexahedral shape (see Figure B.2). The force that works normal to the plane, or pressure, causes a small amount of fluid to have pure dilation and compression. If the shear force by the viscosity of fluid is negligibly small compared to the fluid pressure, the net force applied to surface $dy\,dz$ in Figure B.2 can be written as:[2]

$$p\,dy\,dz-\left(p+\frac{\partial p}{\partial x}dx\right)dy\,dz=-\frac{\partial p}{\partial x}dx\,dy\,dz.\tag{B7}$$

When the force applied to the surface in Equation (B7) is applied to $dz\,dx$, $dx\,dy$ faces in the same manner, the net force that is applied to a small volume $dx\,dy\,dz$ can be expressed as

$$-\left(\frac{\partial p}{\partial x}\mathbf{e}_x+\frac{\partial p}{\partial y}\mathbf{e}_y+\frac{\partial p}{\partial z}\mathbf{e}_z\right)dx\,dy\,dz=-\nabla p dV,\tag{B8}$$

where $\mathbf{e}_x$, $\mathbf{e}_y$, $\mathbf{e}_z$ represent unit vectors in the x, y, z directions, respectively.

Another force that is not considered here is the body force, which is the gravitational acceleration applied to an entire fluid. Considering this, the force applied to a small volume is written as:

$$-\rho dV g_a\mathbf{e}_z-\nabla p dV,\tag{B9}$$

where g_a is the gravitational acceleration of Earth.

Here, the balance of momentum without considering gravity can be written as:[3]

$$\left\{\frac{\partial}{\partial x}(u\rho\mathbf{u})+\frac{\partial}{\partial y}(v\rho\mathbf{u})+\frac{\partial}{\partial z}(w\rho\mathbf{u})\right\}+\frac{\partial}{\partial t}(\rho\mathbf{u})=-\nabla p.\tag{B10}$$

[1] The terms in Equation B6 involved with changes of momentum through the surface of small volume are expressed in a similar form ($dx\,dy\,dz=dV$) to the changes in momentum within a small volume; the area integral can be converted into a volume integral based on Gauss' theorem.

[2] The kinematic viscosity of water and air at normal temperature is 1.45×10^{-5} and 1.14×10^{-6} m^2/s, respectively.

[3] As it can be assumed that the body force due to the density change in a small volume is very small compared to the force by sound pressure, the former is not usually included in the momentum balance when deriving governing equations in acoustics.

The terms in the parentheses on the left side of Equation (B10) can be rewritten as:

$$\rho\mathbf{u}\frac{\partial u}{\partial x} + u\frac{\partial(\rho\mathbf{u})}{\partial x} + \rho\mathbf{u}\frac{\partial v}{\partial y} + v\frac{\partial(\rho\mathbf{u})}{\partial y} + \rho\mathbf{u}\frac{\partial w}{\partial z} + w\frac{\partial(\rho\mathbf{u})}{\partial z}$$

$$= \rho\mathbf{u}(\nabla\cdot\mathbf{u}) + (\mathbf{u}\cdot\nabla)\rho\mathbf{u}. \tag{B11}$$

The final term of the left side in Equation (B10) can be rearranged to yield

$$\frac{\partial}{\partial t}(\rho\mathbf{u}) = \rho\frac{\partial\mathbf{u}}{\partial t} + \mathbf{u}\frac{\partial\rho}{\partial t} \tag{B12}$$

And, by using the conservation of mass,

$$\frac{\partial}{\partial t}(\rho\mathbf{u}) = \rho\frac{\partial\mathbf{u}}{\partial t} - (\mathbf{u}\cdot\nabla)\rho\mathbf{u} \tag{B13}$$

is finally obtained. The left side in Equation (B10) can be rewritten

$$\rho\mathbf{u}(\nabla\cdot\mathbf{u}) + (\mathbf{u}\cdot\nabla)\rho\mathbf{u} + \rho\frac{\partial\mathbf{u}}{\partial t} - (\mathbf{u}\cdot\nabla)\rho\mathbf{u}$$

$$= \rho\frac{\partial\mathbf{u}}{\partial t} + \rho(\mathbf{u}\cdot\nabla)\mathbf{u}$$

$$= \rho\frac{D\mathbf{u}}{Dt} \tag{B14}$$

where $\frac{D}{Dt}$ represents total derivative or material derivative. Consequently, the balance between the force and momentum can be defined as

$$\rho\frac{D\mathbf{u}}{Dt} = -\nabla p. \tag{B15}$$

This equation is called the three-dimensional Euler equation.

B.1.3 Equation of State

Equations (B4) and (B15) represent the relationship between the density of the fluid (ρ) and the velocity of the fluid particles ($\mathbf{u}$), and the relationship between fluid particle velocity and pressure. Therefore, another equation is needed to identify the relationship between the three physical variables of density, fluid particle velocity, and pressure. This is the equation of state that determines the relationship between fluid pressure and density. As sound pressure in a fluid is generally thought to be governed by density and entropy, it can be expressed as

$$p = p(\rho, s), \tag{B16}$$

where s is entropy. A small change of pressure can occur due to changes in density and entropy (or $d\rho, ds$), respectively. This can be expressed as

$$dp = \left.\frac{\partial p}{\partial\rho}\right|_s d\rho + \left.\frac{\partial p}{\partial s}\right|_\rho ds. \tag{B17}$$

For sounds smaller than 10^9 Hz, it can be assumed that during compression and expansion of a fluid it undergoes an isentropic reversible process. Equation (B17) can then be written as

$$p' = \left.\frac{\partial p}{\partial\rho}\right|_{s0} \rho', \tag{B18}$$

where p' and ρ' represent small changes in pressure and density respectively. Moreover,

$$\left.\frac{\partial p}{\partial \rho}\right|_{s_0} = c^2, \quad \text{(B19)}$$

where c is speed of sound. In other words, the square of the speed of sound is equal to the pressure change with respect to the density change.

Alternatively, we can obtain Equation (B19) by considering wavefront propagation in a duct. Suppose that we make a disturbance which induces a small volume change in the one-dimensional duct as illustrated in Figure B.3. This disturbance, which is assumed to compress the fluid by a small volume (dV), will introduce an infinitesimal increase in pressure and density (Figure B.3). As a result of this increase, we can postulate that a wavefront will propagate with the speed of sound c (Figure B.3). We now want to find the relation between the speed of sound propagation c and other physical variables, such as pressure and density.

Conservation of mass implies the identity:

$$(\rho + d\rho)(c - dv)S = \rho c S. \quad \text{(B20)}$$

The left-hand side simply represents the amount of mass change per unit time due to the disturbance. The right-hand side is the mass flux of the fluid at rest. These two have to be balanced, and can be written as

$$\rho dv = c d\rho. \quad \text{(B21)}$$

We next apply Newton's second law to the fluid of interest. The force difference will be $pS - (p + (\partial p/\partial x)dx)S$, and the corresponding momentum change under consideration is $\{(cS - Sdv) - cS\}\rho c$ which neglects higher order terms induced by $d\rho$. We can therefore write

$$dp = \rho c dv. \quad \text{(B22)}$$

Equations (B21) and (B22) lead to the following relation which describes the relation between the speed of propagation and other physical variables of fluid:

$$c^2 = \frac{dp}{d\rho}. \quad \text{(B23)}$$

Equation (B23) states that the square of the speed of sound depends on the rate of compression with respect to density, that is, the amount of pressure requires to generate a unit change in density. Note, however, that the change in pressure and density of the fluid also depends on temperature

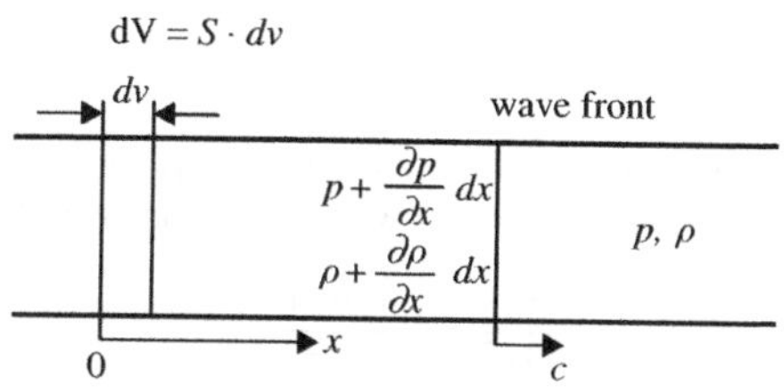

Figure B.3 S is cross-sectional area of the duct, x is the coordinate that measures the distance from the disturbance, and v is the disturbance velocity

or entropy. Therefore, Equation (B23) has to be rewritten as

$$c^2 = \frac{\partial p}{\partial \rho} + \frac{\partial p}{\partial s}. \tag{B24}$$

For isentropic process, Equation (B24) can be written as

$$c^2 = \left.\frac{\partial p}{\partial \rho}\right]_{s_c}. \tag{B25}$$

Equation (B25) simply implies that the speed of sound can be predicted if we know the relation between pressure and density. For example, if the fluid can be assumed to be an ideal gas in isentropic process, then we can obtain the relations between pressure and density (the ideal gas law and the isentropic relation) as

$$\frac{p}{\rho} = nR\Theta, \tag{B26}$$

$$\frac{\partial p}{\partial \rho} = \gamma \frac{p}{\rho} \left(\text{or } \frac{p}{\rho^\gamma} = \text{constant}\right) \tag{B27}$$

where n is the number of moles defined as mass M (kg) per unit molar mass (kg/mol), R is the universal gas constant (= 8.314 J/(K mol) in standard air), Θ is the absolute temperature (K) and γ is the heat capacity ratio which is defined as the ratio of the specific heat capacity under constant pressure to the specific heat capacity under constant volume. Consequently, we can predict the speed of sound for an ideal gas as

$$c = \sqrt{\gamma n R \Theta} \tag{B28}$$

under isentropic (i.e., no change in entropy) conditions.

B.1.4 Velocity Potential Function

When a viscous force is negligibly small compared to other forces, the angular deformation or velocity of the fluid (Figure B.2) will satisfy the relation

$$\nabla \times \mathbf{u} = 0. \tag{B29}$$

The velocity can then be expressed as

$$\mathbf{u} = \nabla \Phi, \tag{B30}$$

because $\nabla \times \nabla = 0$. In other words, when sound waves propagate toward a compressible fluid, we can consider a certain function which expresses the velocity vectors of the fluid particles. This function is called the velocity potential function. Substituting Equation (B30) into the linearized Euler equation yields:

$$p = -\rho_0 \frac{\partial \Phi}{\partial t}. \tag{B31}$$

This means that sound pressure can be obtained from the potential function. In addition, when Equation (B31) is substituted into a linear sound wave equation, we obtain

$$\frac{\partial^2 \Phi}{\partial x^2} = \frac{1}{c^2}\frac{\partial^2 \Phi}{\partial t^2}. \tag{B32}$$

From the above equation, we can conclude that the velocity potential satisfies the wave equation just as sound pressure does. Note also that the velocity and sound pressure can be directly obtained by differentiating the velocity potential function with respect to space or time, as indicated in

Equations (B30) and (B31). As also indicated by Equation (B.30), the velocity along an equipotential line (the line along which the potential is constant) is zero.

B.1.5 *Complex Intensity*

We examine basic concepts underlying complex intensity by using one-dimensional sound waves of single frequency. Instant intensity is represented by multiplying the sound pressure

$$p(x,t) = p_0(x)\cos(\omega t - \phi_p(x)) \tag{B33}$$

by the particle velocity,

$$u(x,t) = u_0(x)\cos(\omega t - \phi_u(x)). \tag{B34}$$

where $p_0(x)$ and $u_0(x)$ are real numbers and $\phi_p(x)$ and $\phi_u(x)$ represent the phase of the sound pressure and of the particle velocity, respectively. For simplicity, x dependency is omitted. Therefore, the instant intensity is expressed as

$$I(x,t) = p_0 u_0 \cos(\omega t - \phi_p)\cos(\omega t - \phi_u), \tag{B35}$$

and can also be written

$$\begin{aligned} I(x,t) &= \frac{1}{2} p_0 u_0 [\cos(2\omega t - \phi_p - \phi_u) + \cos(\phi_p - \phi_u)] \\ &= \frac{1}{2} p_0 u_0 [\cos(2(\omega t - \phi_p) + (\phi_p - \phi_u)) + \cos(\phi_p - \phi_u)] \\ &= \frac{1}{2} \mathrm{Re}\left\{ p_0 u_0 e^{i(\phi_p - \phi_u)} \left[1 + e^{2i(\omega t - \phi_p)}\right]\right\}. \end{aligned} \tag{B36}$$

Let us consider the result of Equation (B36) in conjunction with Equation (1.54). First, we define complex intensity as

$$\begin{aligned} \hat{C}(x) &= I_{avg}(x) + i I_r(x) \\ &= \frac{1}{2} p_0 u_0 e^{i(\phi_p - \phi_u)}. \end{aligned} \tag{B37}$$

The magnitude of complex intensity is then

$$|\hat{C}(x)| = \frac{1}{2} p_0 u_0. \tag{B38}$$

The average intensity or active intensity is found to be

$$I_{avg}(x) = \frac{1}{2} p_0 u_0 \cos(\phi_p - \phi_u) \tag{B39}$$

and the reactive intensity is expressed

$$I_r(x) = \frac{1}{2} p_0 u_0 \sin(\phi_p - \phi_u). \tag{B40}$$

Note also that the phase difference between Equations (B39) and (B40) is 90° (or −90°). Furthermore, when the phase of sound pressure is identical to that of particle velocity (for instance, plane waves), only active intensity exists since reactive intensity becomes 0. On the other hand, when the phase difference between sound pressure and particle velocity is ±90° (for instance, an interior sound field in a rigid boundary), the active intensity becomes 0 while only reactive intensity exists.

We now express the intensity using complex magnitudes and relate them to Equations (B37), (B39), and (B40). The sound pressure and particle velocity can be written as

$$\hat{p}(x,t) = \hat{p}(x,\omega)e^{-i\omega t} \tag{B41}$$

$$\hat{u}(x,t) = \hat{u}(x,\omega)e^{-i\omega t}, \tag{B42}$$

where $\hat{p}(x)$ and $\hat{u}(x)$ represent the complex magnitude of the sound pressure and particle velocity, respectively. The complex intensity defined by Equation (B37) can be written as

$$\hat{C}(x,\omega) = \frac{1}{2}\hat{p}(x,\omega)\hat{u}(x,\omega)^*, \tag{B43}$$

and the active or reactive intensity can be readily written as

$$I_{avg}(x,\omega) = \frac{1}{2}\mathrm{Re}\left[\hat{p}(x,\omega)\,\hat{u}(x,\omega)^*\right], \tag{B44}$$

$$I_r(x) = \frac{1}{2}\mathrm{Im}\left[\hat{p}(x,\omega)\,\hat{u}(x,\omega)^*\right]. \tag{B45}$$

Note that the complex variables $\hat{p}$ and $\hat{u}$ are denoted by p and u in other chapters.

B.1.6 Singular Sources

B.1.6.1 Monopole Sources

The sound pressure of a monopole source can be expressed as follows:

$$p(r) = q\frac{e^{ikr}}{4\pi r}, \tag{B46}$$

where the complex variable q is the monopole amplitude. Here, the amplitude of the sound source is the inertial force of fluid mass that the sound source exerts on the surface surrounding it. In other words, it can be expressed as

$$q = \rho_0(-i\omega v), \tag{B47}$$

where v represents the volume velocity which is the volume radiated by a monopole sound source per unit time (m^3/s).

From Euler equation, the velocity for the r direction can be written as

$$u_r(r) = \frac{1}{ik\rho_0 c}\frac{\partial p}{\partial r} = \frac{q}{\rho_0 c}\frac{e^{ikr}}{4\pi r}\left\{1 + i\left(\frac{1}{kr}\right)\right\}. \tag{B48}$$

The velocity in nearfield and farfield sound fields are approximately obtained as

$$\text{Near field}(kr \ll 1): \quad u_r(r) \approx \frac{q}{\rho_0 c}\frac{e^{ikr}}{4\pi r}i\left(\frac{1}{kr}\right) \tag{B49}$$

$$\text{Far field}(kr \gg 1): \quad u_r(r) \approx \frac{q}{\rho_0 c}\frac{e^{ikr}}{4\pi r} \tag{B50}$$

From Equations (B46) and (B48), the impedance can also be obtained as

$$Z_r(r) = \frac{p(r)}{u_r(r)} = \rho_0 c \frac{1}{1 + i\left(\frac{1}{kr}\right)}. \tag{B51}$$

The impedance in the nearfield and farfield sound fields is defined:

$$\text{Near-field}: \quad Z_r(r) \approx -i\rho_0 c(kr) \tag{B52}$$

$$\text{Far-field}: \quad Z_r(r) \approx \rho_0 c \tag{B53}$$

Equation (B53) implies that it becomes a plane wave in a farfield.

The average intensity in the r direction can be calculated from Equations (B46) and (B48) as follows:

$$I_{avg}(r) = \frac{1}{2}\text{Re}\{p(r)u_r(r)^*\} = \frac{1}{2\rho_0 c}\frac{|q|^2}{(4\pi r)^2}. \tag{B54}$$

For a monopole sound source, the average acoustic power can be obtained by integrating Equation (B54) over an area with a radius r, that is,

$$\Pi_{avg} = \int_0^{\pi}\int_0^{2\pi} I_{avg}(r) r^2 \sin\theta \, d\phi \, d\theta = \frac{|q|^2}{8\pi\rho_0 c}. \tag{B55}$$

B.1.6.2 Dipole Source

A dipole sound source can be constructed by two monopole sound sources that are closely located with a 180° phase difference. It can be mathematically written as:

$$p(r, \theta, \phi) = \lim_{\Delta \to 0} \frac{q}{4\pi}\left(\frac{e^{ikr_1}}{r_1} - \frac{e^{ikr_2}}{r_2}\right), \tag{B56}$$

where

$$r_1^2 = x^2 + y^2 + (z - \Delta/2)^2, \tag{B57}$$

$$r_2^2 = x^2 + y^2 + (z + \Delta/2)^2, \tag{B58}$$

$$q\Delta = q_{di}. \tag{B59}$$

Here, (x, y, z) is a point midway between two sound sources and q_{di} represents the dipole-moment amplitude vector.[4] Note that the dipole moment is a different quantity from the monopole amplitude, although the similar notation is used. When Equation (B56) is expanded with a Taylor series, and higher order terms are assumed to be negligible, Equation (B56) can be written as:

$$p(r, \theta, \phi) = \lim_{\Delta \to 0} \frac{q}{4\pi}\left\{\left(\frac{e^{ikr_1}}{r_1} - \frac{e^{ikr_2}}{r_2}\right)\Bigg|_{\Delta=0} + \frac{d}{d\Delta}\left(\frac{e^{ikr_1}}{r_1} - \frac{e^{ikr_2}}{r_2}\right)\Bigg|_{\Delta=0}\Delta\right\}. \tag{B60}$$

[4] When a dipole sound source is expressed using two sound sources with 180° phase difference, their sound fields change if the locations of the two sound sources change. Representing the directions of the two sound sources with vectors is therefore a more accurate means of representation. It is desirable to represent them as $\mathbf{\Delta}$, $\mathbf{q}_{di}$; however, in this chapter, it is assumed that the two sound sources are located on the z axis for convenience.

Equations (B57) and (B58) can be rewritten as

$$\lim_{\Delta\to 0} r_1 = r, \tag{B61}$$

and

$$\lim_{\Delta\to 0} r_2 = r. \tag{B62}$$

Then the first term of Equation (B60) can therefore be neglected. The second term can be rewritten as

$$\begin{aligned} p(r,\theta,\phi) = \lim_{\Delta\to 0} \frac{q\Delta}{4\pi} \Bigg\{ & \frac{d}{dr_1}\left(\frac{e^{ikr_1}}{r_1}\right)\Bigg|_{\Delta=0} \frac{dr_1}{d\Delta}\Bigg|_{\Delta=0} \\ & - \frac{d}{dr_2}\left(\frac{e^{ikr_2}}{r_2}\right)\Bigg|_{\Delta=0} \frac{dr_2}{d\Delta}\Bigg|_{\Delta=0} \Bigg\}. \end{aligned} \tag{B63}$$

Note that

$$\frac{d}{dr}\left(\frac{e^{ikr}}{r}\right)\Bigg|_{\Delta=0} = \frac{ikre^{ikr} - e^{ikr}}{r^2}. \tag{B64}$$

From Equations (B56) and (B58), we have

$$\frac{dr_1}{d\Delta}\Bigg|_{\Delta=0} = -\frac{1}{2}\frac{z}{r} = -\frac{1}{2}\cos\theta, \tag{B65}$$

and

$$\frac{dr_2}{d\Delta}\Bigg|_{\Delta=0} = \frac{1}{2}\frac{z}{r} = \frac{1}{2}\cos\theta. \tag{B66}$$

By substituting Equations (B59), (B64), (B65), and (B66) into Equation (B63) and rearranging them, the sound pressure by the dipole sound source can be written

$$\begin{aligned} p(r,\theta,\phi) &= -ikq_{di}\frac{e^{ikr}}{4\pi r}\frac{z}{r}\left\{1 + i\left(\frac{1}{kr}\right)\right\} \\ &= -ikq_{di}\frac{e^{ikr}}{4\pi r}\cos\theta\left\{1 + i\left(\frac{1}{kr}\right)\right\}. \end{aligned} \tag{B67}$$

In the nearfield and farfield, the pressure can be estimated as

$$\text{Near field}: \quad p(r,\theta,\phi) \approx q_{di}\frac{e^{ikr}}{4\pi r^2}\cos\theta \tag{B68}$$

$$\text{Far field}: \quad p(r,\theta,\phi) \approx -ikq_{di}\frac{e^{ikr}}{4\pi r}\cos\theta. \tag{B69}$$

From Euler equation, the velocity of direction r (shown in Figure B.4) can be written as

$$u_r(r,\theta,\phi) = -i\frac{kq_{di}}{\rho_0 c}\frac{e^{ikr}}{4\pi r}\cos\theta\left\{1 + 2i\left(\frac{1}{kr}\right) - 2\left(\frac{1}{kr}\right)^2\right\} \tag{B70}$$

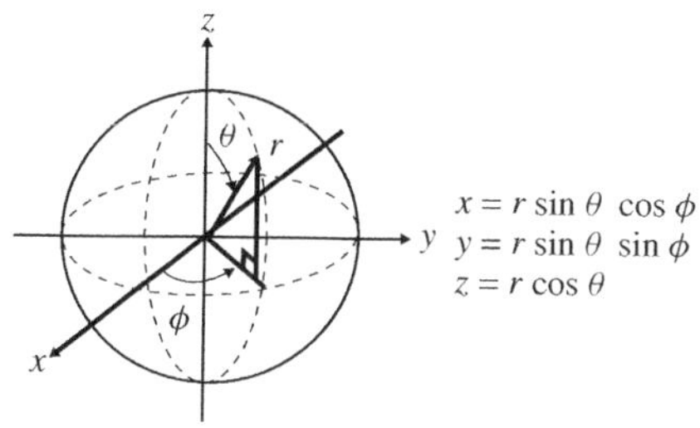

Figure B.4 Spherical coordinate system

and the nearfield and farfield can be written as:

$$\text{Near field}: \quad u_r(r,\theta,\phi) \approx i\frac{2q_{di}}{k\rho_0 c}\frac{e^{ikr}}{4\pi r^3}\cos\theta \tag{B71}$$

$$\text{Far field distance}: \quad u_r(r,\theta,\phi) \approx -i\frac{kq_{di}}{\rho_0 c}\frac{e^{ikr}}{4\pi r}\cos\theta. \tag{B72}$$

Impedance in the r direction can be obtained from Equations (B67) and (B70) as

$$Z_r(r,\theta,\phi) = \rho_0 c\frac{1+i\left(\frac{1}{kr}\right)}{1+2i\left(\frac{1}{kr}\right)-2\left(\frac{1}{kr}\right)^2} \tag{B73}$$

and the nearfield and farfield can be estimated:

$$\text{Near field}: \quad Z_r(r,\theta,\phi) \approx -\frac{i}{2}\rho_0 c(kr) \tag{B74}$$

$$\text{Far field}: \quad Z_r(r,\theta,\phi) \approx \rho_0 c. \tag{B75}$$

The average intensity in the r direction can be written (see Equation (B54)):

$$I_{avg}(r,\theta,\phi) = \frac{k^2|q_{di}|^2}{2\rho_0 c}\frac{1}{(4\pi r)^2}\cos^2\theta. \tag{B76}$$

As examined in, the power obtained by the area integral of the intensity is:

$$\Pi_{avg} = \frac{k^2|q_{di}|^2}{24\pi\rho_0 c}. \tag{B77}$$

Radiation characteristics of monopole and dipole sound sources can be easily understood by comparing Equation (B54) with Equation (B76) or (B55) with Equation (B77).

B.1.6.3 Quadrupole Source

Quadrupole sources can be made by having very closely located two dipole sound sources with opposite phases. It can be mathematically expressed:

$$p(r,\theta,\phi) = \lim_{\Delta\to 0}\left(\frac{-ikq_{di}}{4\pi}\right)\left[\frac{e^{ikr_1}}{r_1}\cos\theta_1\left\{1+i\left(\frac{1}{kr_1}\right)\right\} - \frac{e^{ikr_2}}{r_2}\cos\theta_2\left\{1+i\left(\frac{1}{kr_2}\right)\right\}\right. \tag{B78}$$

B.2 Wavenumber Domain Representation of the Rayleigh Integral

B.2.1 Fourier Transform of Free-Field Green's Function (Weyl's Identity)

Consider the Rayleigh's first integral for the finite planar radiator on S.

$$p(\mathbf{r}) = (-2i\omega\rho_0)\int_S \frac{e^{ikR_s}}{4\pi R_s}u(\mathbf{r}_s)dS. \tag{B101}$$

In Cartesian coordinates, Equation (B101) is given by

$$p(x, y, z) = (-2i\omega\rho_0)\int_{-\infty}^{\infty}\int_{-\infty}^{\infty} \frac{e^{ikR_s}}{4\pi R_s}u(x_s, y_s, 0)dx_s dy_s, \tag{B102}$$

where the distance R_s is a function of relative distance: $x - x_s$ and $y - y_s$, given by

$$R_s(x - x_s, y - y_s, z) = \sqrt{(x - x_s)^2 + (y - y_s)^2 + z^2}.$$

Equation (B102) takes the form of the spatial convolution

$$\begin{aligned} p(x, y, z) &= (-2i\omega\rho_0)\int_{-\infty}^{\infty}\int_{-\infty}^{\infty} g_f(x - x_s, y - y_s, z)u(x_s, y_s, 0)dx_s dy_s \\ &= (-2i\omega\rho_0)\left[g_f(x, y, z) *_x *_y u(x, y, 0)\right], \end{aligned} \tag{B103}$$

where g_f denotes the free-field Green's function.

Applying spatial Fourier transform on both sides of Equation (B103), we obtain the representation of Rayleigh integral in wavenumber domain:

$$P(k_x, k_y, z) = (-2i\omega\rho_0)\left[G_f\left(k_x, k_y, z\right)U(k_x, k_y, 0)\right]. \tag{B104}$$

The spatial Fourier transform of the monopole source $G_f(k_x, k_y, z)$ can be indirectly derived as follows. First, consider a pressure distribution on the planar radiator $S(z = 0)$. The spatial Fourier transform of $p(x, y, 0)$ will decompose the pressure field into many plane waves:

$$P(k_x, k_y, 0) = \int_{-\infty}^{\infty}\int_{-\infty}^{\infty} p(x, y, 0)e^{-i(k_x x + k_y y)}dxdy \tag{B105}$$

The wavenumber spectrum represents the complex amplitude of each plane wave. A single plane wave of wavenumber (k_x, k_y) should propagate in z direction with wavenumber $k_z = \sqrt{k^2 - k_x^2 - k_y^2}$, so we know that the plane wave at z is given by

$$P(k_x, k_y, z) = P(k_x, k_y, 0)e^{ik_z z}. \tag{B106}$$

In addition, from the Euler equation, the pressure at $z = 0$ is related to the surface normal velocity as

$$\begin{aligned} P(k_x, k_y, 0) &= \frac{\rho_0\omega}{k_z}U(k_x, k_y, 0) \\ &= (-2i\rho_0\omega)\frac{i}{2k_z}U(k_x, k_y, 0). \end{aligned} \tag{B107}$$

Then, the wavenumber spectrum at z (Equation (B106)) can be related to U using Equation (B107) as

$$P(k_x, k_y, z) = (-2i\rho_0\omega)\frac{ie^{jk_z z}}{2k_z}U(k_x, k_y, 0). \tag{B108}$$

By comparing Equations (B104) and (B108), the spatial Fourier transform of the monopole sound field ($G_f(k_x, k_y, z)$) can be written as

$$G_f(k_x, k_y, z) = \left(\frac{ie^{ik_z|z|}}{2k_z}\right) \tag{B109}$$

The monopole sound field is the same for $\pm z$, hence it depends on the absolute value of z, as shown in Equation (B109).

B.2.2 *High Frequency Approximation (Stationary Phase Approximation)*

The stationary phase approximation is a useful tool that can be used in the high frequency region. As its name implies, it approximates an integral only using the contribution around the point of stationary phase. If the phase-related term of the integral changes rapidly on the integral surface and if the spatial variation of the magnitude-related term is slow, then most of the integral would converge to zero because of the cancellation of the rapidly changing components. What only remains is the contribution from the area near the point in which the phase is stationary. This point is called *stationary phase point* and useful to convert the surface integral into a line integral.

The derivation is as follows. First, we consider a full integral with respect to an argument z

$$I = \int_{-\infty}^{\infty} f(z)e^{i\zeta(z)}dz. \tag{B110}$$

In the case in which the phase $\zeta(z)$ is quickly changing with respect to z, most of the contribution from the integral is determined by the integration around the stationary phase point. The stationary phase point z_0 can be identified from the derivative of phase on the surface. That is,

$$\zeta'(z_0) = 0. \tag{B111}$$

Then $\zeta(z)$ can be expanded in a Taylor series near the stationary phase point as

$$\zeta(z) \approx \zeta(z_0) + \zeta'(z_0)(z - z_0) + \frac{1}{2}\zeta''(z_0)(z - z_0)^2 \tag{B112}$$

where the second term on the right-hand-side is zero from Equation (B111). Furthermore, if the magnitude variation is slow near the stationary phase point, the magnitude term can be assumed to be $f(z) \approx f(z_0)$. Substituting Equations (B111) and (B112) to Equation (B110) yields

$$I \approx f(z_0)e^{i\zeta(z_0)}\int_{-\infty}^{\infty} e^{i\zeta''(z_0)(z-z_0)^2/2}dz \tag{B113}$$

To evaluate the integral, the argument of exponential is replaced by $\eta^2 = \zeta''(z_0)(z - z_0)^2/2$. Then

$$\sqrt{2/\zeta''(z_0)}d\eta = dz \tag{B114}$$

and Equation (B113) can be rewritten as

$$I \approx f(z_0)e^{i\zeta(z_0)}\sqrt{\frac{2}{\zeta''(z_0)}}\int_{-\infty}^{\infty} e^{i\eta^2}\,d\eta$$

$$\approx \sqrt{\frac{2\pi i}{\zeta''(z_0)}} f(z_0)e^{i\zeta(z_0)} \quad . \tag{B115}$$

B.3 Separation of Variables in Spherical Coordinates

Consider the problem of finding solutions of the wave equation in spherical coordinates.

$$[\nabla^2 + k^2]p(r, \theta, \phi, \omega) = 0$$

$$\frac{1}{r^2}\frac{\partial}{\partial r}\left(r^2\frac{\partial p}{\partial r}\right) + \frac{1}{r^2\sin\theta}\frac{\partial}{\partial\theta}\left(\sin\theta\frac{\partial p}{\partial\theta}\right) + \frac{1}{r^2\sin^2\theta}\frac{\partial^2 p}{\partial\phi^2} + k^2 p = 0 \tag{B116}$$

The solution satisfying Equation (B116) can be found by introducing separation of variables. If p is represented by the factored quantity

$$p(r, \theta, \phi, \omega) = R(r)\Theta(\theta)\Phi(\phi)e^{-i\omega t}, \tag{B117}$$

and if we consider the harmonic basis function for azimuthal angle $\phi(\Phi = e^{im\phi})$, then the wave equation can be separated as

$$\begin{aligned}
&Radial: && \frac{1}{r^2}\frac{d}{dr}\left(r^2\frac{dR}{dr}\right) + \left(k^2 - \frac{C}{r^2}\right)R = 0\\
&Azimuthal: && \frac{d^2\Phi}{d\phi^2} + m^2\Phi = 0\\
&Elevational: && \frac{1}{\sin\theta}\frac{d}{d\theta}\left(\sin\theta\frac{d\Theta}{d\theta}\right) + \left(C - \frac{m^2}{\sin^2\theta}\right)\Theta = 0
\end{aligned} \tag{B118}$$

The set of functions satisfying each of the separated equations in (r, θ, ϕ) directions can be derived independently.

B.3.1 Angle Functions: Associated Legendre Functions

B.3.1.1 Definition

First, the function Φ describing the azimuthal change of a sound field is of concern. We can use either the exponential function ($\Phi = e^{im\phi}$) or sinusoidal function ($\Phi = \sin(m\phi)$ or $\cos(m\phi)$) to satisfy Equation (B118). In either case, the number m should be an integer, in order that Φ is continuous at $\phi = 0, 2\pi$. We denote the harmonic function depends only on ϕ as *circular harmonics*.

For the elevational function Θ, substitution of $\eta = \cos\theta$ yields

$$(1 - \eta^2)\frac{d^2\Theta}{d\eta^2} - 2\eta\frac{d\Theta}{d\eta} + \left(C - \frac{m^2}{1 - \eta^2}\right)\Theta = 0. \tag{B119}$$

The function satisfying the differential equation can be derived from the series expansion of Θ in terms of η. However, it can be shown [1] that the series is finite at $\eta = \pm 1$ only for the special combinations of C:

$$C = n(n + 1). \tag{B120}$$

The differential equation (Equation (B119)) for this case is called as Legendre equation, and the series of $\eta(P_m^n)$ satisfying the Legendre equation is denoted as the *associated Legendre functions.*

$$P_n^m(\eta) = (-1)^m (1-\eta^2)^{m/2} \frac{d^m}{d\eta^m} P_n(\eta) \tag{B121}$$

where $P_n(\eta)$ is the *Legendre polynomial* that is defined as the associated Legendre function of a special case $m = 0$ (Rodrigues' formula).

$$\begin{aligned} P_n(\eta) &\equiv P_n^0(\eta) \\ &= \frac{1}{2^n n!} \frac{d^n}{d\eta^n} (\eta^2 - 1)^n. \end{aligned} \tag{B122}$$

The variable $\eta = \cos\theta$ is real, and hence the Legendre polynomial and associated Legendre function are real-valued functions. Since the highest power of the η in $P_n(\eta)$ is η^n, the mth derivative is always zero for $m > n$. Consequently, the associated Legendre functions are nonzero only for $m \le n$. The Legendre polynomial can also be defined by the following generating function:

$$\frac{1}{\sqrt{1 - 2\eta t + t^2}} = \sum_{n=0}^{\infty} P_n(\eta) t^n. \tag{B123}$$

B.3.1.2 Special Value at End Points

The special value of the associated Legendre polynomial at $\eta = 1$ or $\theta = 0$ is given by

$$P_n^m(1) = \begin{cases} 0 & for\ m \neq 0 \\ 1 & for\ m = 0 \end{cases}. \tag{B124}$$

For $\eta = -1$ or $\theta = \pi$,

$$P_n^m(-1) = \begin{cases} 0 & for\ m \neq 0 \\ (-1)^n & for\ m = 0 \end{cases}. \tag{B125}$$

At the center ($\eta = 0$ or $\theta = \pi/2, 3\pi/2$),

$$P_n^m(0) = \begin{cases} (-1)^{(n-m)/2} \dfrac{(n+m-1)!!}{(n-m)!!}, & n+m \text{ even} \\ 0, & n+m \text{ odd.} \end{cases} \tag{B126}$$

B.3.1.3 Orthogonality

The associated Legendre functions of different orders n are orthogonal to each other over $[-1, 1]$.

$$\int_{-1}^{1} P_{n'}^m(\eta) P_n^m(\eta) d\eta = \frac{2}{2n+1} \frac{(n+m)!}{(n-m)!} \delta_{nn'}. \tag{B127}$$

For Legendre polynomial with $m = 0$, this leads to

$$\int_{-1}^{1} P_{n'}(\eta) P_n(\eta) d\eta = \frac{2}{2n+1} \delta_{nn'}. \tag{B128}$$

B.3.1.4 Angular Step Function

$$a_n = \int_0^1 P_n(\eta)d\eta = \frac{2n+1}{2} \times \begin{cases} 1 \text{ for } n = 0, \ 0 \text{ for } n : \text{even} \\ (-1)^{(n-1)/2}\dfrac{(n-2)!!}{(n+1)!!} \text{ for } n : \text{odd} \end{cases} \tag{B129}$$

B.3.1.5 Parity, Negative Order

The parity relation is useful for calculating the sound field in the opposite angle $(-\eta = -\cos(\theta) = \cos(\theta + \pi))$. The odd order of the Legendre polynomial is an odd function. That is,

$$P_n(-\eta) = (-1)^n P_n(\eta). \tag{B130}$$

From the definition of the associated Legendre function, we can see that

$$P_n^m(-\eta) = (-1)^{n+m} P_n^m(\eta). \tag{B131}$$

For negative order m, the polynomial becomes

$$P_n^{-m}(\eta) = (-1)^m \frac{(n-m)!}{(n+m)!} P_n^m(\eta). \tag{B132}$$

B.3.1.6 Recurrence Relations

The Legendre polynomials have the following recursive relations:

$$\begin{aligned}(2n+1)\eta P_n(\eta) &= (n+1)P_{n+1}(\eta) + nP_{n-1}(\eta) \\ (2n+1)P_n(\eta) &= \frac{d}{d\eta}[P_{n+1}(\eta) - P_{n-1}(\eta)].\end{aligned} \tag{B133}$$

B.3.2 Angle Functions: Spherical Harmonics

B.3.2.1 Definition

The associated Legendre function, representing the variation in θ direction, now can be combined with the circular harmonics for the coordinate ϕ. The combination called spherical harmonics can be defined as

$$\begin{aligned}Y_n^m(\theta, \phi) &= v_{nm} P_n^m(\cos\theta)e^{im\phi} \\ &= \sqrt{\frac{(2n+1)}{4\pi}\frac{(n-m)!}{(n+m)!}} P_n^m(\cos\theta)e^{im\phi}\end{aligned} \tag{B134}$$

The coefficient v_{nm} consisting of n and m is the normalization coefficient that gives the orthonormal property described below.

B.3.2.2 Orthogonality

The spherical harmonics form a complete orthonormal system on Ω

$$\int_\Omega Y_n^m(\theta, \phi) Y_{n'}^{m'}(\theta, \phi)^* d\Omega = \delta_{nn'}\delta_{mm'} \tag{B135}$$

where $\int_\Omega d\Omega = \int_0^{2\pi} \int_0^{\pi} \sin\theta d\theta d\phi$. From the orthogonality and closure relation, we can express arbitrary angular function $f(\theta, \phi)$ as a summation of spherical harmonics.

$$f(\theta,\phi) = \sum_{n=0}^{\infty} \sum_{m=-n}^{n} a_{nm} Y_n^m(\theta,\phi) \tag{B136}$$

with a_{nm} being the coefficient of spherical harmonics. Arbitrary harmonics coefficient a_{nm} can be extracted from $f(\theta, \phi)$ using the orthogonality relation. Integration of $f(\theta, \phi)$ over Ω with $(Y_n^m)^*$ gives

$$\begin{aligned}\int_\Omega f(\theta,\phi) Y_{n'}^{m'}(\theta,\phi)^* d\Omega &= \sum_{n=0}^{\infty} \sum_{m=-n}^{n} \left[a_{nm} \int_\Omega Y_n^m(\theta,\phi)\, Y_{n'}^{m'}(\theta,\phi)^* d\Omega \right]. \\ &= a_{n'm'} \end{aligned} \tag{B137}$$

B.3.2.3 Closure Relation

A counterpart of the orthogonality relation is the closure relation given by

$$\sum_{n=0}^{\infty} \sum_{m=-n}^{n} Y_n^m(\theta,\phi) Y_n^m(\theta',\phi')^* = \delta(\phi-\phi')\delta(\cos\theta - \cos\theta'). \tag{B138}$$

B.3.2.4 Parity

$$Y_n^m(\pi-\theta, \pi+\phi) = (-1)^n Y_n^m(\theta,\phi) \tag{B139}$$

$$Y_n^m(\theta,\phi)^* = (-1)^m Y_n^{-m}(\theta,\phi). \tag{B140}$$

The complex conjugation of the spherical harmonics changes the sign of its harmonic coefficient. Let us consider the spherical harmonics expansion of $f(\theta, \phi)$

$$f(\theta,\phi) = \sum_{n=0}^{\infty} \sum_{m=-n}^{n} \overline{a}_{nm} Y_n^m(\theta,\phi)^*, \tag{B141}$$

then from Equations (B136) and (B140), the coefficients $\overline{a}_{nm}$ and a_{nm} have the following relation:

$$a_{nm} = \overline{a}_{n,-m}(-1)^m. \tag{B142}$$

B.3.2.5 Addition Theorem

A generalization of the trigonometric identity

$$\cos(\theta-\theta') = \cos\theta' \cos\theta + \sin\theta \sin\theta', \tag{B143}$$

for the two-dimensional case is given by the addition theorem of spherical harmonics.

$$P_n\left(\frac{\mathbf{r}\cdot\mathbf{r}'}{rr'}\right) = \frac{4\pi}{2n+1} \sum_{m=-n}^{n} Y_n^m(\theta,\phi) Y_n^m(\theta',\phi')^* \tag{B144}$$

with $\mathbf{r} = (r, \theta, \phi)$, $\mathbf{r}' = (r', \theta', \phi')$. This also can be simply expressed as

$$\sum_{m=-n}^{n} Y_n^m(\theta,\phi) Y_n^m(\theta',\phi')^* = \frac{2n+1}{4\pi} P_n(\cos\gamma) \tag{B145}$$

where γ denotes an angle between (θ, ϕ) and (θ', ϕ').

B.3.2.6 Reciprocity

Since $P_n(\cos\gamma)$ is real, the following relation holds:

$$\sum_{m=-n}^{n} Y_n^m(\theta,\phi)Y_n^m(\theta',\phi')^* = \sum_{m=-n}^{n} Y_n^m(\theta,\phi)^* Y_n^m(\theta',\phi'). \quad \text{(B146)}$$

B.3.3 Radial Functions

Now let us go back to the differential equation for the radial direction. From Equations (B118) and (B120), we have

$$\frac{d}{dr}\left(r^2\frac{dR}{dr}\right) + \left[(kr)^2 - n(n+1)\right]R = 0. \quad \text{(B147)}$$

Substituting $x = kr$ to get rid of k dependency yields the *spherical Bessel differential equation.*

$$x^2\frac{d^2R}{dx^2} + 2x\frac{dR}{dx} + [x^2 - n(n+1)]R = 0. \quad \text{(B148)}$$

The differential equation with $n = 0$ becomes 1D Helmholtz equation

$$\frac{d^2(xR)}{dx^2} + (xR) = 0 \quad \text{(B149)}$$

and thus we can have the solution given by $R = e^{ix}/x = e^{ikr}/kr$.

B.3.4 Radial Functions: Spherical Bessel and Hankel Functions

B.3.4.1 Spherical Bessel Functions

The general functions that satisfy Equation (B148) are given by [2]

$$j_n(x) = 2^n x^n \sum_{s=0}^{\infty} \frac{(-1)^s(s+n)!}{s!(2s+2n+1)!}x^{2s}$$

$$y_n(x) = \frac{(-1)^{n+1}}{2^n x^{n+1}} \sum_{s=0}^{\infty} \frac{(-1)^s(s-n)!}{s!(2s-2n)!}x^{2s} \quad \text{(B150)}$$

The functions j_n and y_n are called spherical Bessel functions and spherical Neumann functions of order n, respectively. The differential equation (Equation (B148)) equals to the Bessel differential equation of order $n + 1/2$, and hence the spherical Bessel function is related to the Bessel function as

$$j_n(x) = \sqrt{\frac{\pi}{2x}}J_{n+1/2}(x),$$

$$y_n(x) = \sqrt{\frac{\pi}{2x}}N_{n+1/2}(x), \quad \text{(B151)}$$

where $J_{n+1/2}$ and $N_{n+1/2}$ are cylindrical Bessel and Neumann functions. Figure B.5 depicts the spherical Bessel and Nuemann functions of order $n = 1, 2, 3$.

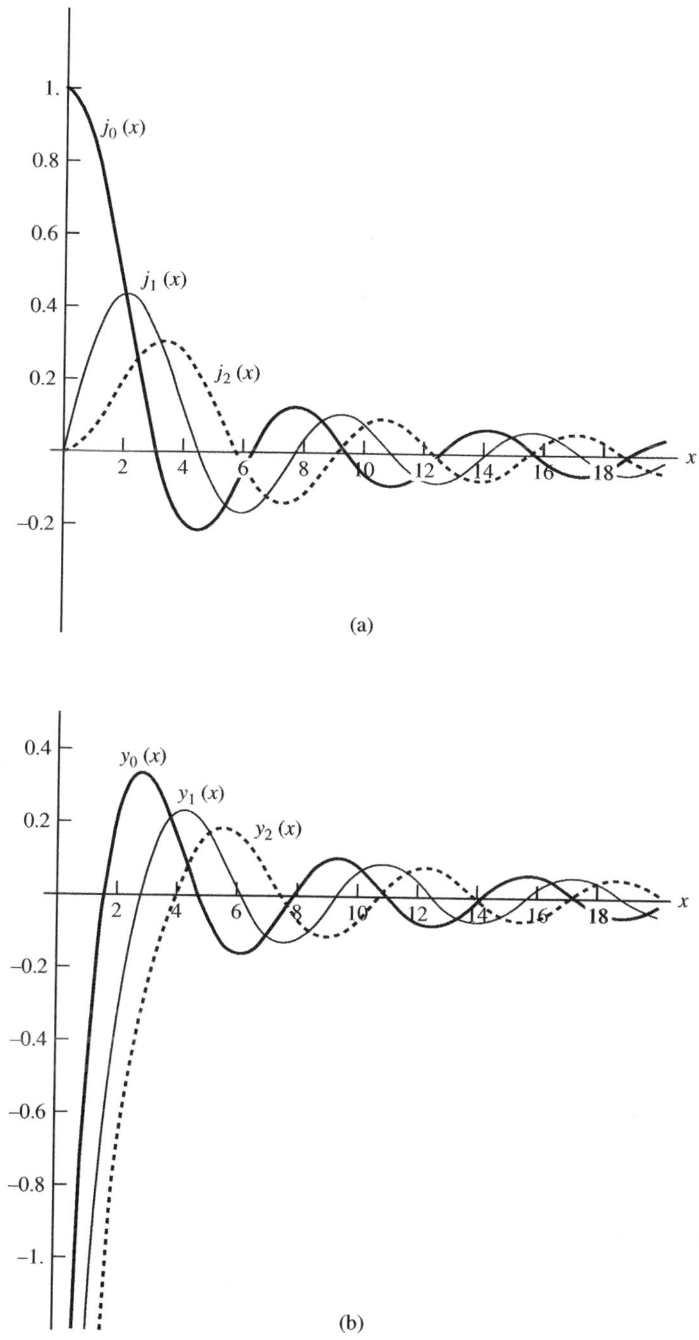

Figure B.5 Spherical Bessel (a) and spherical Neumann (b) functions

B.3.4.2 Spherical Hankel Functions

The complex functions defined by the linear combination of these two functions

$$h_n^{(1)} = j_n + i\,y_n,\quad h_n^{(2)} = j_n - i\,y_n \tag{B152}$$

are known as spherical Hankel function of the first and second kind of order n. The first and second kinds are related by complex conjugation ($h_n^{(1)} = h_n^{(2)*}$). Like the spherical Bessel functions, they are related to the cylindrical Hankel functions as follows:

$$\begin{aligned} h_n^{(1)}(x) &= \sqrt{\frac{\pi}{2x}}H_{n+1/2}^{(1)}(x), \\ h_n^{(2)}(x) &= \sqrt{\frac{\pi}{2x}}H_{n+1/2}^{(2)}(x). \end{aligned} \tag{B153}$$

The series expansion of the spherical Hankel functions can be derived as [2]

$$\begin{aligned} h_n^{(1)}(x) &= (-i)^{n+1}\frac{e^{ix}}{x}\sum_{s=0}^{n}\frac{i^s}{s!(2x)^s}\frac{(n+s)!}{(n-s)!} \\ h_n^{(2)}(x) &= (i)^{n+1}\frac{e^{-ix}}{x}\sum_{s=0}^{n}\frac{(-i)^s}{s!(2x)^s}\frac{(n+s)!}{(n-s)!}. \end{aligned} \tag{B154}$$

B.3.4.3 Special Values of Spherical Bessel Functions

For the special case $n = 0$, the spherical Bessel function is equivalent to the sinc function.

$$j_0(x) = \frac{\sin(x)}{x} = \mathrm{sinc}(x) \tag{B155}$$

whereas the Neumann function is given by

$$y_0(x) = -\frac{\cos x}{x}. \tag{B156}$$

From the definition of the spherical Hankel function, we also have

$$\begin{aligned} h_0^{(1)}(x) &= -\frac{i}{x}e^{ix} \\ h_0^{(2)}(x) &= \frac{i}{x}e^{-ix}. \end{aligned} \tag{B157}$$

B.3.4.4 Generating Function

The spherical Bessel function and Legendre polynomial are related to each other by

$$j_n(x) = \frac{1}{2i^n}\int_0^{\pi} e^{ix\cos\theta}P_n(\cos\theta)\sin\theta d\theta. \tag{B158}$$

B.3.4.5 Small Argument Approximation

For the radial function with small argument x, the first-order approximation can be made from the series expansion of Equations (B150) and (B154).

$$j_n(x) \approx \frac{2^n n!}{(2n+1)!} x^n, \quad y_n(x) \approx -\frac{(2n)!}{2^n n!} \frac{1}{x^{n+1}} \tag{B159}$$

However, for this approximation, the second term in the series should be negligible compared to the first one. Such condition for j_n is given by

$$x << 2\sqrt{\frac{(2n+2)(2n+3)}{n+1}}. \tag{B160}$$

The small argument approximation is used to analyze the nearfield behavior of a sound field. It is noteworthy that, for $x < 1$, low order spherical Bessel functions dominate, whereas higher order Bessel functions' contribution are greater for $x > 1$.

The spherical Neumann function diverges near $x = 0$ with the rate of $1/x^{n+1}$. Therefore, it is much greater than the spherical Bessel function, and hence, the magnitude of the spherical Hankel function follows that of the spherical Neumann function for small x (Figure B.6).

B.3.4.6 Large Argument Approximation

The large argument approximation is useful when we derive a formula for farfield radiation (large kr). For large argument, however, we first consider the series expansion of the spherical Hankel function given in Equation (B154). If x is large enough, then the first term ($s = 0$) of the series

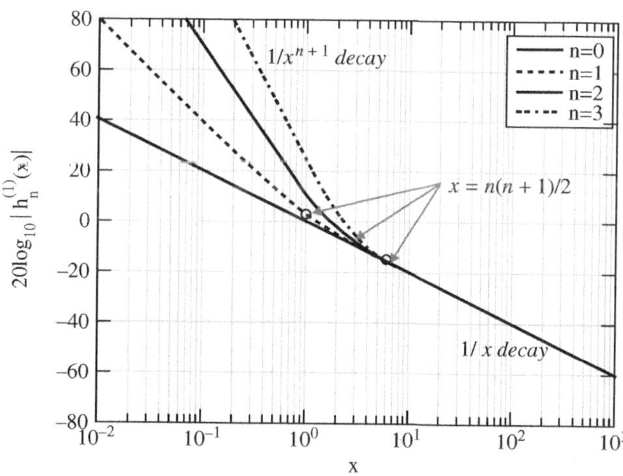

Figure B.6 Magnitude variation of spherical Hankel function in decibel scale

will give the most of contribution, hence it can be rewritten as

$$h_n^{(1)}(x) \sim (-i)^{n+1}\frac{e^{ix}}{x}$$
$$h_n^{(2)}(x) \sim (i)^{n+1}\frac{e^{-ix}}{x}. \tag{B161}$$

Accordingly, we can also have the approximation for the spherical Bessel and Neumann functions.

$$j_n(x) \sim \frac{1}{x}\sin\left(x - \frac{n\pi}{2}\right)$$
$$y_n(x) \sim -\frac{1}{x}\cos\left(x - \frac{n\pi}{2}\right). \tag{B162}$$

However, the approximation is valid only when the first term of the series is much greater than the second or sum of the others. From Equation (B154), this condition can be summarized as follows:

$$x \gg n(n+1)/2. \tag{B163}$$

Figure B.6 depicts the decay of spherical Hankel function in decibel scale. The spherical Hankel function of the first kind has $1/x$ decay at farfield irrespective of the order n. The large argument approximation hence enables us to express the radial decay rate of a farfield sound by a constant factor.

B.3.4.7 Wronskian Relations

The radial functions can be related to each other in terms of Wronskian relations:

$$j_n(x)h_n^{(1)'}(x) - j_n'(x)h_n^{(1)}(x) = \frac{i}{(x)^2}, \tag{B164}$$

$$j_n(x)y_n'(x) - j_n'(x)y_n(x) = \frac{i}{(x)^2} \tag{B165}$$

$$[h_n^{(1)}(x)h_n^{(1)'}(x)^* - h_n^{(1)'}(x)h_n^{(1)}(x)^*] = -\frac{2i}{(x)^2}. \tag{B166}$$

B.3.5 Description of Sound Fields by Spherical Basis Function

Combination of the radial function, such as spherical Bessel or Hankel functions, and the spherical harmonics can efficiently describe a sound in spherical coordinates. Now consider two different types of combinations of the radial function and spherical harmonics:

$$\alpha_n^m(\mathbf{r}, \omega) = j_n(kr)Y_n^m(\theta, \phi), \quad \beta_n^m(\mathbf{r}, \omega) = h_n^{(1)}(kr)Y_n^m(\theta, \phi). \tag{B167}$$

According to their radial functions, α_n^m represents a standing wave in radial direction, whereas β_n^m denotes the radiating function. Because α_n^m is the solution of the Helmholtz equation for arbitrary $\mathbf{r}$,

$$[\nabla^2 + k^2]\alpha_n^m(\mathbf{r}, \omega) = 0,$$

the function α_n^m is useful for describing the sound propagating over a region that does not include any sound source. In other words, any sound field within a volume propagating from the outside of the volume can be expressed in terms of α_n^m.

$$p(\mathbf{r}, \omega) = \sum_{n=0}^{\infty} \sum_{m=-n}^{n} c_{nm}(\omega)\alpha_n^m(\mathbf{r}, \omega). \tag{B168}$$

In contrast, the function β_n^m has a singularity at $r = 0$, hence homogeneous only for $r \neq 0$. The Hankel function of the first kind represents a radiating function, and therefore, can be used to describe the exterior sound field generated from a sound source inside a volume.

$$p(\mathbf{r}, \omega) = \sum_{n=0}^{\infty} \sum_{m=-n}^{n} c_{nm}(\omega)\beta_n^m(\mathbf{r}, \omega). \tag{B169}$$

B.3.6 Representation of the Green's Function

B.3.6.1 Free-Field Green's Function

The expansion of the free-field Green's function explained below follows the derivation of Colton and Kress [3]. Consider an exterior sound field generated by integrating the elementary functions (Equation (B167)) over a spherical surface S of radius r_s. From the property of surface integral (Table 6.1), the elementary function α_n^m satisfying the Helmholtz equation also obeys the homogeneous integral equation. That is,

$$\int_S \left[\frac{\partial \alpha_n^m(\mathbf{r}_s)}{\partial r_s} g(\mathbf{r}|\mathbf{r}_s) - \alpha_n^m(\mathbf{r}_s) \frac{\partial g(\mathbf{r}|\mathbf{r}_s)}{\partial r_s} \right] dS(\mathbf{r}_s) = 0, \qquad r > r_s. \tag{B170}$$

For the radiating function β_n^m that has a singularity at $r = |\mathbf{r}| = 0$, the integral equation gives

$$\beta_n^m(\mathbf{r}) = \int_S \left[\frac{\partial \beta_n^m(\mathbf{r}_s)}{\partial r_s} g(\mathbf{r}|\mathbf{r}_s) - \beta_n^m(\mathbf{r}_s) \frac{\partial g(\mathbf{r}|\mathbf{r}_s)}{\partial r_s} \right] dS(\mathbf{r}_s), \qquad r > r_s. \tag{B171}$$

From the definition of Equation (B167) and with the aid of Wronskian relation, the integrals (Equations (B170) and (B171)) can be arranged to show the following relation:

$$\int_{\Omega_s} \left[g_f(\mathbf{r}|\mathbf{r}_s)\, Y_n^m(\theta_s, \phi_s) \right] d\Omega_s = (ik) h_n^{(1)}(kr) j_n(kr_s) Y_n^m(\theta, \phi), \quad r > r_s \tag{B172}$$

where $r_s^2 \int_{\Omega_s} d\Omega_s = \int_S dS$. Equation (B172) is the spherical harmonic coefficient of the Green's function. From Equation (B141) and the orthogonality of spherical harmonics, we have

$$g_f(\mathbf{r}|\mathbf{r}_s) = ik \sum_{n=0}^{\infty} \left[h_n^{(1)}(kr) j_n(kr_s) \sum_{m=-n}^{n} Y_n^m(\theta, \phi) Y_n^m(\theta_s, \phi_s)^* \right], \quad r > r_s. \tag{B173}$$

From the reciprocity of the Green's function and addition theorem, the harmonics expansion of the Green's function can be summarized as follows:

$$g_f(r, \theta, \phi | r_s, \theta_s, \phi_s) = \begin{cases} \dfrac{ik}{4\pi} \displaystyle\sum_{n=0}^{\infty} h_n^{(1)}(kr) j_n(kr_s)(2n+1) P_n(\cos\gamma) & \mathit{for}\ r > r_s \\ \dfrac{ik}{4\pi} \displaystyle\sum_{n=0}^{\infty} j_n(kr) h_n^{(1)}(kr_s)(2n+1) P_n(\cos\gamma) & \mathit{for}\ r < r_s \end{cases} \tag{B174}$$

where γ is the angle between two position vectors $\mathbf{r}, \mathbf{r}_s$.

B.3.6.2 Spherical Harmonic Expansion of a Plane Wave

A plane wave is characterized by the constant phase along a planar surface. Denoting the surface normal vector of a planar surface as $\mathbf{e}_k$, the plane wave of unit amplitude can be written as

$$p(\mathbf{r}, \omega) = e^{i\mathbf{k}\cdot\mathbf{r}}, \quad \mathbf{k} = k\mathbf{e}_k \tag{B175}$$

at the observation point $\mathbf{r} = r\mathbf{e}_r$. In terms of the relative angle between $\cos\gamma = \frac{\mathbf{e}_k\cdot\mathbf{e}_r}{|\mathbf{e}_k||\mathbf{e}_r|} = \mathbf{e}_k \cdot \mathbf{e}_r$, the pressure field can be rewritten as

$$p(\mathbf{r}, \omega) = e^{ikr(\mathbf{e}_k\cdot\mathbf{e}_r)} = e^{ikr\cos\gamma}. \tag{B176}$$

The angular dependency of a plane wave can be expanded in terms of the Legendre polynomial $P_n(\cos\gamma)$. Since the plane wave source satisfies the Helmholtz equation, the radial expansion of a sound field within a sphere of finite radius can be regarded as an interior problem. Therefore, radial dependency of a plane wave can be expressed in terms of spherical Bessel functions.

$$p(\mathbf{r}, \omega) = \sum_{n=0}^{\infty} C_n j_n(kr) P_n(\cos\gamma). \tag{B177}$$

From the orthogonality of the Legendre polynomial, we can construct an equation to find C_n.

$$\begin{aligned}&\int_0^{\pi} p(\mathbf{r}, \omega) P_n(\cos\gamma) \sin\gamma \, d\gamma \\ &= \int_{-1}^{1} e^{ikr\eta} P_n(\eta) d\eta = C_n \frac{2}{2n+1} j_n(kr).\end{aligned} \tag{B178}$$

Since this equation for C_n holds for every kr, we can obtain $C_0 = 1$ by inserting $kr = 0$ to Equation (B178). The higher order coefficients can be obtained by differentiating Equation (B178) with respect to r and evaluating at $r = 0$. As a result, we obtain

$$C_n = i^n(2n+1). \tag{B179}$$

Consequently, the plane wave can be expanded in Legendre polynomials as

$$p(\mathbf{r}, \omega) = \sum_{n=0}^{\infty} i^n (2n+1) j_n(kr) P_n(\cos\gamma). \tag{B180}$$

Using addition theorem of spherical harmonics (Equation (B145)), the expansion of a plane wave is given by

$$p(\mathbf{r}, \omega) = 4\pi \sum_{n=0}^{\infty} i^n j_n(kr) \sum_{m=-n}^{n} Y_n^m(\theta, \phi) Y_n^m(\theta_k, \phi_k)^* \tag{B181}$$

where (θ_k, ϕ_k) represents the propagating direction of a plane wave ($\mathbf{e}_k$) in spherical coordinates.

The same result can be derived from the farfield approximation of the monopole sound. Consider a monopole located at $\mathbf{r}_s$. If we take the limit $|\mathbf{r}_s| \to \infty$, we can make the substitution $|\mathbf{r} - \mathbf{r}_s| \simeq r_s - \mathbf{e}_s \cdot \mathbf{r}$, where $\mathbf{e}_s$ is a unit normal pointing in the direction of $\mathbf{r}_s$. The free-field monopole's sound field can then be reduced to

$$g_f(\mathbf{r}|\mathbf{r}_s) = \frac{e^{ik|\mathbf{r}-\mathbf{r}_s|}}{4\pi|\mathbf{r}-\mathbf{r}_s|} \approx \frac{e^{ikr_s}}{4\pi r_s} e^{-ik\mathbf{e}_s\cdot\mathbf{r}}. \tag{B182}$$

On the other hand, if we apply the large argument approximation of the spherical Hankel function to the spherical harmonic expansion of Equation (B174), the monopole sound field can be written as

$$g_f(r,\theta,\phi|r_s,\theta_s,\phi_s) \approx \frac{e^{ikr_s}}{4\pi r_s}\sum_{n=0}^{\infty}(-i)^n(2n+1)j_n(kr)P_n(\cos\gamma) \quad \textit{for } r < r_s. \tag{B183}$$

From Equations (B182) and (B183), we can get the same expansion as Equation (B180):

$$e^{ik\mathbf{e}_s\cdot\mathbf{r}} = \sum_{n=0}^{\infty} i^n(2n+1)j_n(kr)P_n(\cos\gamma). \tag{B184}$$

References

1. Morse, P.M. and Ingard, K.U. (1986) *Theoretical Acoustics*, Princeton University Press, Princeton, NJ, pp. 333–334.
2. Arfken, G.B. and Weber, H.J. (2005) *Mathematical Methods for Physicists*, 6th edn, Academic Press, pp. 726–729.
3. Colton, D. and Kress, R., (1998) *Inverse Acoustic and Electromagnetic Scattering Theory*, 2nd edn, Springer-Verlag, Berlin, pp. 30–32.

Index

Sound Visualization and Manipulation, First Edition. Yang-Hann Kim and Jung-Woo Choi.

Printed and bound by CPI Group (UK) Ltd, Croydon, CR0 4YY
16/09/2026
14963390-0004